# GAP JUNCTIONS

# Gap Junctions

Proceedings of the 8th
International Gap Junction Conference
Key Largo, Florida

Edited by

Rudolf Werner
*Department of Biochemistry and Molecular Biology,*
*University of Miami School of Medicine, Miami, Florida, USA*

IOS Press

Ohmsha

Amsterdam • Berlin • Oxford • Tokyo • Washington, DC

ISBN 90 5199 374 9 (IOS Press)
ISBN 4 274 90197 1 C3047 (Ohmsha)
Library of Congress Catalog Card Number 97-75195

*Publisher*
IOS Press
Van Diemenstraat 94
1013 CN Amsterdam
Netherlands

*Distributor in the UK and Ireland*
IOS Press/Lavis Marketing
73 Lime Walk
Headington
Oxford OX3 7AD
England

*Distributor in Germany*
IOS Press
Spandauer Strasse 2
D-10178 Berlin
Germany

*Distributor in the USA and Canada*
IOS Press, Inc.
5795-G Burke Centre Parkway
Burke, VA 22015
USA

*Distributor in Japan*
Ohmsha, Ltd.
3-1 Kanda Nishiki-cho
Chiyoda-ku
Tokyo 101
Japan

*Cover illustration*: Projection density map at 7Å resolution of a recombinant gap junction channel formed by a C-terminal truncation mutant of rat heart $\alpha_1$ (Cx43) connexin. The recombinant connexin was expressed in BHK cells under control of the inducible mouse metallothionin promoter. The expressed protein assembles into recognizable gap junction plaques, and 2D crystals are grown by detergent treatment to extract lipids and concentrate the protein in the membrane plane. The map was derived by electron cryo-microscopy and image analysis of frozen-hydrated specimens. This end-on view shows that each connexon is formed by a hexameric cluster of subunits with three major features: (i) a ring of circular densities centered at a radius of 17Å , interpreted as α-helices that line the channel, (ii) a ring of densities centered at a radius of 33Å, interpreted as α-helices that are most exposed to the lipid and (iii) a continuous band of density at a radius of 25Å, separating the two groups of helices. From Unger, V.M., Kumar, N.M., Gilula, N.B. and Yeager, M. Projection Structure of a Gap Junction Channel at 7Å Resolution. *Nature Struct. Biol.* **4**: 39–43 (1997). Graphics by Brian Sheehan and Mark Yeager using AVS software.



PRINTED IN THE NETHERLANDS

1997 International Gap Junction Conference at Ocean Reef, Key Largo, Florida

# PREFACE

The 8th International Gap Junction Conference was held July 12-17, 1997 at Ocean Reef Club, Key Largo, Florida. A total of 192 scientists from 23 countries participated in the conference presenting 175 papers that included 59 platform presentations and 116 poster presentations. This volume includes 66 of these papers, mostly platform presentations.

As previous gap junction meetings, this conference was a true working conference. Every lecture was attended by almost all participants. Since all platform presentations were chosen by a committee on the basis of the submitted abstracts, there was much new information presented. This, in turn, produced extensive discussion, which usually continued beyond the lecture hall into the dining hall Everglades National Park, and even Pennekamp Coral Reef Park. If the number of new collaborations that originated from this meeting is any indication of the success of a meeting, then this meeting was scientifically very rewarding for most participants.

The beautiful tropical surroundings of the Florida Keys probably also contributed to the generally happy feeling of all participants. One hundred scientists visited the Everglades National Park, and 120 people went on a snorkeling trip to Pennekamp Coral Reef Park, lying just offshore of Ocean Reef Club. I was certainly very relieved when all 120 made it back into harbor safe and sound.

The proceedings of several previous gap junction conferences have been published in monographs that have proven a valuable source of information for scientists in the field of gap junction research. Publishing conference presentations in book form, however, raises a potential problem. Peer-reviewed scientific journals require that results submitted to them for publication not be published previously. To avoid this problem, I suggested to the authors that they include only their results and conclusions but keep the actual experimental data for the peer-reviewed publication. Most people followed my suggestions and submitted manuscripts but a few still had reservations and chose not to submit. I would like to suggest to potential reviewers of journal manuscripts, that they dismiss papers included in this monograph as non-peer-reviewed and thus allow the same data to be included in peer-reviewed publications.

I would like to thank the members of the organizing committee, Drs. Mario Delmar, Habo Jongsma, David Paul, David Spray, and Robert Weintraub for their valuable suggestions and help in putting this conference together. I am also indebted to my associates from the Biochemistry department of the University of Miami, Debbie Bradley, Sandy Black, Mark Royer, and Erika Thompson, for their help in organizing this conference and attending to every detail, and to my graduate students Alice Hudder, Elisa Oltra, and Adam Schiavi who helped putting people on the buses at Miami International Airport and who took turns at the slide projectors during the sessions. Finally, I would like to thank my wife Beth for participating in many decisions.

I thank my colleagues in the gap junction field for entrusting me with the organization of the 1997 conference. I feel privileged to have had the opportunity to host this important conference. I hope that the spirit of open collaboration that prevailed at this meeting will continue at the next gap junction conference, which will take place in 1999 in Switzerland.

Rudolf Werner

# Sponsors

**Grants and contributions** from the following sponsors helped making this conference possible.

American Heart Association

Axon Instruments

Bio-Rad Laboratories

Environmental Protection Agency

Merck Research Laboratories

National Institute of Health

Perkin-Elmer

Searle Monsanto

Searle, Skogie Ill.

University of Miami

# Contents

**Section 4. Assembly and Degradation of Gap Junction Channels**



**Section 5. Biosynthesis of Gap Junction Channels**



**Section 6. Function of Gap Junctions in Different Tissues**

**Section 7. Regulation of Gap Junctional Communication**

**Section 8. Intercellular Communication by Calcium Signalling**



**Section 9. Functional Studies of Gap Junctions in Mutants and Knockout Mice**



**Section 10. Connexin Gene Regulation**

**Section 11. Gap Junctions in Development and Cancer**

# Section 1:
# Gap Junction Structure

*Gap Junctions*
*R. Werner (Ed.)*
*IOS Press, 1998*



# Diverse Molecular Mechanisms of Gap Junction Channel Gating

Bruce J. NICHOLSON, Lan ZHOU, FengLi CAO, Hui ZHU and Ye CHEN
*Dept. Biological Sciences, SUNY at Buffalo, Buffalo, NY 14260*

**Abstract:** Gap junction channels are subject to acute regulation by a number of factors, both electrical and chemical. Gating kinetics of the channels reveals two classes of responses: rapid closure in response to voltage; and slower responses associated with the original formation of channels and their closure induced by chemical signals such as lypophillic agents and reduced cytoplasmic pH. Using the paired Xenopus oocyte expression system, we have found that modifications to several domains, including M2 and both extracellular loops, can induce a reversal in the response of gap junctions to transjunctional voltage (Vj). This suggests that voltage gating involves an inherent conformational change of the connexin that is "global" in that many parts of the molecule are involved. In contrast, the kinetically slower closure of Cx43 channels in response to cytoplasmic acidification has been associated with a discrete "ball and chain " mechanism involving the C-terminal domain. Application of a novel oocyte perfusion chamber has revealed that this gating mechanism also requires additional heat labile component(s) of the cytoplasm. Analogous studies of the gating of Cx43 channels by v-src indicate that this response also utilizes a "ball and chain" mechanism. Contrary to previous findings, however, this appears not to depend on phosphorylation of Cx43, but may also require interaction with cytoplasmic factors, possibly by way of SH3 domains in the C-terminal tail. These initial results suggest that kinetically slower chemical gating of gap junctions may occur via interactions between discrete domains of the connexin, mediated in at least some cases by additional soluble factors, while faster, electrically mediated gating events represent global conformational changes of the connexin.

## 1. Molecular Analysis of Channel Gating

The mechanisms that underlie the switch between open and closed states is appropriately portrayed as a holy grail in the structural and functional analysis of ion channels. Despite many innovative studies, this is understood in surprisingly few cases. The inactivation "ball and chain" gate of $K^+$ channels is perhaps the only case where most of the molecular components have been identified (see[1] for review). In other cases certain components have been defined, such as the voltage sensor of $Na^+$ and $K^+$ channels [2], the ligand binding site, or the gating movement within the pore of ACh Receptors [3]. However, the conformational changes that link these steps are ill defined. Gap junction channels are likely to provide a novel perspective to this field as the intercellular nature of these channels results in certain unique properties. Firstly, the default state of the channels appears to be open, so the gating events involve closure of the channel. Secondly, in cases where the gating process involves general conformational changes of the protein, part of the activation energy barrier of this process must include contributions from the docking interface between the two hemichannels. Finally, the high resolution structures of the gap junction channel that are now becoming available (reviewed in [4]) greatly enhance the prospects for definitive conclusions from these molecular studies.

Closure of gap junction channels is induced by either response to cell trauma (e.g. drop in pH, increase in $Ca^{++}$ or voltage differences between the cells - see [5] for review) or physiological regulators such as phosphorylation (e.g. v-src or MAP kinase - reviewed in [6]). Kinetic studies at the single channel level indicate that *gap junction gating can be divided into two general classes: rapid (~2 msec) responses to electrical stimuli (changes in Vj), and; slower (10-40 msec) responses to chemical stimuli (e.g. pH)* that are also associated with initial channel formation [7]. We have applied molecular approaches to address the question of whether these kinetic differences reflect fundamentally different gating mechanisms.

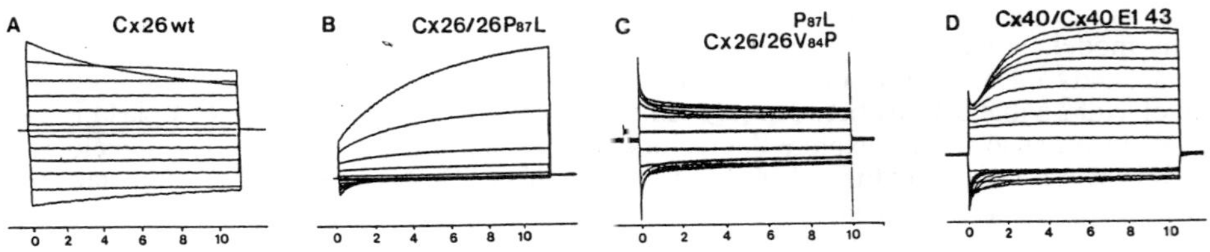


**Figure 1**: Current traces from oocyte pairs injected with cRNA as indicated. The mutation of the conserved proline in M2, as well as the chimera of Cx40 with E1 from Cx43 cause a remarkably similar phenotype by reversing the response to changes in Vj (increase instead of decrease in conductance). The effect of the P87 mutant could be reversed by the reintroduction of P one turn away in the α-helix. (Time shown in seconds)

## 2. Voltage Gating of Gap Junctions

Previous analyses of the molecular determinants of voltage gating in chimera of Cx32 and 26 have associated the voltage sensor with the N-terminus (NT), first transmembrane domain (M1) and extracellular loop (E1) domains of connexins [8], although other portions of the sequence were found to influence the gating parameters [9]. Mutation of a conserved proline in M2 produced a different phenotype in Cx26, reversing the response of the channel so that conductance increased rather than decreased in response to applied Vj (see[10], and Fig 1A&B). This suggested that the distortion introduced in M2 by this proline played a central role in the conformational change that links the original response to voltage by the sensor to eventual movements of the channel lining that would normally lead to channel closure. The critical role of proline in this response is more definitively established by recent studies in which the proline was reintroduced 3,4 or 7 residues away from its original position in Cx26, representing one or two repeat distances in an α-helix. These mutants each served to re-establish the normal polarity of the response, albeit with different kinetics and sensitivity (Fig 1C).

We have noted similar changes in the response of other connexins to applied Vj (i.e. increases in conductance compared to the decreases seen in wt channels) in very different mutant constructs involving chimera between the extracellular loops of Cx43 and the transmembrane and cytoplasmic domains of Cx40 (described in [11]). These mutants were originally constructed for the purposes of defining determinants in the specificity of hemichannel docking. In this regard, the results indicated that specificity does not lie exclusively at the level of primary sequence, but also depends on the tertiary structure of the extracellular loops that is in part determined by the transmembrane and cytoplasmic domains (see abstract by Zhu et al., also [11]). However, most of these constructs also displayed responses to Vj that were in part reminiscent of the proline mutant discussed above (see example in Fig 1D). Although analysis of all of these mutants is needed at the single channel level, this result does implicate the docking domains of connexins in the conformational response leading to voltage gating of the channel This leads to a model of voltage gating, consistent with the earlier studies [9 & 10], as a "global" conformational change of the channel that involves many domains of the connexin sequence other than the sensor and channel lining themselves. As noted above, the energetics of any such "global" rearrangement is likely to be affected by the docking interaction between connexons, as seen here in the responses of these chimeras of Cx40 and 43.

## 3. Chemical Gating of Gap Junctions

### *1. pH Gating*

In sharp contrast to the case with voltage gating described above, the closure of gap junctional channels in response to cytoplasmic acidification has been associated with discrete domains. Demonstration that the truncated C-terminal domain of Cx43 could still mediate gating of the channel led to the proposal that it serves as a "particle" in a "particle-receptor" model of gating [12] analogous to the "ball and chain" inactivation gate of $K^+$ channels [1]. This model, however, may not be universal, as the less sensitive pH responses seen in Cx32, and truncated forms of Cx43, do not appear to require the C-terminal region, but may involve close interactions between its membrane proximal regions and the cytoplasmic loop [13]. In order to further explore the factors that mediate this gating, we have developed a novel, paired oocyte perfusion/voltage clamp chamber. The two oocytes are maintained in electrically isolated chambers, communicating only through a small hole. This allows one of the oocytes to be cut open for perfusion with a variety of pseudo-intracellular solutions, while the other oocyte can remain intact with an independently controlled extracellular environment.

Previous studies using other perfused systems where the connexin composition was unknown had indicated that the gap junctions could respond to changes in pH and $Ca^{++}$ independently [14], but with significantly reduced sensitivity compared to their synergistic action [15]. In the current system, where the connexin composition is controlled and the perfusion conditions can be monitored more readily, we were surprised to find that perfusion resulted in complete loss of responsiveness of Cx43 and Cx32 to pH (Fig. 2A) and $Ca^{++}$ (data not shown) changes. This was demonstrated not to be due to irreversible changes/denaturation of the gap junction upon perfusion, as the responsiveness to reduced pH or elevated $Ca^{++}$ on the intact side was fully retained (Fig 2A). Furthermore, it was found that the response of Cx43 to pH could be restored with the addition of an oocyte cytoplasmic extract (Fig 2B). The somewhat slower kinetics and incompleteness of this response is likely to stem from the dilution of this extract on its reintroduction to the system. The activity appears to be heat labile, and may be different for Cx43 and 32, as the latter appears less stable. Possible candidates for this factor(s) abound ranging from various $Ca^{++}$ binding proteins to kinases, etc. However, even without their definitive identification, it is clear that pH gating is more complex than previously thought, as the connexin cannot be considered in isolation.

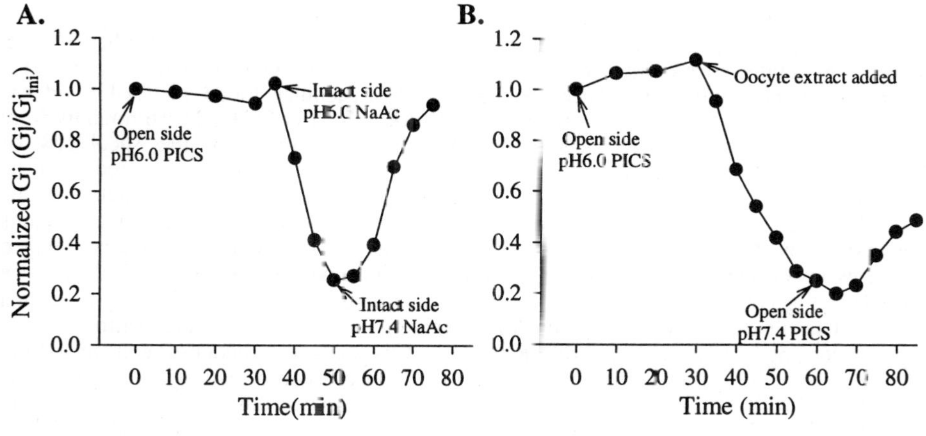


**Figure 2:** (A) A novel chamber that allows perfusion of one of a pair of coupled *Xenopus* oocytes with different pseudo-intracellular solutions (PICS) was used to demonstrate the loss of pH induced closure of Cx43 junctions upon removal of the cytoplasmic milieu, although the reversible uncoupling in response to acidification was retained on the intact side. (B) Reversible pH induced uncoupling could be restored to the perfused side by addition of an oocyte cytoplasmic extract.

## 2. *v-src mediated gating of Cx43*

Although phosphorylation has frequently been associated with the regulation of many channels, it has rarely been implicated in their direct gating. A notable exception that has been described is the closure of Cx43 gap junction channels induced by the *v-src* oncogene, associated with phosphorylation of tyr 265 [16]. Given the implied role of the C-terminal domain, we were interested in a comparison of the mechanism of this gating with that of pH described above. Oocyte pairs injected with Cx43 cRNA were allowed to develop stable conductance levels overnight before injection with cRNA for pp60$^{v\text{-}src}$. Final conductance levels, reached 6 to 12 hours later, were then recorded as a percentage of the original levels in the same oocyte pair. In this way, the effects of *v-src* on gating were largely isolated from any effects on biosynthesis of the connexin protein. Consistent with the previous implication of tyr 265 in this process [16], we found that much of the inhibition by *v-src* could be eliminated by truncation of Cx43 at residue 257 (Table I). However, a more complete loss of inhibition was achieved on further truncations to residue 245 (data not shown), suggesting roles of additional parts of the C-terminal sequence. In an analogous experiment to those of Delmar and colleagues on pH gating, we also found that *v-src* inhibition could be completely restored by the co-expression of the truncated form of Cx43 and the C-tail domain as separate polypeptides (Table I). Thus, this oncogene mediated closure of gap junction channels also appears to use a "ball and chain" mechanism.

A logical extension of this model would be that the signal for triggering interactions of the "ball" with the rest of the channel would be addition of a phosphate by the *v-src* tyrosine kinase. However, contrary to this earlier work, we found that changing tyr 265 to phe had no effect on the inhibition by *v-src* (Table I). This was consistent with our earlier finding that truncations of the C-terminus at residue 257 only partially eliminated *v-src* inhibition of coupling. Preliminary studies on deletions of specific domains within the C-terminus (mutants kindly provided by Taffet and Delmar - SUNY, Syracuse) have supported this contention, and implicated the two putative SH3 target domains in the C-terminus (between residues 240 and 280) as playing a role in this process. This suggests that *v-src* mediated gating of Cx43 may also require interactions with other cytoplasmic factors, presenting even closer analogies to the mechanism deduced for pH gating. Based on initial co-precipitations with the Cx43 tail (Zhou et al, in preparation), pp60$^{v\text{-}src}$ itself represents a possible candidate for such an interacting factor.

To date, the reasons for the discrepancy between these data and earlier studies that implicated tyr 265 in this process[21] remain to be elucidated. One possibility is that cRNA forv-*src* and Cx43 were simultaneously injected in the previous work. In contrast, our results reflect only inhibitory effects of *v-src* after stable expression levels of Cx43 had been reached, and thus should predominantly reflect effects on existing channels. Thus, the greater levels of inhibition that were correlated with tyrosine phosphorylation by Swenson et al.[16] may be related to effects on the biosynthesis of Cx43. Certainly, phosphorylation on serines has already been correlated with regulation of Cx43 assembly [17], so this would not be without precedent.

**Table I:** Inhibition of Various Cx43 Constructs upon Expression of pp60$^{v\text{-}src}$ in *Xenopus* oocytes. Normalized conductance represents the ratio of conductances post- and pre- v-src injection in the same oocyte

| Construct | Normalized conductance | S.E. | Number of oocytes injected |
|---|---|---|---|
| Cx32 | 0.66 | 0.05 | 7 |
| Cx43 | 0.009 | 0.002 | 22 |
| Δ257 | 0.15 | 0.03 | 14 |
| Δ257 + C-TERM$_{258}$ | 0.01 | 0.005 | 10 |
| Cx43Y265F | 0.01 | 0.006 | 13 |
| Δ241-260 | 0.15 | 0.05 | 9 |
| Δ261-280 | 0.22 | 0.05 | 8 |

## 4. Conclusions

The compilation of data presented here, and published previously from a number of other laboratories, suggest some generalized models that, while far from proven, pose useful hypotheses for testing. There is now good evidence that the "ball and chain" model of pH gating of Cx43 may be a general motif utilized by other forms of chemical gating such as that induced by *v-src*. It is also apparent that this model needs to include the effects of accessory cytoplasmic factors that are likely to be unique to each form of gating. The availability of the oocyte perfusion chamber described here should greatly facilitate the identification and analysis of such factors. It is also possible that the slower kinetics associated with chemical gating of gap junctions could to some extent derive from the reliance on intermolecular interactions. In contrast, the more rapid response of gap junctions to electrical stimuli such as changes in Vj appears to involve a more "global" conformational change of the connexin itself that involves many domains within the molecule, predominantly those within the membrane and in the extracellular docking surface. Clearly major challenges lie ahead to refine and reinvent these crude models through further mutagenic and biochemical approaches that will be increasingly enhanced by the higher resolution images of the structure of the channel that are emerging [4].

#### Acknowledgments

We would like to express our thanks to LieXian Xu, Steve Taffet, Mario Delmar, Jurgen Schwarz and Klaus Willecke for providing several of the mutant constructs used in these studies. We would also like to thank Heather Kuruvilla and Joe Smith for providing preliminary data and Charles Fourtner for discussions on all of this work. Support was provided by NIH awards from the NCI, HLBI and GMS to BJN, a Max Planck Prize to BJN and Klaus Willecke, and an AHA Investigator award to BJN

## References

[1] C. Miller, 1990: annus mirabilis of potassium channels. *Science* **252** (1991) 1092-1096.

[2] E. Perozo *et al.*, S4 mutations alter gating currents of Shaker $K^+$ channels. *Biophys. J.* **66** (1994) 345-354.

[3] N. Unwin, Acetylcholine receptor channel imaged in the open state. *Nature* **373** (1995) 37-43.

[4] M. Yeager and B.J. Nicholson, Structure of gap junction intercellular channels. *Curr. Opin. Struct. Biol.* **6** (1996) 183-192.

[5] R. Bruzzone *et al.*, Connections with connexins: the molecular basis of direct intercellular signaling. *Eur. J. Biochem.* **238** (1996) 1-27.

[6] A.F. Lau *et al.*, Regulation of connexin43 function by activated tyrosine protein kinases. *J. Bioenerg. and Biomemb.* **28** (1996) 359-368.

[7] F.F. Bukauskas *et al.*, Biophysical properties of gap junction channels formed by mouse Cx40 in induced pairs of transfected human HeLa cells. *Biophys. J.* **68** (1995) 2289-2298

[8] V.K. Verselis *et al.*, Opposite voltage gating polarities of two closely related connexins.*Nature* **368** (1994) 348-351.

[9] J.B. Rubin *et al.*, A domain substitution procedure and its use to analyze voltage dependence of homotypic gap junctions formed by connexins 26 and 32. *Proc. Nat. Acad. Sci. USA* **89** (1992) 3820-3824.

[10] T.M. Suchyna *et al.*, Identification of a proline residue as a transduction element involved in voltage gating of gap junctions. *Nature* **365** (1993) 847-849.

[11] S. Haubrich *et al.*, Incompatibility of connexin 40 and 43 hemichannels in gap junctions between mammalian cells is determined by intracellular domains. *Mol. Biol. Cell* **7** (1995) 1995-2006.

[12] G.E. Morley *et al.*, Intramolecular interactions mediate pH regulation of connexin43 channels. *Biophys. J.* **70** (1996) 1294-1302.

[13] X.G. Wang and C. Peracchia. Positive charges of the initial C-terminus domain of Cx32 inhibit gap junction gating sensitivity to $CO_2$ *Biophysical J.* **73** (1997) 798-806

[14] D.C. Spray *et al.*, Gap junctional conductance: comparison of sensitivities to H and Ca ions. *Proc. Natl. Acad. Sci. USA* **79** (1982) 441-445

[15] R.L. White *et al.*, Gap junctional conductance between pairs of ventricular myocytes os modulated synergistically by $H^+$ and $Ca^{++}$. *J. Gen. Physiol.* **95** (1990) 1061-1075.

[16] K.I. Swenson *et al.*, Tyrosine phosphorylation of the gap junction protein connexin43 is required for the pp60$^{v\text{-}src}$-induced inhibition of communication. *Cell Regulation* **1** (1990) 989-1002.

[17] L.S. Musil and D.A. Goodenough, Biochemical analysis of Connexin43 intracellular transport, phosphorylation, and assembly into gap junctional plaques. *J. Cell Biol.* **115** (1991) 1357-1374.

*Gap Junctions*
*R. Werner (Ed.)*
*IOS Press, 1998*

# A Ball-and-Chain Model for the Chemical Regulation of Connexin43

Mario Delmar, Kathleen Stergiopoulos, Nobuo Homma, Jose F Ek-Vitorin and Steven M Taffet
*SUNY Health Science Center, Syracuse Ny 13210*

Abstract. We have been interested in the characterization of the molecular mechanisms mediating chemical regulation of connexins. In this paper we review previous studies and outline some new ones showing that chemical regulation of connexin43 follows a ball-and-chain model. We also discuss the possibility that other molecular intermediaries are involved in the pH gating process.

Like in many families, connexins may look alike, but they have rather disparate behavior. Different connexins form channels with unique conductance and permselectivity properties;[10] moreover, individual connexins respond differently to phosphorylation[3,6] as well as to acidification of the intracellular space.[20,23,24,27] Given the functional diversity of connexins, the task of studying gap junction regulation seems overwhelming. Our approach has been to focus our attention on the regulation of one specific connexin (Cx43), and by one specific factor (pHi). In the following pages, we briefly review our previous work on pH gating of Cx43, discuss some new hypotheses, and describe recent preliminary data suggesting that molecular models that apply to pH gating may be pertinent to the study of connexin regulation by other factors.

## *Structural bases for the pH gating of Cx43: The ball-and-chain model*

We have shown that truncation of the carboxyl terminal of Cx43 at amino acid 257 drastically reduces the sensitivity of Cx43 to pH gating.[20,21] A similar observation was reported at the meeting regarding the pH sensitivity of Cx50.[18] Interestingly, truncation of the carboxyl terminal of Cx32[26] or Cx46 (unpublished) does not have the same effect, further showing the diversity of connexin function. In the following paragraphs, we summarize the current knowledge on the structure-function relation of pH gating of Cx43.

Summary of Methods - Data were obtained from pairs of Cx43-expressing Xenopus oocytes. The cells were injected with antisense against the endogenous Cx38, as well as with mRNA for wild-type Cx43. The cells were also injected with the dextran form of the pH-sensitive dye SNARF. This fluorophore changes its emission properties depending on the pH of the surrounding solution. After proper calibration, SNARF serves as an excellent indicator of intracellular pH in the range of 8.5 to 6.3.[7,21] SNARF-injected oocyte pairs expressing Cx43 were recorded electrophysiologically for measurements of junctional conductance (Gj); Changes in $G_j$ were measured relative to the maximal conductance recorded during acidification (i.e., $G_{j\ max}$). Acidification was induced by superfusing the oocytes with a bicarbonate-buffered solution gassed with progressively increasing concentrations of $CO_2$. Simultaneously, optical recordings were obtained to measure pHi. In this configuration, when the appropriate acidification protocol was used, we could correlate the pHi value with the degree of cell-cell coupling at each point in time. Details can be found in Morley et al,[21] as well as in Delmar et al.[7]

Role of the carboxyl terminal domain - Earlier studies from our laboratory showed that truncation of the carboxyl terminal of Cx43 impaired pH gating.[20] More recently, we have shown that Cx43 pH gating results from an interaction between the carboxyl terminal domain, and a separate region of connexin.[2] Figure 1 summarizes the results. In all panels, the diagram at the left represents the membrane topology of Cx43, and points to the particular construct being tested. Panel A shows the dependence of Cx43 to intracellular pH. As shown in Panel B, truncation of the carboxyl terminal of Cx43 at amino acid 257 (mutant M257; Cx43 is 382 amino acids long) caused the loss of pH sensitivity within the pHi tested (data labelled M257). Panel C shows an important piece of data: When M257 was co-expressed with mRNA coding only for the carboxyl terminal domain (amino acids 259 to 382; data labelled M257+CT), pH sensitivity was partly restored. In other words, the carboxyl terminal is capable of closing the channels upon acidification even if it is expressed separate from the rest of the protein. The latter indicates that the carboxyl terminal interacts (directly or indirectly) with a separate region of connexin in order to close the channel.

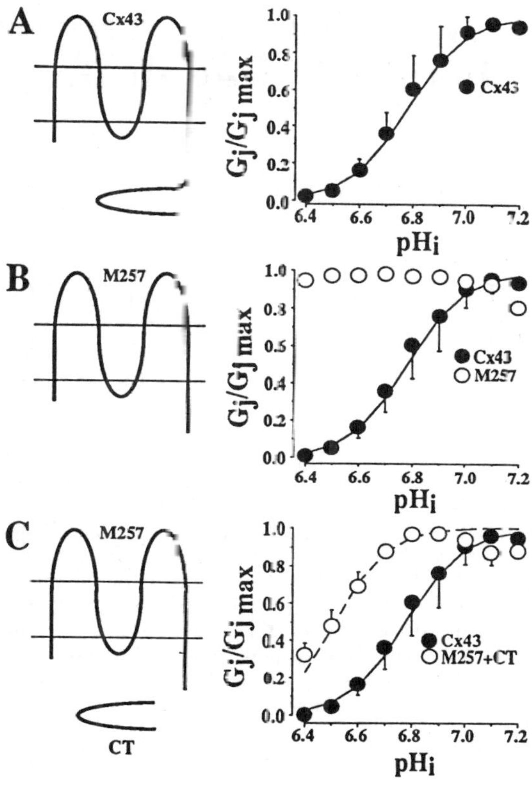


Figure 1

The particle-receptor model - The data presented in Figure 1 led us to a particle-receptor model for Cx43 pH gating, similar to the ball-and-chain model of voltage dependent gating of nonjunctional channels.[2,12] In its simplest version (shown by the diagram in Figure 2), we propose that the carboxyl terminal of Cx43 constitutes a gating particle (labelled "CT" in Figure 2, top). A separate domain of Cx43, related to the pore-forming region, would be a specific receptor for such particle. Upon intracellular acidification (Fig 2, middle), the particle would bind non-covalently to a separate specific domain acting like a receptor site. Protonation of histidine residues may be part of the particle-receptor interaction.[8] The particle-receptor interaction would lead to channel closure. Accordingly, the results presented in Figure 1 can be explained by the fact that the carboxyl terminal fragment can find its receptor even if expressed separately from the rest of the channel (figure 2, bottom).

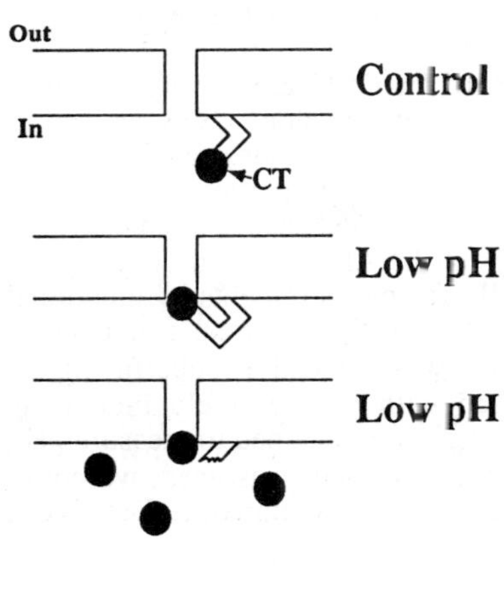


Figure 2

One of the questions that stem from this model is: how many carboxyl terminal domains does it take to close the channel? Indeed, 12 CT domains are supposed to "surround" a gap junction pore. To address this question, we created concatenated constructs where cDNA coding for M257 was cloned in tandem with a full length Cx43. Theresulting protein represented a "dimer" of Cx43, but with only one CT. It was our rationale that 3 of these dimers would form a hexamer, with only 3 CT domains per connexon. Several of these concatenated constructs (with 2, 3 or 4 M257 subunits in tandem) were engineered. They all showed pH sensitivity. To our surprise, though, a concatenated dimer lacking the CT domain (i.e., M257 in tandem with another M257) also resulted to be pH sensitive. The latter indicated that the molecular function of concatenated dimers is not the simple addition of the functions of each independent subunit. Indeed, as shown in Figure 1B, M257 is not pH sensitive (in the range of pH 7.2-6.4). However, two concatenated M257's showed the same pH sensitivity as a normal Cx43. In our interpretation, these results warn against the use of concatenated constructs to study connexin function, since they seriously question the assumption that each molecule in the tandem will behave as though it were expressed separate from the others.

### *The proline-rich region and Cx43 regulation*

We have conducted a thorough survey of the regions of the CT domain that are necessary for acidification-induced uncoupling.[9] Of particular interest was the observation that amino acids within the region 261-300 are essential for normal pH gating.[9] This segment includes a region rich in proline residues. A repeat $(PXX)_4$ is found between amino acids 274 and 285. Proline-rich sequences commonly form left-handed alpha-helices,[1,28] and they are often involved in protein-protein interactions.[1,5,19,28,29] We therefore proposed that the proline-rich region in the Cx43 CT is implicated in a binding reaction that is critical for acidification-induced channel closure.

The proline-rich region is also essential for the phosphorylation of Cx43 by mitogen-activated protein kinase (MAP-K),[25] as well as for the binding of Cx43 to the SH3 domain of v-src.[15,17] Moreover, the interaction between the SH3 domain of v-src and the CT of Cx43 can be prevented by a synthetic peptide formed by the sequence 271-287 of Cx43. In this case, the proline-rich 17-mer peptide competes against the equivalent region of the CT domain for binding to the kinase.[15,17] The same peptide acts as a substrate for MAPK.[25] Given that the proline-rich region is essential for normal pH gating, we asked whether the same 17-mer peptide could prevent the pH gating reaction. The latter would be expected if at least two conditions were met: 1) that the proline-rich region is involved, directly or indirectly, in the particle-receptor binding, and 2) that the binding of the proline-rich region to its target molecule is not sufficient to close the channel. Our recent results (in collaboration with Alan Lau and Martha Kanemitsu) indicate that the 17-mer peptide does interfere with acidification-induced uncoupling of Cx43. The possible role of SH3 domain-containing molecules on pH gating ought to be further studied.

### *Applicability of the particle-receptor model to gap junction regulation by other factors*

Acidification of the intracellular space is only one of the several ways by which the intra- and extracellular environments modulate cell-to-cell communication. We therefore asked whether the ball-and-chain model applies to the regulation of connexins by factors other than pH. To address this question, we developed an experimental model of insulin-induced uncoupling in Cx43-expressing oocytes (in collaboration with Alan Lau). Our preliminary results, communicated at the recent gap junctions meeting, show that the insulin-induced regulation of Cx43 follows the ball-and-chain paradigm. However, the structure-function determinants of insulin-induced uncoupling are not the same as those required for pH gating, thus indicating that these are two independent modes of connexin regulation that share a common gating mechanism.

*The ball-and-chain model and possible intermediary steps*

Our studies show that a) truncation of the CT domain of Cx43 at amino acid 257 prevents pH gating, and b) co-expression of the carboxyl terminal domain as a separate protein rescues the pH sensitivity of the truncated, otherwise pH-insensitive mutant. In view of the current data, at least three models could be proposed (see Figure 3):

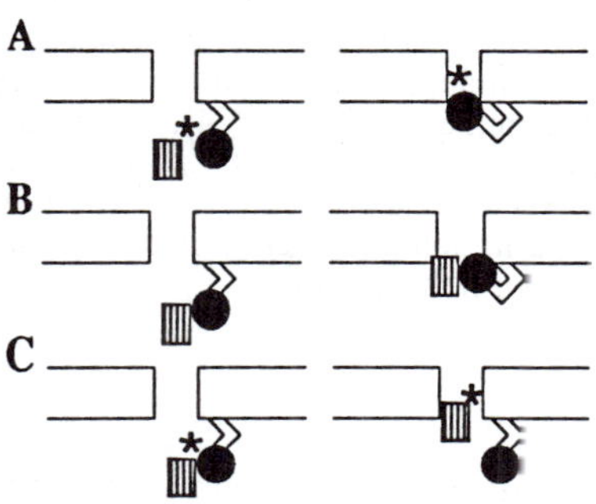


Figure 3

Model A. The CT domain (identified as the solid circle in Figure 3A) is post-translationally modified by an intermediary molecule (solid square). Once modified (asterisk), the CT domain interacts directly with the pore-forming region of the channel.

Model B: The CT domain binds to an intermediary. The intermediary molecule acts as an adapter for the interaction between the CT domain and the pore (Figure 3B).

Model C: The CT domain acts to modify an intermediary molecule. The intermediary then binds to the pore-forming region to close the channel (see Figure 3C).

A combination of the first two models is also possible. In that case, the CT would first be modified. The modification (e.g., phosphorylation) would "prime" the CT, enabling it for interaction with the adapter molecule. While the first two models (A and B) assume close physical proximity between the CT domain and the rest of the connexin, the third one implies that pH gating does not require of direct interaction between the separate connexin domains. Further experiments currently in progress in our laboratory are aimed at discarding model C as a viable possibility.

*Conclusion*

It is the current view that gap junctions are not only passive conduits of electrical charge, but dynamic filters that modulate the passage of molecular messages within the cellular network. Hence, not only the presence, but also the regulation of connexins is important for synchronous tissue function. Our studies have focused on the regulation of Cx43 by pHi. Control of Cx43 by pHi may be critical in the development of lethal cardiac arrhythmias during ischemia.[4,13,14,16] Moreover, our studies have led to molecular models that apply to the regulation of connexin by other factors.[11,22] Understanding the molecular events controlling connexin regulation, and developing means to manipulate their occurrence will allow us to learn more about the importance of intercellular communication both in health and disease.

## References

1. Adzhubei AA, Sternberg MJE: Left-handed polyproline II helices commonly occur in globular proteins. *J Mol Biol* 1993;229:472-493

2. Amstrong CM, Bezanilla F: Inactivation of the sodium channel. II. Gating current experiments. *J Gen Physiol* 1977;70:567-590

3. Bruzzone R, White TW, Paul DL: Connections with connexins: The molecular basis of direct intercellular signaling. *Eur J Biochem* 1996;238:1-27

4. Cascio WE, Johnson TA, Gettes LS: Electrophysiologic changes in ischemic ventricular myocardium: I. Influence of ionic, metabolic and energy changes. *J Cardiovasc Electrophysiol* 1995;6:1039-1062

5. Cohen GB, Ren R, Baltimore D: Modular binding domains in signal transduction proteins. *Cell* 1995;80:237-248

6. Darrow BJ, Laing JG, Lampe PD, Saffitz JE, Beyer EC: Expression of multiple connexins in cultured

neonatal rat ventricular myocytes. *Circ Res* 1995;76:381-387

7. Delmar M, Morley GE, Taffet SM: Molecular analysis of the pH regulation of the cardiac gap junction protein Connexin43, in Spooner PM, Joyer RW, Jalife J (eds): *Discontinuous conduction in the heart.* Armonk,NY., Futura Publishing Co, 1997, pp In Press

8. Ek JF, Delmar M, Perzova R, Taffet SM: Role of His 95 in pH gating of the cardiac gap junction protein. *Circ Res* 1994;74:1058-1064

9. Ek-Vitorin JF, Calero G, Morley GE, Coombs W, Taffet SM, Delmar M: pH regulation of connexin43: molecular analysis of the gating particle. *Biophys J* 1996;71:1273-1284

10. Elfgang C, Eckert R, Lichtenberg-Frate H, Butterweck A, Traub O, Klein RA, Hussler DF, Willecke K: Specific permeability and selective formation of gap junction channels in connexin-transfected HeLa cells . *J Cell Biol* 1995;129:805-817

11. Homma N, Taffet SM, Lau AF, Delmar M: Insulin, and insulin-lige growth factor, induce uncouping in Cx43-expressing oocytes. The gating mechanism follows a particle-receptor model. *Proc Intl Meeting on Gap Junctions* 1997;pp:62

12. Hoshi T, Zagotta WN, Aldrich RW: Biophysical and molecular mechanisms of Shaker potassium channel inactivation. *Science* 1990;250:533-538

13. Janse MJ, Opthof T: Mechanisms of ischemia-induced arrhythmias, in Zipes DP, Jalife J (eds): *Cardiac Electrophysiology: From Cell to Bedside.* Philadelphia, W.B. Saunders Co., 1995, pp 489-496

14. Janse MJ, Wit AL: Electrophysiological mechanisms of ventricular arrhythmias resulting from myocardial ischemia and infarction. *Physiol Rev* 1989;69:1049-1169

15. Kanemitsu M, Loo Y, Simon LWM, Lau AF, Eckhart W: Tyrosine phosphorylation of connexin43 by vsSrc is mediated by SH2 and SH3 domain interactions. *J Biol Chem* 1997;In Press:

16. Kleber AG, Riegger CB, Janse MJ: Electrical uncoupling and increase of extracellular resistance anfter induction of ischemia in isolated, arterially perfused rabbit papillary muscle. *Circ Res* 1987;61:271-279

17. Lau AF, Kurata WE, Kanemitsu MY, Loo LWM, Warn-Cramer BJ, Eckhart W, Lampe PD: Regulation of connexin43 function by activated tyrosine protein kinases. *J Bioenergetics and Biomembranes* 1996;28:359-368

18. Lin JS, Eckert R, Kistler J: calpain cleaves connexins in the mammalian lens. *Proc Intl Meeting on Gap Junctions* 1997;pp:74

19. Linn L, Brasseur R: The hydrophobic effect in protein folding. *FASEB J* 1995;9:535-540

20. Liu S, Taffet SM, Stoner L, Delmar M, Vallano ML, Jalife J: A structural basis for the unequal sensitivity of the major cardiac and liver gap junctions to intracellular acidification. The carboxyl terminal length. *Biophys J* 1993;64:1422-1433

21. Morley GE, Taffet SM, Delmar M: Intramolecular interactions mediate the pH regulation of Connexin43. *Biophys J* 1996;70:1294-1302

22. Nicholson BJ, Zhou L, Cao F, Fourtner C: Diverse molecular mechanisms of gap junctional channel gating to voltage, pH, Ca and phosphorylation. *Proc Intl Meeting on Gap Junctions* 1997;pp:115

23. Steiner E, Ebihara L: Functional characterization of canine connexin 45. *J Membr Biol* 1996;150:153-161

24. Wang X, Li L, Peracchia LL, Peracchia C: Chimeric evidence for a role of the connexin cytoplasmic loop in gap junction channel gating. *Pflugers Archiv* 1996;431:844-852

25. Warn-Cramer BJ, Lampe PD, Kurata WE, Kanemitsu MY, Loo LWM, Eckhart W, Lau AF: Characterization of the mitogen-activated protein kinase phosphorylation sites on the connexin-43 gap junction protein. *J Biol Chem* 1996;271:3779-3786

26. Werner R, Levine E, Rabadan-Diehl C, Dahl G: Gating properties of connexin32 cell-cell channels and their mutants expressed in Xenopus oocytes. *Proc R Soc Lond* 1991;243:5-11

27. White TW, Bruzzone R, Wolfram S, Paul DL, Goodenough DA: Selective interactions among the multiple connexin proteins expressed in the vertebrate lens: The second extracellular domain is a determinant of compatibility between connexins. *J Cell Biol* 1994;125:879-892

28. Williamson MP: The structure and funcion of proline-rich regions in proteins. *Biochem J* 1994;297:249-260

29. Yu H, Chen JK, Feng S, Dalgarno DC, Brauer AW, Schreiber SL: Structural basis for the binding of proline-rich peptides to SH3 domains. *Cell* 1994;76:933-945

*Gap Junctions*
*R. Werner (Ed.)*
*IOS Press, 1998*



# Structural Considerations for Connexon Docking

**Guy A. Perkins[§,‡], Daniel A. Goodenough[¶] and Gina E. Sosinsky[§,◊]**
*§Dept. of Neurosciences 0608 & ◊Dept. of Biology 0322, University of California, San Diego, La Jolla, CA 92093, ‡Dept. of Biology, San Diego State University, San Diego, CA, 92182-4614 ; ¶Dept. of Cell Biology, Harvard Medical School, Boston, MA 02115*

**Abstract.** The intercellular gap junction membrane channel provides a low resistance pathway for cell-cell communication in animal cells. Each of two cells contributes one hemichannel or connexon. Formation of the channel occurs when two connexons recognize each other, pair and seal at their extracellular surfaces. Here we discuss insights into how the structure of an individual connexon which we determined [1], the surface topology of the extracellular domains and the primary amino acid sequence influence the docking of two connnexons to form an intercellular channel.

## Introduction

Gap junctions are unique among membrane channels in that the channels span two cell membranes. These channels are created through the interaction of two hexamers (two hemichannels, called connexons) of connexins which meet and dock to form a sealed cell-cell conduit [2]. Each connexin has two extracellular loops (E1, E2) ~40 amino acid residues per loop which are key components to connexon recognition and docking. Each loop has three cysteines which are strictly conserved [3] and form inter-loop disulfide bonds within connexins [4] that contribute to the structural rigidity [5, 6] of the extracellular domains (see Fig. 1). A fundamental question we are focusing on is how the structure of these domains may control interactions between like and unlike connexins.

## Results and Discussion

The key tool that allowed us to visualize the extracellular domains was the development of reproducible procedures for the mechanical [5, 7] or biochemical splitting [8] of purified gap junctions with high efficiency without losing good structural integrity [1]. By using the atomic force microscope (AFM) cantilever to perform mechanical dissection of gap junction plaques to produce hemichannel plaques Hoh and coworkers [5] showed that the extracellular side had protrusions (height modulations) at the six-fold connexin subunits, giving the first indication that connexons may dock in the same fashion as interdigitating cogs by a 30° rotation with respect to each other. Because urea (a chaotropic agent) is the principal component in the biochemical splitting protocol, it is assumed that hydrogen bonding and hydrophobic interactions play a substantive role in connexon pairing, which is supported by the high percentage of hydrophobic residues in both E1 and E2 [9] (Fig. 1). Moreover, it was determined by Ghoshroy et al. [8] through the use of EGTA (a $Ca^{2+}$ chelator) that divalent cations contribute to the docking stability, and that higher temperatures were necessary for successful splitting (indicative of breaking hydrophobic interactions).

Perkins et al. [1] built on the extracellular domain studies of Hoh and coworkers [5, 7] and Ghoshroy et al. [8] by generating a three-dimensional (3-D) reconstruction of the connexin32 (*Cx32*) connexon from rat liver (see Fig. 1 top left and Fig. 2) produced by biochemical splitting and imaged at low temperature and low electron dose using electron microscope tomography coupled with crystallographic image-processing tools. This reconstruction emphasized the structural asymmetry between the extracellular domains which displayed strong peaks (protrusions) and the cytoplasmic side which had only slight height modulations. With the 3-D structure of a connexon at a resolution of ~2 nm, we were able to address the important question of how the extracellular surface facilitates the recognition of two apposing connexons. Our connexon structure provided further support for the interdigitating model of docking proposed by Hoh et al. [5] and extended the visualization of

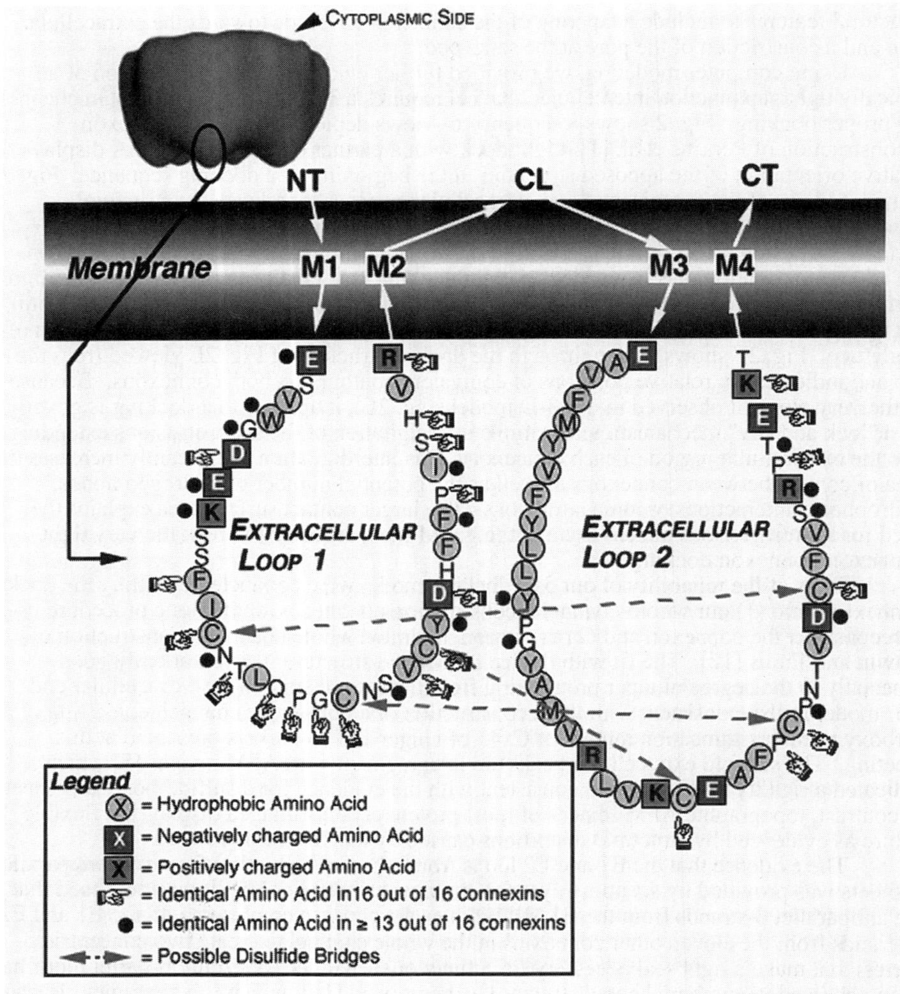


**Fig. 1. Sequence information about structure of extracellular domains.** *Top Left:* A side view of the three-dimensional reconstruction of a connexon. A schematized topology drawing of Cx32 is shown *below*. Abbreviations: **NT** = amino terminus, **CL**=cytoplasmic loop, **CT** = carboxy terminus, **M1, M2, M3, M4** indicate the four transmembrane spanning putative α–helical domains. We have indicated on the reconstruction the extracellular domains of one possible subunit or connexin and the arrow from this structure points to the primary amino acid sequence of the two extracellular loops. On the primary sequence we have indicated all hydrophobic amino acids (strongly hydrophobic residues are W, Y, F, L, I; intermediate hydrophobic residue is V; and weakly hydrophobic resides are A, M, C; [24]). Also indicated are both positively (K, L) and negatively charged (E and D) residues. The disulfide bonds are indicated as per the arrangement of Foote et al. [4]. Note the strong identity of these primary sequences among the connexin family that has been postulated to be a requirement for heterotypic or heteromeric channel formation (sequence alignment according to Peracchia et al. [2]).

structural features to include a tapering of the connexon envelope towards the extracellular side and a constriction of the pore at the same end.

Using computer modeling, we provided further evidence that the formation of an ionically tight gap junction intercellular channel requires a 30° rotation between hemichannels for proper docking. Fig. 2 shows a sequence of views depicting how the connexon reconstruction of Perkins et al. [1] might dock with a partner connexon. Fig. 2A displays the relative orientation of the apposed structures at the beginning of a docking sequence. Fig. 2B provides views of the cytoplasmic and extracellular surfaces to indicate the structural asymmetry between the two sides. Fig. 2C depicts how the peaks on the extracellular surface of one connexon fit into the valleys of the same surface on the apposed connexon. An additional rotation of 8° is needed for tight docking (see misfitting in Fig. 2D) and almost perfect fitting as shown in Fig. 2E. This well-studied 8° skew [10, 11] comes about from the detergent extraction used to isolate gap junction plaques and is a result of compaction of the lattice. Fig. 2F shows a slice through the docked structure of Fig. 2E viewed from the top and indicates the relative positions of equivalent contours on both connexons. Because of the very close fit observed in our 3-D model (Fig. 2E), it appears that docking is governed by a "lock and key" mechanism via a simple interdigitation of the six protrusions extending into the extracellular region of each connexon. This interdigitation significantly increases the area of contact between connexons as well as the potential number of hydrogen bonds, hydrophobic interactions or ionic attractions. The larger contact surface area explains the need for a strong chaotropic environment (e.g. 4M urea) required to break the very tight connexon-connexon docking.

Tests of the reliability of our 3-D docking model were conducted by fitting the docked connexons into (1) our whole-channel reconstruction obtained with the same procedure used to reconstruct the connexon and (2) the frozen, hydrated whole channel reconstruction of Unwin and Ennis [12]. The fit with the ice-embedded structure was remarkably good, especially in the degree of taper progressing from the cytoplasmic to the extracellular end. Our model is also consistent with the reconstruction of a homotypic channel consisting of a carboxy terminus truncation mutant of Cx43 of Unger and coworkers presented at this meeting. The six-fold extracellular peaks are in agreement with AFM images [5] which indicated a rigidity in this domain consistent with the evidence for disulfide-bonded β-sheets. In contrast, topographic AFM images of most protein crystals imaged displayed a flexible nature as evidenced by structural distortions caused by the AFM tip [13].

The evidence that the E1 and E2 loops from adjacent connexins form two antiparallel β-sheets was provided by scanning cysteine mutagenesis studies [4]. It was proposed that the antiparallel β-strands from the E1 and E2 in each connexin combines with the E1 and E2 β-strands from the eleven other connexins in the whole channel to create two concentric β-barrels that make a tight seal necessary for a functional channel. A similar β-barrel motif has been observed in another channel structure, α-hemolysin [14], which is a heptamer. Each α-hemolysin subunit in the transmembrane domain contributes two antiparallel β-strands to create a 14-strand β-barrel. By the use of synthetic peptides and connexin mutagenesis, it was discovered that a large fraction of the E1 and E2 residues are required for channel assembly [15]. It was proposed that once docking is completed an exchange of disulfide bonds stabilizes the open conformation. Using a new mapping scheme with a resolution of ~3 residues, N-glycosylation scanning mutagenesis, it was confirmed that the E2 sequence was in the extracellular domain, yet a considerable fraction was buried [16]. However, which portion of this sequence is buried has yet to be determined.

Recent work demonstrated that subsets of connexons comprised of different connexins will dock and form functional channels [17, 18, 19]. The results of Haubrich et al. [17] using Cx40 and Cx43 extended the work of White et al. [19] by providing evidence that both E1 and E2 from compatible connexin partners are required for proper docking. A further intriguing result was the role the C-terminal domain and cytoplasmic central loops played in determining **functional** incompatibility, *even when both loops were derived from the same connexin type* and could facilitate docking. The combination of E1, E2, C2 and C3 for functional compatibility may reflect the distinction between the channel gate, which can be activated by potential changes and cytoplasmic environment, and the docking gate, which is activated by connexon-connexon binding [20, 21]. Another role implicated for E1 and E2 is the correct integration of the connexin into the plasma membrane and connexon formation. The vast majority of connexons in the plasma membrane are in the closed state [22]. Yet, a

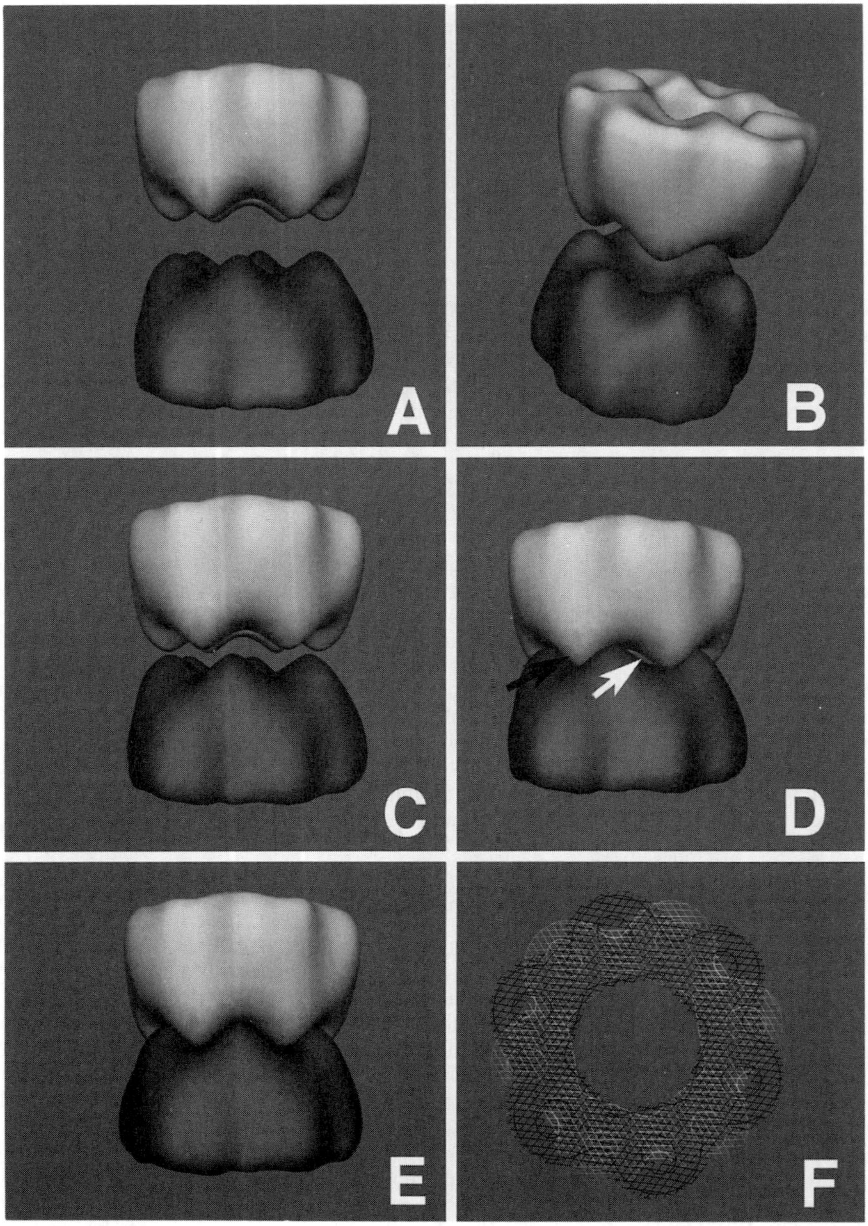


**Fig. 2. Sequence of views showing the docking procedure of two connexons. A.** 3D reconstruction of the Cx32 hemichannel and a partner connexon which was generated by rotating the reconstruction 180° along the X-axis. **B.** Tilted view of the two connexons in (A), showing the structural asymmetry between the cytoplasmic and extracellular surfaces. **C.** The two connexons were moved closer together along the Z axis until there was a close fit as in **D.** Arrows point to places of misfitting at the extracellular surface (black arrow point to an overlap and white arrow indicate a hole). **E.** A perfect fit was obtained after imposing an 8° rotation about the Z-axis due to the 8° skew of the connexon within the crystal lattice. **F.** Section through the center of the docked structure showing the 30° rotation between the two apposing connexons.

chimeric connexin which replaces the Cx32 E1 sequence with the Cx43 sequence transforms the connexon into an open hemichannel [23]. It is not known if this chimera induces an open channel because of interference with loop interactions or because a calcium binding site was eliminated and thereby reduces $Ca^{2+}$ sensitivity such that physiological extracellular $Ca^{2+}$ concentration cannot keep the hemichannel closed. Therefore, since the loops are highly conserved, small changes in the primary structure must cause small tertiary structural changes in the contact surface areas that must have long-range consequences on the physiological state of the channel.

*This work is supported in part by NIH Grants GM43217 (G.E.S.), GM18974 (D.A.G.), RR04050 (M.H. Ellisman) and NSF grant MCB-9629285 (G.E.S.).*

## References

[1] Perkins, G.A., Goodenough, D.A. and Sosinsky, G.E. (1997) Three-dimensional structure of the gap junction connexon. *Biophys. J.*, **72**, 533-544.

[2] Peracchia, C., Lazrak, A. and Peracchia, L.L. (1994) Molecular models of channel interaction and gating in gap junctions *in* Membrane channels. Molecular and cellular physiology. Academic Press, New York. pp. 361-377.

[3] White, T., Paul, D., Goodenough, D., and Bruzzone, R. (1995) Functional analysis of selective interactions among rodent connexins. *Mol. Biol. Cell* **6**, 459-470.

[4] Foote, C.I., Zhu, X., Zhou, L.and Nicholson, B.J. (1997) Pattern of disulfide linkages in the extracellular loop regions of Cx32 suggest a model of the docking interfaces of gap junctions. *J. Cell Biol.*, in press.

[5] Hoh, J., Sosinsky, G. E., Revel, J.-P., and Hansma, P. K. (1993) Structure of the extracellular surface of the gap junction by atomic force microscopy. *Biophys. J.* **65**, 149-163.

[6] Tibbitts, T. T., Caspar, D. L. D., Phillips, W. C., and Goodenough, D. A. (1990) Diffraction diagnosis of protein folding in gap junction connexons. *Biophys. J.* **57**, 1025-1036.

[7] Hoh, J. H., Lal, R., John, S. A., Revel, J.-P., and Arnsdorf, M. F. (1991) Atomic force microscopy and dissection of gap junctions. *Science* **235**, 1405-1408.

[8] Ghoshroy, S., Goodenough, D. A., and Sosinsky, G. E. (1995) Preparation, characterization, and structure of half gap junctional layers split with urea and EGTA. *J. Membr. Biol.* **146**, 15-28.

[9] Paul, D.L. (1986) Molecular cloning of cDNA for rat liver gap junction protein. *J.Cell Biol.* **103**, 123-134.

[10] Baker, T. S., Caspar, D. L. D., Hollingshead, C. J., and Goodenough, D. A. (1983). Gap junction structures. IV. Asymmetric features revealed by low-irradiation microscopy. *J.Cell Biol.* **96**, 204-216.

[11] Baker, T. S., Sosinsky, G. E., Caspar, D. L. D., Gall, C., and Goodenough, D. A. (1985). Gap junction structures VII. Analysis of connexon images obtained with cationic and anionic negative stains. *J. Mol. Biol.* **184**, 81-98.

[12] Unwin, P. N. T., and Ennis, P. D. (1984). Two configurations of a channel-forming membrane protein. *Nature* **307**, 609-613.

[13] Lacapere, J.-J., Stokes, D.L. and Chatenay, D. (1992) Atomic force microscopy of three-dimensional membrane protein crystals. *Biophys. J.* **63**, 303-308.

[14] Song, L., Hobaugh, H.R., Shustak, C., Cheley, S., Bayley, H. and Gouaux, J.E. (1996) Structure of the *Staphylococcal* α-hemolysin, a heptameric transmembrane pore. *Science* **274**, 1859-1866.

[15] Dahl, G. (1996). Where are the gates in gap junction channels? *Clinical Exper. Pharmacol. Physiol.* **23**, 1047-1052.

[16] Dahl, G., Nonner, W. and Werner, R. (1994) Attempts to define functional domains of gap junction proteins with synthetic peptides. *Biophys. J.* **67**, 1816-1822.

[17] Haubrich S., Schwarz H.J., Bukauskas F., Lichtenberg-Frate H., Traub O., Weingart, R., & Willecke K. (1996) Incompatibility of connexin 40 and 43 hemichannels in gap junctions between mammalian cells is determined by intracellular domains. *Mol. Biol. Cell* **7**, 1995-2006.

[18] White, T.W. and Bruzzone, R. (1996) Multiple connexin proteins in single intercellular channels: Connexin compatibility and functional consequences. *J. Bioener. Biomembr.* **28**, 339-350.

[19] White, T. W., Bruzzone, R., Wolfram, S., Paul, D. L., and Goodenough, D. A. (1994) Selective interactions among multiple connexin proteins expressed in the vertebrate lens: the second extracellular domain is a determinant of compatibility between connexins. *J. Cell Biol.* **125**, 879-892.

[20] Brink, P.R., Cronin, K. and Ramanan, S.V. (1996) Gap junctions in excitable cells. *J. Bioener. Biomembr.* **28**, 351-358.

[21] Veenstra, R.D. (1996) Size and selectivity of gap junction channels formed from different connexins. *J. Bioener. Biomembr.* **28**, 327-337.

[22] Paul, D.L., Ebihara, L., Takemoto, L.J., Swenson, K.I. and Goodenough, D.A. (1991) Connexin46, a novel lens gap junction protein, induces voltage-gated currents in nonjunctional plasma membrane of Xenopus oocytes. *J. Cell Biol.* **115**, 1077-89.

[23] Pfahnl, A., Zhou, X.W., Werner, R. and Dahl, G. (1997) A chimeric connexin forming gap junction hemichannels. *Pflugers Archiv.* **433**, 773-779.

[24] Schultz, G.E. and Schirmer, R.H. (1979) Principles of Protein Structure. Springer-Verlag, New York.

*Gap Junctions*
*R. Werner (Ed.)*
*IOS Press, 1998*

# Immunogold Labeling of Dispersed and Ubiquitinated Gap Junction Channels

Dieter F. HÜLSER, Beate REHKOPF, and Otto TRAUB*
*Biologisches Institut, Abt. Biophysik, Universität Stuttgart, D-70550 Stuttgart*
**Institut für Genetik, Abt. Molekulargenetik, Universität Bonn, D-53117 Bonn*

**Abstract.** Replicas of freeze fractured membranes of rapidly frozen cells were explored for gap junction channels and ubiquitinated membrane proteins with an indirect immunogold labeling technique. In BICR/M1R$_k$ cells Cx43 was identified in quasi-crystalline gap junction plaques, the same domains also contained ubiquitinated proteins. In Cx45-transfected HeLa cells we detected immunogold labeling of quasi-crystalline gap junction plaques and of contact areas with dispersed gap junction channels. Plaques were not detected in coupling deficient HeLa wild type cells, but after probing these cells with antibody directed against Cx45, we identified gap junction channels as dispersed particles in non-typical gap junction areas. In double transfected HeLa cells a co-localization of Cx40 and Cx43 was occasionally detected in quasi-crystalline gap junction plaques whereas in contact areas with dispersed particles only one Cx-type was present.

## 1. Introduction

In contact regions of adjacent cells up to several hundreds of gap junction channels are aggregated to gap junction plaques. In freeze fracture replicas of plasma membranes gap junction plaques are identified by distinct particles on the P-face or by pits (extracted particles) on the E-face. Particles and pits represent individual connexons, which are hexamers of connexins (Cx). However, also a non-ordered arrangement of membrane proteins with different particle to particle distances may still be considered as a typical gap junction area. Whether this variation in pattern reflects different physiological states of cells and/or cell-specific differences is controversially discussed [1-8] When only dispersed particles are present, an indisputable identification is impossible. In unfixed and rapidly frozen biological material, membrane patches of attached cells and tight junctions can help to identify close cellular contacts, they do not certify, however, that particles in these contact areas represent gap junction channels. Utilizing a combination of rapid freezing and immunogold labeling [9-11] we have investigated permanently growing cell lines as well as Cx-transfected HeLa cells.

## 2. Materials and Methods

BICR/M1R$_k$ is a neoplastic cell line derived from a spontaneous mammary tumor of the Marshall rat. HeLa is a permanently growing epithelioid cell line derived from a human cervix carcinoma. BICR/M1R$_k$ cells, HeLa wild type cells, and connexin transfectants [12, 13] were cultivated in Dulbecco's modified Eagle's medium. Transfectants were kept with 1 µg/ml puromycin or geneticin, and Cx40/43 transfectants with 0.5 µg/ml puromycin and 0.3 µg/ml geneticin.

For labeling with immunogold [9-11], cells were washed in PBS, collected, pelleted and placed between two gold specimen carriers and plunged into liquid propane. Carrier sandwiches were fractured and replicated by platinum/carbon in a Balzers 301 instrument. Replicas were floated on PBS and treated at least for 60 min in 2.5 % sodium dodecyl sulfate containing 10 mM Tris HCL plus 30 mM sucrose. After washing in PBS, unspecific reactions were blocked with 2 % bovine serum albumin and the replicas were put for 1 hour on a droplet of primary antibody, washed again, treated with BSA, and put on a droplet of secondary antibody for 1 hour. For double labeling the same procedure was followed with a mixture of diluted primary and secondary antibodies. Controls were treated with inappropriate antibodies or secondary antibodies only.

Primary antibodies were prepared by immunizing rabbits with synthetic peptides directed against the last 22 amino acids of Cx40- or Cx43-COOH ends [14] and against a fusion protein of glutathion S transferase and COOH terminus of Cx45 [12]. The following antibodies were purchased from Biotrend (Köln, FRG): Mouse-anti-Cx43 (Nr. CT 1359), rabbit-anti-ubiquitin (Nr. PDR C10), goat-anti-mouse IgG (H+L) 15 nm gold (Nr. 115.022), goat-anti-mouse IgG (H+L) 10 nm gold (Nr. 110.022), goat-anti-rabbit IgG (H+L) 10 nm gold (Nr. 110.011).

HeLa wild type cells and Cx43 transfectants were also explored for ubiquitin-connexin conjugates. For this purpose, cells were harvested after two washes in PBS with a rubber policeman and immediately frozen in liquid nitrogen. Cell lysates (100 µg per lane) were separated on SDS polyacrylamide-gel and analyzed by Western blot analysis using antibodies directed against Cx43 or ubiquitin.

## 3. Results

Fractured membranes of unfixed cells are often not smooth but appear as ruffled surface, however, typical quasi-crystalline gap junction plaques are easily identified. Figure 1 gives an example of a gap junction plaque between BICR/M1R$_k$ cells which was labeled with mouse-anti-Cx43 antibody. The presented area exhibits mainly E-face tagged with 15 nm immunogold. The labeling efficiency, however, does not vary with the replicated membrane half (E- or P-face) of a gap junction plaque or with the size of the immunogold, but with 6 nm gold it is difficult to discriminate immunogold from replicated connexons [11].

Connexin co-localization was investigated with double-transfected (Cx40 and Cx43) HeLa cells. Both homomeric channel types form gap junction plaques in single transfected HeLa cells and these structures were also present in double-transfected cells (Fig. 2). Both channel types can be co-localized in one quasi-crystalline gap junction plaque. However, plaques were often found to be labeled against one Cx-type only, which was always the case for contact areas with dispersed particles.

**Fig. 1** A Cx43 gap junction plaque in BICR/M1R$_k$ cells labeled with 15 nm immunogold. Bar: 200 nm

**Fig. 2** Cx40/Cx43 transfected HeLa cells. The large plaque is double labeled. (Cx40 = 10 nm; Cx43 = 15 nm). Bar: 200 nm

Typical quasi-crystalline gap junction plaques were present in Cx45-transfected HeLa cells (Fig. 3a) and were as clearly labeled as were contact areas with dispersed particles (Fig. 3b), which would not necessarily be identified as connexons without this label.

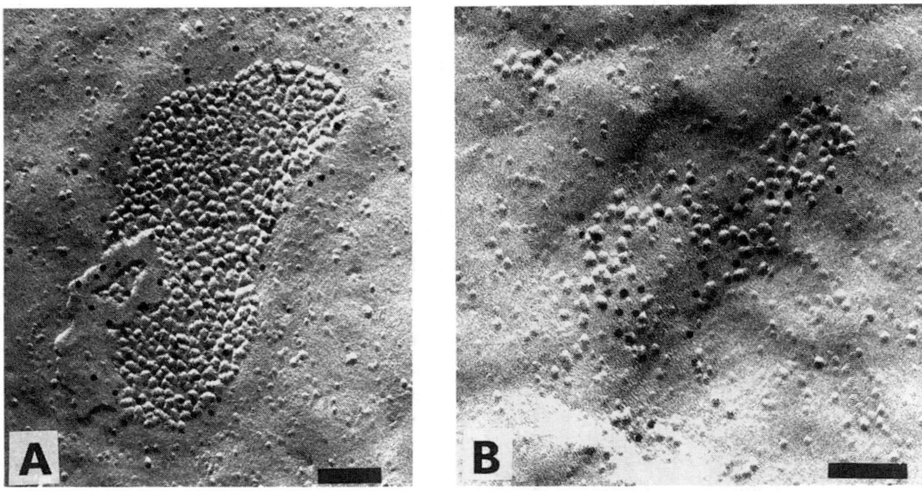


**Fig. 3** HeLa Cx45 transfectants. (A) Gap junction plaque, (B) Contact area. Bar: 100 nm

Gap junction plaques were not detected in HeLa wild type cells but electrophysiological measurements with high resolution revealed that they are deficiently coupled through channels which are characterized by a low single channel conductance [15]. We probed HeLa wild type cells with antibody directed against Cx45 and found label with dispersed particles in tight junctional areas and at attachment sites of filaments (Fig. 4).

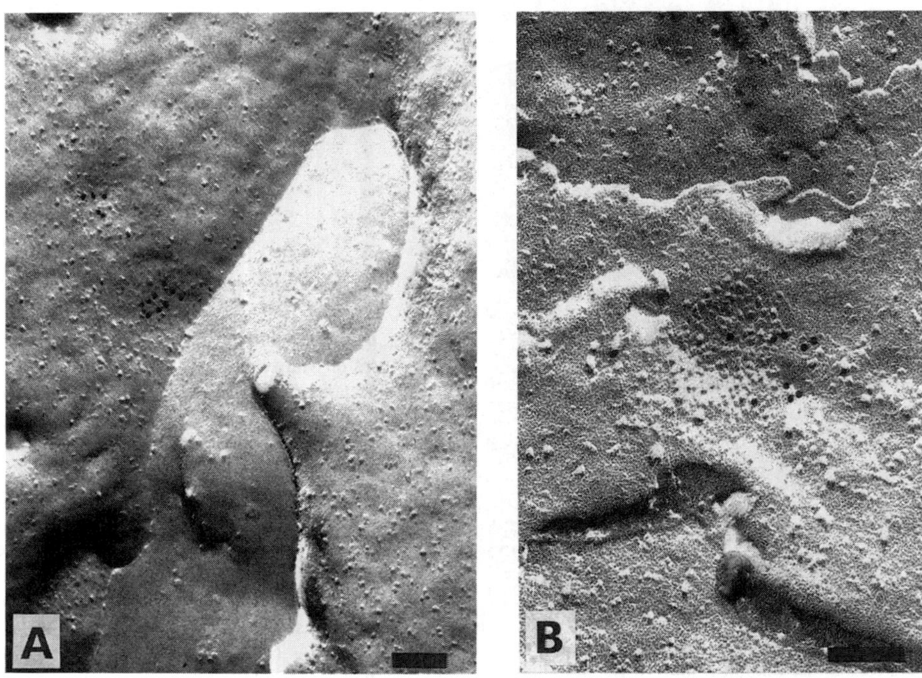


**Fig. 4** HeLa wild type cells labeled for Cx45. (A) Broken filament which extended from the right to the left cell, two attachment sites are labeled. (B) Region of close contact with dispersed particles. Bar: 100 nm

We have also used this immunogold labeling method to elucidate the degradation of gap junction channels by ubiquitination. In BICR/M1$R_k$ cells both labeled (Fig. 5) and unlabeled gap junction plaques were detected but also other areas revealed labeled particles. An identification of these ubiquitinated proteins has not yet been made. In HeLa Cx43 transfectants ubiquitinated Cx43 could be detected by Western blot analysis (Fig. 6, lane 5 and 6). No cross reactivity was detected with HeLa wild type cells (Fig. 6, lane 4 and 4'). The two upper protein bands at the level of the stacking and of the separation gel in lane 5 correspond to the ubiquitinated form of Cx43, since both protein bands are recognized by anti-ubiquitin antibody (lane 6,) as well as by anti-Cx43 antibody in HeLa Cx43 transfectants (lane 5) but not in HeLa wild type cells (lane 7). Presumably other ubiquitinated proteins of the same molecular mass were detected in wild type and transfected cells (lanes 2', 3, 4, 6, 7). When cells were not frozen immediately before lysis, the ubiquitinated Cx43 molecules were degraded so fast that they could be detected only as 43 kDa monomers and their phosphorylated isoforms (Fig. 6, lane 1).

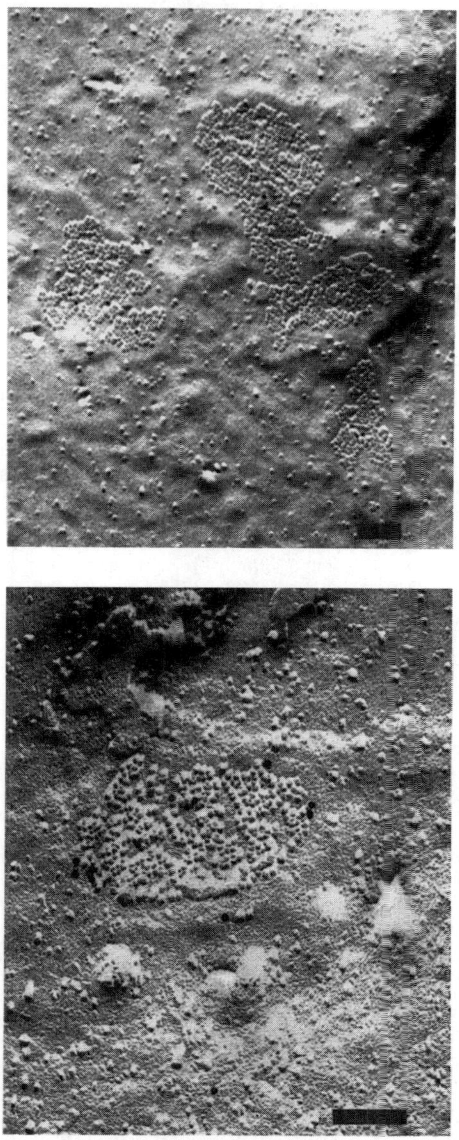

**Fig. 5** BICR/M1$R_k$ cells labeled with antibody directed against ubiquitin and 10 nm immunogold. Bar 100 nm

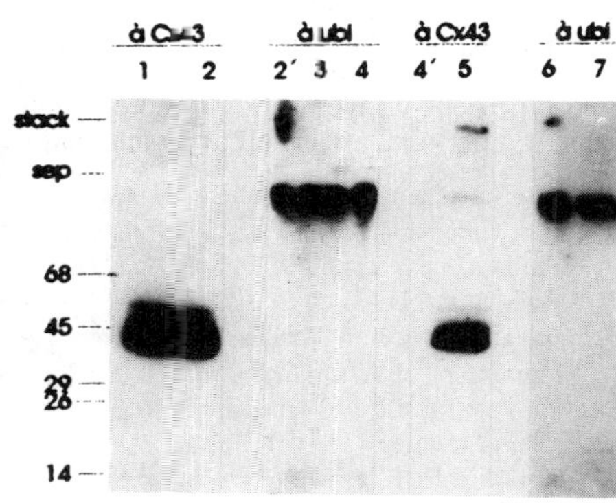


**Fig. 6** Western blot of ubiquitinated HeLa cells. Lysates of wild type cells (lanes 4, 7) and Cx43 transfectants (lanes 1, 2, 3, 5, 6) were analyzed with antibodies directed against Cx43 and ubiquitin as well as with protein A-$^{125}$I for autoradiography. To compare the protein bands detected by the two different antibodies, lane 2 and lane 4 were divided in two halves before incubation with these antibodies. The numbers on the left side indicate the molecular mass in kDa of bovine serum albumin (68), ovalbumin (45), carboanhydrase (29), chymotrypsinogen (26), ribonuclease A (14); stack: beginning of the stacking sep: separation gel.

## 4. Discussion

We have demonstrated that gap junction channels are also present as dispersed channels in contact areas. It must still be clarified, whether labeling reveals contact sites with gap junction channels or non-contacting membranes with connexons. This might be answered with future experiments when extracellular and intracellular epitopes of connexons will be labeled separately. Since no plaques were detected in HeLa wild type cells which are coupled with only a few Cx45 channels, but labeling was observed with dispersed particles, the existence of functional gap junction channels distant from plaques has a high probability.

Contact areas contain only one Cx-type but plaques can consist of more than one type, therefore, these domains may well represent different functional states of channels. Thus far, ubiquitinated connexins could be isolated *in vivo* only from cells with defects in their proteasome complex or in the presence of inhibitors [16]. Now we have shown with two different techniques that gap junction plaques contain ubiquitinated proteins, indicating that connexins may be degraded by an ubiquitin dependent proteolysis. However, also internalization and lysis of gap junction plaques can be observed.

## References


[1] C. Peracchia and A.F. Dulhunty, Gap junctions: structural changes associated with changes in permeability. J. Cell Biol. 63 (1974) (2, Pt 2), 263a

[2] E. Raviola, D.A. Goodenough and G. Raviola, Structure of Rapidly Frozen Gap Junctions. J. Cell Biol. 87 (1980) 273-279

[3] S. Sikerwar and S. Malhotra, Structural correlates of glutaraldehyde induced uncoupling in mouse liver gap junctions. Eur. J. Cell Biol. 25 (1981) 319-323

[4] R.B. Hanna, T.S. Reese, R.L. Ornberg, D.C. Spray and M.V.L. Bennett, Fresh frozen gap junctions: Resolution of structural detail in the coupled and uncoupled states. J. Cell Biol. 91 (1981) 125a

[5] J. Délèze and J.C. Hervé, Effect of Several Uncouplers of Cell-to-Cell Communication on Gap Junction Morphology in Mammalian Heart. J. Membrane Biol. 74 (1983) 203-215

[6] C.R. Green and N.J. Severs, Gap Junction Connexon Configuration in Rapidly Frozen Myocardium and Isolated Intercalated Disks. J. Cell Biol. 99 (1984) 453-463

[7] D.F. Hülser, D. Paschke and J. Greule, Gap Junctions: Correlated Electrophysiological Recordings and Ultrastructural Analysis by Fast Freezing and Freeze-Fracturing. In: H. Plattner (ed.), Electron Microscopy of Subcellular Dynamics, ISBN 0-8493-6079-X. CRC Press, Boca Raton, 1989, pp. 33-49

[8] H. Wolburg and A. Rohlmann, Structure-Function Relationships in Gap Junctions. International Review of Cytology 157 (1995) 315-373

[9] K. Fujimoto, Freeze-fracture replica electron microscopy combined with SDS digestion for cytochemical labeling of integral membrane proteins. Application to the immunogold labeling of intercellular junctional complexes. J. Cell Science 108 (1995) 3443-3449

[10] K. Fujimoto, SDS-digested freeze fracture replica labeling electronmicroscopy to study the two-dimensional distribution of integral membrane proteins and phospholipids in biomembranes: practical procedure, interpretation and application. Histochem. Cell Biol. 107 (1996) 87-96

[11] D.F. Hülser, B. Rehkopf and O. Traub, Dispersed and Aggregated Gap Junction Channels Identified by Immunogold Labeling of Freeze-Fractured Membranes. Exp. Cell Res. 233 (1997) 240-251

[12] A. Butterweck, U. Gergs, C. Elfgang, K. Willecke and O. Traub, Immunochemical Characterization of the Gap Junction Protein Connexin45 in Mouse Kidney and Transfected Human HeLa Cells. J. Membrane Biol. 141 (1994) 247-256

[13] C. Elfgang, R. Eckert, H. Lichtenberg-Fraté, A. Butterweck, O. Traub, R.A. Klein, D.F. Hülser and K. Willecke, Specific Permeability and Selective Formation of Gap Junction Channels in Connexin-transfected HeLa Cells. J. Cell Biol. 129 (1995) 805-817

[14] O. Traub, R. Eckert, H. Lichtenberg-Fraté, C. Elfgang, B. Bastide, K.H. Scheidtmann, D.F. Hülser and K. Willecke, Immunochemical and electrophysiological characterization of murine connexin40 and -43 in mouse tissues and transfected human cells. Eur. J. Cell Biol. 64 (1994) 101-112

[15] R. Eckert, A. Dunina-Barkovskaya and D.F. Hülser, Biophysical characterization of gap-junction channels in HeLa cells. Pflügers Arch. Eur. J. Physiol. 424 (1993) 335-342

[16] J.G. Laing and E.C. Beyer, The Gap Junction Protein Connexin43 Is Degraded via the Ubiquitin Proteasome Pathway. J. Biol. Chem. 270 (1995) 26399-26403

# Section 2:
# Gating Mechanisms

*Gap Junctions*
*R. Werner (Ed.)*
*IOS Press, 1998*



# Gating Properties of Gap Junction Channels in Schwann Cells and Fibroblasts Isolated from the Sciatic Nerve of Neonatal Rats

Feliksas BUKAUSKAS, Peter SHRAGER and Camillo PERACCHIA
*Department of Pharmacology and Physiology*
*University of Rochester, Rochester, NY 14642-8711, USA*

**Abstract.** The goal of this project was to define by double whole-cell patch clamp the characteristics of cell-cell coupling in homo- and heterocellular pairs formed between Schwann cells and fibroblasts. The cells were isolated from 1-2 day old neonatal rat sciatic nerve. Schwann cell pairs exhibited strong and symmetrical dependence of steady state coupling conductance, $g_j(ss)$, on transjunctional potential, $V_j$. In contrast, in fibroblasts the $g_j(ss)$-$V_j$ relation varied from symmetrical to strongly asymmetrical. In heterocellular pairs, asymmetry of $g_j(ss)$-$V_j$ relation was strongly related to $V_j$ polarity: $g_j(ss)$-$V_j$ was more sensitive to $V_j$ when the potential from the side of the Schwann cell was negative. In four out of twelve experiments there was no electrical cell-cell coupling just after patch opening, but a positive potential from the Schwann cell side revealed step-by-step increase of junctional current, $I_j$. A negative potential from the side of Schwann cell closed gap junction (GJ) channels. Based on the $g_j(ss)$-$V_j$ relation in homo- and heterocellular pairs, we suggest that the $V_j$-gating mechanism of Schwann cell GJs is activated by negative potentials from the side of the Schwann cell cytoplasm. An analysis of single channel conductance shows that in Schwann cells $\gamma_j$(main state) and $\gamma_j$(residual) are ~37 and 6 pS, respectively. In fibroblasts, the value of $\gamma_j$(main state) was highly dispersed, ranging from 30 to 130 pS. In heterocellular pairs, $\gamma_j$(main state) displayed $V_j$-dependent rectification, i.e. $\gamma_j$(main state) increased when the Schwann cell was depolarized or the fibroblast was hyperpolarized. Schwann cells are likely to exhibit one type of GJ channels, whereas in fibroblasts GJs are formed by Cx43 and other connexin not yet determined.

## 1. Methods

Schwann cells and fibroblasts were enzymatically isolated from sciatic nerves of 1-2 days old neonatal rats under aseptic conditions [1]. Spontaneously preformed Schwann/Schwann, fibroblast/fibroblast and Schwann/fibroblast cell pairs were studied by double whole cell clamp for monitoring macroscopic ($g_j$), and single GJ channel ($\gamma_j$) conductance. Homo- and heterocellular pairs could be distinguished by the difference in the shape between Schwann cells and fibroblasts. Typically, Schwann cells have a spindle shape, whereas fibroblasts are flat and lack obvious structural asymmetry.

## 2. Results and discussion

*Schwann/Schwann cell pairs*

Macroscopic cell-cell coupling was weak, mean $g_j$ =0.21±0.07 nS (n=20). $g_j(ss)$-$V_j$ relation was symmetrical and demonstrated high $V_j$-sensitivity. $V_o$=15-20 mV (Boltzmann parameter characterizing $V_j$ at which $g_j$ is half maximum).

At the single channel level Schwann cells exhibited two different types of gating activity: 1) $V_j$-sensitive $I_j$ transitions between open state and one of the substates, mostly residual, and 2) $V_j$-insensitive $I_j$ transitions between open and closed states. The frequency of the first type of $I_j$ transitions depended on $V_j$ and the open channel probability decreased by increasing $V_j$. Preliminary data indicate that $I_j$ transitions of the second type occurred most often spontaneously at small $V_j$ values and were activated only by high $V_j$ values. We estimated the single channel conductance to be: $\gamma_j$(main state)=~37 pS, at open state and $\gamma_j$(residual)=~6 pS, at residual state (Fig. 1A). The ratio $\gamma_j$(main state)/$\gamma_j$(residual)=6.2.

Our estimates for $V_o$ and $\gamma_j$(main state) agree with the data of Chanson et al. [1]. In addition, we found that during $V_j$ gating at the single channel level Schwann cells exhibited substates (Fig. 1B). The most frequently observed substate was similar to the residual state that we observed in other types of connexins, e.g., Cx32, Cx37, Cx40, Cx43, Cx46 [2-3, and our unpublished data]. The ratio $\gamma_j$(main state)/$\gamma_j$(residual) for Schwann cells is higher than for other connexins, e.g., this ratio is 5.5 for Cx40 and 5.3 for Sf9 cells [2, 4]. The molecular mechanisms of $V_j$-gating and $\gamma_j$(residual) are not known. Verselis et al. [5] have suggested that N-terminus and extracellular loop E1 of connexins are involved in $V_j$-gating mechanism. However, the sensorial and gating elements of the $V_j$-gating mechanism need to be identified more precisely. Our data indicate that the ratio $\gamma_j$(main state)/$\gamma_j$(residual) is not constant and varies depending on $V_j$ [6]. This variation is maximally expressed in heterotypic GJ channels, e.g., Cx26/Cx32, Cx32/Cx37, where $\gamma_j$(main state) strongly depends on $V_j$ [7]. On the other hand, $\gamma_j$(residual) also depends on $V_j$ in homo- and heterotypic channels of insect [4, 6] and mammalian (Bargiello, personal communication) cells. It is likely that the ratio $\gamma_j$(main state)/$\gamma_j$(residual) varies depending on $V_j$ within a relatively small range in homotypic GJ channels, but within a much larger range in heterotypic channels.

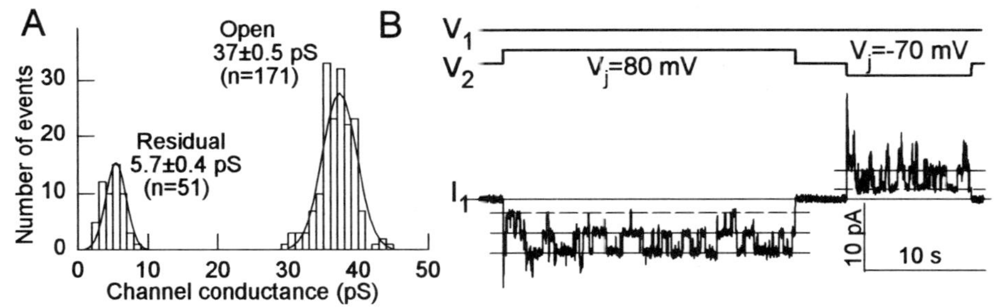


Fig. 1. Single channel characteristics of Schwann/Schwann cell pairs. (A) Frequency histogram of $\gamma_j$(main state) and $\gamma_j$(residual), summarized from 6 experiments. (B) Illustration of unitary single channel events during $V_j$-sensitive gating at $V_j$=80 and -70 mV. $V_1$ and $V_2$ voltage in cell 1 and cell 2, respectively. Junctional current, $I_1$, was measured in cell 1. Dashed and continuous lines indicate residual and open states, respectively.

Schwann cells of sciatic nerve in the neonatal period are likely to express one connexin type, assembling homotypic channels. The value of $\gamma_j$(main state) that we have measured in Schwann cells (~40 pS) is similar to that measured earlier in Schwann cells by Chanson et al. [1] and both are close to those estimated for Cx32 [8] and Cx45 [9]. Schwann cells exhibit a $V_j$-gating sensitivity similar to that of Cx45, $V_o \approx 15$ mV [9], but about twice as small as that of Cx32 [10-11]. Recently, it has been shown that cultured Schwann cells, isolated from neonatal rat sciatic nerve, express Cx46 mRNA and that the level of Cx46 expression correlates with the degree of coupling [12]. However, $\gamma_j$(main state) in HeLa46 (rat Cx46) is about 120-150 pS (our unpublished data). Based on single channel conductance and $g_j$-$V_j$ relation measurements, we can suggest that Schwann cells of sciatic nerve in the neonatal period assemble GJs with functional characteristics similar to those of Cx45. However, more precise estimates of Cx46 single channel parameters may eliminate some of these contradictions.

*Fibroblast/fibroblast cell pairs*

Coupling conductance was higher between fibroblasts than between Schwann cells: mean $g_j$=14.4±6 nS (n=10). The macroscopic $g_j$-$V_j$ relation showed high variability from symmetrical, with $V_o$≈50 mV, to highly asymmetrical, when $V_o$ ranged from 15 to 50 mV.

Single channel conductance measurements were performed in 5 well-coupled cell pairs uncoupled by heptanol and in 2 poorly coupled cell pairs without uncoupling treatments. Under the influence of heptanol we measured $\gamma_j$(main state) during the last channel closing and during the first channel opening when heptanol was washed out. Fig. 2A illustrates the frequency histogram of $\gamma_j$(main state) summarized from 5 experiments. The data show a wide distribution of $\gamma_j$(main state) in the range of 25-140 pS. In two experiments, when only one channel was in operation, we were able to estimate $\gamma_j$(main state) and $\gamma_j$(residual). Fig. 2B shows a frequency histogram summarized from these two experiments. Both $\gamma_j$(main state) and $\gamma_j$(residual) exhibited two separate Gaussian distributions with mean values for $\gamma_j$(main state) of 113±3 pS and for $\gamma_j$(residual) of 28.5±1.5 pS. We believe that in the two poorly coupled cell pairs (Fig. 2B) the GJs were made of Cx43 [3, 13-14]. Most likely, Cx43 was also preferentially expressed in cell pairs in which we measured a symmetrical $g_j$-$V_j$ relation with $V_o$≈50 mV. The wide range of $\gamma_j$(main state) (Fig. 2A) and asymmetrical $g_j$-$V_j$ relation, with $V_o$ in the range of 15-50 mV, indicates that heterotypic and/or heteromeric channels were expressed. Our data suggest that at least two types of connexins are likely to be expressed in fibroblasts. One of them most likely is Cx43 [14], and the other should be a connexin that exhibits a relatively small $\gamma_j$(main state) and high sensitivity to $V_j$-gating.

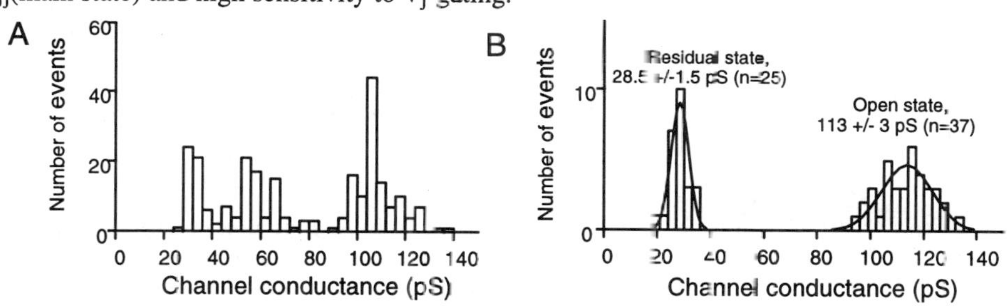


Fig. 2. Frequency histograms of single channel events in fibroblast/fibroblast cell pairs. (A) Frequency histogram of $\gamma_j$(main state) measured under the influence of heptanol in 5 well-coupled cell pairs. In view of the wide range of channel conductance values, it is likely that heterotypic and/or heteromeric channels are formed. (B) Frequency histogram of $\gamma_j$(main state) and $\gamma_j$(residual) in two cell pairs exhibiting only one functional channel.

*Schwann/fibroblast cell pairs*

Cell-cell coupling in heterocellular Schwann/fibroblast pairs was relatively weak, mean $g_j$=0.23±0.08 nS (n=12), corresponding to 4 or 5 functional channels. The macroscopic $g_j$(ss)-$V_j$ relation was symmetrical in 2 out of 12 cell pairs. In both cell pairs the $g_j$(ss)-$V_j$ relation was close to that measured in Schwann/Schwann cell pairs, i.e., $V_o$≈20 mV. In 4 out of 10 cell pairs with asymmetric $g_j$(ss)-$V_j$ relation there was no coupling at $V_j$≈0, but coupling was present when the Schwann cell was made positive. Negative potentials (15-50 mV) from the side of Schwann cell led to cell-cell uncoupling (Figs. 3 and 4A) until the next period of Schwann cell depolarization.

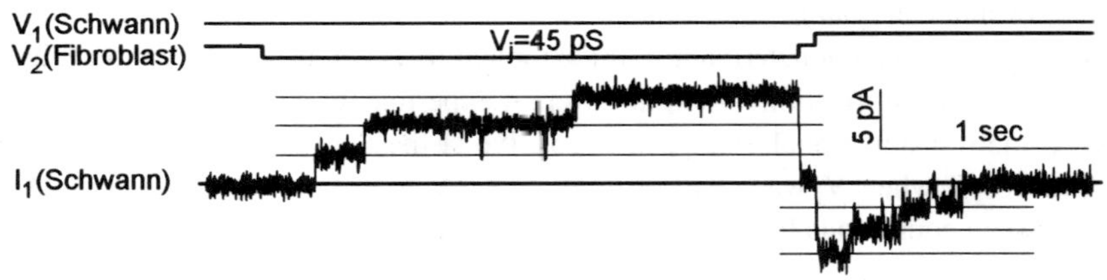


Fig. 3. Asymmetry of $V_j$-sensitive gating in Schwann/Fibroblast cell pair. $V_1$ and $V_2$ indicate the Schwann cell and fibroblast voltages, respectively. Fibroblast hyperpolarization by 45 mV results in step-by-step junctional current, $I_1$, increase indicating the opening of three channels that were in closed state at $V_j \approx 0$. Fibroblast depolarization by 45 mV leads to rapid and complete uncoupling. $\gamma_j$(main state) (distance between continuous lines) is larger when the fibroblast is hyperpolarized instead of depolarized.

All of 10 heterocellular pairs with asymmetrical $g_j(ss)$-$V_j$ relation were more $V_j$-sensitive when the potential from the side of the Schwann cell was relatively negative (Figs. 3 and 4A). These data suggest that the $V_j$-gating mechanism of Schwann cell GJs is activated by a negative potential. Based on the $V_j$-sensitive gating properties of homo- and heterocellular pairs, we suggest that the major connexin of Schwann cells exhibits $V_j$-gating properties similar to those of one of the fibroblast connexin (different from Cx43). It is likely that in two experiments with symmetrical $g_j(ss)$-$V_j$ relation, homotypic GJ channels were formed, and in the other 10 experiments heterotypic and/or heteromeric GJs were present.

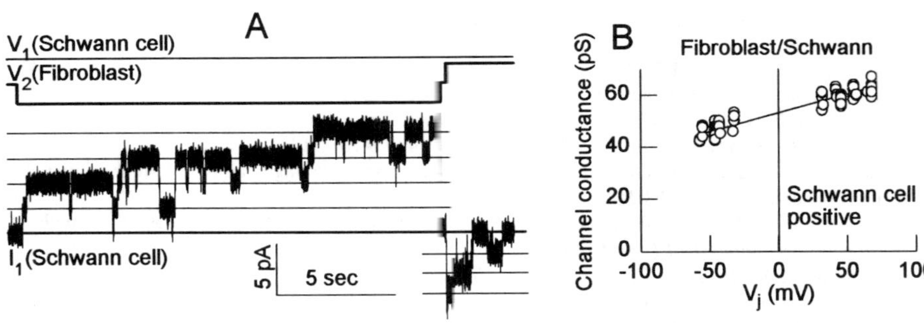


Fig. 4. Asymmetrical $V_j$-sensitive gating and rectification of $\gamma_j$(main state) in Schwann/fibroblast cell pairs. (A) Illustration of asymmetrical $V_j$-gating. Fibroblast hyperpolarization for about 40 mV results in progressive opening of four GJ channels that were in closed state at $V_j$=0. Fibroblast depolarization leads to rapid and complete uncoupling. The continuous lines indicate channel open states. $\gamma_j$(main state) (distance between continuous lines) is larger when the fibroblast is hyperpolarized than depolarized. (B) Illustration of $\gamma_j$(main state) rectification measured in one cell pair: $\gamma_j$(main state) is higher when the Schwann cell is relatively positive.

Our data suggest that the highly asymmetrical $V_j$-sensitive gating of heterotypic junctions between Schwann cells and fibroblasts prevents heterocellular coupling at $V_j \approx 0$ mV, and that a small $V_j$ change can significantly modulate cell-cell coupling. Functional isolation of Schwann cells and fibroblasts from direct exchange by ions and metabolites could be important for the myelination process. The single channel conductance in heterocellular pairs with symmetrical $g_j(ss)$-$V_j$ relation was close to that measured in

Schwann/Schwann cell pairs, i.e., $\gamma_j$(main state) was about 40 pS. In cell pairs with asymmetrical $g_j(ss)$-$V_j$ relation, $\gamma_j$(main state) exhibited $V_j$ dependent rectification (Fig. 4B) and the steepness of rectification varied in different experiments. $\gamma_j$(main state) was always higher when the transjunctional voltage was positive at the Schwann cell side and ranged from ~35 to 70 pS when $V_j$ was changed from -75 mV to +75 mV (Figs. 3 and 4A, see distances between continuous lines for positive and negative $V_j$'s.

## 3. Conclusions

1) Schwann cells of neonatal rat sciatic nerve are likely to form homotypic GJ channels expressing functional characteristics similar to those of channels made of Cx45.
2) The GJs of fibroblasts are likely to be formed by at least two connexin types assembled in heterotypic and heteromeric channels. They contain Cx43 [14] and a connexin with properties similar to those expressed in Schwann cells.
3) Our data suggest that the asymmetrical $V_j$-sensitive gating of Schwann/fibroblast pairs causes a functional uncoupling at $V_j \approx 0$ mV. Therefore, small $V_j$ changes can significantly modulate cell-cell coupling and cellular functions likely to depend on cell communication, such as cells proliferation, myelination, etc. The $V_j$-gating mechanism of Schwann cell GJs appears to be activated from the side of cytoplasm with more negative potential.

## References


[1] Chanson, M., Chandros, K., Rook, M., Kessler, J., Spray, D. 1993. Gating characteristics of a steeply voltage-dependent gap junction channel in rat Schwann cells. J. Gen. Physiol.: 102. 925-946.

[2] Bukauskas, F.F., Elfgang, C., Willecke, K., Weingart, R. 1995. Biophysical properties of gap junction channels formed by mouse connexin 40 in induced pairs of transfected human HeLa cells. *Biophysical J* 68:2289-2298.

[3] Valiunas, V., Bukauskas, F.F., Weingart R. 1997. Conductance and ionic selectivity of connexin43 gap junction channels examined in neonatal rat heart cells. *Circulation Res.* 80:708-719.

[4] Bukauskas, F.F., Vogel, R., Weingart, R. 1997. Biophysical properties of heterotypic gap junctions newly formed between two types of insect cells *J. Physiology* 499/3:701-714.

[5] Verselis, V.K., Ginter, C.S., and Bargiello. T.A. 1994. Opposite voltage gating polarities of two closely related connexins. *Nature* 368:348-351.

[6] Bukauskas, F.F., Weingart, R. 1994. Voltage-dependent gating of single gap junction channels in an insect cell line. *Biophysical J.* 67: 613-625.

[7] Bukauskas, F.F., Elfgang, C., Willecke, K., Weingart, R. 1995. Heterotypic gap junction channels (connexin26-connexin32) violate the paradigm of unitary conductance. *Pflugers Arch.*. 429:870-872.

[8] Bukauskas, F.F., Weingart, R. 1995. Gating properties of homo- and hetero-typic gap junction channels formed by different mouse connexins. In: *Abstr. of Gap Junction Conference* France.

[9] Moreno, A.P., Laing, J.G., Beyer, E.C., Spray, D.C. 1995. Properties of gap junction channels formed of connexin 45 endogenously expressed in human hepatoma (SKHep1) cells. *Am. J. Physiol.* 37: C356-C365.

[10] Wang, X., Li, L., Peracchia L. and C. Peracchia. 1996. Chimeric evidence for a role of the connexin cytoplasmic loop in gap junction channel gating. *Pflugers Arch.* 431: 844-852.

[11] Moreno, A.P., Eghbali, B., Spray, D.C. 1991. Connexin32 gap junction channels in stably transfected cells: Equilibrium and kinetic properties. *Biophys. J.* 60: 1-11.

[12] Chandross, K.J., Spray, D.C., Cohen, R.I., Kumar, N.M., Kremer, M., Dermietzel, R., Kessler, J.A. 1996. TNFa Inhibits Schwann cell proliferation, Connexin46 expression and gap junction communication. *Mol. and Cell. Neuroscience.* 7: 479-500.

[13] Bukauskas, F., Peracchia, C. Two distinct gating mechanisms in gap junction channels: a CO2 sensitive and a voltage sensitive. *Biophys. J.* 72:2137-2142 (1997).

[14] Beyer, E.C., Kistler, D.L., Paul, D.I., Goodenough, D.A. 1988. Antisera directed against connexin43 peptides react with a 43-kDa protein localized to gap junctions in myocardium and other tissues. *J. Cell Biol.* 108: 595-605.

*Gap Junctions*
*R. Werner (Ed.)*
*IOS Press, 1998*

# Gap Junction-Like Currents Detected from Membrane Channels of Novikoff and NRK Cells

Rafiqul Hasan, Ross G. Johnson* and Alonso P. Moreno. *Krannert Institute of Cardiology, Indianapolis IN 46202, and *Department of Genetics and Cell Biology. University of Minnesota, St. Paul MN 55108*

The inward diffusion of charged fluorescent dyes into cells in culture strongly suggests the presence of gap junction hemichannels. [1]. To determine their electrophysiological characteristics in the plasma membrane, a whole cell voltage technique was performed on isolated Novikoff and NRK cells. Most of these cells had a very low input resistance that could be increased 50 times using 2mm halothane. Two major membrane currents were identified. One was a $Cl^-$ like rectifying current, present in 100% of the cells, that activates on hyper-polarization and can be inhibited by IAA-94. The second current was activated after prolonged hyper-polarizations (10 sec at 60mv) and was present in approx. 65% of the cells. The voltage dependence of the inactivation, (adjusted by a mono-exponential function) was not significantly different from that for complete Cx43 gap junction channels. Two main conductive states were recorded, with similar ratios to those found in cell pairs expressing Cx43 and their activity could be blocked by 2 mm halothane. Together, these data strongly suggest that this second type of current is due to the Cx43 gap junction hemichannels, expressed in Novikoff and NRK cells.

## 1. Introduction

Gap junctions have been described mostly in terms of functional channels between adjacent cells. To form one complete gap junction channel, a direct docking of two hemichannels or connexons is required, one provided by each cell in the system. Some of these connexons can also be found at the surface of the membrane. Their function, besides being a precursor of full channels, is not known, but because of their limited selectivity, we presume that in physiological conditions they must be closed, in order to maintain the integrity of the cell.

The process of formation and assembly of these channels is currently under study, and the characterization of their behavior as individual channels is of critical importance in order to understand their function and capabilities.

It has been demonstrated by cell co-cultures [2] and cell-to-cell forced apposition [3] that functional full channels can be formed in a few minutes. These experiments reinforce old theories on how gap junctions are formed from fully pre-assembled precursor hemichannels present in the non-junctional membrane [4][5][6]. The first real description of functional hemichannels was presented by deVries in 1992 [7]using horizontal cells from catfish retina. With their work, they established most of what we actually know about these channels. Later, others have studied the properties of Cx46 hemichannels, which are highly expressed in Xenopus oocyte membranes [8][9]. A large number of connexins have been reported to be involved in the uptake of fluorescent dyes in various cell types [10], but the electrophysiological characteristics of only a few of them have been studied. The list includes Cx46, chick Cx56, mCx30 and mCx50 (see Valiunas, this book).

In this report, we are showing that hemichannel currents can be recorded from the membranes of Novikoff and NRK cells in culture which constitute some of the first recordings obtained from cell lines in culture. We propose that these hemichannels are formed by Cx43 based on: 1) Cx43 is highly expressed in both of these cell lines, and 2) the electrophysiological properties of these currents (i.e. voltage dependence and gating charge and unitary conductance sub-states), resemble those of full channels formed by Cx43.

## 2. Methods

Novikoff hepatoma cells and normal rat kidney (NRK) cells were used in this study. They were grown at 37 °C in supplemented D-MEM media and seeded on 12 mm circular glass coverslips. Junctional conductance ($g_j$) between cell pairs was calculated using the dual whole cell voltage clamp technique. Access to the cytoplasm was achieved using a brief negative pressure pulse, after giga-ohm seal formation between polished glass micropipettes (3-5 Ohm) and the cell membrane. The micropipettes were filled with a cesium- and ATP-containing patch solution (**PCs** in mM; 130 CsCl; 0.5 $CaCl_2$; 2 $Na_2ATP$; 3 MgATP; 10 Hepes; 10 EGTA, pH 7.2). During recordings, cells were kept at room temperature and in cesium-containing bathing solution (**SALHigh** in mM; 160 NaCl, 7 CsCl; 0.1 $CaCl_2$, 0.6 $MgCl_2$, 10 Hepes, pH 8). Low calcium solution contained (**SALLow** in mM; 160 NaCl, 7 CsCl; 5 mM $CaCl_2$, 20 EGTA, 0.6 mM $MgCl_2$, 10 mM Hepes, pH 8).

To determine the voltage sensitivity of junctional channels, transjunctional current ($I_j$) was measured in one of the cells (held at constant voltage) while a protocol of voltage steps (ranging from -80 to 80 mV) was applied to the contiguous cell using pClamp software (Axon Instruments). Time constants for voltage inactivation currents were obtained using a pClamp algorithm for single exponential curves. Single channel currents were measured using freshly split cells that present a small number of functional hemichannels, or where junctional conductance was reduced after either superfusion of Halothane (2 mM) or 2,3Butanedione monoxime (BDM). Unitary currents were obtained during long voltage steps of 10 second voltage ramps (100 to –100 mV) applied to the cells. The unitary opening or closing current events were measured using pClamp software (Axon Instruments) or through a digitizing board (Summagraphics with SigmaScan software [Jandel, Corta Madera CA]).

## 3. Results

***3.1. Dye uploading of cell attached cells***. Cells attached to glass coverslips, also present a high level of LY uptake in low calcium solutions (Li et al., 1996).

***3.2. Input resistance under low calcium solutions.*** Using a whole cell voltage clamping technique on isolated cells, it was possible to determine their input resistance. In control conditions, we found values between 100 and 500 MOhms During the application of low $Ca^{++}$ solution, we found that the input resistance decrease more than 95% of the original values. This change was not totally reversed even after using high calcium solution for 10 minutes. This opens that possibility that the channels were opened at low concentrations of extracellular calcium, but they could not go back to their closed state because the cell has been dialyzed with the attached pipette However, the uptake experiments show that when the cells are intact, the channels can close, not allowing the fluorescent dye to escape. This suggests the participation of an intracellular mechanism in the closure of the channel.

***3.3. Inactivation kinetics of hemichannels.*** To determine the kinetics of inactivation of multiple hemichannels present in Novikoff cells we applied a protocol of pulses shown in Figure 1. In most of the cases, the resulting currents for hyperpolarizing pulses were slowly activated. For depolarizing pulses, these currents inactivated with a voltage dependence and kinetics similar to those of homotypic Cx43.

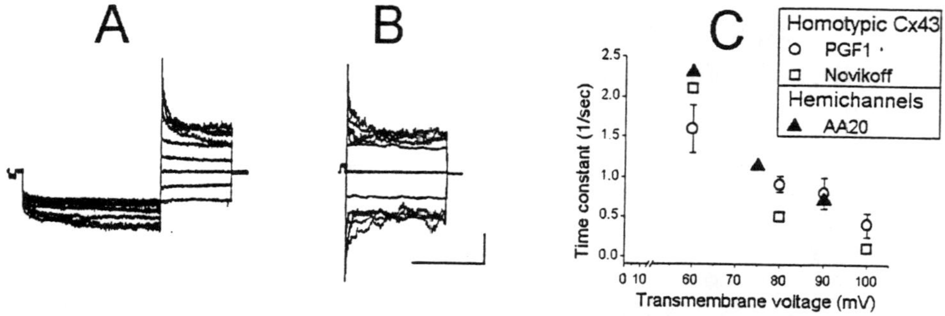


Figure 1. Membrane currents from isolated Novikoff cells resemble those from Homotypic Cx43 channels. **A.** Hyperpolarization slowly opens a conductance related to a $Cl^-$ channel. **B.** Transjunctional currents recorded from full gap junction channels expressed in SKHep1 cells. **C.** The inactivation time constant of AA20 cells (these are Novikoff cells that poorly express Cx43) hemichannels is not significantly different than those calculated from full gap junction channels.

***3.4. Halothane resolves the unitary conductance of the hemichannels.*** Low input resistance can be increased with Halothane. This anaesthetic has been shown effective in blocking almost all gap junctions formed in mammalian cells. When Novikoff cells were exposed to 0.1 mM calcium solution, their input resistance (IR) was significantly reduced. This IR could be increased substantially by the use of Halothane, as shown in Figure 2A. Upon recovery of Halothane, it is possible to record the opening of channels, whose conductances are mainly 48±8 or 75±7 pS as shown in Figure 2B. These unitary currents are the ones associated with the decrease in the input resistance in the presence to low calcium solutions.

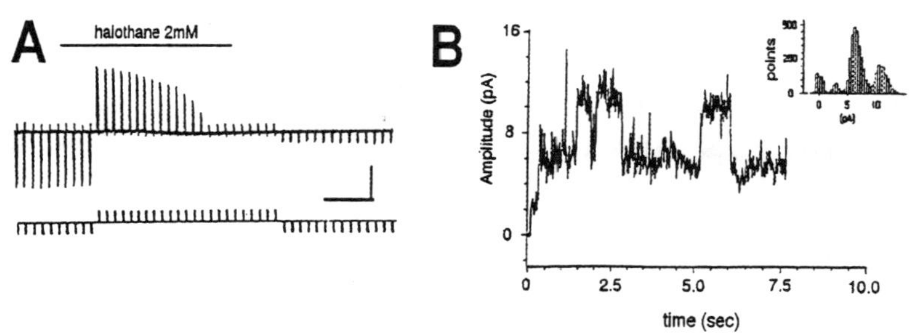


Figure 2. Membrane currents associated with hemichannels are blocked by Halothane. **A.** A 300 ms, 5 mV pulse was applied at one Hz to record membrane conductance. During Halothane perfusion, this conductance is significantly decreased. **B.** Upon recovery, multiple channel recordings show conductances of 40 and 70 pS. The histogram represents the total number of points during the first 2500 milliseconds of recording.

To study more properly these channels, we recorded the opening and closing of the channels during the application of SALLow. When applied slowly it was possible to record unitary events form the surface of NRK cells. Again, the unitary conductances observed here are 40 and 70 pS, consistent with those observed in Novikoff cells. In Figure 3, we show an example of these transitions in a cell that is clamped at various voltages. In the inset, there is a clear presence of these single channel sub-conductances.

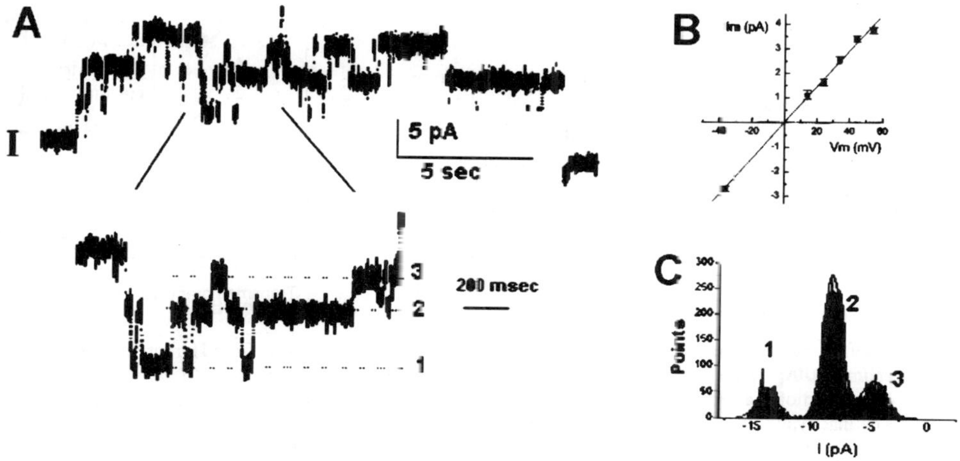


Figure 3. Unitary conductance recordings from NRK membranes after exposure to SALLow solution. **A.** These small transitions, found in most of our records present two main conductances. One of ~48 and the other of ~75pS. The ratio between these conductances is also observed in homotypic gap junction channels (60, 90 pS). **B** I/V relation showing that the unitary current for the large conductance is voltage independent up to −40 mV. **C.** Histogram of all points from the lower trace in A digitized at 2kHz.

***3.5. Other channels present in these cells.*** Other channels are present in the membranes of these tumor cell lines, including large highly voltage dependent chloride channels (unitary conductance of 400pS). This channel activates with hyperpolarizing currents. In Figure 4, we show recordings of single channel during the application and washout of IAA-94, a specific Cl- channel blocker.

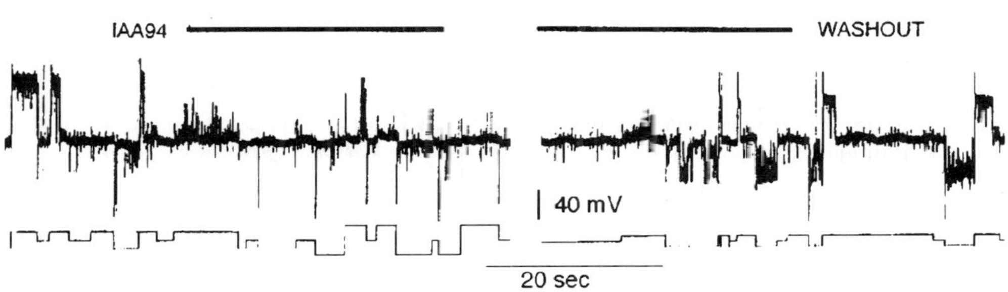


Figure 4. Chloride channels in Novikoff and NRK cells can be blocked by AAI-94. This substance is an anion channel blocker. In an excised patch of an NRK cell, the presence of these channel can be eliminated by using 10 mM of this compound. Note in this sequence, that during IAA-94 application, the activity of the chloride channel disappears and that the effect is reversible.

## 4. Discussion

The presence of gap junction hemichannels has been proposed by experiments that demonstrate fluorescent dye uptake when the cultured cells are treated with low calcium solutions [1]. Cells expressing this behavior have been used to study the electrical behavior of those hemichannels.

In terms of the voltage dependency of these gap junction-like currents, their Inactivation constants are very similar to those of Cx43 homotypic channels [2]. This suggests that the hemichannel gating system is functional and maintains its structure and sensitivity after docking.

The unitary conductance that we see in these experiments is smaller than the one predicted by considering that the full gap junction channel pore is formed by two identical cylinders. If a phosphorylated gap junction channel is close to 60 pS, then the conductance of an hemichannel was expected to be close to 120 pS. In other reports, other hemichannels show sizes from 200 up to 400 pS, somewhat higher than the respective full channels (compare [9] and Tsutsumi, this book). In NRK cells, the largest conductance found so far is 75 pS and remains linear up to –40 mV. This characteristic differs to Cx46 hemichannels expressed in oocytes as well as those from Cx30 and 50 expressed in HeLa cells (Valiunas, this book).

The fact that Cx43 hemichannels have smaller conductances than other connexons suggested that the Cx43 hemichannel change its structure or its permeability after assembly with other connexons, re-establishing thereafter the conductance shown in homotypic Cx43 full channels. This can be observed in the formation of heterotypic Cx45/Cx43 channels. Here it has been strongly suggested that the conductance of the phosphorylated Cx43 hemichannels have a unitary conductance of 120 pS [2].

One of the most attractive feature of the hemichannels expressed in NRK and Novikoff cells, is that they have at least two conductive states, just like the ones observed in Cx43 full channels expressed in SKHep1 and other cell types. It is our goal to determine the significance of these sub-conductance states in the behavior and permeability properties of the hemichannels.

Other channels are also expressed in these cultured cell lines. The most abundant and present in both cells types are chloride channels, which we found blockable by the compound IAA-94. Fortunately, these channels are highly voltage dependent and could be easily detected and their activity eliminated from these cells.

All our data so far obtained strongly suggest that those cell currents and single channel activity observed in the two cell types studied correspond to hemichannels formed by Cx43.

## References

1. Li H. et al. Properties and regulation of gap junctional hemichannels in the plasma membranes of cultured cells. *Journal of Cell Biology*. 134(4):1019-30, 1996 Aug.

2. Moreno A.P., G.I. Fishman, E.C. Beyer and D.C. Spray. Voltage dependence gating and single channel analysis of heterotypic gap junction channels formed of Cx45 and Cx43. Prog .Cell Res. 4; 405, 1995

3. Bukauskas FF. Elfgang C. Willecke K. Weingart R. Biophysical properties of gap junction channels formed by mouse connexin40 in induced pairs of transfected human HeLa cells. Biophys. J. 68(6):2289-98, 1995

4. Loewenstein WR. Junctional intercellular communication: the cell-to- cell membrane channel. Physiological Reviews. 61(4):829-913, 1981.

5. Ramon F. Rivera A. Gap junction channel modulation--a physiological viewpoint. Progress in Biophysics & Molecular Biology. 48(3):127-53, 1986.

6. Johnson R.G., R.A. Meyer and Paul D. Lampe. Gap junction formation: A self assembly model involving membrane domains of lipid and protein. In "Cell interactors and gap Juncitons. Eds Sperelakis and Cole. CDC Press, 1989.

7. deVries S.H. and E.A. Schwartz. Hemi-gap-junction channels in solitary horizontal cells of the catfish retina. J. Physiol. 445; 201-230, 1992

8. Ebihara L. Berthoud VM. Beyer EC. Distinct behavior of connexin56 and connexin46 gap junctional channels can be predicted from the behavior of their hemi-gap-junctional channels. Biophysical Journal. 68(5):1796-803, 1995.

9. Trexler EB. Bennett MV. Bargiello TA. Verselis VK. Voltage gating and permeation in a gap junction hemichannel. PNAS USA. 93(12):5836-41, 1996.

10. Li H., Atkinson M., C. Elfgang, k. Willeke and R.G. Johnson. Gap junction (GJ) hemichannel analysis in HELA cells transfected with different connexins. Mol. Biol. Cell. 7:91a, 1996.

*Gap Junctions*
*R. Werner (Ed.)*
*IOS Press, 1998*



# Domains of Connexin32 Relevant to $CO_2$-Induced Channel Gating

Xiao Guang WANG and Camillo PERACCHIA

*Department of Pharmacology and Physiology, University of Rochester School of Medicine and Dentistry, 601 Elmwood Avenue Rochester, NY 14642-8711, USA*

**Abstract.** To identify Cx32 domains involved in chemical gating, the $CO_2$ sensitivity of Cx38, Cx32 or various chimeras and mutants was studied by double voltage clamp in *Xenopus* oocytes. Cx32 is much less sensitive to $CO_2$ than Cx38. With a 3 min $CO_2$ treatment, the junctional conductances ($G_j$) of channels made of Cx38 or Cx32 dropped to 0% and 85% of initial values, respectively. With a 15 min $CO_2$ treatment, $G_j$ of Cx32 only dropped to ~47% in 12-13 min. With 15 min $CO_2$, $pH_i$ dropped to 6.38 in 5-7 min. In contrast, $Ca^{2+}{}_i$ increased with a time course similar to that of the $G_j$ drop. The chimeras Cx32/38I, Cx32's cytoplasmic loop (CL) replaced with that of Cx38, or Cx32/38I$_2$, second half of Cx32's CL ($CL_2$) replaced with that of Cx38, behaved like Cx38, suggesting that $CL_2$ is relevant to $CO_2$ gating. A 17 res domain ($CT_1$) of Cx32 at the beginning of the C-terminus (CT) chain seems to play an inhibitory role. Whereas CT deletion by 84% (beyond res. 219, Cx32-D219) did not affect $CO_2$ sensitivity, replacement of 5 arginines (R, res. 215, 219, 220, 223 and 224) with asparagines (N) or threonines (T) at $CT_1$, either in Cx32 (Cx32-5R/N) or in Cx32 deleted at res. 225 (D-5R/N), greatly enhanced $CO_2$ sensitivity, such that with 3 min $CO_2$ $G_j$ dropped to ~7%. In contrast, replacement of 5 R with other basic res., K or H, did not increase $CO_2$ sensitivity. Partial R/N replacement partially increased $CO_2$ sensitivity. R215 is a stronger inhibitor than R219-R220, whereas R223-R224 reduce the inhibitory efficiency of both R215 and R219-R220. We propose a model for $CO_2$ gating that involves charge interaction between acidic res. of the beginning of the cytoplasmic loop ($CL_1$) and basic res. of $CT_1$. $CT_1$ may keep the channel open by latching to $CL_2$ and limiting its mobility. By eliminating the $CL_1$-$CT_1$ interaction the gating process may enable $CL_2$ to move toward the channel lumen. The channel may be closed by a radial displacement of 6 conserved tyrosines, one per connexin.

## 1. Introduction

Gap junction channels close in response to various changes in cytosolic homeostasis. As a result of channel closure cells uncouple from each other electrically and metabolically. Although cell uncoupling is primarily a protective mechanism, evidence for channel permeability regulation by nearly physiological changes in $[Ca^{2+}]_i$ [1] indicates that fine

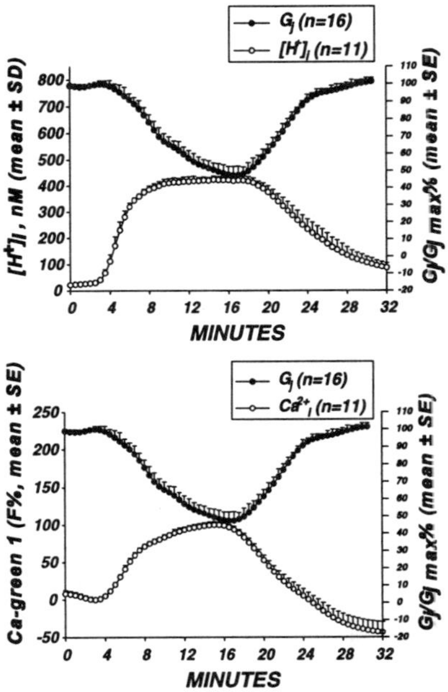


**Figure 1** Time course of changes in $[H^+]_i$, $Ca^{2+}_i$ and $G_j$ (Cx32), in *Xenopus* oocytes, exposed to 100% $CO_2$ for 15 minutes. The saline solution, bubbled with 100% $CO_2$, was perfused at 0.6 ml/min. $[H^+]_i$ (measured with BCECF) and $Ca^{2+}_i$ (monitored with Calcium Green-1) were determined at the oocytes' perifery. Note the similarity between $G_j$ and $Ca^{2+}_i$ curves. $[H^+]_i$ rapidly reaches a plateau ($pH_i$ = 6.38) that is sustained at the same level with longer (30 min) $CO_2$ treatments as well.

modulation of cell communication may play a role in normal cellular functions. In order to understand how cell communication is regulated and how coupling regulation is linked to specific cellular activities it is important to define the nature of the uncoupling agents and the molecular basis of channel gating. The latter requires first a full understand of the connexin domains that participate in the gating mechanism.

## 2. Role of pH and $Ca^{2+}$ in channel gating

Over the years a large body of evidence has emphasized the role of $Ca^{2+}_i$ and $H^+_i$ in cell coupling regulation [2, 3, 4], but it is still unclear whether $H^+$ and $Ca^{2+}$ act independently from each other, and which of them regulates cell coupling under physiological circumstances [rev. in 4].

In *Xenopus* oocytes pairs expressing Cx38, exposure to 100% $CO_2$ for 3 min caused a rapid drop of $G_j$, $pH_i$ and $pCa_i$, but the time course of $G_j$ matched better that of $pCa_i$ than that of $pH_i$ [5]. Similar results were obtained in oocytes expressing Cx32 and exposed to 100% $CO_2$ for 15 min while monitoring $Ca^{2+}_i$ or $pH_i$ with Calcium Green-1 or BCECF, respectively (Fig. 1). These data confirmed previous evidence on Novikoff hepatoma cells showing that gap junction channels are sensitive to nanomolar $[Ca^{2+}]_i$ and are insensitive to $pH_i$ in the range 7.2-6.1 [1]. The poor relationship between $G_j$ and $pH_i$ is consistent with data obtained in other cells as well [rev. in 4].

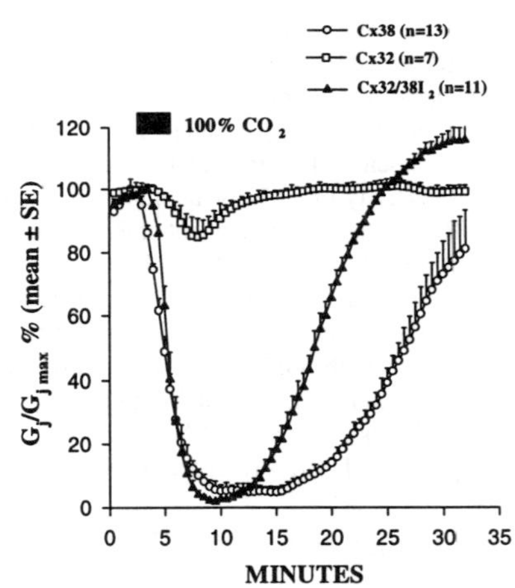


**Figure 2.** Sensitivity to $CO_2$, expressed as normalized $G_j$ ($G_j/G_{j\,max}$; 100% =control, pretreatment value), in oocyte pairs expressing Cx32, Cx38 or the chimera Cx32/38$I_2$. A 3 min exposure to $CO_2$ decreases $G_j$ to nearly 0% with Cx38 and to ~85% with Cx32. The chimera Cx32/38$I_2$ (second half of inner loop, $CL_2$, of Cx32 replaced by that of Cx38), reproduced the uncoupling efficiency of Cx38, indicating that $CL_2$ plays an important role in pH gating sensitivity. From Ref. 10.

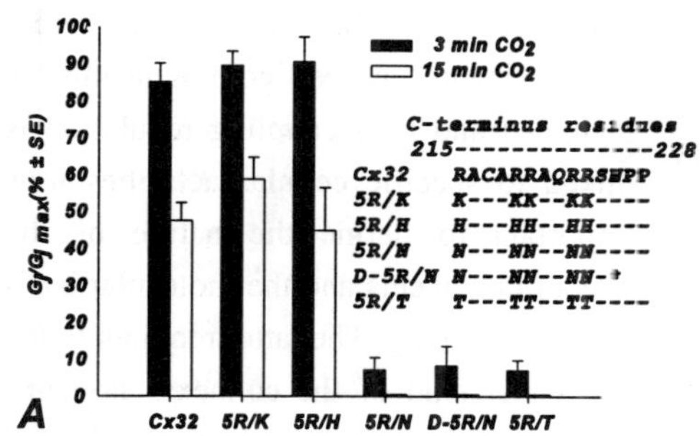


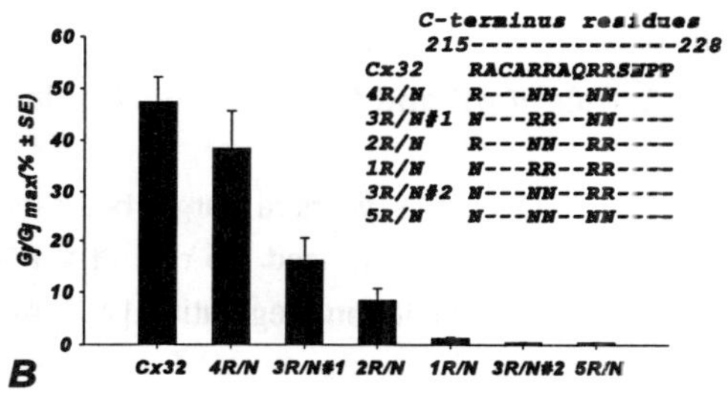


**Figure 3. A.** Summary of the effects of replacing 5 arginine (R) res. with basic (K, H) or neutral (N, T) res., in the initial domain ($CT_1$) of the C-terminus chain of Cx32, on normalized $G_j$ following 3 min or 15 min exposures to $CO_2$. Replacement of R with N or T res. greatly increases the $CO_2$ sensitivity of Cx32 or its mutant deleted at res. 225 (D-5R/N), whereas replacement or R with K or H res. has no effect. **B.** Partial R/N replacement results in intermediate $CO_2$ sensitivities, indicating that the R res. differ in their ability to inhibit the $CO_2$ sensitivity of Cx32. R215 inhibits more than R219-220. In contrast, R223-224 seem to partly counteract the inhibitory activity of R215 and R219-220, as 2R/N and 1R/N are more sensitive to $CO_2$ than 4R/N and 3R/N#1, respectively.

There is evidence that the calcium effects may be mediated by calmodulin (CaM). This hypothesis, based on early studies with CaM inhibitors, has been recently strengthened by data on the effect of CaM expression inhibition in *Xenopus* oocytes [5].

## 3. Domains relevant for pH/Ca gating

The molecular mechanism of $CO_2$-induced gating is still unknown, but data on the potential involvement of certain connexin domains are accumulating. The CT domain has been suggested to play a major role in Cx43 chemical gating and a ball-and-chain-like model for $CO_2$-gating has been proposed [6]. This model, is based on provocative data showing that the reduced $CO_2$ sensitivity of a Cx43 mutant deleted at the CT end is reversed by coexpression of deleted Cx43 and CT end [6]. However, the deleted Cx43 was not insensitive to acidification, but just less sensitive, a finding hard to explain if the postulated gating "ball" (CT end) is missing. Nonetheless, this model may not be relevant for Cx32, because a Cx32 mutant in which 84% of the CT had been deleted (Cx32-D219) was as sensitive to $CO_2$ as wild-type Cx32 [7, 8].

For defining domain(s) of Cx32 involved in $CO_2$-induced gating, we have studied the gating sensitivities to $CO_2$ of channels made of Cx32, Cx38, or various chimeras and mutants, expressed in *Xenopus* oocytes [9, 10]. Cx32 is much less sensitive to $CO_2$ than Cx38 (Fig. 2). Our data show that two chimeras: Cx32/38I, Cx32 containing the cytoplasmic loop (CL) of Cx38 and Cx32/38$I_2$ (Fig. 2), Cx32 containing the second half of CL ($CL_2$) of Cx38, are as sensitive to $CO_2$ as Cx38 [9, 10]. This indicates that $CL_2$ plays a major role in pH gating sensitivity.

Since CT chimeras did not express functional channels, the potential role of CT could not be tested with chimeras, but useful data were obtained by mutating basic residues of its initial 18 res. segment ($CT_1$) and by partially deleting CT. Although much of Cx32's CT

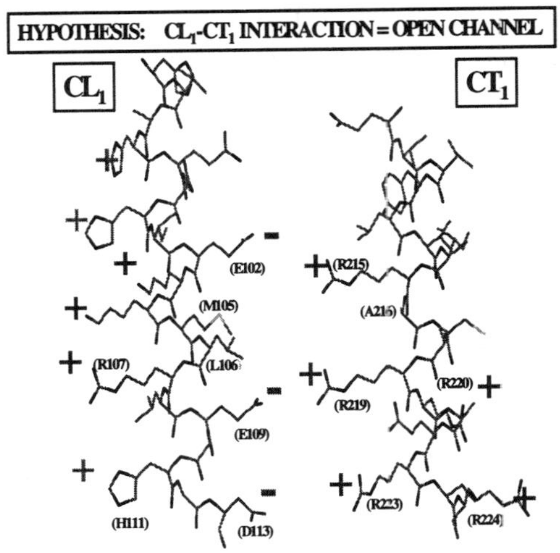


**Figure 4.** Model of potential electrostatic interaction between two domains ($CL_1$ and $CT_1$) of Cx32, displayed in alpha-helical conformation. In view of the fact that: $CT_1$ is positively charged, the inhibitory action of $CT_1$ depends on its positive charges, and the only cytoplasmic domain with negative charges is $CL_1$ (not considering some acidic res. of a CT domain that can be deleted without gating consequences), we propose that $CL_1$ and $CT_1$ interact electrostatically with each other. This interaction may inhibit channel gating by reducing the mobility of the cytoplasmic loop which is likely to participate in the gating mechanism.

seems to be irrelevant for $CO_2$-gating sensitivity, because 84% deletion of it, at res. 219 (Cx32-D219), does not affect $CO_2$ sensitivity [7, 8], some basic residues of $CT_1$ have an inhibitory role. This is suggested by recent data with mutants in which positively charged res. (R) of $CT_1$ were replaced with neutral (polar) res. (N or T) [8], another basic residue (K) or one that becomes positive at low pH (H)(Fig. 3A). Replacement of R with N or T res. greatly increased the $CO_2$ sensitivity of Cx32 or its mutant deleted at res. 225 (D-5R/N) [8], whereas replacement of R with K or H res. had no effect (Fig. 3A) [11], indicating that inhibition is due indeed to positive charges rather than to specific residues. Progressive replacement of R with N res. resulted in progressive increase in Cx32 sensitivity to $CO_2$ (Fig. 3B). R215 has greater inhibitory power than R219-220, whereas R223-224 seem to partly counteract the inhibitory activity of both R215 and R219-220. This is suggested by the fact that 2R/N and 1R/N were more sensitive to 15 min exposures to $CO_2$ than 4R/N and 3R/N#1, respectively.

The inhibitory action of $CT_1$ may involve charge interaction between acidic res. of the N-terminus end of the cytoplasmic loop ($CL_1$) and basic res. of $CT_1$ (Fig. 4). $CT_1$ may function as a latch that keeps the channel open by preventing the second half of the cytoplasmic loop ($CL_2$) from moving toward the channel pore. The gating process may somehow break the $CL_1$-$CT_1$ interaction, enabling $CL_2$ to move toward the channel pore. The channel may be closed by the radial protrusion into the pore of six conserved tyrosines (Y135), one per connexin, because in preliminary experiments a Y135G mutant generated junctions whose $G_j$ paradoxically increased by 15-20% with 15 min $CO_2$ treatments.

This model may also be supported by preliminary evidence for increased gating sensitivity to $CO_2$ with E102R (Fig. 4) mutation at $CL_1$, and by the observation that gating sensitivity to $CO_2$ is increased in oocyte pairs in which one expressed tandem-Cx32 (Cx32 dimers genetically linked NT-to-CT) and the other wild-type Cx32. In connexons made of tandem-Cx32, 3 of the 6 CT and NT chains are linked. This may prevent 3 of the 6 $CT_1$ domains from interacting with $CL_1$ domains, rendering 3 CL more mobile and free for gating.

## 4. Conclusion

The chemical gating of gap junction channels is a complex phenomenon that involves more than one connexin domain. Both an increase in $[Ca^{2+}]_i$ and a decrease in $pH_i$ initiate the cell-cell uncoupling process, but the effect of low $pH_i$ appears to be primarily mediated by an increase in $[Ca^{2+}]_i$. The chain of events that link the increase in $[Ca^{2+}]_i$ and/or $[H^+]_i$ to the channel gating mechanism are unclear, but indirect evidence suggests a role of CaM in the uncoupling process. At the molecular level, two connexin domains have been considered important: the cytoplasmic (inner) loop and the C-terminus chain. However, different functions have been attributed to the latter in Cx32 and in Cx43. In Cx32 the second half of the inner loop seems to be the most relevant gating domain, whereas positive charges of the initial domain of the C-terminus have a strong inhibitory function. This may result from electrostatic interactions between this domain and the first half of the cytoplasmic loop. This interaction may inhibit gating by reducing the mobility of the cytoplasmic loop.

### References


1. A. Lazrak and C. Peracchia, Gap Junction Gating Sensitivity to Physiological Internal Calcium Regardless of pH in Novikoff Hepatoma Cells. *Biophys. J.* 65 (1993) 2002-2012.
2. B. Rose and W.R. Loewenstein, Permeability of a Cell Junction and the Local Cytoplasmic Free Ionized Calcium Concentration: a Study with Aequorin. *J. Membr. Biol.* 28 (1976) 87-119.
3. L. Turin and A.E. Warner, Carbon Dioxide Reversibly Abolishes Ionic Communication between Cells of Early Amphibian Embryo. *Nature,* 270 (1977) 56-57.
4. C. Peracchia, A. Lazrak and L.L. Peracchia, Molecular Models of Channel Interaction and Gating in Gap Junctions. In: C. Peracchia (ed.), Handbook of Membrane Channels. Molecular and Cellular Physiology, ISBN: 0 12 550640 6. Academic Press, San Diego, 1994, pp. 361-377.
5. C. Peracchia, X.G. Wang, L.Q. Li and L.L. Peracchia, Inhibition of Calmodulin Expression Prevents Low-pH-Induced Gap Junction Uncoupling in *Xenopus* Oocytes. *Pflüg. Arch.* 431 (1996) 379-387.
6. G.E. Morley, S.M. Taffet and M. Delmar, Intramolecular Interactions Mediate pH Regulation of Connexin43 Channels. *Biophys. J.* 70 (1996) 1294-1302.
7. R. Werner, E. Levine, C. Rabadan-Diehl and G. Dahl, Gating Properties of Connexin32 Cell-Cell Channels and Their Mutants Expressed in *Xenopus* Oocytes. *Proc. Roy. Soc.* 243 (1991) 5-11.
8. X.G. Wang and C. Peracchia, Positive Charges of the Initial C-ter Domain of Cx32 Inhibit Gap Junction Gating Sensitivity to $CO_2$. *Biophys. J.* 73 (1997) 798-806.
9. X.G. Wang, L.Q. Li, L.L. Peracchia and C. Peracchia, Chimeric Evidence for a Role of the Connexin Cytoplasmic Loop in Gap Junction Channel Gating. *Pflüg. Arch.* 431 (1996) 844-852.
10. X.G. Wang and C. Peracchia, Connexin32/38 Chimeras Suggest a Role for the Second Half of the Inner Loop in Gap Junction Gating by Low pH. *Am. J. Physiol.* 271 (1996) C1743-C1749.
11. X.G. Wang and C. Peracchia, Role of C-Terminus Positive Charges in $CO_2$ Sensitivity of Cx32. *Molec. Biol. Cell* (1997) Abstract in Press.

*Gap Junctions*
*R. Werner (Ed.)*
*IOS Press, 1998*

# Function of Gap Junction Channels Formed in Cells Co-expressing Connexins 40 and 43

Ding Sheng He and Janis M. Burt
*Department of Physiology, University of Arizona, Tucson, Arizona*

Cxs 40 and 43 are co-expressed in a variety of cell types including heart and vascular smooth muscle. It is not clear what advantage is conferred on cells by co-expression of these two connexins. In their homomeric-homotypic forms, the channels formed by these connexins exhibit distinct unitary conductance and voltage-dependent gating properties. These two connexins do not form homomeric-heterotypic junctions. If they form heteromeric channels, the functional properties could be quite different from the homomeric/homotypic parental channel types. The goal of the present study was to determine whether these connexins form heteromeric-(homotypic or heterotypic) channels. Using a cell line that expresses both connexins, A7r5 cells, unitary channel events and voltage-dependent gating behavior were determined and the results compared to published data obtained from homomeric/homotypic Cx43 or Cx40 channels. Unitary events and gating behavior distinct from that observed for the homomeric/homotypic channels and junctions support the conclusion that connexins 40 and 43 form heteromeric channels.

## 1. Introduction

Gap junction channels connect the cytoplasms of adjacent cells and provide a conduit for the exchange of materials between cells. The nature of the materials exchanged is limited by the permselectivity characteristics of the channels. The functional gap junction channel is composed of two hemichannels, called connexons, one contributed by each cell. Each connexon is believed to be a hexamer of protein subunits called connexins (Cx). Of the more than 12 connexin genes found in the mammalian genome, the products of only two, Cx40 and Cx43, have been localized to vascular smooth muscle. It is not clear what advantage might be conferred on cells that co-express these two connexins.

Connexins 40 and 43, in their homomeric/homotypic forms, exhibit distinct conductance, permselectivity and voltage-dependent gating properties [1-4], not to mention potential differences in regulation at the transcriptional, translational, post-translational, etc. levels. Thus co-expression of the homomeric/homotypic forms of Cx40 and Cx43 can be expected to expand the breadth of gap junction function beyond that conferred by either protein alone. Connexins 40 and 43 do not form heterotypic channels (i.e. connexons comprised of Cx40 do not form functional channels with connexons comprised of Cx43) [5-7]. However, it is not yet clear whether these connexins can form heteromeric channels (wherein each connexon contains both connexins), and if they do what the ramifications of heteromeric interaction might be on unitary conductance, permselectivity, gating and regulation.

Working with A7r5 cells, which express both Cx40 and Cx43, Moore and Burt [8] documented the presence of channels with diverse amplitudes. Relative frequency histograms of unitary conductance data collected by hand from strip chart recorder records could be fit equally well by two (68 ± 10 and 119 ± 30pS (S.D.)) or three (70 ± 11, 108 ± 14 and 141 ± 18pS) Gaussian distributions (via Peakfit); however, each peak had greater breadth than expected based on the precision of data collection (± 6pS). It was not clear whether this breadth resulted from true biological variability or technical limitations in data collection and analysis. In the current study we return to the question of whether Cxs 40 and 43 form heteromeric channels with better data collection and analysis strategies.

## 2. Materials and methods

A7r5 cells were obtained from American Type Culture Collection (ATCC). Cells were grown until confluent in DMEM supplemented with 10% fetal bovine serum. Single channel and voltage-dependent behavior were assessed by dual whole-cell voltage clamp techniques on pairs of A7r5 cells approximately one hour after plating. Single channel events were recorded from naturally occurring low conductance pairs after establishing a transjunctional potential difference of 40mV. Data were filtered at 2kHz, digitized at 22kHz (Neurocorder) and stored on tape for future computer-assisted analysis. Taped data were reacquired at 500Hz filtering, 2kHz digitization frequency and analyzed with the analysis program of Ramanan and Brink [9]. All points histograms were constructed and fit to provide channel amplitude, Po, MOT and MCT. Voltage dependent behavior was assessed using a protocol that stepped one cell of a pair to ± 20 mV for ~10s while holding the other cell at 0mV. The polarity of the stepped cell was subsequently reversed after a brief (~4s) return to 0mV. The step protocol was then repeated with larger steps (20mV increment) up to a maximum of 100mV.

## 3. Results

Multiple discrete channel events ranging in amplitude from 38 to 196 pS were observed in single channel recordings and all-points histograms (Figure 1). These channels had open probabilities, Po, ranging from 0.08 to 0.93 with differences in mean open time accounting for most of the variation in open probability. Examples of mode shifting behavior [10] were also prevalent in these records (Figure 2).

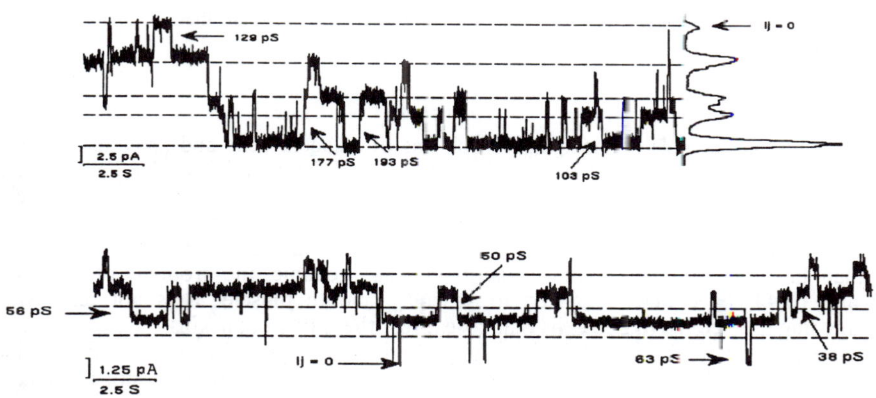


Fig.1. Top panel: A 24 second multi-channel trace from a naturally low conductance pair. The transjunctional voltage was 40 mV. Downward deflections represent channel openings. Note the diversity of event amplitudes, some of which are specifically noted. MOT= 873msec, MCT = 824msec and Po = 0.51. Lower panel: A second cell pair that exhibited channel events of smaller amplitude - 38 to 63 pS. $V_j$=40mV MOT= 993msec, MCT = 420msec and Po = 0.7.

Fig.2. This 220 second multiple-channel trace shows mode shifting behavior of gap junction channels. The transjunctional voltage is 40 mV. Upward deflections represent channel opening.

Diversity of response to voltage was also observed in the junctions formed by A7r5 cells. The plots in Fig 3 illustrate this diversity as well as the "average" behavior. In both plots, steady-state current normalized to instantaneous current is plotted as a function of transjunctional voltage. Results from fifteen cell pairs are illustrated in A, "average" data are displayed in B. When the average data were fit (Sigma Plot - Boltzmann equation), a $V_0$ of 70mV and $G_{min}$ of 0.6 were revealed (Fit not shown).

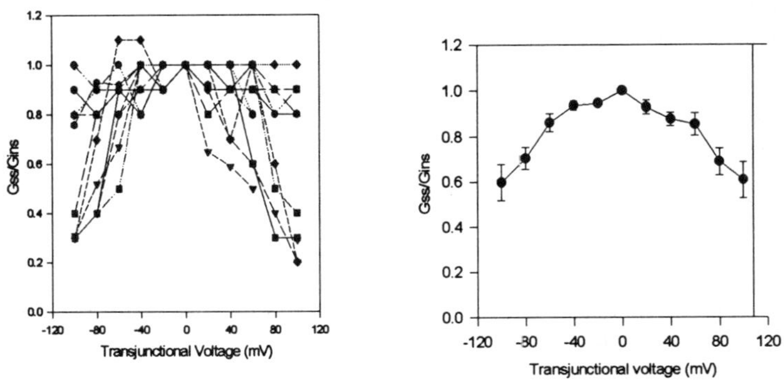


A B

Fig 3.. Voltage-dependent gating of gap junctions in A7r5 cells. Plot A is presented from 15 individual cell pairs that describe the relationship of transjunctional voltage and conductance. Same data is presented in Plot B and conductance is plotted in M± SE.

## 4. Discussion

Connexins 40 and 43 in their homomeric/homotypic forms exhibit distinct unitary conductances and voltage dependent gating behavior. These properties of channel function would be expected to differ in heteromeric channels, if such channels form [10]. In this study we evaluated these parameters of channel function in A7r5 cells, a cell line that co-expresses these connexins, and compare, in the text that follows, the results with published data on the behavior of the homomeric/homotypic forms of Cx40 and Cx43.

Baukauskas et al. [4] examined the biophysical properties of Cx40 channels expressed in HeLa cells. They found that Cx40 exhibited a 198 pS main state, a 36 pS residual state, and infrequent substates of several amplitudes. In contrast, Valiunas et al. [2] found that Cx43 channels exhibited a main state of 96pS, a residual state of 23pS, and infrequent substates of several amplitudes. Based on these and other data, we would predict that if the only type of channel formed in cells that co-express these connexins is the homomeric/homotypic channel, then we would observe channel events with the following amplitudes: 198 and 36 pS (Cx40), and 96 and 23pS (Cx43) as the channels transition from the closed state to either the residual or main states, and 162pS (Cx40) and 73pS (Cx43) events as the channels transition between the main and residual states. Occasional events of amplitudes other than these would also be expected as the channels enter or leave substates. Our records show numerous transitions with amplitudes other than the expected amplitudes; indeed transitions of the expected size were rare. Openings from the zero transjunctional current level included events with the following amplitudes: 170, 160, 144, 135, 120, 100, 95, 63, 55pS.

Several investigators have examined the voltage dependence of Cx40 channels in their homomeric/homotypic form. In all cases, voltage dependence is evident with $V_0$ ranging between 38 and 50mV and $G_{min}$ between 0.18 and 0.33 [3,4,11]. In comparable studies on Cx43, weaker voltage dependence was observed, with $V_0$ between 50 and 70mV and $G_{min}$ between 0.27 and 0.46 [2,3,12]. In cells co-expressing these connexins, if only homomeric/homotypic channels form, then one might observe a biphasic response of junctional conductance to voltage - i.e. at moderate vs. large transjunctional voltages, the decrease in junctional conductance might resemble voltage dependent closure of Cx40 vs. Cx43 containing junctions. The magnitude of response to moderate vs. large voltages would reflect the relative contributions of the two channel types to the overall junction. Regardless of relative contribution, $G_{min}$ would certainly be less than ~0.5, the largest of reported $G_{min}$ values for either Cx43 or Cx40 in their homomeric forms. In our experiments we observed a huge diversity of voltage dependent behavior. Some cell pairs exhibited essentially no voltage dependent decrease in conductance over ± 100mV. Other cell pairs exhibited voltage dependence behavior that resembled Cx40-containing junctions while still others resembled Cx43-containing junctions.

Thus, our single channel and voltage dependence data are at odds with expected behavior if the A7r5 cell gap junction were composed solely of homomeric/homotypic channels. It is unlikely that gap junctions in A7r5 cells are comprised of homomeric/heterotypic junctions, as several investigators [5-7] have demonstrated that Cxs 40 and 43 do not form such channels (i.e. connexons comprised of Cx40 do not form functional channels with connexons comprised of Cx43). Haubrich et al [7] demonstrated that the failure of these connexins to form homomeric/heterotypic channels reflects their inability to dock with one another in the extracellular space. They also demonstrated that Cx40*43E1,2 form functional homotypic channels as well as heterotypic channels with Cx43 or Cx40. These data suggest that the membrane spanning domains of these two connexins do not prevent these connexins from interacting to form connexons or functional channels. Thus available data are consistent with the possibility that connexins 43 and 40 form heteromeric connexons. Haubrich et al's data would further predict that such heteromeric connexons would only dock with themselves to form heteromeric/homotypic channels. If these predictions are correct, then 64 different channel types could be formed in cells co-expressing Cxs 40 and 43. The probability of observing the activity of either homomeric/homotypic channel (if all 64 channel types are formed with equal probability) is, 0.0156 (or 1/64). Thus, if heteromeric/homotypic channels form they would constitute the vast majority of channels in the junction and their unitary sizes and voltage dependence would predominate in single channel and macroscopic recordings.

The diversity of channel events in our single channel recordings and of voltage dependence in our macroscopic recordings are expected results if Cxs 40 and 43 form heteromeric/homotypic channels. Consequently, we conclude that connexins 40 and 43 form heteromeric channels with functional properties distinct from either Cx40 or Cx43 homomeric/homotypic channels. The diversity of channel and gap junction function conferred by co-expression of connexins 40 and 43 may provide such cells with considerable flexibility in regulating the function of their intercellular communication pathway, flexibility that may facilitate reponsiveness to a broad range of physiological stimuli. Such flexibility could be of considerable importance to cells in the cardiovascular system as they exploit the full dynamic range of this system.

## References


[1] R. D. Veenstra, H. Z. Wang, D. A. Beblo, M. G. Chilton, A. L. Harris, E. C. Beyer, and P. R. Brink, Selectivity of connexin-specific gap junctions does not correlate with channel conductance *Circ. Res.* vol. 77, pp. 1156-1165, 1995.

[2] V. Valiunas, F. Bukauskas, and R. Weingart, Conductances and selective permeability of connexin43 gap junction channels examined in neonatal rat heart cells *Circ. Res.* vol. 80, pp. 708-719, 1997.

[3] R. Bruzzone, J.-A. Haefliger, R. L. Gimlich, and D. L. Paul, Connexin40, a component of gap junctions in vascular endothelium, is restricted in its ability to interact with other connexins *Mol. Biol. Cell*, vol. 4, pp. 7-20, 1993.

[4] F. F. Bukauskas, C. Elfgang, K. Willecke, and R. Weingart, Biophysical properties of gap junction channels formed by mouse connexin40 in induced pairs of transfected human HeLa cells *Biophys. J.* vol. 68, pp. 2289-2298, 1995.

[5] T. W. White, D. L. Paul, D. A. Goodenough, and R. Bruzzone, Functional analysis of selective interactions among rodent connexins *Mol. Biol. Cell*, vol. 6, pp. 459-470, 1995.

[6] C. Elfgang, R. Eckert, H. Lichtenberg-Frate, A. Butterweck, O. Traub, R. A. Klein, D. Hulser, and K. Willecke, Specific permeability and selective formation of gap junction channels in connexin-transfected HeLa cells *J. Cell Biol.* vol. 129, pp. 805-817, 1995.

[7] S. Haubrich, H. J. Schwarz, F. Bukauskas, H. Lichtenbergfrate, O. Traub,Weingart R, and K. Willecke, Incompatibility of connexin 40 and 43 hemichannels in gap junctions between mammalian cells is determined by intracellular domains *Mol. Biol. Cell*, vol. 7, pp. 1995-2006, 1996.

[8] L. K. Moore and J. M. Burt, Selective block of gap junction channel expression with connexin-specific antisense oligodeoxynucleotides *Am. J. Physiol.* vol. 267, pp. C1371-C1380, 1994.

[9] S. V. Ramanan, S. F. Fan, and P. R. Brink, Model invariant method for extracting single-channel mean open and closed times from heterogeneous multichannel records *J. Neuro. Methods*, vol. 42, pp. 91-103, 1992.

[10] P. R. Brink, K. Cronin, K. Banach, E. Peterson, E. M. Westphale, K. H. Seul, S. V. Ramanan, and E. C. Beyer, Evidence for heteromeric gap junction channels formed from rat connexin43 and human connexin37 *Am. J. Physiol.* vol. (in press), 1997.

[11] D. A. Beblo, H. Z. Wang, E. C. Beyer, E. M. Westphale, and R. D. Veenstra, Unique conductance, gating, and selective permeability properties of gap junction channels formed by connexin40 *Circ. Res.* vol. 77, pp. 813-822, 1995.

[12] P. R. Brink, S. V. Ramanan, and G. J. Christ, Human connexin 43 gap junction channel gating: Evidence for mode shifts and/or heterogeneity *Am. J. Physiol. Cell Physiol.* vol. 271, pp. C321-C331, 1996.

*Gap Junctions*
*R. Werner (Ed.)*
*IOS Press, 1998*



# Anion Permeability of Gap Junctions is Enhanced in Growth Arrested Smooth Muscle Cells

David T. Kurjiaka, Mary V. Olsen and Janis M. Burt
*Department of Physiology, University of Arizona, Tucson Arizona, 85724*

**Abstract**

In a non-diseased vessel, vascular smooth muscle cells are typically growth arrested (GA). In response to a variety of pathophysiologic conditions, however, these cells are stimulated to enter the cell cycle and proliferate. To assess what role changes in intercellular communication might play in maintenance of the growth arrested and proliferating states and the transition between these states, we have examined gap junction coupling and connexin (Cx) expression in a smooth muscle cell (SMC) line that expresses Cx40 and Cx43, A7r5 cells. Treatment of A7r5 cells for 48h in GA media (1% FBS or 0.1% BSA) inhibited cell proliferation and arrested nearly 90% of the cells (from 70% in 10% FBS: determined by FACS) in $G_1/G_0$. This increase in $G_1/G_0$ cells was associated with a 38% enhancement of dye coupling (Lucifer Yellow) in GAed cells. In contrast, GAed cells exhibited a 40% reduction in electrical coupling over their proliferating counterparts. One explanation for the discrepancy between dye and electrical coupling is a change in permselectivity of the junction. To assess this possibility, macroscopic conductance ($g_j$) was measured with patch solutions (135mM) differing only in their anion and cation content (KGlutamate, KCl and TEACl). In GAed cells, $g_j$ remained unchanged for KCl and TEACl patch solutions. However, $g_j$ was reduced in GAed cells with KGlutamate in the patch solution. These results suggest that GAed A7r5 cells exhibit a greater junctional anion permeability than their proliferating counterparts. Enhanced anion permeability could mediate exchange of anionic signaling molecules that facilitate maintenance of the GAed state.

## 1. Introduction

Gap junction communication may play an important role in growth control. Cells proliferating in an uncontrolled fashion frequently exhibit altered or absent cell-cell communication [1]. When such cells have their native gap junction protein reintroduced via transfection, controlled growth properties are restored [2]. In addition, cells stimulated to proliferate with growth factors (PDGF, EGF, etc.) exhibit an acute but transient reduction in gap junction communication [3,4]. In the vasculature, smooth muscle cell proliferation occurs in several pathophysiological conditions. The goal of this study was to evaluate cell-cell communication in growth arrested vascular smooth muscle cells and assess the mechanisms underlying any observed differences with proliferating cells.

## 2. Hypotheses

1) Gap junction communication is enhanced in growth arrested A7r5 cells, an aortic smooth muscle cell line.
2) Junctional permeability to anions is enhanced in growth arrested A7r5 cells, cation permeability remains unchanged.

## 3. Materials and methods

The A7r5 cell line, derived from embryonic rat aortic smooth muscle cells, was chosen for study because they express both Cx40 and Cx43, a feature of some vascular smooth muscle cells *in vivo* [5]. Cells were propagated in DMEM with 10% FBS [lipoprotein deficient serum (LPDFBS) was used in some experiments]. Growth arrest (GA) was induced by serum reduction (1% FBS) or serum deprivation (0.1% BSA) for 48h or 72h. Cell cycle position (cells in $G_1/G_0$) was assessed on propidium iodide stained cell nuclei via fluorescence activated cell sorting.

Lucifer Yellow dye was microinjected into confluent proliferating (in 10% LPDFBS) and GAed (in 1% LPDFBS for 48h) cells and the number of neighboring cells to which dye diffused (within one minute) was quantified. Macroscopic junctional conductance ($g_j$) of proliferating and GAed cells was determined by dual whole-cell voltage clamp. For the proliferating state, cells were plated at low density on gelatin (1%) coated coverslips in 10% FBS and $g_j$ determined 18-24h later. To initiate the growth arrest protocol, subconfluent proliferating cells were washed three time with PBS and growth arrest media (1% FBS or 0.1% BSA) was added. After 24h, the PBS wash was repeated to eliminate any autocrine growth factors and 1% FBS returned. Initially macroscopic conductance was determined with a patch solution containing 50:50 KGlutamate and CsCl. In addition, three different patch solutions were used to determine $g_j$. These patch solutions varied by the primary current carrying ions (135 mM; KGlutamate, KCl or TEACl) to test for differences in permselectivity. Single channel events were observed by treating cell pairs with halothane (5mM) to decrease $g_j$. A driving force of ±40 mV was applied in all experiments.

## 4. Results

Treatment of subconfluent A7r5 cells for 48h in either GA media (1% FBS or 0.1% BSA) induced ~ 90% of the cells into $G_1/G_0$. Increasing the duration of GA to 72h elevated the proportion of cells in $G_1/G_0$ by 2%. When these cells (1% FBS) were tested for dye coupling, GAed SMCs exhibited a 38% increase in the number of neighboring cells with dye (Fig. 1). In contrast, when GAed cells were tested for electrical coupling, they were ~40% less well coupled than their proliferating counterparts (Fig. 2).

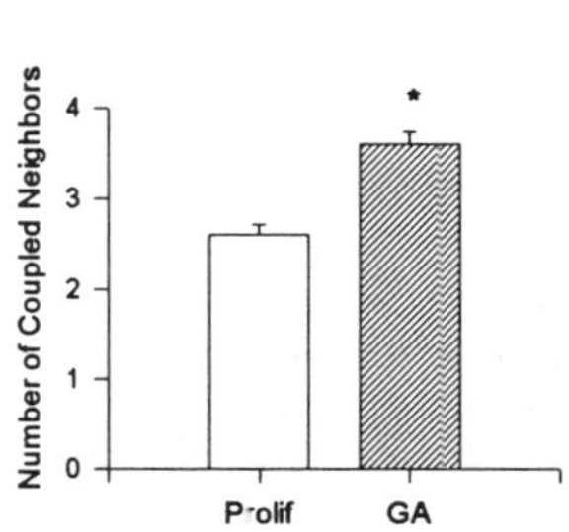


**Figure 1.** Dye coupling is enhanced in GAed vs proliferating A7r5 cells. Dye coupling was determined as the number of neighboring cells to which Lucifer Yellow diffused following dye microinjection. LPDFBS (1% and 10%) was used in these experiments. Dye injections were performed on 70 proliferating and 212 GAed cells. * - significantly different from proliferating cells at $p < 0.05$.

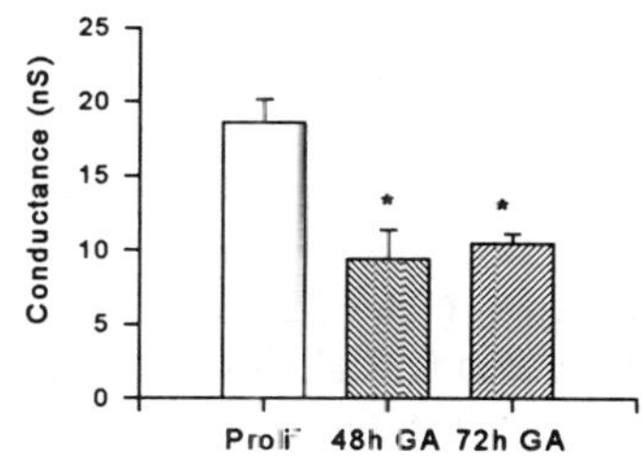


**Figure 2.** Electrical coupling (macroscopic conductance in nS) is reduced in GAed vs proliferating A7r5 cells. Cells were GAed in 0.1% BSA and their conductance measured in 50:50 KGlutamate and CsCl pipette solution. Sample sizes were as follows: proliferating n=23, 48h GA n=6, 72h GA n=13.
* - significantly different from proliferating cells at $p < 0.05$.

To explain this discrepancy, the permselectivity of these channels was assessed by changing the primary current carrying ion in the patch solution (135mM). In GAed SMC, conductance with KGlutamate in the pipette was significantly reduced (Fig. 3). At the same time, no differences were apparent with KCl or TEACl in the patch solution (Fig. 3). Treatment of the cells from Figure 2 with halothane revealed an increase in the frequency of single channel events with amplitudes less than 90pS in GAed SMC (Fig. 4). The observed changes in permselectivity and unitary conductance could be explained by changes in Cx expression. As an initial test of this possibility, Cx40 and 43 mRNA levels were determined in GAed and proliferating cells. Cx40 was reduced in GAed cells, Cx43 was unchanged.

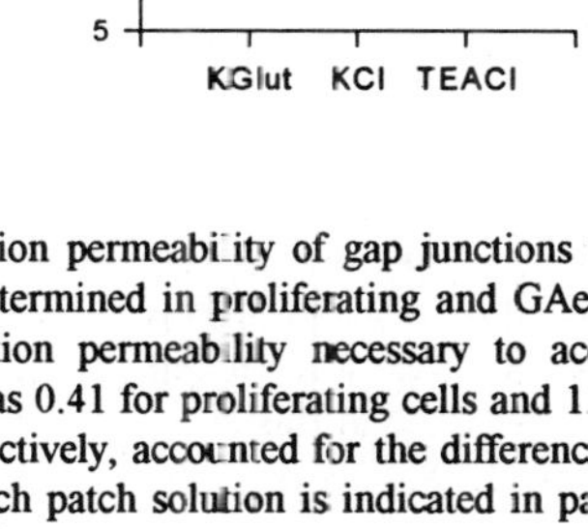


**Figure 3.** Relative anion to cation permeability of gap junctions was enhanced in GAed A7r5 cells. Macroscopic conductance was determined in proliferating and GAed cells with 3 patch solutions. The change in relative anion to cation permeability necessary to account for differences in junctional conductance between solutions was 0.41 for proliferating cells and 1.71 for GAed cells in KGlutamate vs KCl whereas 0.62 and 1.37 respectively, accounted for the differences observed in TEACl vs KCl. The number of pairs sampled with each patch solution is indicated in parentheses. * - significantly different from proliferating cells at $p < 0.05$.

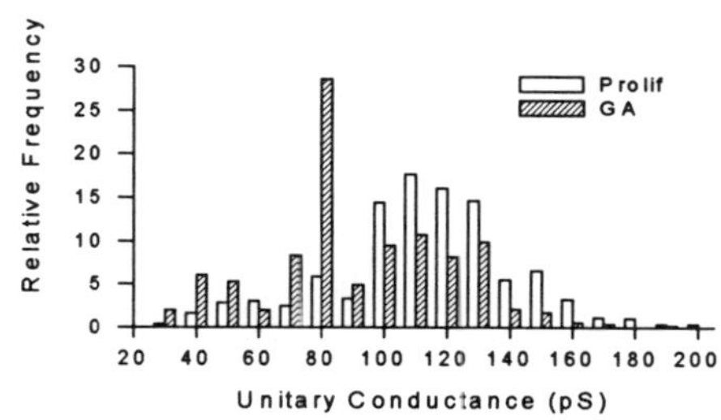


**Figure 4.** In GAed cells the frequency of events with amplitudes <90pS was enhanced while the frequency of events with amplitudes >90pS was decreased (50:50 KGlutamate and CsCl). All cell pairs were treated with halothane to decrease macroscopic conductance. The sum of the weighted conductance [Σ (bin conductance • relative frequency)] for growth arrested and proliferating cells were calculated as 93pS and 113pS.

## 5. Discussion

Treatment of A7r5 cells for 48h and 72h with serum-deficient or deprived medium induces GA in ~90% of smooth muscle cells. In these GAed cells, the diffusion of Lucifer Yellow dye between cells was augmented. Despite enhanced dye coupling electrical coupling, determined with KGlutamate/CsCl patch solution, was reduced. The apparent discrepancy between dye and electrical data are explained by an increased anion:cation permeability ratio in the GAed vs. proliferating cells (from 0.41 in Proliferating to 1.71 GA for KCl vs. KGlutamate: Fig. 3). That $g_j$ did not differ between GAed and proliferating cells when determined with KCl (14.56 and 14.89 nS) and TEACl (10.29 and 11.56 nS), indicates that gap junction cation permeability was not different between proliferating and GAed cells. Therefore, the change in anion:cation permeability ratio must reflect an increase in anion permeability. From these data we can also infer that the number of active gap junction channels (or their open probability) must be considerably less in the GA cells (see below). These conclusions are consistent with the significant enhancement of dye coupling observed in the GAed cells, since Lucifer Yellow is an anionic dye.

To identify the mechanisms underlying the observed changes in anion permeability, we investigated single channel events in GAed vs. proliferating cells. $g_j = \text{Perm} \cdot \gamma_j \cdot N \cdot P_o$ where Perm is the relative anion:cation permeability, $\gamma_j$ the single channel conductance, N the number of channels and $P_o$ the open probability. Since $g_j$ in GAed and proliferating cells was equal when measured with KCl, we know that the product of these parameters must be equal. Using the calculated mean channel conductance (Fig. 4) and Perm for the GA vs. proliferating cells ($\gamma_j$ GA 93pS; $\gamma_j$ proliferating 113pS; Perm in GA 1.71; Perm in proliferating 0.41), we calculated that the ratio of the product of N and $P_o$ in GAed vs. proliferating cells is 0.29. If the difference in this ratio stemmed entirely from a change in N, and if, for simplicity, we assume $P_o$ equals one, then the number of active channels in the GA cells would be ~70% less than in the proliferating cells or 93 vs. 314 channels. It is interesting to compare this calculated number of active gap junction channels to the 56% reduction in Cx40 mRNA level and to consider the greater unitary conductance of the full open state of Cx40 homotypic vs. Cx43 homotypic channels (~155pS vs. ~90pS: 6).

In previous studies, individual Cx40 channels were shown to be relatively impermeable to anions (0.29 relative permeability: [7]) whereas Cx43 channels were found to be permeated equally well by anions and cations [8]. That the junctional anion:cation permeability ratio for GAed cells exceeds one is inconsistent with these previously measured permselectivities. To explain these findings, we propose the following: 1) Cx40 and Cx43 form heteromeric channels with properties, in particular permselectivity, not predicted by their homomeric/ homotypic behavior, or 2) relative to the fully open states of the homomeric/homotypic forms of these channels, other conductance states (perhaps as induced by phosphorylation) exhibit significantly different permselectivities. With either possibility, a change in proportion of Cx40 to Cx43 would be expected to alter gap junction behavior. Further experiments will be necessary to determine the mechanism whereby anion permeability of gap junctions is enhanced.

## 6. Conclusion

Gap junctions in GAed smooth muscle cells exhibit enhanced anion permeability relative to junctions in their proliferating counterparts. While the mechanism(s) underlying this enhanced anion permeability are unclear, the observed decrease in Cx40 mRNA in the GA cells may play a role. The enhanced permeation of gap junctions by anions in GA cells is expected to support enhanced capacity for intercellular exchange of anionic regulatory molecules (e.g., cyclic nucleotides, metabolites, $IP_3$, etc.) which could facilitate maintenance of the growth arrested state of vascular smooth muscle cells.

## Acknowledgments

This work was supported by funding from an NRSA Institutional Training Grant (2T32 HL07249-21) and Arizona Disease Control Research Commission (5-090).

## References

1. Yamasaki, H and CCG Naus. Role of connexin genes in growth control. *Carcinogenesis* 17:1199-1213, 1996.

2. Mesnil, M, V Krutovskikh, C Piccoli, C Elfgang, O Traub, K Willecke and H Yamasaki. Negative growth control of HeLa cells by connexin genes: connexin species specificity. *Cancer Res.* 55:629-39, 1995.

3. Pelletier, DB and AL Boynton. Dissociation of PDGF receptor tyrosine kinase activity from PDGF-mediated inhibition of gap junctional communication. *J. Cell. Physiol.* 158:427-434, 1994.

4. Lau, AF, MY Kanemitsu, WE Kurata, S Danesh and AL Boynton. Epidermal growth factor disrupts gap-junctional communication and induces phosphorylation of connexin43 on serine. *Mol. Biol. Cell.* 3:865-874, 1992.

5. Little, TA, EC Beyer and BR Duling. Connexin 43 and connexin 40 gap junctional proteins are present in arteriolar smooth muscle and endothelium in vivo. *Am. J. Physiol.* 268:H729-H739, 1995.

6. Traub, O, R Eckert, H Lichtenberg-Frate, C Elfgang, B Bastide, KH Scheidtmann, Hulser DF, K Willecke. Immunochemical and electrophysiological characterization of murine connexin40 and -43 in mouse tissues and transfected human cells. *Eur. J. Cell Biol.* 64:101-12, 1994

7. Beblo, DA, H-Z Wang, EC Beyer, EM Westphale and RD Veenstra. Unique conductance, gating and selective permeability properties of gap junction channels formed by connexin40. *Circ. Res.* 77:813-822, 1995.

8. Veenstra, RD, H-Z Wang, DA Beblo, MG Chilton, AL Harris, EC Beyer and PR Brink. Selectivity of connexin-specific gap junctions does not correlate with channel conductance. *Circ. Res.* 77:1156-1165, 1995.

*Gap Junctions*
*R. Werner (Ed.)*
*IOS Press, 1998*

# Selectivity of rat Connexin-43 and Connexin-40 Gap Junction Channels: Insights into Connexin Pore Structure and Mechanisms of Ion Permeation

*R.D. Veenstra, D.A. Beblo and H.-Z. Wang*
*Department of Pharmacology, SUNY Health Science Center at Syracuse, NY 13210, USA*

## 1. Abstract

Channel conductances and ionic permeabilities were determined on homotypic rat Cx43 and Cx40 gap junction channels using conventional double whole cell patch clamp methods. Varying KCl concentration produced nonlinear conductance changes indicative of single site saturation kinetics with a $K_d$ of 127 mM for Cx40 and 143 mM for Cx43. Cation permeabilities did not vary significantly from conductance or mobility values and both channels exhibited an 8:1 K/Cl selectivity ratio. Anion permeabilities were not consistent with Goldman-Hodgkin-Katz theory, but were consistent with models derived for two distinct cation/anion permeable channels requiring direct counterion binding to the primary permeant ion. In one case, anions still produce a counter flux (Cx43) and in the other example, anions are cotransported with the permeant cation (Cx40). Thus $K^+$ flux is predicted to have significant effects on second messenger permeability since most are anionic (cAMP, $IP_3$). These hypotheses need to be examined by direct experimentation.

## 2. Introduction

The primary physiological function of the connexins is believed to be the formation of intercellular channels that permit the exchange of electrical and chemical signals between adjacent cells. In mammalian cells, this translates to an ability to effectively conduct electrical signals at intracellular $K^+$ concentrations of 140-150 mM and second messengers upwards of 900 daltons in molecular mass and 10-14 Å in diameter at much lower concentrations (e.g. µM to mM). Typically, ion channels are characterized by their ionic selectivity, limiting pore size, channel conductance, and modulatory gating mechanisms. These properties of gap junction channels remain largely unexplored despite the prevailing interpretation of gap junction channels as being large aqueous pores with only modest ionic selectivities and molecular selectivities based primarily on steric hindrance due to the relative molecular and pore diameters. In this brief report, I would like to convey to you our recent findings on the ionic permeability, channel conductance, and ionic affinity properties of two homotypic connexin channels formed by rat connexin43 (Cx43) and connexin40 (Cx40) [1,2,3]. To conclude, I will compare our findings to published accounts of the ionic permeation mechanisms reported for three different cation and anion permeable channels as a means of gaining insight into how ions actually traverse the gap junction channel pore and the consequences of these mechanisms on molecular permeation.

## 3. Results

The two most commonly reported means of functionally identifying the connexin channel expressed in a particular cell type is the determination of the channels transjunctional voltage ($V_j$) sensitivity and the unitary channel conductance in a 120-140 mM potassium salt solution. We will not be presenting any evidence about the gating of gap junction channels at this time, but I would like to address some of the channel conductance and selectivity properties of the Cx43 and Cx40 channels that we have investigated in our laboratory over the past two years.

### *3.1 Channel Conductance Sequences*

The easiest experimental permutation to perform is the equimolar substitution of one ion for another of like valence. For an aqueous channel, the change in conductance should mimic the relative differences in their aqueous mobilities. We used a series of 115 mM XCl salts including the five alkali earth cations ($Cs^+$, $Rb^+$, $K^+$, $Na^+$, and $Li^+$) and two small organic cations, tetramethylammonium (TMA) and tetraethylammonium (TEA). The unitary conductance ($\gamma_j$) sequences for both Cx43 and Cx40 followed the aqueous mobility sequence with only modest variations [1,2]. This is most evident in Figure 1 (left panel) where the ionic channel conductance ratios relative to $K^+$ are plotted relative to their aqueous mobility ratios ($\mu_{ion}/\mu_K$). The solid line indicates unity, or $\gamma_{ion}/\gamma_K = \mu_{ion}/\mu_K$. The deviations from unity are more pronounced for the less mobile cations and the only sequence transposition occurs with $Li^+$ which has a higher $\gamma_j$ than $TMA^+$ despite a lower $\mu_{ion}$. These data do not provide any indication of ionic affinity or selectivity and are consistent with the general interpretation of an essentially aqueous channel that is permeable to monovalent cations.

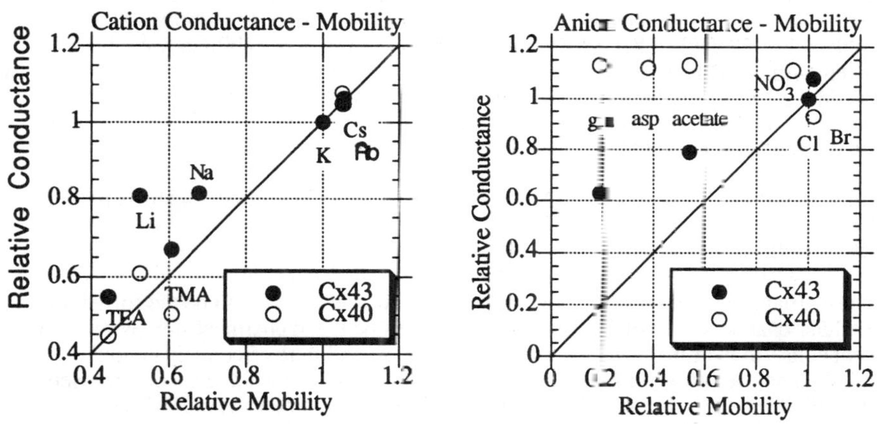


Figure 1. Unitary channel conductance ($\gamma_j$) of Cx43 and Cx40 gap junction channels in 115 mM $Cl^-$ salt of the indicated cation (left panel) and 115 mM $K^+$ salt of the indicated anion (right panel) were normalized to their relative value in KCl [1,2]. $\gamma_j$ (KCl) was 97 pS for Cx43 and 142 pS for Cx40. The diagonal line indicates equality to their ionic aqueous mobilities [4] also plotted relative to $K^+$.

Additional information was obtained by performing similar conductance and permeability measurements using 115 mM $K^+$ salts of various anions. Figure 1 (right panel) also illustrates the channel conductances and mobilities relative to $Cl^-$ for Cx43 and Cx40. The Cx43 channel again did not produce any striking results, but the lack of an effect on Cx40 $\gamma_j$ by a variety of organic anions including nitrate, acetate, aspartate, and glutamate results does not follow conventional constant field theory.

### *3.2 Channel Conductance - Concentration Relations*

The above experiments were performed using a constant salt concentration and the change in gj depends on the apparent mobility of the cation within the channel pore. When one varies the concentration of a single salt, KCl, one obtains a conductance plot that depends on the kinetics of transport through the pore. For simple diffusion, this relation is a straight line and saturation of the channel conductance is indicative of saturable enzyme binding site. The general assumption is that gap junction channels pass intracellular $K^+$

ions with relative ease, thereby creating a pathway for rapid transmission of electrical signals. When we varied the KCl concentration across a Cx43 or Cx40 gap junction channel by adding an impermeant polysaccharide (raffinose) to the internal pipette solution, we obtained a curvelinear relationship relative to KCl concentration that became sublinear at higher concentrations [3]. The conductance and concentration values are listed in Table 1.

Table 1. Channel Conductance- Concentration.

| [KCl] (mM) | Cx43 (pS) | Cx40 (pS) |
|---|---|---|
| 30 | 70±3 | 78±3 |
| 58 | 102±1 | 112±4 |
| 87 | 137±7 | 141±2 |
| 115 | 158±6 | 185±6 |
| 130 | 164±7 | --- |
| 147 | 198±6 | 187±4 |
| 190 | --- | 199±7 |
| 230 | 203±1 | 210±7 |

This indicated to us that the Cx43 and Cx40 channels were indeed beginning to saturate and extrapolation of the Hill equation fit provided an estimate of the limiting $\gamma_j$ in KCl. The Hill coefficients were ≈1.1 for both channels and the limiting conductances were 233 pS and 253 pS for Cx40 and Cx43 respectively. By definition, the $K_d$ of an enzyme substrate equals the concentration at half maximum velocity which was 127 and 143 mM KCl for Cx40 and Cx43 respectively. These data indicate that Cx40 and Cx43 gap junction channels are indeed designed to conduct $K^+$ at physiological intracellular concentrations.

### *3.3 Relative Ionic Permeabilities*

Ionic selectivity is best determined by measuring the diffusion potential that develops when two ions are distributed across the channel with oppositely directed Nernst potentials. Conventionally, relative permeability ratios can be calculated using the Goldman-Hodgkin-Katz voltage equation. These experiments were performed using all of the above 115 mM $Cl^-$ salt internal pipette solutions. One might expect the ionic conductance ratios to be accurate predictors of the ionic permeability ratios for an essentially aqueous channel. The results for both channels are illustrated in Figures 2 and 3. While the Cx40 channel ionic permeability ratios closely approximate the conductance ratios (slope = 1.04), the ionic permeabilities for the Cx43 channel follow a steeper linear relationship (slope = 1.6) even though the permeability and conductance sequences for both channels remain unchanged.

Additional interactions must be occurring to reduce permeability more than conductance, particularly for the larger ions. Conductance is a kinetic measurement of how many ions are transported per unit time while permeability is a measure of how often one ion is transported relative to another. Hence, there may be ionic interactions occurring at a site or sites within the pore that were not anticipated from conductance measurements.

The relationship between relative ionic permeability and ionic radius was utilized to provide estimates of pore size based on the hydrodynamic equation [5]:

$$P_{ion} = C \bullet (1-\alpha^2) \bullet \left( \frac{1 - 2.105\alpha + 2.0865\alpha^3 - 1.7068\alpha^5 + 0.72603\alpha^6}{1 - 0.75857\alpha^5} \right)$$

where α is the ratio of the ion/pore radii and C is a constant used to calculate the area at the mouth of the pore. Fits of the relative ionic permeability - (hydrated) radius values provided estimates of pore radius of 6.3 ± 0.4 Å for Cx43 and 6.6 ± 0.7 Å for the Cx40 channel [1,2]. Using published permeability data on the Cx46 hemichannel, we obtained an estimate of 5.5 ± 0.6 Å for its pore radius [3,6].

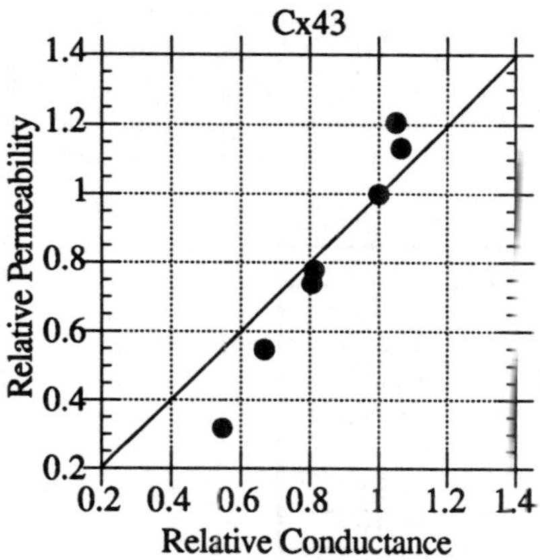


Figure 2. Ionic permeability versus ionic conductance relative to potassium for the Cx43 channel.

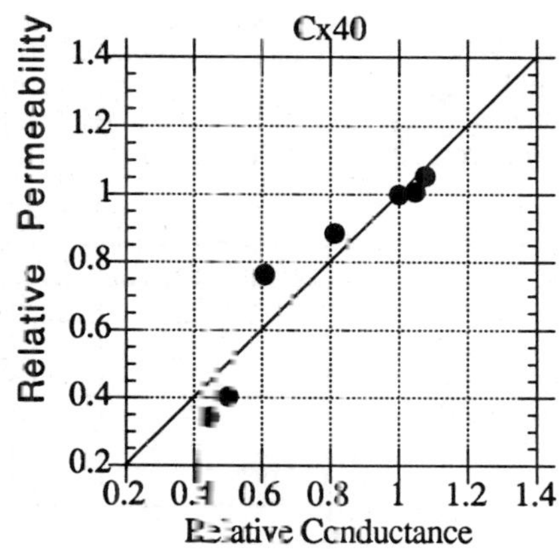


Figure 3. Ionic permeability versus ionic conductance relative to potassium for the Cx40 channel.

An asymmetric single XCl salt gradient permits direct measurement of the $P_x/P_{Cl}$ ratio. We used LiCl since it was the standard we used for cation permeability measurements and again used an impermeant sugar to replace LiCl as required. The reversal potentials for 115/30 mM LiCl were -16.5 and -16.1 mV for Cx43 and Cx40 [1,2], which translate to a nearly equivalent Li/Cl permeability ratio of 0.18 for both channels (≈ 8 to 1 selectivity for $K^+$ over $Cl^-$). Thus, despite quantitative differences in the cationic conductances, conductance ratios, and permeability ratios, their cation/anion selectivity appears to be identical.

The anion permeability values for the Cx43 channel did not conform to constant field theory either even though the permeability -conductance ratio plot again appears to be close to unity for the tested anions (data not shown,[1]). Not only were the anion $\gamma_j$ values higher for less mobile anions, but the anion permeability values for the Cx40 channel similarly were higher for the larger anions relative to $Cl^-$ and $Br^-$ based on the GHK voltage equation calculations (data not shown [2]). These seemingly paradoxical results require new interpretations based on direct cation-anion interactions within the pore. As such, the assumption of ionic independence inherent to the GHK equation is not valid here.

At least three different papers describing permeation through cation and anion permeable channels also demonstrated that the reversal potentials did not follow conventional Goldman-Hodgkin-Katz theory [7,8,9]. This was true for pores three times larger than the connexin pore estimates as well as smaller pores of 7-8 Å in diameter. There are several similarities between our data on the Cx43 and Cx40 channels and two of these published mechanisms for cation/anion permeation that led us to hypothesize specific mechanisms for cation and anion permeability for the Cx43 and Cx40 channels.

## 4. Mechanisms of Cation and Anion Permeability

One major distinguishing characteristic of the two models that best describe our results for the Cx43 and Cx40 channels is that the permeability of the less permeant ions (anions) is dependent on the presence of a permeant cation. In the case of the Cx43 channel, the $P_{anion}/P_{Cl}$ ratios were readily calculated from the experimental reversal potentials once the cation concentration was divided by the total concentration of all cations in the solution. This derivation was performed by Borisova et al. [8] for the amphotericin B channel and its interpretation assumes that the counterflux produced by the less permeant ion requires binding to the more permeant ion at a specific site within the pore. The same can be assumed for the Cx40 channel except that the anion does not produce a countercurrent flux, but rather is cotransported with a cation in proportion to their respective cation/anion permeability ratio. This explains why the more permeant anions ($Br^-$, $Cl^-$) reduce $\gamma_j$ while the less permeant anions have a negligible effect on $\gamma_j$. This interpretation can also explain why the experimental reversal potentials shift in the opposite direction than expected from constant field theory. This mechanism for cation/anion permeability was derived by Franciolini and Nonner for the neuronal anion channel [9]. Given the similarity of our results on two different connexin channels to these two mechanisms for ion permeation, we hypothesize that there is a cation binding site within the Cx43 and Cx40 pores where the anions must obligatorily bind to a cation in order to pass.

Figure 4 illustrates the proposed mechanism of ion permeation for the Cx43 channel. A putative fixed anionic site within the pore attracts cations which bind weakly (energy well) while transiting through the restricted space of ≈ 12 Å diameter. A lone anion approaching this site would be repelled (energy barrier) unless a cation resides at the site in question. This permits the anion to interact with the resident cation and pass through the channel (lower energy barrier). Such interactions are likely to occur since the $K^+$ and $Cl^-$ ions have a combined diameter of ≈7 Å, thereby leaving space for only two water molecules at the site. The same general mechanism is proposed for the Cx40 channel except that $Cl^-$ is assumed to be cotransported as an ion pair through the site, presumably due in part to a narrower pore diameter at the cation binding site.

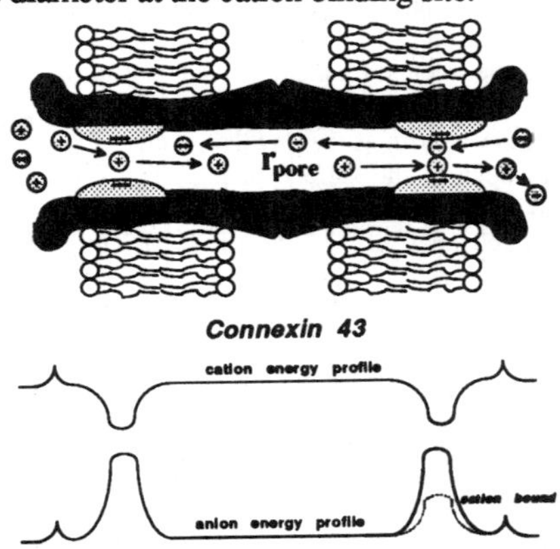


Figure 4. Diagrammatic representation of the Cx43 channel pore. Each hemichannel is presumed to possess a cation binding site that accounts for the 8:1 $K^+/Cl^-$ selectivity. The basis for this cationic selectivity is modeled here as an electrostatic mechanism. Anions are effectively repelled unless a cation is present which facilitates their passage by transiently interacting with the cation during translocation through the pore.

## References

[1] Wang H-Z and Veenstra RD. Monovalent ion selectivity sequences of the rat connexin43 gap junction channel. *J Gen Physiol* 1997;109:491-507.

[2] Beblo DA and Veenstra RD. Monovalent cation permeation through the connexin40 gap junction channel. Cs, Rb, K, Na, Li, TEA, TMA, TBA, and effects of anions Br, Cl, F, acetate, aspartate, glutamate, and $NO_3$. *J Gen Physiol* 1997;109:509-522.

[3] Veenstra, R. D. and H.-Z. Wang. Biophysics of gap junction channels. In: *Heart Cell Communication in Health and Disease*, W. C. DeMello and M. Jansen (ed.), Kluwer Academic, Norwell, MA, 1997, in press.

[4] Hille B. *Ionic Channels of Excitable Membranes*, 2nd edition, Sunderland, MA, Sinauer Associates Inc., 1992, 607 pp.

[5] Dwyer TM, Adams DJ, and Hille B.. The permeability of the endplate channel to organic cations in frog muscle. *J Gen Physiol* 1980;75:469-492.

[6] Trexler EB, Bennett MVL, Bargiello TA, and Verselis VK. Voltage gating and permeation in a gap junction hemichannel. *Proc Natl Acad Sci USA* 1996;93:5836-5841.

[7] Zambrowicz EB and Colombini M. Zero-current potentials in a large membrane channel: A simple theory accounts for complex behavior. *Biophys J* 1993;65:1093-1100.

[8] Borisova MP, Brutyan RA, Ermishkin LN. Mechanism of anion-cation selectivity of amphotericin B channels. *J Membr Biol* 1986;90:13-20.

[9] Franciolini F and Nonner W. A multi-ion mechanism in neuronal background chloride channels. *J Gen Physiol* 1994;104:725-746.

*Gap Junctions*
*R. Werner (Ed.)*
*IOS Press, 1998*


# Studies of Gating in Cx46 Hemichannels

E.B. Trexler, M.V.L. Bennett, T.A. Bargiello and V.K. Verselis
*Department of Neuroscience, Albert Einstein College of Medicine, Bronx, NY 10461*

**Abstract.** Cx46 forms functional hemichannels in the absence of contact by an apposed hemichannel and we have used these hemichannels to study voltage gating and chemical gating by $H^+$ at the single channel level with high time resolution. In both cell-attached and excised patch configurations, we find that single Cx46 hemichannels exhibit some gating properties expected of half of a gap junction channel, as well as novel properties. At inside positive voltages, Cx46 hemichannels close in a manner consistent with gating by $V_j$ in cell-cell channels; the hemichannels close to a long-lived residual conductance and the transitions between open and residual states are fast, typical of ion channel gating. At inside negative voltages, Cx46 hemichannels also close, but by a mechanism that has unusual characteristics; the transitions between fully open and fully closed states take 20-100 ms and appear as combinations of complex sets of transitions between several short-lived substates and slow continuous transitions that we attribute to rapid gating between additional, unresolved substates. We interpret this latter mechanism as the one that confers the ability of unapposed Cx46 hemichannels to open in the surface membrane. For chemical gating, unapposed Cx46 hemichannels, like cell-cell channels, are closed by elevated levels of $H^+$ on the cytoplasmic channel face. Ultrafast (< 0.5 ms) application of $H^+$ using solution switching and excised patches revealed two effects, an immediate small (10 - 15%) reduction in unitary conductance, when added to either side, followed by complete closure when added to the cytoplasmic side. That sensitivity to $H^+$ remains in excised patches indicates that the actions of $H^+$ do not require a soluble cytoplasmic factor. We interpret the reduction in unitary conductance as due to occupancy of $H^+$ in the pore and subsequent closure as due to binding of $H^+$ to a cytoplasmic site(s) that, in turn, causes conformational changes that result in complete hemichannel closure. In summary, the unapposed hemichannel is a valuable tool in addressing fundamental questions concerning the operation of gap junctions channels that are difficult to answer by existing methods.

## 1. Introduction

Gating of gap junctions by transjunctional voltage, $V_j$, and by "chemical" agents such as alkanols and cytoplasmic $H^+$ and $Ca^{2+}$ ions now is regarded as a characteristic of all members of the connexin family. Sensitivities to these agents vary among members of the connexin family [1] and this variability, along with tissue specific connexin expression, suggests that differential regulation of cell-cell coupling likely serves different functions. For example, highly voltage sensitive connexins, such as Cx37 and Cx45, are now being found in neural tissue suggesting a physiological role for $V_j$ dependent uncoupling in neural signaling. Low $V_j$ sensitivity may also be of value in maintaining coupling in the presence of voltage differences between excitable cells. Likewise, a physiological role for $H^+$ - and/or $Ca^{2+}$-induced uncoupling is suggested by the significant effects on coupling that occur with modest changes in the cytoplasmic concentrations of these ions. The complete uncoupling characteristic of chemical gating at higher concentrations of $H^+$ and/or $Ca^{2+}$, while not likely to be physiological, can be of protective value in limiting the spread of cell injury within a normally well-coupled population of cells.

Studies of the mechanisms of gap junction channel gating, whether by chemical agents or voltage, at the single channel level have been somewhat hampered by the inaccessibility of junctional membranes to direct patch recording. The double whole cell patch technique has been widely applied to isolated cell pairs, but in many cases junctional conductance is too large to visualize discrete unitary events and/or to permit quantitative analysis of channel open and closed dwell times. Investigations of the mechanisms of chemical gating are additionally hampered by difficulties in rapidly changing intracellular $H^+$ and/or $Ca^{2+}$ concentrations to avoid slower secondary effects and non-uniform changes in concentration. By utilizing the Cx46 hemichannel in excised patch configurations, we present a means of studying voltage and chemical gating of gap junctions with high-time resolution.

## 2. Results and discussion

### *2.1 Voltage gating in Cx46 hemichannels*

#### *2.1.1 Identification of two distinct voltage gating mechanisms*

In cell-cell channels formed of connexins, the principal form of voltage dependence, termed transjunctional voltage ($V_j$) dependence, reflects a sensitivity to the potential difference between the cells and is insensitive to the absolute membrane potential, $V_m$. In homotypic junctions, in the absence of $V_j$, the open state is favored. When $V_j$ is imposed, junctional conductance, $g_j$, declines symmetrically for either polarity of $V_j$ and it has been proposed that this symmetric reduction in $g_j$ results from the separate gating actions of the two component hemichannels paired in a head-to-head fashion [2,3]. Thus, each polarity of closure is associated with one or the other hemichannel and the expectation for voltage gating of hemichannels in isolation would be that there would be only a single polarity of closure.

Macroscopic recordings of Cx46-induced hemichannel currents in *Xenopus* oocytes originally did show only a single gating polarity with currents shutting-down at hyperpolarized potentials and slowly activating with depolarization [4,5]. At lower levels

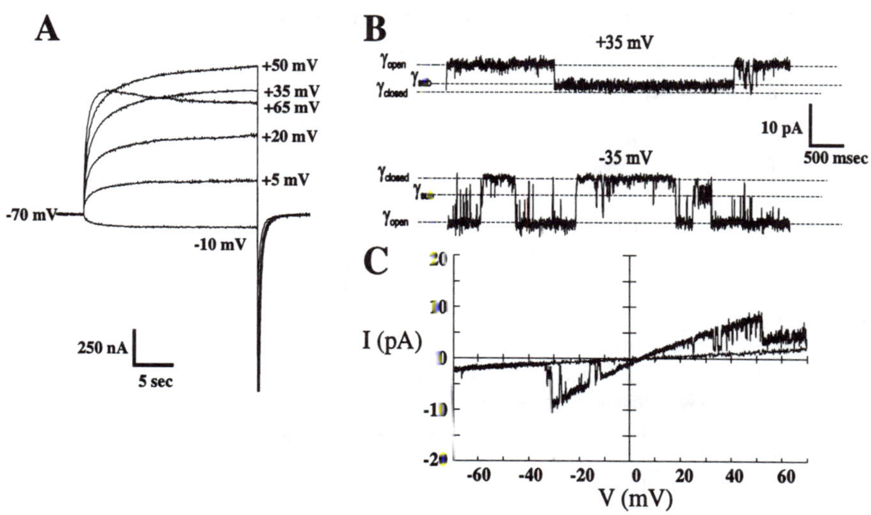


*Figure 1. Examples of Cx46 hemichannel gating. A) Macroscopic Cx46 mRNA injected Xenopus oocyte currents obtained in K⁺ bathing solution. Cx46 hemichannels activate with depolarization and start to close at higher positive potentials. B) Single channel recordings in a cell-attached patch configuration. Shown are representative currents from a patch at +35mV and -35mV. At +35mV the channel mainly gates to a long-lived (residual) subconductance state, while at -35mV, the channel gates mainly to a fully closed state, and transitions between open and closed states involve numerous short-lived substates. C) Voltage ramp showing differential closure to fully closed and a residual conductance and negative and positive voltages, respectively.*

of expression and in $K^+$ salt solutions used to minimize interference from endogenous sodium currents, we demonstrated that Cx46-induced currents not only activate with depolarization, but that they also show relaxation with larger depolarizations [6]. The reduction in current at larger inside positive voltages is only partial and not nearly as strong as when the cell is hyperpolarized to inside negative voltages (compare relaxation at +65 mV with tail currents in Fig. 1A). The ability of Cx46 hemichannels to close at both large negative and positive voltages could also be demonstrated at the single channel level in cell-attached or excised patches. Figs. 1C & D show recordings from cell-attached patches containing a single Cx46 hemichannel. At inside negative voltages, the hemichannel transits between the open and a fully closed (non-conducting) state, whereas at inside positive voltages, the transitions occur between the open and a partially closed (subconducting) state.

### *2.1.2 Gating at inside negative voltages in Cx46 hemichannels is complex and resembles gating associated with cell-cell channel formation*

Activation of Cx46 hemichannels with depolarization is characterized by complex transitions between fully closed and open states. En route between these states, the hemichannel appears to necessarily traverse several short-lived substates. The time course of transiting between fully closed and open states is variable, ranging from 20 - 100 msec, and variable in the apparent number of substates traversed; sometimes there are as many as 6 or 7 identifiable susbtates and sometimes there are fewer interspersed with apparently slow, continuous transitions containing no clearly resolvable substates.

An example of a complex gating transition of this type (at -40 mV) is shown in Fig. 2A along with an idealized trace (see Fig. 2 legend). Opening of a single Cx46 hemichannel from the fully closed state (down arrow) is characterized by a series of transitions to short-lived substates. In all, 7 substates were identified and the total transition

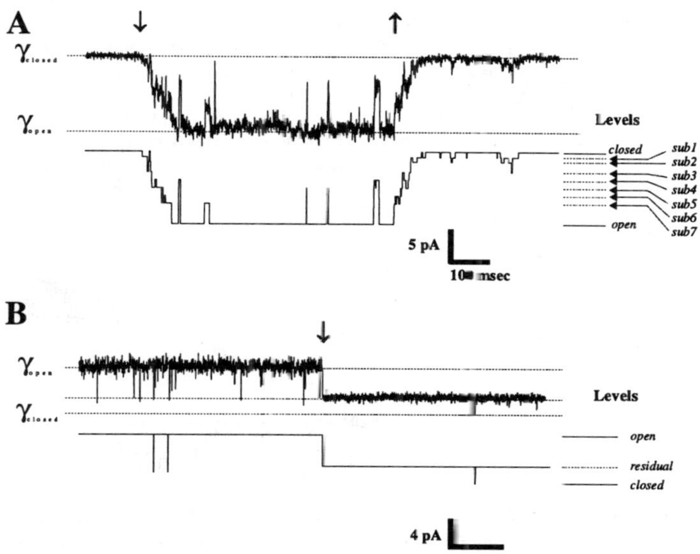


*Figure 2. Examples of single Cx46 hemichannel currents and corresponding idealized traces at (A)-50mV and (B)+50mV. Traces were idealized with the Sublevel Hinkley Detector [7]. A) An example of gating at inside negative voltages showing multiple substates traversed between open and fully closed states. In (B), the gating transition between open and the long-lived residual subconductance states involves no discernible intermediate substates, and appears instantaneous.*

time from fully closed to the fully open states took ≈70 ms. Following a long open epoch, interspersed with a few brief transitions to susbtates, the hemichannel closed fully, again by traversing a series of substates over a time course of 60 - 70 ms. For comparison, a gating transition from open to residual states, which occurs at inside positive voltages, is shown in Fig. 2B. Typically at these voltages, the hemichannel remains open for long periods of time and closes to the residual subconducting state for long periods of time. The transition between open and residual states is rapid (unresolvable within the bandwidth of the amplifier) and is typical of voltage gating in ion channels. This gating behavior resembles gating by $V_j$ in cell-cell channels in that transitions occur between an open and a stable (long-lived) substate. Furthermore, the polarity of this gating mechanism is the same as that seen for component Cx46 hemichannels when incorporated into cell-cell channels.

*2.2 Chemical gating by $H^+$ in unapposed Cx46 hemichannels*

Block of gap junctional communication by cytoplasmic $H^+$ has been demonstrated in many cell types expressing various connexins [1]. A controversial aspect $H^+$-induced uncoupling is whether $H^+$ acts through intermediates or directly on the junctional channel. Also the mechanism and site(s) of action remain unknown. We used the accessibility of the Cx46 hemichannel preparation to examine $H^+$-induced closure.

Shown in Fig. 3A is an example of $H^+$ sensitivity of an excised inside-out patch containing several Cx46 hemichannels. Upon rapidly switching from a pH 7.5 to a pH 6.0 HEPES buffered solution the hemichannels rapidly closed and remained closed until the pH was restored to 7.5. An important conclusion of this simple experiment is that pH sensitivity does not require a soluble intermediate as closure readily occurs in an excised patch configuration. If $H^+$ is applied to the cytoplasmic side of the channel when it is open, Fig 3B, there is an immediate reduction in unitary conductance (arrow) followed by

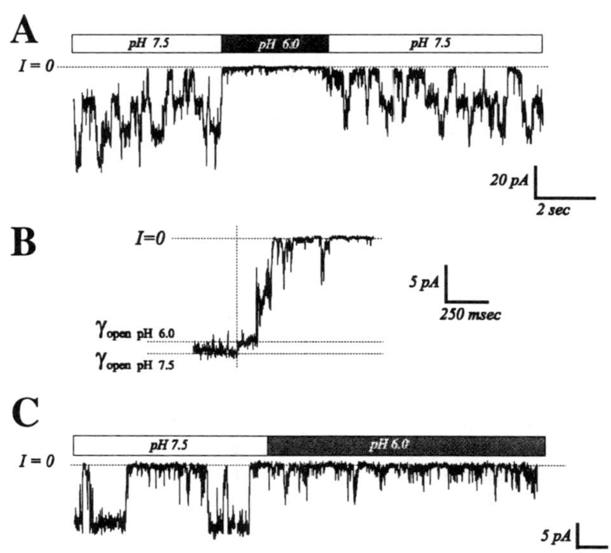


*Figure 3. Shown are examples of excised inside-out patches with Cx46 hemichannel activity. Bathing solutions of lower pH lower hemichannel open probability. A) Rapid switching from pH 7.5 bathing solution to pH 6.0 closes multiple channels in the patch, which reopen with re-application of pH 7.5 bathing solution. B) Single channel conductance decreases at the instant of pH 6.0 application, followed by a slow closure. C) Rapid application of pH 6.0 lowers the open probability of the channel even if it was previously closed, suggesting that $H^+$ entry into the hemichannel is not necessary for closure.*

complete closure of the channel. Interestingly, the closing transition induced by $H^+$ appears very similar to the closing transitions at induced negative voltages, i.e., it takes ≈50 msec to reach the fully closed state and traverses several short-lived substates.

To determine whether $H^+$ acts by eventually blocking the pore, $H^-$ was rapidly applied when the hemichannel was closed by voltage (Fig. 3C). When closed, $H^+$ can still act because the channel never reopens as long as a high concentration of $H^+$ is present. These data indicate that the site(s) of action of $H^+$ that renders the channel closed is not likely to be in the pore, but rather located on the cytoplasmic aspect of the channel. Mutations of His residues in the cytoplasmic loop of Cx43 were shown to significantly affect the sensitivity to acidification in Cx43 [8] suggesting the possibility that of titration of these residues mediates a direct action of $H^+$ on connexins.

## 3. Conclusions

It had always been assumed that a hemichannel is closed in the plasma membrane when unapposed by another hemichannel to prevent loss of cytoplasmic solutes and entry of extracellular ions. This view has changed as a number of connexins are now believed to be capable of forming functional hemichannels. By what mechanism do the hemichannels open? We propose that the ability of Cx46, and perhaps all connexins, to function as hemichannels is related to the normal process of hemichannel opening upon docking and cell-cell channel formation. The energy barrier to hemichannel opening can be surmounted by conformational changes that occur with docking, and similarly can be surmounted, in some connexins, by depolarization and low extracellular $Ca^{2+}$ without docking. Gating related to cell-cell channel formation [9] and opening of Cx46 hemichannels (see Fig. 2 in [6]) is very similar in appearance, being characterized by a slow, complex transitions between open and fully closed states. Since docking must involve one or both of the extracellular loop domains (E1, E2), we provisionally term this mechanism 'loop' gating. It is distinguished from $V_j$ gating typically associated with cell-cell channels.

As we demonstrate here, the Cx46 hemichannel has many properties ascribed to cell-cell channels, such as $V_j$ gating and pH sensitivity. The take-home lesson is that the unapposed hemichannel is likely to have a configuration that is similar to that when it is incorporated into a cell-cell channel. Thus, the unapposed hemichannel can be a valuable tool in the study of gap junction channels due to its accessibility to direct patching.

## References

1. Bennett, M.V. and V.K. Verselis *Biophysics of gap junctions.* Semin.Cell Biol., 1992. **3**: p. 29-47.

2. Harris, A.L., D.C. Spray, and M.V. Bennett, *Kinetic properties of a voltage-dependent junctional conductance.* J.Gen.Physiol., 1981. **77**: p. 95-117.

3. Verselis, V.K., C.S. Ginter, and T.A. Bargiello, *Opposite voltage gating polarities of two closely related connexins.* Nature, 1994. **368**: p. 348-351.

4. Paul, D.L., *et al.*, *Connexin46, a novel lens gap junction protein, induces voltage-gated currents in nonjunctional plasma membrane of Xenopus oocytes.* J.Cell Biol., 1991. **115**: p. 1077-1089.

5. Ebihara, L. and E. Steiner, *Properties of a nonjunctional current expressed from a rat connexin46 cDNA in Xenopus oocytes.* J.Gen.Physiol., 1993. **102**: p. 59-74.

6. Trexler, E.B., *et al.*, *Voltage gating and permeation in a gap junction hemichannel.* Proc Natl Acad Sci U S A, 1996. **93**: p. 5836-5841.

7. Draber, S. and R. Shultze, *Detection of jumps in single-channel data containing subconductance levels.* Biophys J, 1994. **67**: p. 1404-1413.

8. Ek, J.F., *et al.*, *Role of histidine 95 on pH gating of the cardiac gap junction protein connexin43.* Circulation Research, 1994. **74**: p. 1058-1064.

9. Bukauskas, F.F. and R. Weingart, *Voltage-dependent gating of single gap junction channels in an insect cell line [see comments].* Biophys J, 1994. **67**: p. 613-625.

*Gap Junctions*
*R. Werner (Ed.)*
*IOS Press, 1998*

# Molecular Selectivity of Homomeric and Heteromeric Connexin Channels

**Andrew L. Harris and Carville G. Bevans**
***Department of Biophysics, Johns Hopkins University, 3400 North Charles St. Baltimore, MD 21218***

**Abstract.** The selective permeability of immunoaffinity-purified connexin channels among large tracer molecules and cyclic nucleotides was studied by a liposome-based Transport-Specific Fractionation (TSF) system. It was demonstrated that Cx26 contributes to a narrowed pore diameter when in heteromeric Cx32/Cx26 channels. Connexin pore diameters were defined and bracketed using a set of uncharged fluorescent oligosaccharide tracers. Radiolabeled cAMP and cGMP both permeated homomeric Cx32 channels. However, most liposomes containing heteromeric Cx32/Cx26 channels were impermeable to both nucleotides. The fraction permeable to cGMP was greater than that permeable to cAMP. This suggests that the subunit composition of connexin channels determines their selectivity for cyclic nucleotides.

## 1. Introduction

In non-excitable tissues the primary signaling mediated by gap junctions is likely to be molecular in nature. Whereas for most ion channels biologically important function can be directly assessed by current carried by specific atomic ions, for gap junctions the flux of likely biologically important permeants cannot be directly assessed by single-channel current measurements. This, and other features of gap junction channels, have denied investigators many of the tools and approaches available for study of other channels, and forced the development of specialized techniques and analytic approaches to obtain fundamental structural, biochemical and physiological information.

We present here a set of studies using an experimental technique that enables some of the fundamental questions to be addressed. Utilizing purified connexin channels incorporated into unilamellar liposomes, this technique permits exploration of the flux of arbitrary labeled molecules. It also permits testing of molecules for direct modulatory effects [1-5].

## 2. Transport-Specific Fractionation: What It Is, How It Works, What It Reveals

The technique used is a Transport Specific Fractionation (TSF) of liposomes containing functional channels [6,7]. The key processes in the TSF are: (a) liposomes that contain functional channels with specific permeability properties become more dense than liposomes that do not, due to exchange of osmolytes through the channels, and (b) because of this change in density the two populations of liposomes physically separate in a density gradient. A functional property of the channels - molecular permeability - produces the change in density. The fractionation is thus "transport-specific".

To achieve this, liposomes are formed containing a channel-permeable osmolyte and centrifuged through an iso-osmolar density gradient formed by the same osmolyte and another of greater density. Liposomes that do not contain functional channels have a density that is a volume-weighted average of the densities of the entrapped aqueous volume and the lipid; they form a band in the upper part of the gradient. Liposomes containing functional channels, because they continually exchange osmolytes with the external medium, effectively have the density of lipid (which is greater than that of the solution originally trapped inside); they form a band in the lower part of the gradient.

The osmolytes are used at concentrations that create an appropriate range of density in the gradient. The density gradient is iso-osmolar, unlike classical density gradients, to avoid shrinking or swelling of the liposomes. The channels to be studied are incorporated into uniformly-sized unilamellar phospholipid liposomes [8]. For reasons made clear below, it is useful to use a channel/lipid ratio that yields liposomes with an average of less than 1 functional channel per liposome.

For connexin channels, urea and sucrose are a useful pair of osmolytes, both being permeable through connexin channels (sucrose has the greater density). In the experiments described here, liposomes were formed in a solution containing 459mM urea (plus 10mM KCl, 10mM HEPES, 0.1mM EDTA). The TSF gradient was formed from this solution and one in which the urea was osmotically replaced by sucrose (at 400mM). The corresponding specific gravities ($D^{20}_{4}$) were 1.0056 for the urea solution and 1.0511 for the sucrose solution. The densities of the liposomes were 1.0196 for the liposomes without functional channels (875Å diameter liposomes of composition 2:1::PC:PS), and 1.04 (lipid density) for the same liposomes with functional channels.

The liposomes were centrifuged in the gradient at 300,000 x *g* for several hours. Liposomes without permeability to urea/sucrose ideally band 31% into the gradient, and liposomes with channels permeable to urea/sucrose at 76% into the gradient [5,7,9,10]. Incorporated channels do not significantly affect liposome density; the protein required to produce the same change in density corresponds to ~80 connexin hemichannels.

For the liposomes used, equilibration of osmolytes across the liposome membrane through a connexin channel is in the millisecond range. Therefore a channel need be open only for very short durations during the centrifugation for the liposome to move fully to the lower position.

This method for characterizing the activity of reconstituted channels is useful for several types of experiments:

- Success/Efficiency of Reconstitution: The fraction of liposomes in the lower band is the fraction of the liposomes that contain functional channels. It thus demonstrates successful functional reconstitution.
- Permeability and Binding Studies: Spatial separation of the two liposome populations permits the recovery and identification of ligands, protein labels or tracers that specifically associate with active or inactive channels.
- Modulation Studies: Changes in the fraction of liposomes in the lower band (with exposure to pharmacological agents, for example) reflect changes in the fraction of the channels that are active.
- Planar Bilayer Reconstitution: The lower band is composed of liposomes that each contain at least one functional channel.

The TSF system has several advantages for study of channel function, including: (a) being a bulk assay for channel activity, (b) having the ability to select for channels permeable only to large molecules, reducing contamination by other channels with narrower pores (*e.g.*, AChR, potassium channels); an advantage when dealing with only partially purified preparations, (c) purifying channels on the basis of a desired channel property; permeability to sucrose in the above example, but also for channels active/inactive in the presence of arbitrary ligands or toxins.

There are several limitations, including: (a) the disadvantages inherent to all reconstituted systems, (b) the inability to record single channel currents and thus gain kinetic data on gating or block, and (c) the test for activity is all-or-none; any but extreme changes in gating will not be seen.

To explore the structural basis for selectivity, two preparations of connexin channels purified from native tissue were used. Homomeric Cx32 channels were purified from rat liver. A mixture of Cx32/Cx26 was purified from mouse liver (which contains proportionally more Cx26 than rat liver [11,12]) [5,13-15]. The protein compositions of the preparations were established by immunoblotting. By comparing the permeability properties of these two populations of channels one may infer the influence of Cx26 on the molecular selectivity.

## 3. Molecular Permeability: Cx26 Contributes to a Narrowed Pore

The first studies to reveal differences in the limiting pore diameter of channels formed by different connexin isoforms took advantage of the fact that liposomes permeable to the small osmolyte but not the larger behave differently in TSF gradients. Different permeabilities to the two osmolytes result in an osmotically-driven shrinkage of liposomes that causes them to band at an intermediate density/position [15].

In addition to sucrose, the osmolytes used were raffinose (a trisaccharide, 596Da), metrizamide (789Da) and nycodenz (821Da) (the latter two are commonly used in density gradients for cell fractionation studies [16]). TSF gradients were formed using urea and these compounds, all of which are uncharged.

Liposomes containing homomeric Cx32 formed a normal lower band in the urea/sucrose and urea/raffinose gradients, but formed a band at the intermediate density in urea/metrizamide and urea/nycodenz gradients. This indicates that the permeant size cut-off for homomeric Cx32 channels is between the sizes of raffinose and metrizamide. In contrast, liposomes containing the Cx32/Cx26 formed intermediate bands in all three compounds that were larger than sucrose. These results indicate that channels containing Cx26 have a narrower limiting pore diameter than those containing only Cx32. Since the amino acid sequences of rat and mouse Cx32 are identical [17], this difference in permeability can be most readily attributed to the presence of Cx26 in the channels. Furthermore, the presence of *only* an intermediate band for the mouse liposomes in urea/raffinose indicated that all the channels had narrowed pores (*i.e.*, all were impermeable to raffinose), and therefore all were heteromeric; any homomeric Cx32 channels present would have formed a band at the normal lower position.

## 4. Molecular Permeability: Permeability to Uncharged Tracers & Cyclic Nucleotides

The TSF system is well-suited for studies of permeability using labeled tracers. Arbitrary labeled tracer (or a mixture of tracers) can be loaded into the liposomes using DMSO [18,19]. Fluorescent or radiolabeled tracers may be utilized. Tracers that permeate the channels will rapidly leave the liposomes. When the liposomes are then analyzed by the TSF system, those in the upper band (not containing active channels, as defined by sucrose permeability) will contain the tracer(s) as originally loaded. Liposomes in the lower band will retain only the tracer(s) that could not permeate functioning channels. By comparing the tracers in each population the permeable tracers can be identified - the lower band will be missing the permeable tracers. The upper band functions as an internal control for quantitation of tracer trapping, controlling for variations in efficiency of tracer loading and nonspecific leakage. Because the measurement relies on comparison of tracers retained in the two populations, and not the fraction of liposomes that contain functional channels, variations in the number of active channels have no effect on the ability to identify permeable tracer(s). Since the tracers equilibrate rapidly, any non-zero permeability through the channel will result in complete loss of the tracer.

Using a mixture of oligosaccharides labeled with a small uncharged fluorescent moiety (an aminopyridyl group) it was shown that Cx32 channels had a larger limiting pore diameter than the Cx32/Cx26 channels [14,19,20]. Specifically, trisaccharide tracer was completely lost from the lower band Cx32 liposomes, whereas the smaller tracer was lost and the trisaccharide fully retained in the lower band Cx32/Cx26 liposomes. The complete, stoichiometric retention of this tracer from the Cx32/Cx26 liposomes also shows that all these channels were heteromeric (*i.e.*, homomeric Cx32 channels in this population would have selectively reduced the amount of retained trisaccharide). These results independently confirm those from the studies utilizing the osmotically-driven density change. They are being extended by the use of the same oligosaccharides derivatized with positively and negatively charged fluorescent moieties. In this way, the roles of size and charge in connexin channel selective permeability may be independently assessed.

Radiolabeled cyclic nucleotides were used as tracers in the same manner. The data show that homomeric Cx32 channels were permeable to both cAMP and cGMP, but that most of the (heteromeric) Cx32/Cx26 channels were permeable to neither [2,20]. For the heteromeric channels, the proportion of liposomes permeable to cGMP was greater than that permeable to cAMP. The data suggest that the narrowed pore of the Cx32/Cx26 channels can select against second messengers. This is the first evidence for such a selectivity for connexin channels. It also strongly suggests that the arrangement or stoichiometry of the components of the heteromeric channels can determine whether cyclic nucleotides can permeate.

## 5. Discussion

By two experimental protocols, the data show that Cx26, when in heteromeric Cx32/Cx26 channels, contributes to formation of pores of narrower cross-sectional dimension. They further show that the narrowed pore correlates with a greater degree of selectivity against cyclic nucleotides. These findings are consistent with previous work suggesting that homomeric Cx26 forms narrower pores than homomeric Cx32 [21], with less direct experiments in which size and charge were varied together [22], and with access resistance measurements [23].

These results indicate (1) permeability of cyclic nucleotides can be connexin-specific, and (2) some stoichiometries and/or arrangements of connexin isoforms within a channel can distinguish between cAMP and cGMP. This is the first evidence for connexin channel selectivity among second messengers. These studies do not define a basis for this selectivity, but it is likely due to the different chemistries of the permeants (*e.g.*, cAMP can form two hydrogen bonds whereas cGMP can form three), suggesting the presence of selective binding sites within the connexin pore.

**Acknowledgments:** Supported by NIGMS grant RO1-36044. Special thanks to Joshua Zimmerberg at NICHD for contributions to the development of the TSF and for ongoing intellectual and research support.

**References**

[1] Bevans, C. and A.L. Harris. 1996. Cyclic nucleotides, pH and salt modulate connexin channel activity in liposomes. Biophys. J. 70:A32

[2] Bevans, C.G. and A.L. Harris. (in press) Interaction of cAMP, cGMP & taurine with connexin channels. Mol. Biol. Cell

[3] Bevans, C.G. and A.L. Harris. 1997. Connexin channel activity modulation by cyclic nucleotide analogs, pH buffers and magnesium ions. Biophys. J. 72:A272

[4] Bevans, C.G. and A.L. Harris. 1996. Connexin channel modulation by nucleotides and polysaccharides. Mol. Biol. Cell 7:281a

[5] Rhee, S.K., C.G. Bevans and A.L. Harris. 1996. Channel-forming activity of immunoaffinity-purified connexin32 in single phospholipid membranes. Biochemistry 35:9212-9223.

[6] Goldin, S.M. and V. Rhoden. 1978. Reconstitution and "transport specificity fractionation" of the human erythrocyte glucose transport system. A new approach for identification and isolation of membrane transport proteins. J. Biol. Chem 253:2575-2583.

[7] Harris, A.L., A. Walter and J. Zimmerberg. 1989. Transport-specific purification of large channels incorporated into lipid vesicles. J. Membr. Biol. 109:243-250.

[8] Mimms, L.T., G.A. Zampighi, Y. Nozaki, C. Tanford and J.A. Reynolds. 1981. Phospholipid vesicle formation and transmembrane protein incorporation using octyl glucoside. Biochemistry 20:833-840.

[9] Harris, A.L., A. Walter, D.L. Paul, D.A. Goodenough and J. Zimmerberg. 1992. Ion channels in single bilayers induced by rat connexin32. Mol. Brain Res. 15:269-280.

[10] Harris, A.L. 1994. Connexin32: A Protein That Can Form Channels Through One or Two Membranes. In: Biomembrane Electrochemistry. M. Blank and I. Vodyanoy, editors. American Chemical Society, Washington, D.C. 197-224.

[11] Traub, O., J. Look, R. Dermietzel, F. Brümmer, D.F. Hulser and K. Willecke. 1989. Comparative characterization of the 21-kD and 26-kD gap junction proteins in murine liver and cultured hepatocytes. J. Cell Biol. 108:1039-1051.

[12] Nicholson, B.J. and J.-T. Zhang. 1988. Multiple protein components in a single gap junction: Cloning of a second hepatic gap junction protein ($M_r$ 21,000). In: Modern Cell Biology, Volume 7: Gap Junctions. E.L. Hertzberg and R.G. Johnson, editors. Alan R. Liss, Inc. New York. 207-218.

[13] Kordel, M., B.J. Nicholson and A.L. Harris. 1993. Crosslinking studies of rat and mouse liver connexins in purified form and in plasma membranes. Biophys. J. 54:A192

[14] Bevans, C.G. and A.L. Harris. 1995. Connexin-specific permeability and modulation of connexin channel activity in liposomes. Mol. Biol. Cell 5S:190A

[15] Bevans, C., R.A. Brutyan, C. DeMaria and A.L. Harris. 1995. Reconstituted connexin channels: Selective permeation by large molecules. Biophys. J. 68:A204

[16] Rickwood, D. (ed.) 1976. Biological Separations in Iodinated Density Gradient Media. IRL Press, London.

[17] Hennemann, H., G. Kozjek, E. Dahl, B.J. Nicholson and K. Willecke. 1992. Molecular cloning of mouse connexins26 and -32: Similar genomic organization but distinct promoter sequences of two gap junction genes. Eur. J. Cell Biol. 58:81-89.

[18] Fraser, S.E., C.R. Green, H.R. Bode and N.B. Gilula. 1987. Selective disruption of gap junctional communication interferes with a patterning process in hydra. Science 237:49-55.

[19] Bevans, C.G., M. Kordel, S.K. Rhee and A.L. Harris. (submitted) Connexin channels show differential selectivity among second messengers and uncharged molecules.

[20] Bevans, C.G. and A.L. Harris. 1996. Direct measurement of selective permeability among uncharged tracers and cyclic nucleotides by connexin channels. Mol. Biol. Cell 7:91a

[21] Cao, F., R. Eckert, C. Elfgang, D.F. Hulser, K. Willecke and B.J. Nicholson. 1994. Differential pemeabilities of gap junctional studied in *Xenopus* oocytes and HeLa transfectants. Mol. Biol. Cell 5:198a

[22] Elfgang, C., R. Eckert, H. Lichtenbergfrate, A. Butterweck, O. Traub, R.A. Klein, D.F. Hulser and K. Willecke. 1995. Specific permeability and selective formation of gap junction channels in connexin-transfected HeLa cells. J. Cell Biol. 129:805-817.

[23] Oh, S., M.V.L. Bennett, E.B. Trexler, V.K. Verselis and T.A. Bargiello. (in press) Changes in permeation caused by a connexin 32 mutation underlie X-linked Charcot-Marie-Tooth disease. Neuron

# Section 3:

# Channels Conductance and Voltage Dependence

*Gap Junctions*
*R. Werner (Ed.)*
*IOS Press, 1998*



# Molecular Modeling of Gap Junction Pores

Anja MINDERMANN, Jürgen PLEISS*, and Dieter F. HÜLSER
*Institute of Biology, Biophysical Dept., University of Stuttgart, 70550 Stuttgart, Germany*
**Institute of Technical Biochemistry, University of Stuttgart, 70569 Stuttgart, Germany*

**Abstract.** Gap junction channels differ in ionic conductance and in permeability for large molecules. To understand the structural basis for this selective diffusion of charged molecules, we modeled the structure of the transmembrane pore with 24 α-helices (M3 model). In our first approach we excluded extracellular and cytoplasmic domains and the modeled pores revealed only very small differences for connexin(Cx)26, Cx32, and Cx43. The pore lining amino acids of M3 are similar and, therefore, the same pore diameter of about 1 nm and rings of positive (arginine/lysine) and negative (glutamic acid) charges were found for all three connexins. Since this model also excludes membrane lipids and water, it represents the closest possible packing of helices in this arrangement. Including the C-terminal domain and the cytoplasmic loop (M3CD model), the model reveals different cytoplasmic openings ranging from 1.6 nm for Cx26 to 3.2 nm for Cx43. At the extracellular side, the pore diameters are smaller (minimum: 1 nm) because the extracellular loops are relatively short and may not fold into separate domains. The diffusion properties of several fluorescent dyes with different molecular weights and charges were investigated with monolayers of Cx transfected HeLa cells. In addition, the utilized dye molecules were modeled.

## 1. Introduction

Many attempts have been made to experimentally determine the structure of the gap junction pore [1-8]. However, even the best resolved structure [9] is by far not detailed enough to assign the amino acid (AA) sequence to the secondary structure elements. Other membrane proteins, like the nicotinic acetylcholine receptor have been modeled [10-12] on the basis of their AA sequence and similarly resolved structural data [13 14]. Some ion channels have even been modeled with almost no experimental data on their structure [15-20]. We took advantage of these methods and modeled the gap junction pore on the basis of the AA sequences of several connexins.

Since selective diffusion of larger molecules has been reported for different connexins [21-24], a selectivity filter might be located inside the pore. We started modeling with the transmembrane (TM) part, which is the narrowest part of the pore and, moreover, is the only part of the connexon of which the secondary structure (α-helices) is known. Our models of Cx26, Cx32, and Cx43 are very similar in the TM part, but further analysis of their AA sequences reveals that cytoplasmic domains may explain differences in the permeability of gap junction channels for larger molecules.

## 2. Models

For the secondary structure of connexins a rough model consisting of four α-helices and corresponding loops is generally accepted. We modeled a gap junction inner pore on the basis of the Cx32 amino acid sequence (taken from the Internet SWISS-PROT database [25]). With hydrophobicity and amphiphilicity analyses [26,27] assignation of the transmembraneous residues is possible. Our computed models are based on the most amphiphilic helix consisting of nineteen AAs of the M3 region (3rd box in Fig. 1). Six of these membrane spanning helices line the inner pore. They were arranged in the corners of a hexagon (sidelength: 1 nm) with the most hydrophilic sidechains pointing inside the pore. The outer eighteen helices, necessary for the stability of the pore, were modeled with leucine and were arranged on a trigonal lattice with an axis distance of 1 nm. No water, lipids, or other parts of the protein were added for modeling. Energy minimization and molecular dynamics were performed using GROMOS [28]. Cx26 and Cx43 were generated on the basis of the Cx32 structure by changing the sidechains of the helices.

A stable condition is reached where the axes of opposite, parallel helices are 2 nm apart, so that the effective pore diameter is about 1 nm. Pore openings were only slightly different for Cx26, Cx32, and Cx43, since the M3 sequences are similar for most connexins. A ring of positive charge can be seen inside the pore which is due to arginine in Cx26- and Cx32-pores and due to lysine in Cx43-pores. Next to this positively charged ring, all connexins also formed a ring of negative charge by glutamic acid in the center of the pore.

```
rCx26    1 mDwgtLqsIL ggVnKHstsI gKIwLtVLFI FRImILVVaa KEVwgDEqaD FVCntLqpgC KnVCyDHyFp
rCx32    1 mnwtgLytLL sgVnRHstaI gRVwLsVIFI FRImVLVVaa EsVwgDEKss FICntLqpgC nsVCyDHFFp
rCx43    1 gDwsaLgKLL DKVqaystag gKVwLsVLFI FRILLLgtaV EsawgDEqsa FRCntqqpgC EnVCyDKsFp
              :*  * .:* . *: :**:  *::**:*:** ***::* .*. :..****::  * *** **** :.****: **

rCx26   71 IsHIRLwaLq LImVstpaLL VamHVayRRH EKK------- --RKFmKgEI KnEFKDIEEI Kt--------
rCx32   71 IsHVRLwsLq LILVstpaLL VamHVaHcqH IEK------- ---KmLRLEg HgDpLHLEEV KR--------
rCx43   71 IsHVRFwVLq IIFVsVptLL yLaHVFyVmR KEEKLnKKEE EL-KVaqtDg VnVEmHLKqI EIKKFKygIE
           ***:*:* ** :*:**.*:**     ** :   :  ::            *. : :    .   .:::: :

rCx26  124 --qKVRIEgs LwwtyttsIF FRVIFEaVFm yVFyImyngF FmqRLVKCna wpCpntVDCF IsRptEKtVF
rCx32  123 --HKVHIsgt LwwtyVIsVV FRLLFEaVFm yVFyLLypgy amVRLVKCEa FpCpntVDCF VsRptEKtVF
rCx43  140 EHgKVKmRgg LLRtyIIsIL FKsVFEVaFL LIqwyIy-gF sLsaVytCKR DpCpHqVDCF LsRptEKtIF
              **:: *   *  **  *:. *: :**..*:   :: :* *:   :   : .*:   ***: ****  :*******:*

rCx26  192 tVFmIsVsgI CILLnItELCy LFIRyCsgK sKR----pV   226
rCx32  191 tVFmLaasgI CIILnVaEVVy LIIRaCaRR aqRRsnppsR KgsgFgHRLs pEyKqnEInK LLsEqDgsLK
rCx43  209 IIFmLVVsLV sLaLnIIELFy VFFKgVKDR VKgRsD-pyH attgp---Ls psKDCgspKy ayFngCsspt
            :**: .* : .: **: *: * ::::     :   :      *

rCx32  261 DILRRspgtg agLaEKsDRC saC   283
rCx43  275 apLspmsppg yKLVtgDRnn ssCRnynKca sEqnwanysa EqnRmgqags tIsnsHaqpF DFpDDnqnaK

rCx43  345 KVaagHELqp LaIVDqRpss RassRassRp RpDDLEI   381
```

**Fig. 1** Alignment of sequences for Cx26, Cx32 and Cx43 with Clustal [29]. TM parts are boxed, dividing the sequences in NT, M1 (1st box), E1, M2 (2nd box), CL, M3 (3rd box), E2, M4 (4th box), and CT. K, R: pos., D, E: neg.; H: neutral or pos.; F, I, L, V: hydrophobic AAs. *: identical; :: very similar; .: similar AAs.

**Tab. 1** Data for CL and CT derived from AAs in Fig. 1 (5th and 9th segment). Charge of His was assumed as +0.5 e.

| | Volume / nm³ | | Charge / e | | Pore diameter / nm | |
|---|---|---|---|---|---|---|
| | CT | CL | CT | CL | max. | min. |
| **Cx26** | 1.8 | 4.6 | 4.0 | 5.5 | 2.0 | 1.0 |
| **Cx32** | 8.7 | 4.5 | 8.5 | 3.5 | 2.6 | 1.0 |
| **Cx43** | 18.4 | 6.7 | 6.5 | 5.0 | 3.2 | 1.0 |

An extension of this M3 model considers also cytoplasmic domains (M3CD model). For this model we assumed a spherical shape of the COOH-terminus (CT) and the cytoplasmic loop (CL) and calculated their radii and volumes

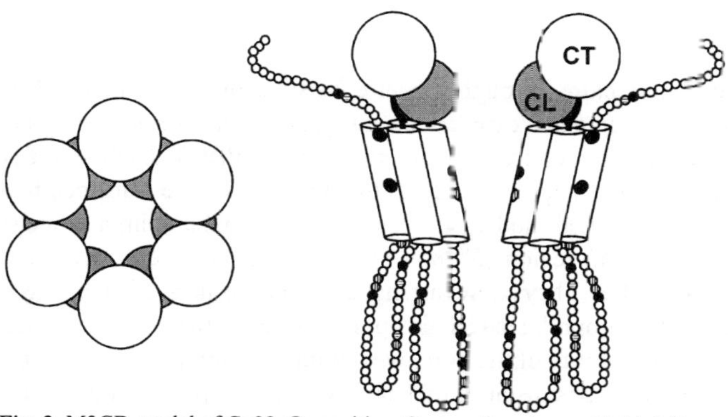


Fig. 2 M3CD model of Cx32. ●: positive, ◍: negative charge; ◒: histidine.

(Tab. 1). Assuming a spherical arrangement of the CT and the CL, one can show that these cytoplasmic domains further open the intracellular entry of the pore and cause a tilted arrangement of the helices (Fig. 2). Different connexins can have different tilt angles. These cytoplasmic domains contain many charged amino acids which might also explain different pore permeabilities. Freeze fracture pictures of Cx26 gap junction plaques do not reveal significant differences in particle packing when compared with Cx43 plaques. This indicates that the volume of the CT might play a minor role in particle spacing, but it might well be that the CL influences the pore's intracellular diameter. The extracellular loops are not as big and do not fold into a separate domain, so that the extracellular opening at the docking site is smaller, furthermore, the conserved cysteine residues as well as the high amount of other conserved residues indicate that no major structural variety may be expected on the extracellular side of the pore.

## 3. Experiments

Four fluorescent dyes {lucifer yellow (LY), methoxypyrene trisulfonic acid (MPTS), carboxyfluorescein (CF) and calcein} have been tested for their diffusion properties by iontophoretic injection into monolayers of HeLa cells transfected with connexin genes (Fig. 3). Between 5 and 22 experiments were carried out. In addition, these dye molecules have been modeled with MOPAC [30] and their properties are shown in Tab. 2.

No apparent differences in selectivity can be seen from these data. However, positively charged dyes {ethidium bromide (EB), propidium iodide (PI), and diamidinophenylindole (DAPI)} partly show a decreased permeability in Cx31, Cx32, and Cx43 channels [31]. We used only negatively charged dyes, which did not show a charge dependent diffusion.

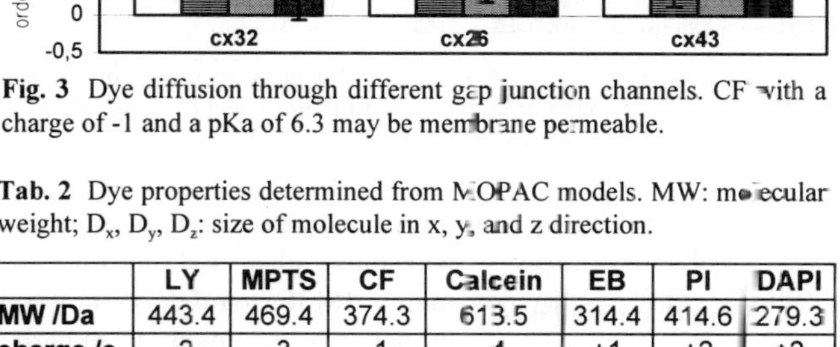


Fig. 3 Dye diffusion through different gap junction channels. CF with a charge of -1 and a pKa of 6.3 may be membrane permeable.

Tab. 2 Dye properties determined from MOPAC models. MW: molecular weight; $D_x$, $D_y$, $D_z$: size of molecule in x, y, and z direction.

| | LY | MPTS | CF | Calcein | EB | PI | DAPI |
|---|---|---|---|---|---|---|---|
| MW /Da | 443.4 | 469.4 | 374.3 | 613.5 | 314.4 | 414.6 | 279.3 |
| charge /e | -2 | -3 | -1 | -4 | +1 | +2 | +2 |
| $D_x$ /nm | 0.35 | 0.31 | 0.72 | 0.71 | 0.43 | 0.79 | 0.30 |
| $D_y$ /nm | 0.91 | 0.71 | 0.94 | 1.60 | 0.92 | 1.14 | 0.70 |
| $D_z$ /nm | 1.22 | 1.07 | 0.94 | 0.93 | 1.08 | 1.25 | 1.48 |

## 4. Conclusions

The M3 model gives a minimal pore diameter of about 1 nm. Since loops, membrane lipids, and water are not included it represents the closest possible packing of helices in this arrangement. Further modeling in a more realistic environment will certainly show a somewhat higher diameter. M3 models for Cx26, Cx32 and Cx43 are very similar with rings of positive (arginine/lysine) and negative (glutamic acid) charges. Whether these charged rings may act as a selectivity filter is an open question.

Because the sequences of the TM and extracellular parts are very conserved (Fig. 1), the differences in structure and properties will be found in the cytoplasmic parts [32]. The CL has a very high density of charged AAs, and the CT varies strongly in the number of AAs. Up to now there are no structural data available for these parts. Furthermore, there is no homology to other proteins of known structure.

Nevertheless, CT and CL are included in the M3CD model (Fig. 2). So far, only the mass and the according volume (assuming a spherical shape) are considered. It explains different tilting of the 24 helices which form a connexon and it matches the electron crystallographic data of Perkins et al. [8] who found a cone shaped Cx32 pore with a minimum diameter of 16 Å at the extracellular side and a maximum diameter of 26 Å at the cytoplasmic entry, which matches our data (Tab. 1). With similar methods, however, Unger et al. [9] revealed the projection structure of a cylindrical pore with an overall diameter of about 24 Å for Cx43 channels deficient in COOH-domains. Influences of other properties (i. e. charge) remain to be shown.

The extracellular parts of the connexin proteins are very conserved. Especially three cysteines in E1 and E2 which form intramolecular disulfide bonds [33] and two proline containing sections (QPGC in E1 and PCP in E2) indicate strongly conserved structures.

Our model of a gap junction pore does not yet allow predictions on docking and gating processes. It will be refined by correlating diffusion properties and chemical structure of diffusing molecules with pore structure and charge and it might help to explain and predict different diffusion properties of physiologically relevant large molecules.

Acknowledgments: These investigations were supported by grants from Landesgraduiertenförderung Baden-Württemberg.

## References

[1] P.N.T. Unwin, G. Zampighi, Structure of the junction between communicating cells. *Nature* 283 (1980) 545-549

[2] P.N.T. Unwin, P.D. Ennis, Two configurations of a channel-forming membrane protein. *Nature* 307 (1984) 609-613

[3] T.T. Tibbitts, D.L.D. Caspar, W.C. Phillips, D.A. Goodenough, Diffraction diagnosis of protein folding in gap junction connexons. *Biophys. J.* 57 (1990) 1025-1036

[4] K.A. Stauffer, N. Unwin, Structure of gap junction channels. *Seminars in Cell Biology* 3 (1992) 17-20

[5] G.E. Sosinsky, Image Analysis of Gap Junction Structures. Electron Microsc. Rev. 5 (1992) 59-76

[6] M. Yeager, *In situ* Two-Dimensional Crystallization of a Polytopic Membrane Protein: the Cardiac Gap Junction Channel. *Acta Cryst.* D50 (1994) 632-638

[7] M. Yeager, B.J. Nicholson, Structure of gap junction intercellular channels. *Curr. Op. Struct. Biol.* 6 (1996) 183-192

[8] G. Perkins, D. Goodenough, G. Sosinsky, Three-Dimensional Structure of the Gap Junction Connexon. *Biophys. J.* 72 (1997) 533-544

[9] V.M. Unger, N.M. Kumar, N.B. Gilula, M. Yeager, Projection structure of a gap junction membrane channel at 7 Å Resolution. *Nature Struct. Biol.* 4 (1997) 39-43

[10] R. Sankararamakrishnan, M.S.P. Sansom, Modelling packing interactions in parallel helix bundles: pentameric bundles of nicotinic receptor M2 helices. *Biochim. Biophys. Acta* 1239 (1995) 122-132

[11] R. Sankararamakrishnan, M.S.P. Sansom, Structural features of isolated M2 helices of nicotinic receptors. Simulated annealing via molecular dynamics studies. *Biophys. Chem.* 55 (1995) 215-230

[12] R. Sankararamakrishnan, C. Adcock, M.S.P. Sansom, The Pore Domain of the Nicotinic Acetylcholine Receptor: Molecular Modeling, Pore Dimensions, and Electrostatics. *Biophys. J.* 71 (1996) 1659-1671

[13] N. Unwin, Nicotinic Acetylcholine Receptor at 9 Å Resolution. *J. Mol. Biol.* 229 (1993) 1101-1124

[14] N. Unwin, Acetylcholine receptor channel imaged in the open state. *Nature* 373 (1995) 37-43

[15] J.C. Bradley, W.G. Richards, Potassium channels: a computer prediction of structure and selectivity. *Prot. Eng.* 7 (1993) 859-862

[16] S. Bogusz, A. Boxer, D.D. Busath, An SS1-SS2 β-barrel structure for the voltage-activated potassium channel. *Prot. Eng.* 5 (1992) 285-293

[17] K.R. Courtney, A model for the $K^+$ channel pore based on close packing of helices with substantial tryptophan bulk. *Proc. West. Pharmacol. Soc.* 35 (1992) 171-175

[18] H.R. Guy, P. Seetharamulu, Molecular model of the action potential sodium channel. *Proc. Natl. Acad. Sci. USA* 83 (1986) 508-512

[19] S. Tani, A. Imamura, K. Kanda, Kinetic Approach with *ab initio* MO Method on Ionic Selectivity and Size in Sodium Channel. *J. theor. Biol.* 140 (1989) 519-536

[20] A. Grove, J.M. Tomich, M. Montal, A molecular blueprint for the pore-forming structure of voltage-gated calcium channels. *Proc. Natl. Acad. Sci. USA* 88 (1991) 6418-6422

[21] R. Eckert, D.F. Huelser, Single channel properties of cell-cell channels in transfected and non-transfected cultured human cell lines. In: Y. Kanno et al. (ed.), Progress in Cell Research, Vol. 4. Intercellular communication through gap junctions. Elsevier Science Publishers B.V.: Amsterdam, New York, 1995, pp. 423-426.

[22] R.D. Veenstra, H.-Z. Wang, E.C. Beyer, P.R. Brink, Selective Dye and Ionic Permeability of Gap Junction Channels Formed by Connexin45. *Circulation Research* 75 (1994) 483-490

[23] R.D. Veenstra, H.-Z. Wang, E.C. Beyer, S.V. Ramanan, P.R. Brink, Connexin37 Forms High Conductance Gap Junctioon Channels with Subconductance State Activity and Selective Dye and Ionic Permeabilities. *Biophys. J.* 66 (1994) 1915-1928

[24] R.D. Veenstra, H.-Z. Wang, D.A. Beblo, M.G. Chilton, A.L. Harris, E.C. Beyer, P.R. Brink, Selectivity of Connexin-Specific Gap Junctions Does Not Correlate With Channel Conductance. *Circulation Research* 77 (1995) 1156-1165

[25] R.D. Appel, A. Bairoch, D.F. Hochstrasser, A new generation of information retrieval tools for biologists: the example of the ExPASy WWW server. *Trends Biochem. Sci.* 19 (1994) 258-260

[26] J. Kyte, R.F. Doolittle, A simple method for displaying the hydropathic character of a protein. *J. Mol Biol.* 157 (1982) 105-132

[27] F. Jähnig, Structure predictions of membrane proteins are not that bad. *Trends Biochem. Sci.* 15 (1990) 93-95

[28] W.F. van Gunsteren, H.J.C. Berendsen, Computer Simulation of Molecular Dynamics: Methodology, Applications and Perspectives in Chemistry. *Angew. Chem. Int. Ed. Engl.* 29 (1990) 992-1023

[29] J.D. Thompson, D.G. Higgins, T.J. Gibson, CLUSTAL W: improving the sensitivity of progressive multiple sequence alignment through sequence weighting, position specific gap penalties and weight matrix choice. *Nucleic Acids Res.* 22 (1994) 4673-80

[30] J.J.P. Stewart, MOPAC MANUAL (Sixth Edition, 1990) (Available at: http://www.psc.edu/general/software/packages/mopac/mopac6_manual.txt)

[31] C. Elfgang, R. Eckert, H. Lichtenberg-Frate, A. Butterweck, O. Traub, R.A. Klein, D.F. Hülser, K. Willecke, Specific Permeability and Selective Formation of Gap Junction Channels in Connexin-transfected HeLa Cells. *J. Cell Biol.* 129 (1995) 805-817

[32] S. Haubrich, H.-J. Schwarz, F. Bukauskas, H. Lichtenberg-Fraté, O. Traub, R. Weingart, K. Willecke, Incompatibility of Connexin 40 and 43 Hemichannels in Gap Junctions between Mammalian Cells is Determined by Intracellular Domains. *Mol. Biol. Cell* 7 (1996) 1995-2006

[33] S. Rahman, W.H. Evans, Topography of connexin32 in rat liver gap junctions. Evidence for an intramolecular disulphide linkage connecting the two extracellular peptide loops. *J. Cell Sci.* 100 (1991) 567-578

*Gap Junctions*
*R. Werner (Ed.)*
*IOS Press, 1998*

# Increase of the Gap Junction Conductance of Adult Mammalian Heart Myocytes by Intracellular Calcium Ions

Bruno DELAGE and Jean DÉLÈZE
*Laboratoire de Physiologie Cellulaire, Unité Mixte de Recherche du CNRS 6558, Faculté des Sciences, Université de Poitiers, France*

**Abstract.** In myocyte pairs isolated from adult rat hearts cultured for 2 to 5 days, either decreases of the cytosolic concentration of calcium ions to very low levels or increases to 1-3 μM promote a definite rise of junctional conductance. $Ca^{2+}$ concentrations at least one order of magnitude higher are requested to impair the cell-to-cell conductance.

## 1. Introduction

Since the early observations of the necessity of calcium ions for triggering the healing-over phenomenon in mammalian heart [1], several reports [e.g. 2,3,4] have concluded that a rise of intracellular calcium concentration decreases the gap junction conductance of cardiac cells.

A knowledge of $Ca^{2+}$ effects on heart cell junctions is obviously important for an understanding of the physiology and pathology of cardiac conduction. A contribution of a rise of $Ca^{2+}$ concentration to the arrhythmias observed during cardiac ischemia has been postulated [5]), and according to other work [6], a decrease of junctional conductance might occur even at the calcium concentrations expected during normal heart function.

In the present investigation, the junctional conductance ($g_j$) was recorded on pairs of mammalian cardiac myocytes from adult rats by the dual whole-cell voltage-clamp method and the cytosolic $Ca^{2+}$ level was modified in one cell by dialysis from the last positioned patch pipette.

## 2. Materials and Methods

**a. Cell preparation.** Adult rat hearts (age 6 to 8 weeks) were perfused in a Langendorff apparatus and submitted to enzymatic dissociation by means of collagenase (Worthington) according to standard methods [7]. The released cells were kept for 1 to 4 days in culture dishes containing Ham F10 (Gibco) supplemented with 2% fetal calf serum (Boehringer) and antibiotics. The dishes were maintained in a water saturated atmosphere (5% $CO_2$ in air) at 37 °C. The cardiac myocytes kept their typical adult rod-shaped morphology. Despite some loss of cross-striation, they did nor revert to the undifferentiated apearance of cultured neonatal heart cells. Experiments were performed at any time after the first day of culture with similar results.

**b. Junctional conductance measurement.** The electrical conductance of the cell junctions was measured by the dual voltage-clamp method, which employs two feed-back amplifiers,

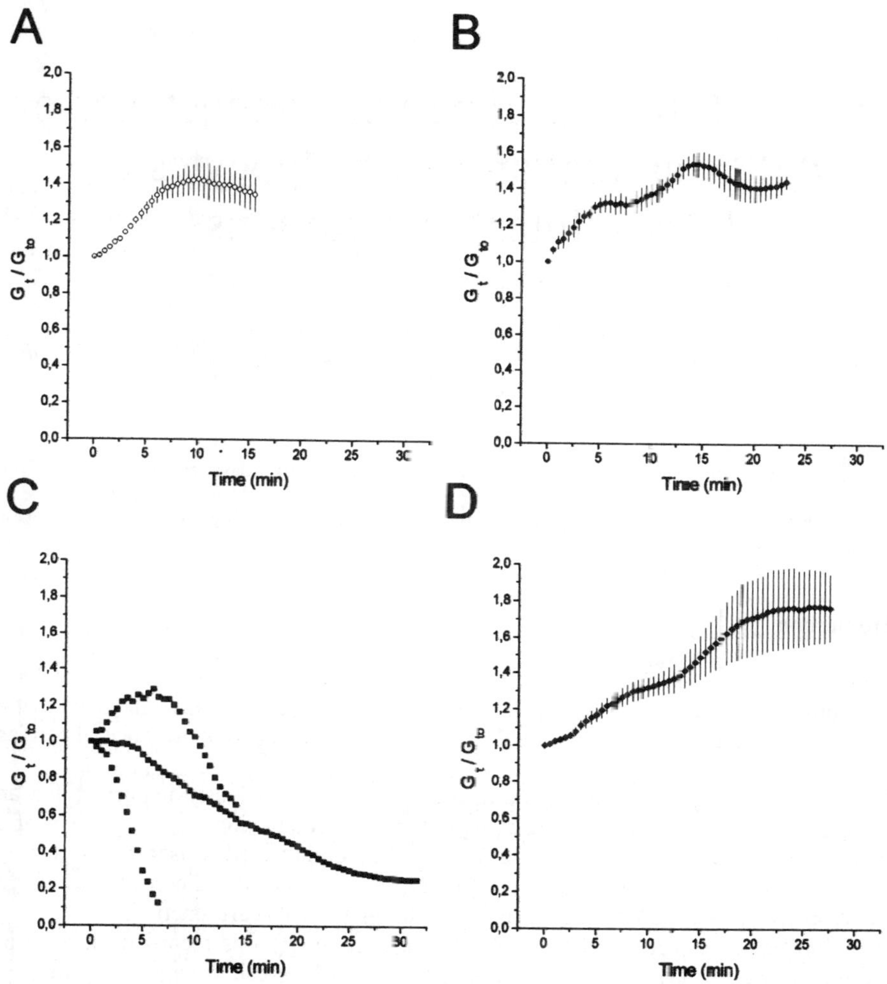


Figure 1. Normalized values (± SE in panels A,B,D) of the junctional conductance recorded in pairs of ventricular myocytes while the cytosolic $Ca^{2+}$ concentration is altered in one cell of the pair by dialysis from the last positioned patch pipette.

A: When no calcium is added to the pipette solution, a substantial increase of $g_j$ is recorded. The mean values of 4 experiments are shown.

B: With 1.8 μM $Ca^{2+}$ in the patch pipette, $g_j$ reaches a mean value 50% above controls within 10 to 15 minutes (range +27 to + 68% in 3 experiments). At later times $g_j$ usually tends to decrease but stays higher then the initial values. Uncoupling has never been observed with these $Ca^{2+}$ concentrations.

C: With 50 μM $Ca^{2+}$, a transitory increase of $g_j$ that can reach 30% of controls can be recorded during the first 10 minutes of dialysis. Note the case of partial uncoupling to about 25% of the initial value of $g_j$ at 30 minutes.

D: Calmodulin inhibition by addition of W7 (100 μM) to a pipette solution containing 2.2 μM $Ca^{2+}$ promotes a further increase of $g_j$ (cf. panels B and D) to an average 80% above initial values (mean of 3 experiments).

each connected in whole-cell configuration to one myocyte of a pair by a patch pipette [8]. Both myocytes are first clamped to the same membrane potential ($V_{h}$.), close to the resting potential, then one myocyte is pulsed to a different potential ($V_h + \Delta V_h$) while the other is maintained at $V_h$ by its feedback circuit compensating the junctional current ($I_j$). Conductances are obtained from $I_j$ and $\Delta V_h$ (mean control value 49 ± 8 nS, n = 200).

**C. Cell dialysis.** Patch pipettes, filled with a solution containing (in mM) KCl 140, Mg-ATP 5, glucose 10, HEPES 10, BAPTA 5 and buffered to pH 7.2 by KOH, had resistances of 2-3 MΩ. In order to alter the cytosolic $Ca^{2+}$ -concentration in one cell of the pair, the last positioned patch pipette contained variable amounts of $CaCl_2$, calculated by means of the software Max Chelator V. 6.63 (a gift of Chris Patton, Stanford University, CA, USA), using a $K_D$ of 107.15 nM for BAPTA-Ca, 436.5 nM for BAPTA-$H_{D1}$, 0.01 M for BAPTA-Mg, 0.1 mM for ATP-Ca, 295 nM for ATP-$H_{D1}$ and 61 μM for ATP-Mg.

## 3. Results

With a nominally $Ca^{2+}$ -free solution in the patch pipette, (Figure 1**A**), a substantial increase of $g_j$ was recorded (mean +40% of controls, range +33 to +71%). Inhibition of calmodulin by W7 and of protein kinases by H7 prevented this increase of $g_j$ at low $Ca^{2+}$ concentations.

When [$Ca^{2+}$] was close to the cytosolic concentration of the resting myocytes (about 0.1 μM, [9]), moderate oscillations of $g_j$ around the initial control value were observed (range -30 to +39%).

With [$Ca^{2+}$] between 0.9 and 2.6 μM, $g_j$ increased to a mean value 40% above controls (range +27 to +68%) after 10 to 15 minutes of dialysis (Figure 1**B**). This increase was prevented by the protein kinase inhibitor H7. Uncoupling has not been observed in this range of $Ca^{2+}$ concentrations. The rise of gj was further enhanced to + 80% (n = 3) above the initial values in presence of the calmodulin inhibitor W7 (cf. panels **D** and **B** in Figure 1).

Impairment of junctional conductance first occurred with $Ca^{2+}$ -concentrations that were at least one order of magnitude higher, and complete or partial uncoupling was regularly obtained with 50 μM $Ca^{2+}$ in the pipette (Figure 1**C**). With this [$Ca^{2+}$], a transitory increase of $g_j$, which could reach 30% above controls, was often observed during the first 10 minutes of dialysis. Partial uncoupling, $g_j$ levelling off to 20 to 30% the initial value within 10-30 min, was frequently observed.

## 4. Discussion

The increased junctional conductance at subphysiological cytosolic calcium levels can be tentatively accounted for by a decreased activity of phosphodiesterase and a subsequent rise of cAMP, which has been shown to upregulate $g_j$ in heart myocytes. Of all the calmodulin-regulated enzymes, phosphodiesterase has the highest affinity for the $Ca^{2+}$-calmodulin complex and is already activated at physiological cytosolic calcium levels. Inhibition of the increase in $g_j$ in this $Ca^{2+}$ range by compound H7 is consistent with this interpretation.

The increase of $g_j$ observed when the dialysis solution contains calcium in the concentration range 0.9 to 2.6 μM possibly depends on the activation of the $Ca^{2+}$ -dependent protein kinase C, since it was inhibited by H7. In this range of calcium concentration, the junctional conductance was further increased by the presence of the calmodulin inhibitor W7 (cf. Figures 1**B** and **D**). This effect may be tentatively accounted for by a decreased rate of calcium transport to the extracellular medium or into the sarcoplasmic reticulum, and consequent further rise of the cytosolic calcium concentration. It is known that elevation of the cytosolic calcium level activates the $Ca^{2+}$ -ATPase by a calcium/calmodulin-dependent protein kinase that phosphorylates phospholamban [10].

The finding of a substantial increase of $g_j$ when the cytosolic calcium concentrations of ventricular myocytes is increased by dialysis with solutions containing calcium ions, in a concentration range corresponding to excitation-contraction coupling or slightly above it, is new. The possible functional significance of this $Ca^{2+}$ effect may be perceived when considering that an increased $g_j$ facilitates cell-to-cell transmission of the action potential. This would contribute to the well-known increase of conduction velocity observed during exercise, at a time when the cytosolic $Ca^{2+}$ concentration is expected to rise with cardiac frequency and strength of contraction.

These regulations of the gap junctional conductance by calcium ions in ventricular myocytes appear very different from those observed in other cell types [11]. It must therefore be concluded that.regulation of the gap junction conductance by calcium ions varies widely with the cell type.

## References


[1] J. Délèze, The recovery of resting potential and input resistance in sheep heart injured by knife or laser, *Journal of Physiology* **208** (1970) 547-562.

[2] W.C. DeMello, Effect of intracellular injection of calcium and strontium in cell communication in heart, Journal of Physiology, **250** (1975) 231-245.

[3] P. Maurer and R. Weingart, Cell pairs isolated from adult guinea pig and rat hearts: effects of [Ca2+]i on nexal membrane resistance, Pflügers Archiv **409** (1987) 394-402.

[4] L. Firek and R. Weingart, Modification of gap junction conductance by divalent cations and protons in neonatal rat heart cells, Journal of Molecular and Cellular Cardiology **27** (1995) 1633-1643.

[5] A.G. Kléber, C.B. Riegger and M.J. Janse, Electrical uncoupling and increase of extracellular resistance after induction of ischemia in isolated, arterially perfused rabbit papillary muscle, *Circulation Research* **61** (1987) 271-279.

[6] A. Noma and N. Tsuboi, Dependence of junctional conductance on proton, calcium and magnesium ions in cardiac paired cells of guinea-pig, Journal of . Physiology **382** (1987) 193-211.

[7] T. Powell, D.A. Terrar and V.W. Twist, Electrical properties of individual cells isolated from adult rat ventricular myocardium, Journal of Physiology **302** (1980) 131-153.

[8] D.C. Spray, A.L. Harris and M.V.L Bennett, Equilibrium properties of a voltage-dependent junctional conductance, Journal of General Physiology **77** (1981) 77-93.

[9] W.G. Wier, M.B. Cannell, J.R. Berlin, E. Marban and W.J. Lederer, Cellular and subcellular heterogeneity of [Ca2+]i in single heart cells revealed by Fura-2, *Science* **235** (1987)325-328.

[10] A.P. Braun and H. Schulman, The multifunctional calcium/calmodulin-dependent protein kinase: from form to function, Annual Review of Physiology, **57** (1995) 417-445.

[11] A. Lazrak and C.Peracchia, Gap junction gating sensitivity to physiological internal calcium regardless of pH in Novikoff hepatoma cells, *Biophysical Journal* **65** (1993) 2002-2012.

*Gap Junctions*
*R. Werner (Ed.)*
*IOS Press, 1998*

# Homotypic hCx37 and rCx43 and their Heterotypic Form

K. Banach, S. V. Ramanan and P.R.Brink
*Dept. of Physiology and Biophysics, SUNY at Stony Brook, Stony Brook, New York 11794-8661*

**Abstract.** Homotypic hCx37 channel conductance shows saturation with increased salt concentration while homotypic rCx43 does not. The data for the former case are consistent with fixed surface charges near the mouth of the pore which attract counterions . Dilute salts also affected the kinetics of the voltage dependent decline in $I_j$ for rCx43. This could arise from the presence of fixed surface charges in the protein distant from the pore mouth which, however, affect the voltage sensor of rCx43. The behavior of the heterotypic rCx43-hCx37 gap junctions can be explained on the basis of two independently gating hemichannels.

## 1. Introduction

In cells expressing only one connexin type, the gap junction channel is composed of 12 identical connexins (homomeric/homotypic) (1,2). For any two adjacent cells expressing different connexins there are additional possiblilites. For two closely apposed cells, each expressing only one type of connexin, a heterotypic gap junction channel is possible. If either of the cells is coexpressing connexins, then intercellular channels with heteromeric connexins are possible. Both heterotypic and heteromeric channels have been demonstrated (3,4). Much of the heterotypic behavior has been documented using oocyte cell pairs (5).
The potential for such a diverse pool of gap junction channels raises a number of questions, one of which is whether these channels also exhibit functional diversity. Under physiological conditions this could mean that the transfer of second messenger, the spread of electrical excitation and the dynamic regulation of intercellular communication may depend on the type or the combination of channels expressed in interacting cells.
In order to understand the physiological significance of the diversity of connexins, it is first necessary to learn more about the characteristics of the individual homotypic channels. In this chapter we present and compare some characteristics of the hCx37 and the rCx43 gap junction channels. Further, we compare the features of the individual homotypic channels with those of their heterotypic form (hCx37/rCx43).

## 2. Methods

The experiments were carried out on cell pairs with the voltage clamp technique in the double whole cell configuration (6). A communication deficient Neuroblastoma cell line (N2a), transfected with cDNA encoding for either hCx37 or rCx43 served as a source of cell pairs (7,3). All experiments were performed on an inverted microscope. The standard extracellular solution contained (in mM): 110 KCl, 10 Hepes, 1.8 $CaCl_2$, 2 $MgCl_2$, pH=7.2. The standard pipette solution contained (in mM): 110 KCl, 10 Hepes, 0.09 $CaCl_2$, 1 EGTA, 0.1 $MgCl_2$, pH=7.1. To obtain concentration conductance relationships for either of the connexins, the main salt concentration of the pipette solution was varied to 30, 55,110 and 270mM KCl. In every case, the KCl concentration in the bath solution was changed accordingly to maintain osmotic equilibrium. In the low salt concentrations (30 and 55 mM), sucrose was added to the solutions to maintain the osmolarity at 230 mOsm.
For experiments on heterotypic cell pairs, N2a cells expressing either hCx37 or rCx43 were

cocultured in the same dish. To enable the identification of heterotypic cell pairs, one cell type was labeled in advance with a fluorescent cell tracker (3). The experimental conditions, data acquisition and analysis are described elsewhere (8, 3).

## 3. Results and discussion

Cx37 and Cx43 are widely distributed in the body (9). Cx43 represents the major connexin in the myocardium and Cx37 is expressed mainly in endothelial cells, where it is coexpressed with Cx43. In the homotypic configuration, the two channel types exhibit differences in their gating behaviour as well as in their single channel conductance.

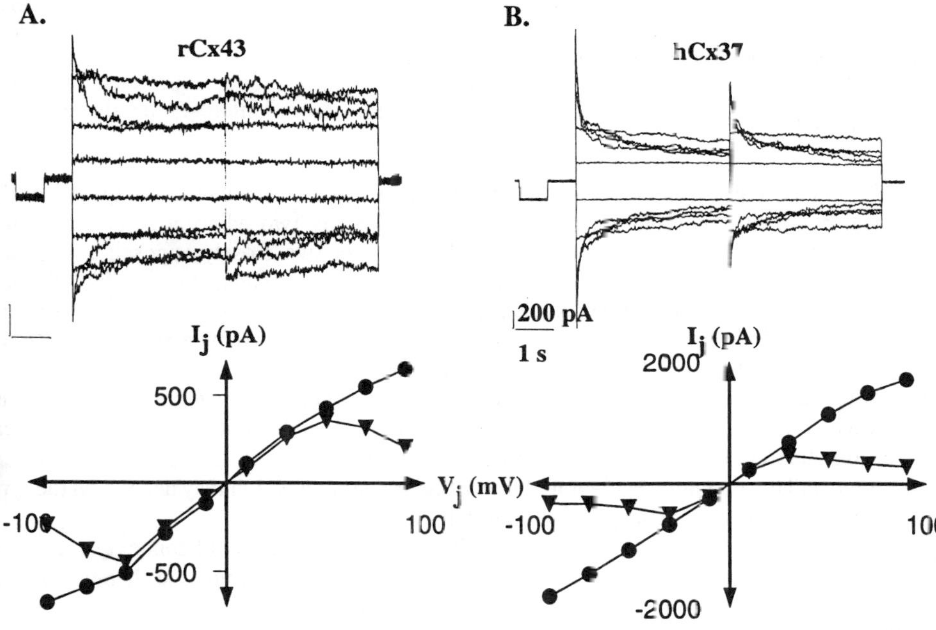


**Fig.1:** Current voltage relationship obtained in N2a cell pairs either expressing rCx43 (A) or hCx37(B). Original current traces and their corresponding $I_j/V_j$ curves are shown. Points represent the instantaneous current, triangles the steady state current.

## 4. Voltage dependence

The gating of all gap junction channels depends on the intercellular voltage gradient ($V_j$). The decline of the intercellular current ($I_j$) with time becomes more pronounced as $V_j$ is increased. We examined the voltage dependent gating of the homotypic channels by monitoring the macroscopic $I_j$ generated in response to voltage pulses from ±10 to ±110 mV in 20 mV increments. After 4s, the polarity of the pulse was reversed and maintained for another 4 s before the holding potential of 0 mV was again established in both cells. Fig. 1 (top) shows original current traces recorded in the non-stepped cell of a cell pair. In the case of rCx43 (A, top) as well as for hCx37 (B, top) low voltage gradients induce a current that remains constant over the duration of the pulse. Increased $V_j$ results in a decline of the current to a new steady state level. In the current voltage relationships (Fig.1 A, B, bottom) this voltage dependent inactivation of $I_j$ is reflected in a deviation of the steady state current from linearity. Human

Cx37 has a half maximal inactivation of the current at $V_0$=±30mV; in rCx43 the $V_0$ is ±70 mV. The instantaneous I-V relationship is linear for both hCx37 and rCx43 in the voltage range examined ($-110 \leq V_j \leq 110$ mV).

## 5. Single channel conductance

We varied electrolyte concentrations to probe the conductance paths of homotypic rCx43 and hCx37 gap junction channels. In Fig. 2 (left), original current traces recorded in hCx37 are shown. Human Cx37 unitary conductance is not linear with regard to concentration. The data indicate that the conductance of the channel tends toward saturation with increasing salt concentration. This can be explained via the rate theory of permeation (10). In low salt concentrations, the unitary conductance of the channel remains high. This is consistent with an aggregation of charges near the orifice of the channel which attracts counterions and repels like ions.
The rCx43 conductance concentration relation differs from that of hCx37. In this case the conductance through the channel depends linearly on the intracellular salt concentration and no tendency toward saturation can be observed over the concentration range utilized.

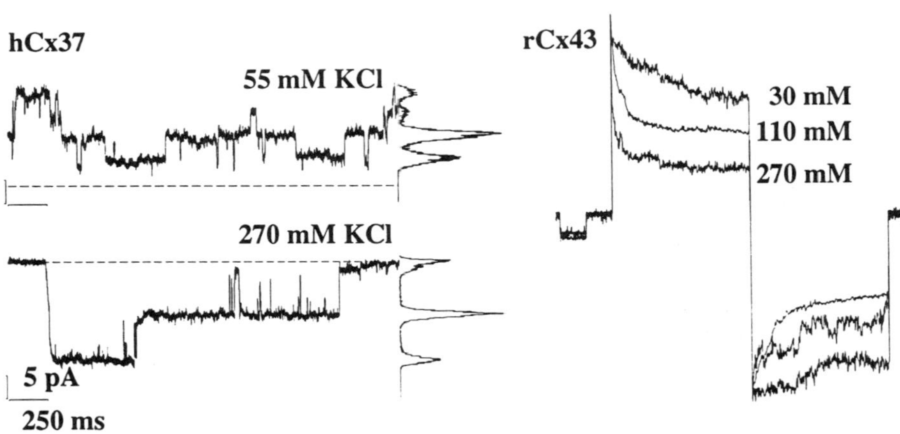


**Fig2:** Effects of the intracellular salt concentration on the current through hCx37. Right panel: Single channel current transitions recorded at $V_j$=-25 (55 mM, 194pS) and $V_j$=+20 mV (270 mM, 480pS) respectively. Left panel: Macroscopic current traces obtained at different salt concentrations at $V_j$=±110 mV. An increase in the salt concentration shifts $V_0$ to lower voltages.

## 6. Voltage dependence of rCx43 depends on the salt concentration

The properties of the rCx43 gap junction channel were affected in another way by the change of the intracellular salt concentration. Fig. 2 (right) shows superimposed current traces recorded at $V_j$=±110mV in 270, 110 or 30 mM KCl. Decreasing salt concentrations resulted in a diminished voltage dependent inactivation of $I_j$.
A possible explanation for this feature is, again, effects due to a fixed titratable surface charge. Since the conductance-concentration relationship for rCx43 is linear, this charge would not lie at the channel orifice. However, since surface charges produce fields inside the membrane(11), screening of these charges by counterions could alter the fields felt by the voltage sensor and thereby change the apparent $V_0$.
From our experiments, we conclude that hCx37 as well as rCx43 are influenced by screenable surface charges in two different ways. Whereas the conductance of hCx37 is increased by

charges in the entrance region of the channel, a surface potential induced by charges alters the apparent $V_0$ of the voltage sensing mechanism of rCx43.

If the fixed charges postulated above are intrinsic to the hemichannels (as opposed to being a manifestation of hemichannel-hemichannel interaction), their effects should be felt in heterotypic channels as well. We therefore performed experiments with hCx37/rCx43 heterotypic gap junction channels.

## 7. Characteristics of the heterotypic channel

In Fig.3, junctional current traces recorded in a heterotypic cell pair in response to the standard protocol (±10 to ±150 mV) are shown. In contrast to recordings from homotypic cell pairs, the current response is highly asymmetric for positive and negative potentials of $V_j$. The current voltage relationship obtained from those current traces is shown in Fig. 3 (right). The instantaneous current is no longer a linear function of the transjunctional voltage but exhibits rectification. Transjunctional voltages elicit an increased instantaneous current when the hCx37 containing cell is more positive with respect to the rCx43 side. The voltage dependent gating does not resemble that of either of the homotypic connexins. When the hCx37 side is positive, the half maximal inactivation of $I_j$ is reached when $V_0$~=70 mV. When the rCx43 side is positive, there appears to be little or only slight voltage dependent gating.

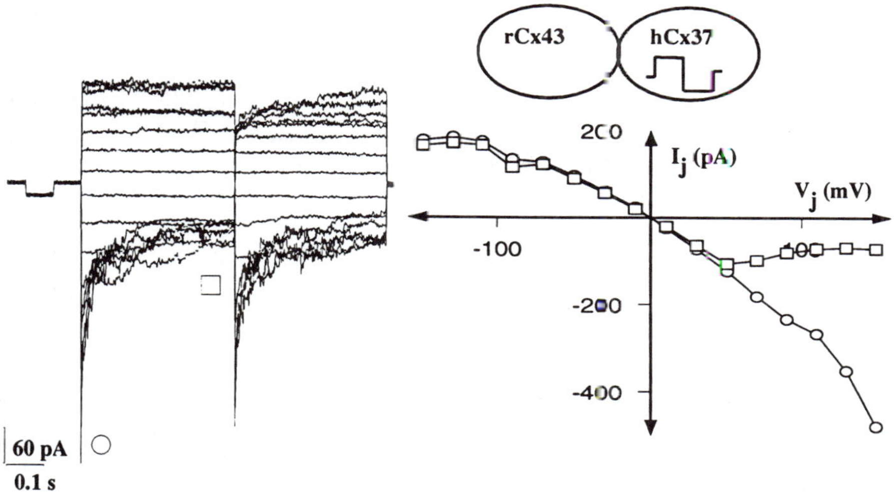


Fig.3: Macroscopic current traces recorded in a heterotypic hCx37/rCx43 cell pair, in response to the standard voltage protocol applied to the hCx37 expressing cell. Duration of the voltage pulse was 400ms for each polarity. The corresponding current voltage relationship for the instantaneous (circle) and the steady state (square) is shown on the right.

Data obtained in oocytes indicate that hCx37 gates on relative positivity (5), whereas for rCx43, it would appear that the connexon on the more negative side of the channel is involved in the voltage dependent gating (12). For heterotypic channels, therefore, the gates on both hemichannels would respond to the same voltage gradient. Voltage dependent gating occurs when the hCx37 side is more positive; when rCx43 is more positive, neither of the voltage sensing mechanisms is activated and the channel stays open (3).

Assume that the gating of the hemichannels in the heterotypic junction is independent and

identical to their respective behavior in the homotypic junction.Then, when both hemichannels are deactivated at the same time (with hCx37 positive), the prediction is that the kinetics and the $V_0$ of the voltage dependent inactivation of $I_j$ should reflect the features of hCx37, since it is most sensitive to $V_j$. This is not the case, as the $V_0$ of the heterotypics is shifted by ~30 mV relative to that of homotypic hCx37. The shift can, however, be explained by the surface charges and the resulting surface potential already postulated for the rCx43 homotypic channel. If this surface potential is additive to any applied transjunctional voltage gradient, then it would shift the assumed $V_j$. A shift of $V_0$ to higher voltages would therefore indicate a positive surface potential on the rCx43 side of ~30 mV relative to hCx37.
Further, the rectification of the instantaneous current through the heterotypic gap junction channel can be explained by the features of hCx37 in the homotypic configuration. Charges in the hCx37 channels entranceway attract counterions and repel like ions, increasing the concentration of counterions and decreasing that of like ions from their bulk value. In the heterotypic configuration, this results in a locally different ion concentration at the opposite entrances of the pore. An increased conductance is observed when the hCx37 side is more positive, i.e., when there is a cation flow from the hCx37 side to the rCx43 side. This indicates that the cation concentration is locally increased in the mouth of hCx37 and therefore the surface charges lining the hCx37 pore entrance are negative.

## 8. Conclusion

While, at first glance, it would appear that the heterotypic hCx37/rCx43 data is quite different from that anticipated from the corresponding homotypic behavior, both (a) the rectification of the instantaneous current and (b) asymmetric gating in the heterotypics can indeed be explained in terms of the (different) surface charge effects postulated to explain the behavior of the homotypic junctions. Since such surface charge effects are most readily exposed by varying the salt concentration, probing changes in gating and conductance with ion activity would seem to be a powerful tool in exploring the behavior of gap junction channels.

## References

[1] Kumar, N. and B. Gilula The gap junction communication channel. Cell 84:384-388. 1996.
[2] Bennett, M.V.L., X.Zheng and M.L.Sogin. The connexin family tree. *Prog. Cell Res.* 4: 3-8, 1995.
[3] Brink, P.R., K. Cronin, K. Banach, E. Peterson, E.M. Westphale, K.H. Seul, S.V. Ramanan, and E.C. Beyer Evidence for heteromeric gap junction channels composed of hCx37 and rCx43. Amer. J. Physiol.( in press).
[4] Jiang, J.X. and D.A. Goodenough. Heteromeric connexons in lens gap junction channels PNAS 93:1287-1291. 1996.
[5] White, T.W., D. Paul, D.A. Goodenough and R. Bruzzone. Functional analysis of selective interactions among rodent connexins. Mol. Biol. Cell 6:459-470. 1995.
[6 ] Neyton, J, and A. Trautmann. Single channel currents of an intercellular junction. Nature 317:331-335 1985.
[7] Reed , K.E., E.M. Westphale, D.M. Larson, H.Z. Wang, R.D. Veenstra and E.C. Beyer. Molecular cloning and functional expression of human Cx37. *J.Clin.Invest.* 91:997-1004 1993.
[8] Brink, P.R., S.V. Ramanan and G.J. Christ. Human connexin43 gap junction channel gating evidence for mode shifts and/or heterogeneity. *Am.J.Physiol.* 271: C321-C331, 1996.
[9] Beyer, E. C., Gap Junctions. Int. Rev. Cytol.137:1-37.
[10] Hille, B. (1992) Ion channels in exitable cells. Sinauer Ass. Inc. Sunderland Mass.
[11] Mathias R. T., G. J. Baldo, K. Manivannan and S. McLaughlin. Discrete charges on biological mambranes. In: Electrified Interfaces in Physics, Chemistry and Biology. 473-490. Kluwer Academic Publishers. 1992.
[12] Moreno, A.P., G.I Fishman, E.C. Beyer and D.C. Spray. Voltage dependent gating and single channel analysis of heterotypic gap junctions formed of Cx45 and Cx43. In: Progress in Cell Research: Intercellular communciation through gap junctions. editied by K. Kanno, K. Kataoka, Y. Shib, Y. Shibata, and T. Shimazu., Vol 4:405-408. Elsevier Science Pub. 1993.

*Gap Junctions*
*R. Werner (Ed.)*
*IOS Press, 1998*



# Voltage Dependent Gating of Gap Junction Channels: Influence of the Second Extracellular Loop

Algimantas KRISCIUKAITIS* and Dieter F. HÜLSER
*Biologisches Institut, Abt. Biophysik, Universität Stuttgart, D-70550 Stuttgart*
**Kaunas Medical Academy, Eiveniu 4, LT-3007 Kaunas, Lithuania*

**Abstract.** Patch clamp measurements with connexin transfected HeLa cells revealed two groups of connexin types which differ by their voltage dependent gating after changing the polarity of the transjunctional voltage. Positive and negative transjunctional voltage steps higher than the half maximum inactivation voltage closed individual channels, which was observed as an exponential decay of the initial to a steady state current. Repetitive pulses of the same polarity gave the same decay and when the polarity of the pulses was reversed, a similar decay was observed only for Cx37- and Cx43- but not for Cx32-, Cx40-, and Cx45-channels. The latter respond upon the first reversed pulse in a way which might be explained by a slow re-opening of closed channels and their subsequent closing. Further pulses of the same polarity resulted in the fast exponential decay as for Cx37- and Cx43-channels. The number and sequence of amino acids in the first extracellular loop is highly conserved for all investigated connexins, whereas a significant difference is obvious for the second extracellular loop, which is longer in Cx37 and Cx43. With the HeLa connexin transfectant Cx40*43E1,2 where both extracellular loops from Cx40 were replaced by Cx43, this slow re-opening was no longer observed. Since intracellular domains and membrane integrated helices of this transfectant were part of Cx40 and the first extracellular loop is highly conserved, we hypothesize that the second extracellular loop is involved in regulatory processes of gating.

## 1. Introduction

Voltage dependent gating is common for gap junction channels of vertebrate cells [1-12]. The voltage of half maximal inactivation and the single channel conductance are determined by the connexin type which assembles the gap junction channel. Our double whole cell voltage clamp measurements with connexin transfected HeLa cells revealed two groups of gap junction channels which differ by their voltage dependent gating after changing polarity of the transjunctional voltage. Cx37- and Cx43-channels always respond with an exponential decay of the junctional conductance from an instantaneous to a steady state conductance; Cx32-, Cx40-, and Cx45-channels respond with a rising and a subsequent decay of the junctional conductance during the first pulse, indicating that a re-opening and a subsequent closing can be resolved for these connexins.

## 2. Materials and Methods

For our study, HeLa cells were used which were transfected with Cx32, Cx37, Cx40, Cx43, Cx45 [13,14] and with a chimeric construct Cx40*43E1,2 where both extracellular loops of Cx40 were exchanged for the extracellular loops of Cx43 [15]. For patch clamp measurements, cell pairs were prepared by trypsinizing monolayer cells for 5-15 min at 37°C and agitation with an Eppendorf pipette. After reseeding on glass cover slips, cells were allowed to attach for at least 30 min before performing the experiments. Cover slips with attached cells were washed with PBS and transferred into a microscope chamber. All experiments were carried out in PBS at room temperature.

Intercellular current recordings were performed with the double whole cell voltage clamp method [2,12]. Patch pipettes were filled with a solution containing 120 mM CsCl, 2,7 mM $MgCl_2$, 2,6 mM $CaCl_2$, 10 mM HEPES, 10 mM EGTA (pH = 7,4) and their resistance ranged from 1-5 MΩ. Voltage of half maximal inactivation of gap junction channels was evaluated by stepping the transcellular voltage in a range between ±100 mV. For analyzing the reopening of channels, repetitive rectangular voltage pulses were applied and their polarity was changed after every fourth pulse. Pulse durations varied between 3 and 20 seconds and amplitudes were at least 10% higher than the voltage of half maximal inactivation.

## 3. Results

In Cx43- and in Cx37-transfectants always an exponential decay of current was registered, regardless of the pulse polarity (Fig. 1). A different current relaxation was seen in Cx32-, Cx40-, and Cx45-transfectants, where after polarity change the junctional current always started with a rising phase during the first pulse before it also declined (Fig. 2). Only after three to four voltage pulses, each lasting at least 5 seconds, the exponential decay occurred regularly (Fig. 2). When the polarity was changed, the decay of the junctional current was again preceded by a rising phase during the first pulse.

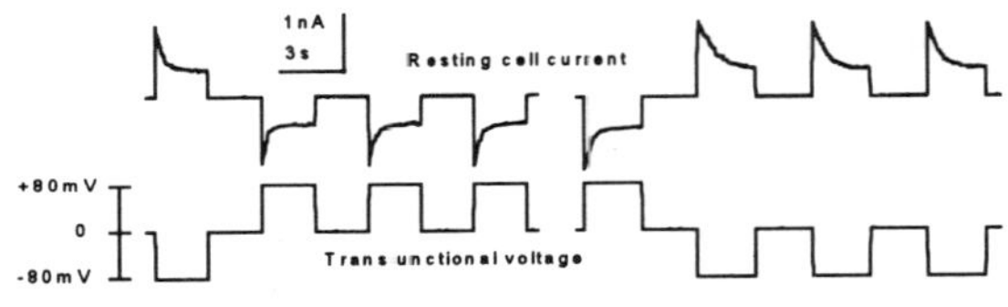


**Fig.1** Junctional current response to rectangular voltage pulses in Cx43-transfectants

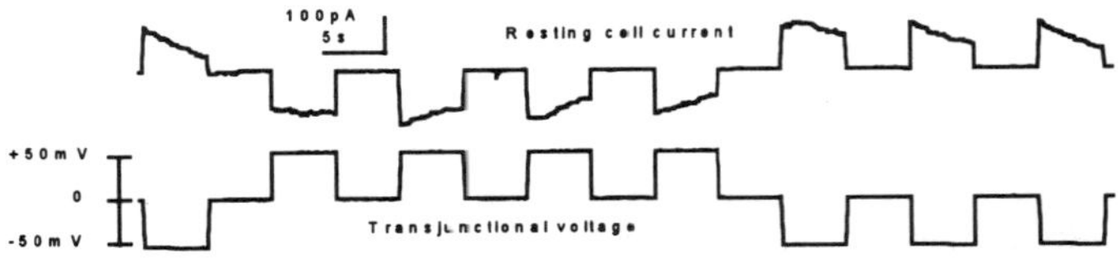


**Fig.2** Junctional current response to rectangular voltage pulses in Cx45-transfectants

When the duration of the pulse was about 20 seconds long, the transjunctional current reached a steady state. The different responses of the gap junction channels are best seen, when the current response after polarity switch is copied into the preceding current pulse. This is demonstrated for Cx45 channels in figure 3 where one can see that the instantaneous current after repetitive pulses is of the same value as the maximum current of the first pulse. This effect is independent of the total junctional conductance, it can be observed when only a few channels are active as well as when many channels connect adjacent cells. Since the maximum amplitude of the transjunctional current is always similar, the same number of active channels must be involved.

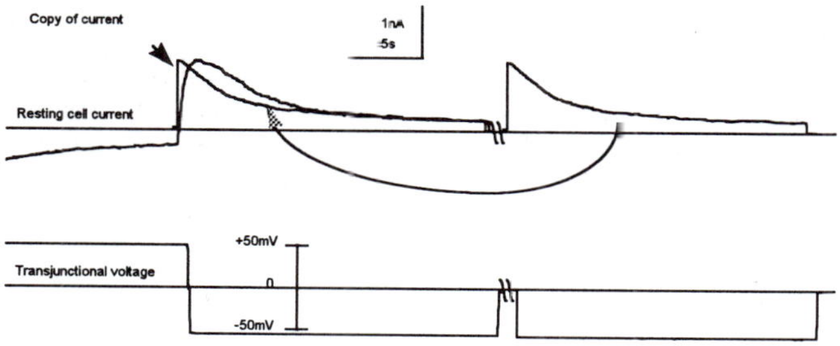


**Fig. 3** Junctional current responses to 20 s voltage pulses in Cx45 transfectants. Arrows indicate a copy of the junctional current response after several pulses with unchanged polarity.

We compared the amino acid sequences of these tested connexins [16] in order to detect differences which might account for the observed behavior in response to transjunctional voltage polarity changes. Especially the domains for the second extracellular loops differed in length of their sequences, which were longer in Cx37 and Cx43 than in those connexins where the re-opening was observed (Fig. 4). The first extracellular loops, however, showed highly conserved sequences for all five connexins.

```
*Cx32                 RLVKCEAFPCPNTVDCFVSRPTEKT
 Cx37 GQWRLYGWTMEPVFVCQRAPCPHIVDCYVSRPTEKT
*Cx40                     CRRSPCPHPVNCYVSRPTEKN
 Cx43           YGFSLSAVYTCKRDPCPHQVDCFLSRPTEKT
*Cx45                     CSRLPCPHKIDCFISRPTEKT
```

**Fig. 4** Amino acid sequences of the 2[nd] extracellular loops of the investigated connexins

Whether the observed re-opening could be attributed to the extracellular loops was tested with the HeLa connexin transfectant Cx40*43E1,2 where both extracellular loops from Cx40 were replaced by Cx43. For clarification, the sequences of amino acids which were exchanged [15] are shown in figure 5

```
                 <--------- 1st ex.loop --------->
   mCx40:  42   SSWGDEQADFRCDTIQPGCQNVCYDQAFPISHI  74
   mCx43:  42   SAWGDEQSAFRCNTQQPGCENVCYDKSFPISHV  74

                            <---- 2nd ex.loop ---->
mCx40  167  IVGQYLLYGIFLDTLHVCRRSPCPHPVNCYVSRPTEKN  203
mCx43  169  LLIQWYIYGFSLSAVYTCKRDPCPHQVDCFLSRPTEKT  205
                 <-------- 2nd ex.loop ---------->
```

**Fig. 5** Amino acids which were exchanged in the chimeric construct Cx40*43E1,2

Re-opening was no longer observed with channels of this chimeric connexin but a relaxation of the transjunctional current occurred as was observed for Cx43- or Cx37-channels (Fig. 6).

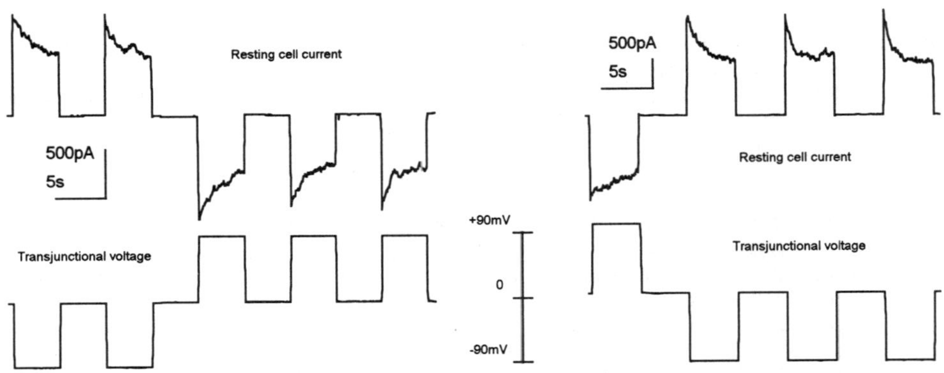


**Fig. 6** Junctional current responses to voltage pulses in Cx40*43E1,2 transfectants

## 4. Discussion

Voltage dependent gating is a well known phenomenon of gap junction channels [1-12] and several models have been proposed to explain this behavior [1,10,12,17]. When both connexons contribute one voltage sensitive gate to a gap junction pore, three different closed states can occur but only one open state is possible. The open probability is maximal when no transjunctional voltage is present and is minimal at transjunctional voltages higher than $U_0$. When the polarity of the transjunctional voltage also influences the gating, one can assume that only one connexon gate closes at a positive voltage whereas the other stays open. When the polarity is reversed, the closed gate will open and the gate in the other connexon senses a positive voltage and will close. When this switch from one gate to the other is fast and the transition time between open and closed states is short, a current response might be seen which corresponds to that of Cx37- and Cx43-channels, but when the switch from one gate to the other is slow, the behavior of Cx32-, Cx40- and Cx43-channels might result, even for the same transition time between open and closed states. Our experiments indicate that not only intracellular and/or pore domains can play a role in voltage sensitive channel gating, but also the second extracellular loop, which might be involved when polarity changes occur.

## References

[1] D.C. Spray, R.L. White, A. Campos de Carvalho, A.L. Harris and M.V.L. Bennet, Gating of gap junction channels. Biophys. J. 45 (1984) 219-230

[2] J. Neyton and A.Trautmann, Single-channel currents of an intercellular junction. Nature 317 (1985) 331-335.

[3] R.D. Veenstra and R.L. DeHaan, Measurement of single channel currents from cardiac gap junctions. Science 233 (1986) 972-974

[4] M.B. Rook, H.J. Jongsma and A.C.G. van Ginneken, Properties of single gap junction channels betwen isolated neonatal rat heart cells. Am. J. Physiol. 255 (1988) H770-H782

[5] R.L. DeHaan, Dynamic behavior of cardiac gap junction channels. In: Modern Cell Biology - Gap Junctions. E.L. Hertzberg and R.G. Johnson eds., B.H. Satir series ed., Alan R. Liss Inc., New York (1988) pp. 303-320

[6] J.M. Burt and D.C. Spray, Single-channel events and gating behavior of the cardiac gap junction channel. Proc. Natl. Acad. Sci. USA 85 (1988) 3431-3434

[7] R.D. Veenstra, Voltage-dependent gating of gap junction channels in embryonic chick ventricular cell pairs. Am. J. Physiol. 258 (1990) C662-C672

[8] H.-Z. Wang, J. Li, L.F. Lemanski and R.D. Veenstra, Gating of mammalian cardiac gap junction channels by transjunctional voltage. Biophys. J. 62 (1992) 139-151

[9] J.B. Rubin, V.K. Verselis, M.V.L. Bennet and T.A. Bargiello, Molecular analysis of voltage dependence of herotypic gap junctions formed by connexins 26 and 32. Biophys. J. 62 (1992) 183-195

[10] J.B. Rubin, V.K. Verselis, M.V.L. Bennet and T.A. Bargiello, A domain substitution procedure and its use to analyze voltage dependence of homotypic gap junctions formed by connexins 26 and 32. Proc. Natl. Acad. Sci. USA 89 (1992) 2820-2824

[11] Y. Chen and R.L. DeHaan, Multiple channel conductance states and voltage regulation of embryonic chick cardiac gap junctions. J. Membr. Biol. 127 (1992) 95-111

[12] R. Eckert, A. Dunina-Barkovskaya and D.F. Hülser, Biophysical characterization of gap-junction channels in HeLa cells. Pflügers Arch. Eur. J. Physiol. 424 (1993) 335-342

[13] A. Butterweck, U. Gergs, C. Elfgang, K. Willecke and O. Traub, Immunochemical Characterization of the Gap Junction Protein Connexin45 in Mouse Kidney and Transfected Human HeLa Cells. J. Membrane Biol. 141 (1994) 247-256

[14] C. Elfgang, R. Eckert, H. Lichtenberg-Fraté, A. Butterweck, O. Traub, R.A. Klein, D.F. Hülser and K. Willecke, Specific Permeability and Selective Formation of Gap Junction Channels in Connexin-transfected HeLa Cells. J. Cell Biol. 129 (1995) 805-817

[15] S. Haubrich, H.-J. Schwarz, F. Bukauskas, H. Lichtenberg-Fraté, O. Traub, R. Weingart and K. Willecke, Incompatibility of Connexin 40 and 43 Hemichannels in Gap Junctions between Mammalian Cells is Determined by Intracellular Domains. Mol. Biol. Cell 7 (1996) 1995-2006

[16] SWISS-PROT database, Geneva University Hospital and University of Geneva. Internet (http://www.embl-heidelberg.de/srs/srsc)

[17] V.K. Verselis, C.S. Ginter and T.A. Bargiello, Opposite Voltage Gating Polarities of Two closely related Connexins. Nature 368 (1994) 348-351.

Gap Junctions
R. Werner (Ed.)
IOS Press, 1998

# Expression of Heteromeric Lens Connexons in *Xenopus* Oocytes

L. Ebihara*, C. Oberti* and V.M. Berthoud**
*Department of Physiology and Biophysics,
The Chicago Medical School, North Chicago, IL, USA
**Department of Pediatrics, Washington,
University School of Medicine, St. Louis, MO, USA

**Abstract**

It has been previously shown biochemically that lens connexins can form heteromeric connexons. However, the functional consequence of heteromer assembly of lens connexins is still not well understood. To investigate the biophysical properties of heteromeric lens connexons, gap junctional proteins cloned from chick(Cx56) and mouse(Cx50) were coexpressed in *Xenopus* oocytes that had been pretreated with an antisense Cx38 oligonucleotide. Whole cell and single channel currents were measured in single oocytes by conventional 2-microelectrode voltage clamp and patch clamp methods. Cx56 expressed as a slowly activating, nonselective cation current that activated on depolarization to potentials > -10 mV; No nonjunctional current was observed in Cx50 cRNA-injected oocytes in the presence of .7 mM $Ca^{2+}$ and .8 mM $Mg^{2+}$. Coinjection of Cx50 cRNA with Cx56 cRNA into oocytes resulted in the development of a nonjunctional current which activated on depolarization but had a more negative threshold for activation(-70 mV) than Cx56. The reversal potential was not altered. These results indicate that coinjection of Cx56 and Cx50 cRNAs give rise to a novel channel which must contain subunits derived from the 2 cRNAs.

## 1. Introduction

The lens is an avascular, transparent tissue that functions to focus light on the retina. It is composed of a solid mass of cells that grows throughout life with the addition of new cells at the surface. Two cell types are found in the lens: epithelial cells and fiber cells. The epithelial cells form a monolayer on the anterior surface of the lens. The bulk of the lens is composed of elongated fiber cells that are closely packed together in an orderly array. Mature fiber cells are known to lack intracellular organellles and have high concentrations of soluble proteins known as crystallins. The accumulation of these crystallins along with the loss of organelles allows the lens to achieve a high refractive index while maintaining transparency. As the the lens lacks blood vessels, the fiber cells are dependent on gap junctions for transport of electrolytes, water and small metabolites between the inner fiber cells and the epithelial cells.

Gap junctional channels are integral membrane proteins connecting the cytoplasm of adjacent cells. These channels are composed of two connexons or hemichannels, one from each cell, which dock via their extracellular domains to form a whole gap junctional channel. Each connexon is formed from six connexins arranged as a hexamer around a central pore. At least 13 rodent connexins have been identified, of which 3 are found in the lens( Cx43[1], Cx50[2] and Cx46[3]). Immunofluorescent studies indicate that Cx43 localizes to the anterior epithelial cell layer while Cx46 and Cx50 colocalize to fiber cell gap junctions[3]. Moreover, recent biochemical studies indicate that Cx46 and Cx50 form heteromeric channels[4]. In the present study, we have taken advantage of the unusual ability of Cx46( and its chicken homolog, Cx56)[5] to form open hemichannels in the nonjunctional membrane of single oocytes to investigate whether Cx50 can form heteromeric connexons with Cx56.

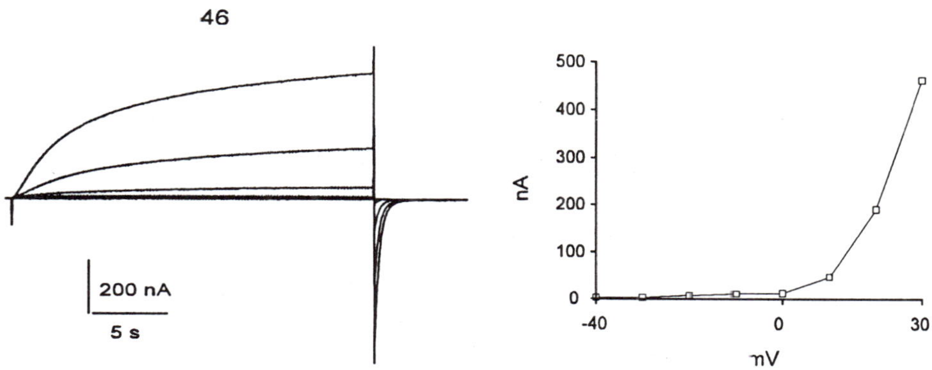


Figure 1. (A) Membrane currents in an oocyte injected with Cx46 cRNA and antisense oligonucleotides for Cx38. The whole cell currents were measured using a two-electrode-voltage-clamp. A series of depolarizing steps were applied in 10 mV increments between -40 and 30 mV from a holding potential of -40 mV. (B) I-V relations obtained from data shown in A. The current was measured at the end of the 24 s pulse and plotted as a function of voltage.

## 2. Results and discussion

We began to investigate the electrophysiological properties of the lens connexins by expressing rat Cx46 in *Xenopus* oocytes[6]. We found that expression of Cx46 in solitary oocytes resulted in the development of a large time- and voltage-dependent current(which will be denoted as $I_h$) that was not observed in non-injected control oocytes or in oocytes injected with cRNA for Cx43 or Cx32.

$I_h$ activated at potentials positive to -20 mV and reversed polarity around -10 mV. Ion substitution experiments suggested that this current was nonselectively permeable for small monovalent cations such as sodium and potassium. Removal of external calcium caused a marked increase in the amplitude of $I_h$, shifted the steady-state activation curve to more negative potentials, and altered the kinetics of activation and deactivation. The ability of Cx46 to induce the formation of cell-to-cell channels was tested in the oocyte pair system. 1 mM cobalt was added to the incubation solution to block current flow through nonjunctional hemi-channels and prevent the oocyte pairs from lysing. Oocyte pairs injected with Cx46 cRNA + antisense oligonucleotides for *Xenopus* Cx38 were strongly coupled. In contrast, oocyte pairs injected with antisense alone showed no coupling. The inactivation kinetics of the gap junctional channels resembled the deactivation kinetics of $I_h$ recorded in external solutions containing low calcium. These results suggest that Cx46 is forming open hemichannels in the nonjunctional membrane of oocytes. The physiological significance of $I_h$ remains unclear although a current resembling $I_h$ has recently been reported in isolated mammalian lens fiber cells in extracellular solutions containing reduced $Ca^{2+}$.

We have also expressed chick Cx56( the chicken homolog of Cx46)[7] and mouse Cx50 in single oocytes and oocyte pairs. Single oocytes injected with Cx56 cRNA developed a slowly activating, time-dependent, outward current that activated on depolarization to potentials greater than -10 mV. In contrast, oocytes injected with Cx50 cRNA showed little or no connexin-induced current. Nonjunctional currents induced by Cx56 differed from those induced by Cx46 in their degree of initial rectification.

To test whether Cx56 and Cx50 could form gap-junctional channels, the connexins were expressed in pairs of oocytes and assayed for coupling using a dual two microelectrode voltage clamp technique. Background endogenous coupling was reduced

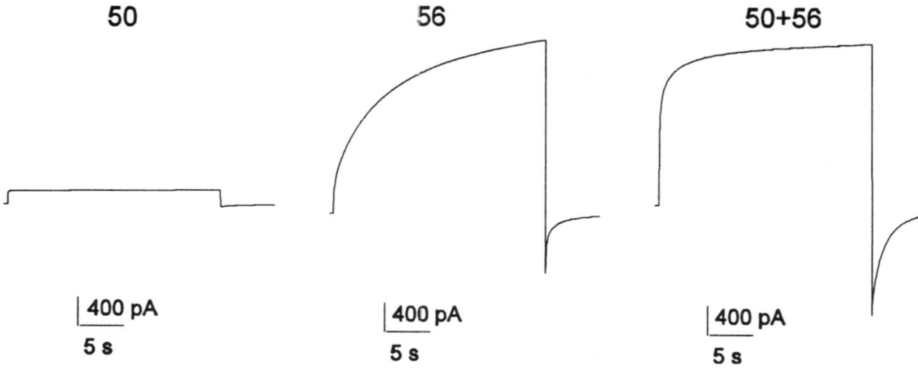


Figure 2. Membrane currents recorded from oocytes injected with Cx50, Cx56 or Cx56+Cx50. Voltage clamp steps were applied from -40 mV to +30 mV.

by preinjecting the oocytes the antisense oligonucleotides for *Xenopus* Cx38, 24 hours prior to pairing. Oocyte pairs expressing either Cx56 or Cx50 were strongly coupled while oocyte pairs treated with antisense alone were not coupled. The steady-state $G_j$-$V_j$ relation of the gap junctions formed from Cx56 had a reduced voltage sensitivity and the time course of channel closure was much slower than for Cx50.

The effect of coexpression of Cx50 with Cx56 was examined by coinjecting oocytes with equal concentrations of cRNA for Cx56 and Cx50 and studying membrane currents in single oocytes using a two-microelectrode-voltage-clamp technique. Coexpression of Cx50 and Cx56 led to a negative shift in the threshold for activation of $I_h$, accelerated the rate of activation and slowed the rate of deactivation as compared to currents in oocytes expressing only Cx56. It also decreased the ratio between the activated outward current and the tail current amplitude. These results differ from those expected from the sum of the two homomeric channels and provide the first functional evidence for heteromeric association of lens connexins. It is likely that in addition to causing alterations in voltage dependent properties, heteromeric interactions between lens connexins may also lead to changes in other properties such as permeability for small metabolites and thus play a critical role in maintaining normal lens homeostasis.

## References

[1] **Beyer, E. C., D. L. Paul, and D. A. Goodenough.** Connexin43: a protein from rat heart homologous to a gap junction protein from liver. *J. Cell Biol.* 105: 2621-2629, 1987.

[2] **White, T. W., R. Bruzzone, D. A. Goodenough, and D. L. Paul.** Mouse cx50, a functional member of the connexin family of gap junction proteins, is the lens fiber protein MP70. *Molecular Biology of the Cell* 3: 711-720, 1992.

[3] **Paul, D. L., L. Ebihara, L. J. Takemoto, K. I. Swenson, and D. A. Goodenough.** Connexin46, a novel lens gap junction protein, induces voltage-gated currents in nonjunctional plasma membrane of *Xenopus* oocytes. *J. Cell Biol.* 115: 1077-1089, 1991.

[4] **Jiang, J. X., T. W. White, D. A. Goodenough, and D. L. Paul.** Molecular cloning and functional characterization of chick lens fiber connexin 45.6. *Molecular Biology of the Cell* 5: 363-373, 1994.

[5] **Rup, D. M., R. D. Veenstra, H. Wang, P. R. Brink, and E. C. Beyer.** Chick connexin-56, a novel lens gap juncion protein. *J. Biol. Chem.* 268: 706-712, 1993.

[6] **Ebihara, L. and E. Steiner.** Connexin46 forms gap junctional hemichannels in *Xenopus* oocytes. In: *Gap Junctions,* edited by J. E. Hall, G. A. Zampighi, and R. M. Davis. Amsterdam: Elsevier Science Publishers, 1993, p. 75-77.

[7] **Ebihara, L., V. M. Berthoud, and E. C. Beyer.** Distinct behavior of connexin56 and connexin46 gap junctional channels can be predicted from the behavior of their hemi-gap-junctional channels. *Biophys. J.* 68: 1796-1803, 1995.

# Section 4:

# Assembly and Degradation of Gap Junction Channels

*Gap Junctions*
*R. Werner (Ed.)*
*IOS Press, 1998*



# Heteromeric Connexon Formation: Analyses Using Transfected Cells

Eric C. BEYER*, Kyung H. SEUL*, Kathrin BANACH†, David M. LARSON‡, and Peter R. BRINK†

**Washington University, St. Louis, MO; †State University of New York, Stony Brook, NY; and ‡Mallory Institute of Pathology, Boston, MA*

**Abstract.** Since many cells contain multiple connexins, there is a potential for mixing of those connexins within the same channel. The greatest complexity of mixing would derive from the formation of heteromeric gap junction channels (which contain different connexins within the same hemichannel or connexon). We used stable transfection of cells to examine the potential mixing of connexin37 and connexin43. In the transfected cells, both connexin proteins were abundantly produced, and they localized to identical distributions as detected by immunofluorescence. Double whole cell patch clamp studies showed that co-expressing cells exhibited unitary channel conductances and gating characteristics that could not be explained by hemi-channels formed of either connexin alone. These observations suggest that connexin37 and connexin43 can readily mix to form heteromeric channels. The ability of different connexins to form heteromeric channels and the novel properties of those mixtures may have a major impact on the regulation of intercellular communication.

## 1. Introduction

Gap junctions are formed by subunit proteins that are members of the multi-gene family of connexins. We have previously used stable transfection of non-communicating cells with individual connexins to demonstrate that they each form gap junction channels with differing gating and permeability/selectivity properties [1,2].

However, determining the characteristics of channels formed of individual connexins may not be sufficient to explain the properties of intercellular communication in most cells, since it has been demonstrated that in many, if not all, tissues (including liver, heart, lens, bone, etc) more than one connexin is expressed in the same cell. Indeed, in osteoblastic cell lines (which express connexin43 and connexin45) alterations in the relative amounts of the two co-expressed connexins lead to changes in the molecular permeability of communicating cells [3,4]. Those observations suggest that the co-expressed connexins might be interacting.

Most prior studies have assumed that gap junction hemichannels (connexons) are composed uniformly of a single connexin (i.e. homomeric connexons). The only kind of connexin mixing that has been thoroughly considered is the formation of heterotypic channels (when a channel is formed by two homomeric connexons containing different connexins) [5,6].

## 2. Heteromeric mixing of connexins

Co-expression of multiple connexins in a single cell leads to the possibility of mixing of connexins within a single gap junction hemi-channel, forming "heteromeric" connexons. There have been some biochemical data to support the existence of heteromeric connexons in the lens [7] and in cells overexpressing the liver connexins Cx26 and Cx32 [8].

There has been little electrophysiological analysis to examine the properties of heteromeric gap junction channels nor even to support their existence. The existence of heteromeric hemichannels would allow the formation of a large number of channels beyond that possible if only homomeric connexons containing a single connexin formed. (See Fig. 1).

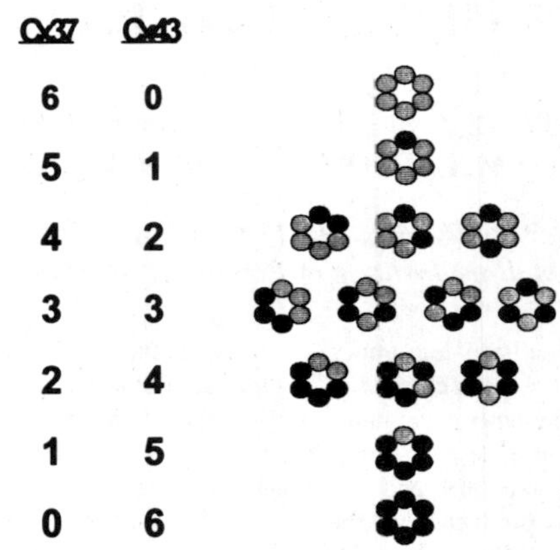


Figure 1. Schematic representation of the types of hemichannels possible in a cell that equally co-expressed connexin37 (Cx37) and connexin43 (Cx43), assuming that the two connexins are freely capable of mixing and that there is six-fold symmetry. There are 12 heteromeric hemichannel types and 2 homotypic types (6/0 and 0/6). For any two such co-expressing cells that were coupled, there would be 14X14 possible combinations or 196 gap junction channel types possible. The 196 forms would include one each of homomeric/homotypic Cx37 and Cx43 channels and two heterotypic Cx37/Cx43 channels (one in each polarity). The remaining 192 channels would all include heteromeric hemichannels (more than 97% of all channels). Figure adapted from [9].

## 3. Co-expression of connexin43 and connexin37

For the present studies, we chose to study the co-expression of Cx43 and Cx37. These two connexins are endogenously co-expressed in vascular endothelial cells, and the relative abundances of these two connexins vary within these cells, and can be modulated according to different growth conditions or in response to TGF-beta [10]. However, the physiological properties of channels formed exclusively of Cx43 or of Cx37 are very distinctive. They differ dramatically in channel unitary conductance and in gating in response to transjunctional voltage.

We used stable transfection of established cell lines with the rat Cx43 and the human Cx37 cDNAs to produce cells that co-expressed the two connexins. In some experiments, mouse Neuro2A cells (a non-communicating cell line) were sequentially transfected with human Cx37 (using G418 selection) followed by rat Cx43 (using Zeocin selection) [9]. In other studies, normal rat kidney (NRK) cells (which abundantly express Cx43) were transfected with Cx37 (using G418 selection). The transfected cells abundantly expressed Cx43 and Cx37 as demonstrated by RNA blot hybridization and immunoblotting [9]. The co-expressed Cx37 and Cx43 localized to identical distributions as demonstrated by immunofluorescence (Fig. 2).

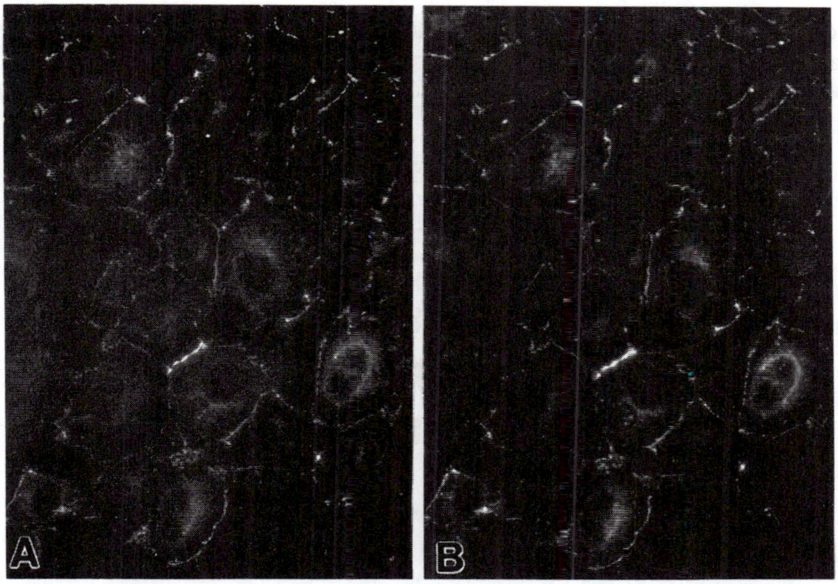


**Figure 2.** Double label immunofluorescent localization of Cx43 (A) and Cx37 (B) in NRK cells transfected with human Cx37. Cells were reacted with simultaneously reacted with mouse anti-Cx43 and rabbit anti-Cx37 antibodies followed by FITC-conjugated goat anti-mouse IgG and CY3-conjugated goat anti-rabbit IgG. The introduced Cx37 is detected in an identical distribution to the endogenous Cx43, suggesting that gap junctions in these cells contain both connexins.

## 4. Electrophysiological evidence for heteromeric mixing

N2A cell pairs co-expressing Cx37 and Cx43 were examined by the double whole cell patch clamp technique. Data obtained from these cells were compared to data derived from Cx43 or Cx37 expressing cell pairs or heterotypic Cx43-Cx37 cell pairs.

Weakly coupled cell pairs allowed the observation of single channel events. In experiments using a pipette solution containing 180 mM CsCl, the conductance of Cx37 channels was 360 pS and that of Cx43 was 115 pS. However, in the Cx37/Cx43 co-transfected cells no single channel size was predominant; rather, multiple unitary conductances were observed (including 280, 220, 200, 150. 70, and 35 pS)[9].

The voltage-dependent gating behavior also differed between single connexin-expressing cells and the co-transfectants. Cx37/Cx37 pairs showed a strong voltage-dependence. Cx43/Cx43 pairs showed a weak voltage dependence. Heterotypic Cx37/Cx43 pairs showed a voltage dependent asymmetry similar to previous observations in oocyte pairs [9]. However, pairs of cells co-expressing Cx37 and Cx43 were much more scattered, and the voltage-dependent decrease of the normalized conductance was shifted to higher voltages than observed with the homotypic or heterotypic combinations ([9] and Fig. 3).

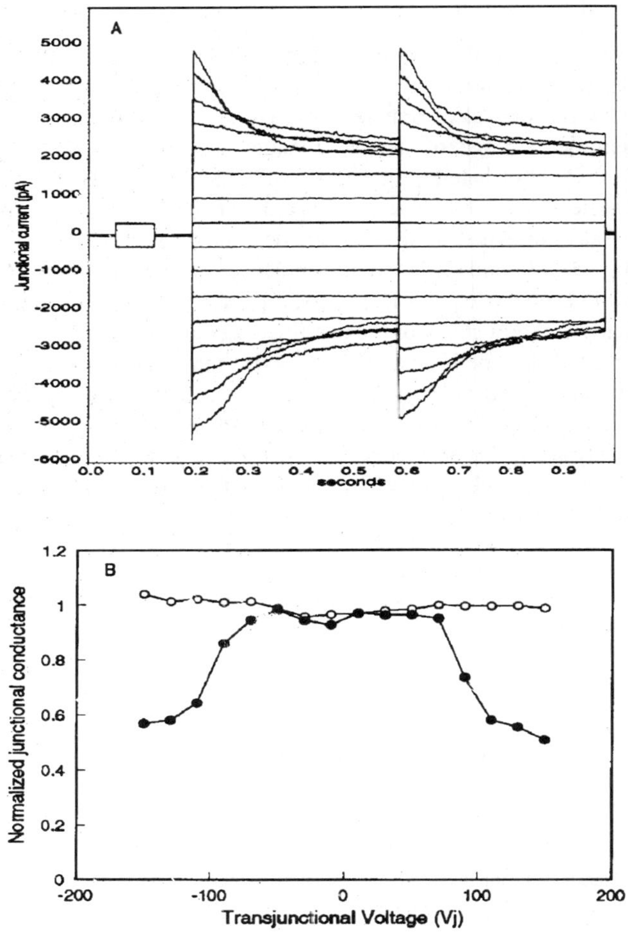


Fig 3. Voltage-dependence of coupling in a pair of N2A cells both co-transfected with Cx37 and Cx43. A. Junctional currents. The polarity of the 10 mV test pulse was flipped from negative to positive for half of the step potentials used. The kinetics of voltage-dependent inactivation of junctional current for the cotransfected cell pair appears slower or elongated relative to homotypic rat Cx43 or human Cx37 [9]. B. Graphs of the instantaneous (○) and non-instantaneous (●) normalized junctional conductance plotted against transjunctional voltage ($V_j$). To calculate instantaneous junctional conductance, the first five data points after the onset of a step in the junctional current records were used. The last five data points before the termination of a step were used to determine non-instantaneous junctional conductance. The instantaneous junctional conductance is the same regardless of the amplitude of $V_j$ of either polarity. $G_{min}$ or $G_{non\text{-}inst}$ is 0.58 in this case. This is very different from homotypic rat Cx43 ($G_{min}$ = 0.3) or homotypic human C37 ($G_{min}$ = 0.2 or less). The elevated $G_{min}$ is a typical feature of the co-transfected N2A cell pairs. Only a minority of the cell pairs showed $G_{min}$s below 0.5 [9].

## 5. Conclusions

Stable transfection of cell lines has allowed us to derive clones that co-express Cx43 and Cx37. Data derived from these cells can be readily compared to that derived from cells transfected with either connexin alone or to heterotypic Cx37/Cx43 cell pairs. The Cx37/Cx43 co-transfectants exhibit a range of unitary conductances that cannot be explained by homomeric Cx37 or Cx43 or heterotypic Cx37/Cx43 channels. Furthermore, gating in the Cx37/Cx43 co-transfectants exhibits less voltage-dependence than in the homomeric or homotypic pairs. Taken together, these data provide strong electrophysiologic evidence for heteromeric mixing of Cx37 and Cx43. While evidence has been presented that some other connexins will form heteromers (e.g. Cx46 and Cx50 [7]), it is unclear whether all connexins are capable of mixing and how many of the possible combinations can occur.

The existence of heteromeric mixing of connexins may have significant consequences for the properties of intercellular communication between coupled co-expressing cells. Such cells would be able to produce a spectrum of channel types with each form having unique permeability/selectivity and gating characteristics. Alternatively, the large number of different combinations might "average each other out", minimizing the divergent properties of channels formed from different connexins.

**Acknowledgements.** These studies were supported by NIH grants HL52697 (to DML), HL45466 (to ECB), and HL31299 (to PRB). Eileen Westphale provided invaluable assistance with transfection and cloning of the connexin-expressing cell lines.

**References**

[1] R. D. Veenstra *et al.*, Connexin37 forms high conductance gap junction channels with subconductance state activity and selective dye and ionic permeabilities. *Biophys. J.* 66 (1994) 1915-1928.
[2] R. D. Veenstra *et al.*, Selectivity of connexin-specific gap junctions does not correlate with channel conductance. *Circ Res* 77 (1995) 1156-1165.
[3] T. H. Steinberg *et al.*, Connexin43 and connexin45 form gap junctions with different molecular permeabilities in osteoblastic cells. *EMBO J.* 13 (1994) 744-750.
[4] M. Koval *et al.*, Transfected connexin45 alters gap junction permeability in cells expressing endogenous connexin43. *J. Cell Biol.* 130 (1995) 987-995.
[5] T. W. White *et al.*, Selective interactions among the multiple connexin proteins expressed in the vertebrate lens: the second extracellular domain is a determinant of compatibility between connexins. *J. Cell Biol.* 125 (1994) 879-892.
[6] T. W. White *et al.*, Functional analysis of selective interactions among rodent connexins. *Mol. Biol. Cell* 6 (1995) 459-470.
[7] J. X. Jiang and D. A. Goodenough, Heteromeric connexons in lens gap junction channels. *Proc. Natl. Acad. Sci. USA* 93 (1996) 1287-1291.
[8] K. A. Stauffer, The gap junction proteins beta(1)-connexin (connexin-32) and beta(2)-connexin (connexin-26) can form heteromeric hemichannels. *J. Biol. Chem.* 270 (1995) 6768-6772.
[9] P. R. Brink *et al.*, Evidence for heteromeric gap junction channels formed from rat connexin43 and human connexin37. *Am. J. Physiol. (Cell. Physiol.)* (in press)
[10] D. M. Larson *et al.*, Differential regulation of connexin43 and connexin37 in endothelial cells by cell density, growth, and TGF-beta1. *Am. J. Physiol. (Cell. Physiol.)* 272 (1997) C405-C415.

*Gap Junctions*
*R. Werner (Ed.)*
*IOS Press, 1998*

# Expression of Retroviruses Containing Connexins in Chick Embryonic Lens In Situ

Jean X. Jiang* and Daniel A. Goodenough
*Department of Cell Biology*
*Harvard Medical School, Boston, MA 02115*
*Present address: *Department of Biochemistry*
*The University of Texas Health Science Center at San Antonio*
*7703 Floyd Curl Drive, San Antonio, Texas 78284-7760, USA*

**Abstract.** To understand the function of connexin-mediated gap junction intercellular communication in lens cell biology, an *in vivo* model system was developed in which exogenous proteins can be expressed in embryonic chick lens. RCAS(A) is a replication-competent chicken retrovirus which infects dividing cells. Retroviral constructs were prepared containing alkaline phosphatase (AP) and FLAG-tagged connexins (Cx). Chick lenses were infected *in situ* by injecting virus into the lumen of lens vesicles at stage 18, followed by culture for varying periods. The lenses were then dissected and the expressed proteins visualized by AP histochemistry and immunostaining. 72 hours after viral infection, AP staining could still be observed in epithelial cells and in differentiating fibers in the bow region, but not in mature lens fibers. FLAG-tagged exogenous lens connexins were also abundantly expressed by viral infection. The exogenous connexins localized at the cell surfaces in junctional maculae and showed the same cell-type specific distribution as their endogenous connexin counterparts. Together, this retroviral expression model system provides opportunities to study the expression of wild-type and mutant proteins during lens differentiation. Expression of wild-type connexins has revealed that the distribution of the three different lens connexins may be translationally or post-translationally regulated.

## 1. Introduction

The lens is dependent for osmotic and metabolic homeostasis on a highly developed network of gap junction-mediated intercellular communication to facilitate exchange of ions and metabolites throughout the organ [1,2]. The chick lens is known to express three different connexins [3]. Cx43 is a component of gap junctions present between lens epithelial cells and differentiating fibers in the lens bow region [4]. Cx56 and Cx45.6 are abundantly expressed in lens fiber junctions [5,6] although they are also found to a lesser degree in epithelial cell to epithelial cell junctions [7].

In order to experimentally probe the functions played by connexin in chick lens physiology and development, an *in vivo* model system is required, since the systems of cultured lens cells which have been developed [8-10] only recapitulate the initial stages in lens fiber development.

While the αA crystallin promoter has been used to drive high-level, non-lethal expression of a variety of transgenes in the mouse lens [11,12], transgenic chicken approaches have had only limited success due to the long generation time and poor efficiency of achieving stable transgenic lines by microinjection. The RCAS(A) chicken retrovirus has been successfully utilized for expression of AP gene in developing chick embryos [13,14] and for misexpression of the *Hox-4.6* gene in chick limb bud [15].

Lens development offers a unique time window in which to specifically infect lens cells with retrovirus. During development, the lens differentiates within the embryonic ectoderm overlying the optic cup by first thickening to form the lens placode which then invaginates to form the lens pit which separates from the ectoderm as the lens vesicle [16]. At this stage, the lens is a hollow sphere of cells, and virions can be injected directly into the vesicle lumen. Following this transient stage, the posterior cells of the lens vesicle elongate anteriorly to form the primary lens fibers, eventually occluding the lumen of the vesicle upon contact with the anterior lens epithelium.

In this report, we have injected RCAS(A) virions carrying the AP gene directly into the lens vesicle of stage 18 chick lenses and measured AP expression during subsequent stages of lens development. We have observed that the lenses formed normally following virus infection, and that AP was expressed in all lens cell types. However, AP activity was measurable only in the epithelium and more superficial differentiating cells since the AP activity was destroyed as lens fibers matured. Virions were also constructed containing cDNA coding for epitope-tagged chick lens connexins, which were injected into the lumen of lens vesicles. We describe here the pattern of expression of the different lens connexins during subsequent lens development.

## 2. Results

The expression of exogenous proteins in the embryonic lens was initially examined by microinjection of RCAS(A) containing the AP reporter gene into the centrallumen of embryonic chick lens vesicles as illustrated in Fig. 1. Expression of AP in retrovirally infected lens cells was detected by the colorimetric AP reaction [17]. Fig. 2 shows photographs of chick lenses 72 hours after infection. Fig. 2A shows whole mounts of infected (left) and control contralateral (right) lenses following the AP reaction. Large regions of the lens surface are stained. Cryosections of this specimen (Fig. 2B) revealed that epithelial cells and bow cells were positive. However, the infected early lens fibers which were visibly stained 24 h after viral infection (data not shown) could no longer be seen, indicating that the exogenous AP was processed by the developing lens cells according to their normal developmental program.

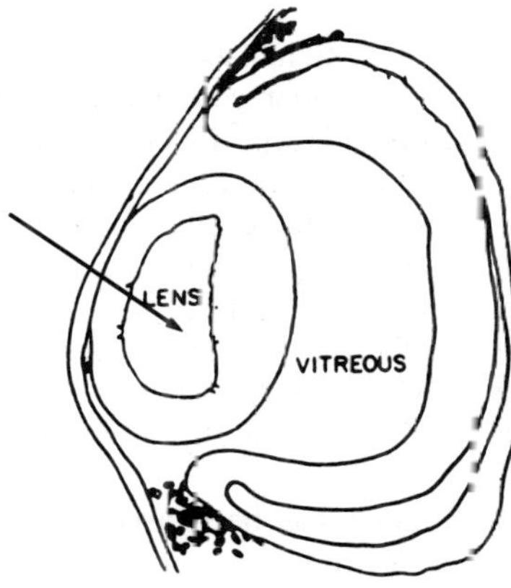


**Figure 1. Microinjection of retrovirus into stage 18 embryonic chick lens.** The virions are directly injected into the central lumen of lens vesicle (arrow).

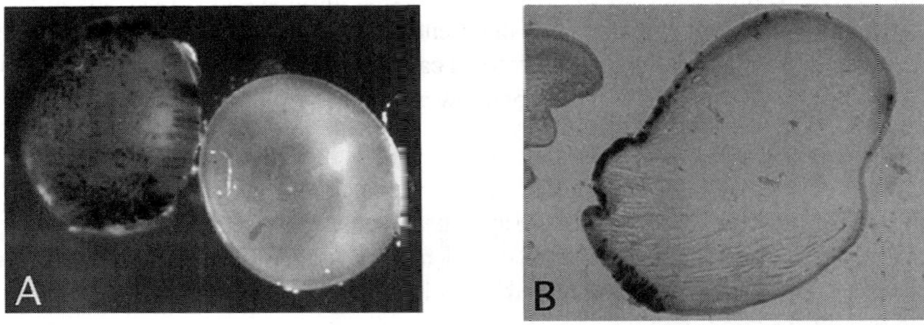


**Figure 2. AP expression in lens after viral injection.** 72 hours after viral infection, the expression of AP was detected either in whole lens (A) or in lens sections (B).

Patterns of retroviral expression of AP in the chick lens infected at stage 18 showed that lens cells could express AP. The AP expression patterns revealed a high concentration of infected cells near the lens equator, where stem cells are known to be mitotically active. Lens epithelial cells, alone or in clusters, expressed AP, indicating continued cell division within the epithelium during the developmental stages examined. Differentiating lens fibers in the equatorial region, seen to be expressing AP at 24 hours, were negative at 72 hours, indicating that previously expressed AP was proteolytically degraded during lens fiber maturation.

The viral constructs were prepared containing each of the three full length lens connexins with each C-terminus tagged with the FLAG sequence to distinguish them from endogenous connexins. RCAS(A) preparations containing each of three lens connexins were individually injected into stage 18 embryonic chick lens as illustrated in Fig.1. 72 hours after viral infection, RCAS(A) viruses containing connexins were abundantly expressed in embryonic chick lens and a portion of the signal was punctate at the cell surface (Fig. 3), typical of gap junction plaques. Expression of exogenous connexins in large quantity had no detectable effect on lens morphology and did not interfere with lens developmental process.

The expression of FLAG-tagged Cx43 in the lens was found abundantly in the bow region (A and B) where lens cell mitosis and differentiation are active. Cx43 was also found in the lens epithelium (C and D). However, there was no staining in the lens fibers. FLAG-tagged Cx45.6 (E and F) was expressed in the lens bow region as well, although the staining extended deeper into the cortex than that seen for Cx43. In addition, the exogenous Cx45.6 was also expressed in both superficial (E and F) and in deep fibers (G and H). The Cx45.6 staining was not detectable in lens epithelium. FLAG-tagged Cx56 was also abundantly expressed in the lens and it localization was similar to that for Cx45.6 (data not shown).

In summary, all three exogenous lens connexins introduced by viral infection were found at the same specific locations as their endogenous counterparts. This cell type-specific localization of the gap junction proteins cannot be transcriptionally regulated, since the RCAS(A)-driven AP expression did not show any selectivity.

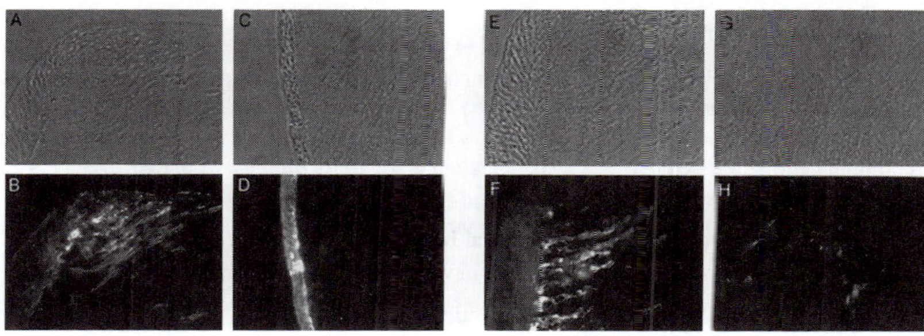


**Figure 3. Expression of exogenous Cx43 and Cx45.6 in chick embryonic lens.** RCAS(A)-Cx43 (A-D) and Cx45.6 (E and H) were microinjected into the lumen of stage 18 lens *in situ*. After 72 h incubation, lenses were isolated and frozen sections were prepared and stained with anti-FLAG mAb and subsequently with rhodamine-conjugated anti-mouse IgG.

## 3. Discussion

RCAS(A), a replication-competent avian retrovirus [18-20] has been used to express exogenous proteins in the embryonic chick lens *in situ*. Specificity of viral targeting to the lens was achieved by injection of the virus into the lumen of the lens vesicle at developmental stage 18. Viral infection and spread was found not to visibly alter lens development over a period of 1 week, the longest time studied.

In this study, we have introduced two types of proteins by retroviral injection into embryonic chick lens *in situ*: AP, a soluble cytosolic protein, and three lens connexins, insoluble integral membrane proteins used as structural proteins in gap junctions. 24 hours following viral infection at stage 18, AP was found in all lens cell types except the primary lens fibers at the posterior surface of the lens. Since the primary fibers are very slowly growing or postmitotic [21, 22], they were refractory to viral infection. The distribution of AP was seen in all lens cell types, indicating that there was no cell type-specific selectivity in the regulation of expression. However, AP signal was lost as lens fibers matured, indicating that the AP was proteolytically removed from the fibers as part of the maturation program.

In contrast to these AP data, lens connexins introduced by retroviral infection, accumulated only in the same cellular locations as their endogenous counterparts: Cx43 in lens epithelium, but not in lens fibers; Cx56 and Cx45.6 mainly in lens fibers and not in the epithelium. These observations clearly suggested cell- and connexin-specific mechanisms for controlling protein accumulation, not only in terms of synthesis, but also in terms of escaping the proteolytic processing program paradigmatic of fiber differentiation. It is possible that there are targeting signals residing in the coding region of lens fiber connexins, since exogenous retroviral constructs containing cDNA of connexins which lack their non-coding regulatory regions nonetheless show endogenous cellular distributions.

The control of connexin expression in the lens may involve both translational and post-translational mechanisms, both of which have been documented in lens biology

[23,24]. Studies have reported that there is proteolytic processing of ovine Cx50 [25-27], notably involving calpain [28]. The experimental model described here will allow the design and expression of dominant negative mutant connexins to address the function of lens gap junction in lens physiology and development.

## References


[1] R.T. Mathias *et al.*, Physiological properties of the normal lens. *Physiol Rev* 77 (1997) 21-50.

[2] K. Cooper *et al.*, The physiology of lens junctions. In: Peracchia C, ed. *Biophysics of Gap Junction Channels*. Boca Raton: CRC Press; 1991 pp. 57-74.

[3] D.A. Goodenough, The crystalline lens: a system networked by gap junctional intercellular communication. In: Gilula NB, ed. *Seminars in Cell Biology*. London: Sauders Scientific Pub./Academic Press; 1992 pp. 49-58.

[4] L.S. Musil *et al.*, Expression of the gap junction protein connexin43 in embryonic chick lens: molecular cloning, ultrastructural localization, and post-translational phosphorylation. *J Membr Biol* 116 (1990) 163-175.

[5] J.X. Jiang *et al.*, Molecular cloning and functional characterization of chick lens fiber connexin45.6. *Mol Biol Cell* 5 (1994) 363-373.

[6] V.M. Berthoud *et al.*, Characterization of the gap junction protein connexin56 in the chicken lens by immunofluorescence and immunoblotting. *Invest Ophthalmol Vis Sci* 35 (1994) 4109-4117.

[7] J.X. Jiang *et al.*, Changes in connexin expression and distribution during chick lens development. *Dev Biol* 168 (1995) 649-661.

[8] A.S. Menko *et al.*, Chicken embryo lens cultures mimic differentiation in the lens. *Dev Biol* 103 (1984) 129-141.

[9] P.G. Fitzgerald and D.A. Goodenough, Rat lens cultures: MIP expression and domains of intercellular coupling. *Invest Ophthalmol Vis Sci* 27 (1986) 755-771.

[10] A.S. Menko *et al.*, Junctions between lens cells in differentiating cultures: structure, formation, intercellular permeability, and junctional protein expression. *Dev Biol* 123 (1987) 307-320.

[11] P.A. Overbeek *et al.*, Lens-specific expression and developmental regulation of the bacterial chloramphenicol acetyltransferase gene driven by the murine aA-crystallin promoter in transgenic mice. *Proc Natl Acad Sci (USA)* 82 (1985) 7815-7819.

[12] C.E. Egwuagu *et al.*, Ectopic expression of gamma interferon in the eyes of transgenic mice induces ocular pathology and MHC class II gene expression. *Invest Ophthalmol Vis Sci* 35 (1994) 332-341.

[13] D.M. Fekete and C.L. Cepko, Replication-competent retroviral vectors encoding alkaline phosphatase reveal spatial restriction of viral gene expression/transduction in the chick embryo. *Mol Cell Biol* 13 (1993) 2604-2613.

[14] D.M. Fekete and C.L. Cepko, Retroviral infection coupled with tissue transplantation limits gene transfer in the chicken embryo. *Proc Natl Acad Sci (USA)* 90 (1993) 2350-2354.

[15] B.A. Morgan *et al.*, Targeted misexpression of Hox-4.6 in the avian limb bud causes apparent homeotic transformations [see comments]. *Nature (Lond)* 358 (1992) 236-239.

[16] G.W. Philpott and A.J. Coulombre, Lens development. II. The differentiation of embryonic chick lens epithelial cells in vitro and in vivo. *Exp Cell Res* 38 (1965) 635-644.

[17] S.C. Fields-Berry *et al.*, A recombinant retrovirus encoding alkaline phosphatase confirms clonal boundary assignment in lineage analysis of murine retina. *Proc Natl Acad Sci (USA)* 89 (1992) 693-697.

[18] S.H. Hughes *et al.*, Adaptor plasmids simplify the insertion of foreign DNA into helper-independent retroviral vectors. *J Virol* 61 (1987) 3004-3012.

[19] S. Hughes and E. Kosik, Mutagenesis of the region between env and src of the SR-A strain of Rous sarcoma virus for the purpose of constructing helper-independent vectors. *Virology* 136 (1984) 89-99.

[20] C.J. Petropoulos and S.H. Hughes, Replication-competent retrovirus vectors for the transfer and expression of gene cassettes in avian cells. *J Virol* 65 (1991) 3728-3737.

[21] J. Piatigorsky and S.S. Rothschild, Loss during development of the ability of chick embryonic lens cells to elongate in culture: inverse relationship between cell division and elongation. *Dev Biol* 28 (1972) 382-389.

[22] J.W. McAvoy, Cell division, cell elongation and the co-ordination of crystallin gene expression during lens morphogenesis in the rat. *J Embryol Exp Morphol* 45 (1978) 271-281.

[23] J.X. Jiang *et al.*, Posttranslational phosphorylation of lens fiber connexin46: a slow occurrence. *Invest Ophthalmol Vis Sci* 34 (1993) 3558-3565.

[24] M. Kantorow *et al.*, Conversion from oligomers to tetramers enhances autophosphorylation by lens αA-crystallin - specificity between αA- and αB-crystallin subunits. *J Biol Chem* 270 (1995) 17215-17220.

[25] J. Kistler and S. Bullivant, Protein processing in lens intercellular junctions: cleavage of MP70 to MP38. *Invest Ophthalmol Vis Sci* 28 (1987) 1687-1692.

[26] J. Kistler *et al.*, MP38 contains the membrane-embedded domain of the lens fiber gap junction protein MP70. *J Biol Chem* 265 (1990) 13357-13361.

[27] G.N. Rao *et al.*, Bovine lens membrane proteins: MP70, MP64, and MP38 are products of the same gene. *Ophthalmic Res* 22 (1990) 166-172.

[28] J.S. Lin *et al.*, Processing of the gap junction protein connexin50 in the ocular lens is accomplished by calpain. *Eur J Cell Biol* 73, (1997) 141-149.

*Gap Junctions*
*R. Werner (Ed.)*
*IOS Press, 1998*

# Cell-Specific Differences in Transport and Assembly of Connexin46

Michael KOVAL*, Elizabeth HICK#, James E. HARLEY+, and Thomas H. STEINBERG+
**Department of Physiology and Institute for Environmental Medicine, University of Pennsylvania School of Medicine, Philadelphia, PA 19104, #Department of Biology, Massachusetts Institute of Technology, Boston, MA 02139, and +Department of Internal Medicine, Washington University School of Medicine, St. Louis, MO 63110*

**Abstract.** Connexin46 (Cx46) is most frequently associated with lens. We have found that both transformed and primary osteoblastic cells also express Cx46. Unlike lens cells that assemble Cx46 into gap junctions at the cell surface, Cx46 expressed by osteoblastic cells was retained in an aspect of the *trans* Golgi network (TGN). Also, Cx46 expressed by osteoblasts was found to be in a monomeric, rather than oligomeric form. Osteoblasts did not process Cx46 into the 68 kD phosphorylated form. To confirm that our results were not a general phenomenon of Cx46 expression by non-lentoid cells, we stably transfected Cx46 into Hela cells (Hela/Cx46). Hela/Cx46 cells showed assembly of Cx46 into functional gap junctions at the plasma membrane and Cx46 phosphorylation. This suggests that osteoblasts specifically regulate the expression of Cx46 using mechanisms that involve connexin transport, oligomerization and phosphorylation.

## 1. Introduction

Functional gap junctions consist of two hexameric membrane protein hemichannels, one in each cell, that join to form an aqueous pore that interconnects the cytoplasm of two adjacent cells [1-5]. There is considerable evidence that the connexins are assembled into hemichannels in an intracellular compartment prior to transport to the plasma membrane. Both endoplasmic reticulum (ER) [6,7] and the trans Golgi network (TGN) [8,9] have been implicated as sites for the assembly of connexin monomers into hexameric hemichannels.

The mechanisms that control connexin assembly and transport have remained elusive. One possibility is that connexin assembly into hemichannels is constitutive. While this has not been ruled out, there are examples from lens [10], thyroid [11] and uterine luminal epithelia [12] where multiple connexins expressed by the same cell form distinct gap junctions in distinct regions of the plasma membrane. These examples suggest that connexin hemichannel assembly is a regulated process. Of course, the information that controls whether or not two different connexins interact may be encoded in the primary sequence of the connexin itself [6], somewhat analogous to the control of heterotypic channel formation [13,14]. However, other mechanisms such as differential targeting and association with accessory proteins or chaperones remain possibilities for the regulation of gap junction hemichannel assembly.

We have found that rat osteoblastic cells express at least two connexins, Cx43 and Cx46 [9,15]. These two gap junction proteins have distinct fates in osteoblasts. While Cx43 is assembled into gap junction channels at the plasma membrane, Cx46 is retained as a monomeric protein in an aspect of the trans Golgi network (TGN). This suggests that gap junction assembly is a regulated process in osteoblasts.

Typically, Cx46 expression has been associated with lens. As lens epithelial cells differentiate into fiber cells, they express decreasing amounts of Cx43 and increasing amounts of Cx46 and Cx50 [10,16]. Both Cx46 and Cx50 are located at the cell surface as gap junctions, as determined by immunohistochemical analysis [10]. Lens fiber cells contain roughly equimolar amounts of Cx46 and Cx50 [17] and by sucrose gradient fractionation these two connexins have been found to associate into mixed hemichannels [18]. Interestingly, in lens epithelia that express Cx43 in addition to Cx46 and Cx50, Cx43 is found to form distinct gap junctions adjacent to Cx46/Cx50 containing channels [10], consistent with the notion that gap junction proteins are not randomly co-mingling in these cells.

## 2. Expression of Cx46 in osteoblastic and transfected Hela cells

Immunofluorescence was used to analyze the intracellular distribution of Cx43, Cx45 and Cx46 in three model systems, primary osteoblasts derived from rat calvaria, and two osteosarcoma cell lines, ROS 17/2.8 (ROS) and UMR 109-01 (UMR) [9,15,19,20]. All three cell types showed localization of Cx43 at the cell surface where cells are in close contact, in punctate regions corresponding to gap junctions (see figure 1a). Further, Cx45 was also found assembled into gap junctions, both in cells that express endogenous Cx45 (UMR) [19] or transfected Cx45 (ROS/Cx45; figure 1b) [15].

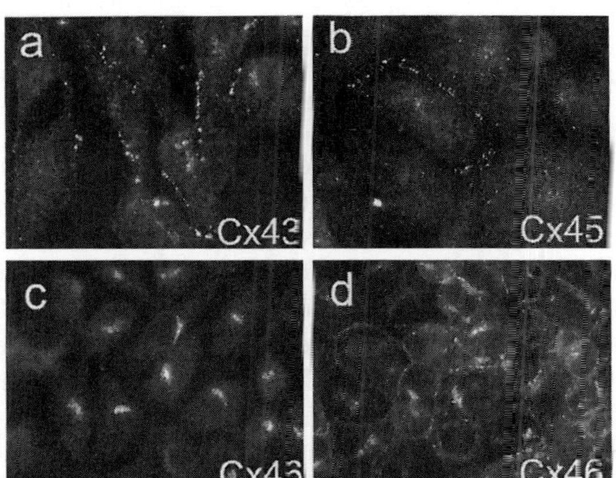


**Figure 1**. Immunofluorescence localization of connexins. ROS (a,c) , ROS/Cx45 (b) or Hela/Cx46 (d) cells were cultured on glass coverslips, fixed, permeabilized and then incubated with antiserum to either Cx43 (a), Cx45 (b) or Cx46 (c,d). The cells were labeled with rhodamine –conjugated goat anti-rabbit IgG and then visualized by fluorescence microscopy.

In contrast to Cx43 and Cx45, osteoblastic cells retain Cx46 exclusively in an intracellular, perinuclear compartment by indirect immunofluorescence localization (figure 1c). Primary osteoblasts, UMR and ROS/Cx45 cells showed a similar immunostaining pattern for Cx46. The perinuclear compartment containing Cx46 was found to be an aspect of TGN as determined by co-localization TGN-38 [9]. Neither endocytosed Lucifer yellow nor ER membranes labeled with rhodamine-ConA showed significant co-localization with Cx46 [9]. Also, using immunofluorescence, we cannot rule out the presence of low, undetectable levels of Cx46 at the cell surface. For instance, a surface pool of Cx43 in communication-defective cell lines that is not visualized by immunofluorescence can be readily detected biochemically by surface biotinylation [21,22].

While lens expresses Cx50, we found that ROS cells did not. Thus, one possible concern was that Cx46 was trapped in the TGN due to an essential interaction with Cx50. However, channels composed of Cx46 alone have been observed in transfected Xenopus oocytes [14,23,24]. To test whether mammalian cells were capable of forming gap junction channels consisting of Cx46 alone, we transfected Hela cells with rat Cx46 cDNA (Hela/Cx46). As seen by immunofluorescence, Cx46 was found to be present at the cell surface (figure 1d). The formation of functional gap junction channels by Hela/Cx46 cells was confirmed by the ability of these cells to transfer microinjected Lucifer yellow between adjacent cells [9].

Connexin expression was also biochemically examined. Cell homogenates were solubilized in 1% Triton X-100 for 30 min at 4°C, then centrifuged at 100,000g for 30 min to obtain Triton X-100 soluble supernatant (sol) and insoluble pellet (ppt) fractions, which were analyzed by immunoblot (Figure 2).

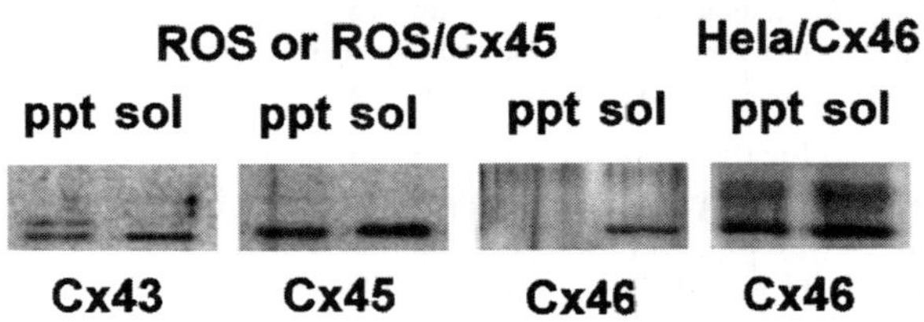


**Figure 2.** Triton X-100 solubility of connexins. ROS (Cx43,Cx46) , ROS/Cx45 (Cx45) or Hela/Cx46 cells were harvested and then solubilized in 1% Triton X-100 at 4°C as described in the text. Note the presence of two Cx46 bands in the sample obtained from Hela/Cx46 cells, as opposed to ROS cells which only show the 53 kD form. Similar results were obtained from primary osteoblastic cells and UMR cells (not shown).

In good agreement with previous studies [21], we found that there was a correlation between the appearance of a Triton X-100 insoluble fraction and immunolocalization of a given connexin at the cell surface. In contrast, virtually all of the Cx46 associated from ROS cells was soluble in 1% Triton X-100 at 4°C. However, Cx46 analyzed from Hela/Cx46 cells showed a significant Triton X-100 insoluble fraction. Thus, Triton X-100 insolubility appears to be a good marker for the assembly of Cx46 into gap junction channels. Further work will be required to understand the biochemical basis for differential connexin detergent solubility.

Another aspect of the biochemistry of Cx46 in osteoblasts can be seen in figure 2. Cx46 expressed by Hela/Cx46 cells shows two dominant bands, corresponding to Mr of 53 kD and 68 kD. This is similar to lens, where the 68 kD band corresponds to a phosphorylated form of Cx46. Other post-translational modifications may also be involved in generating the 68 kD form of Cx46 [17,24,25]. In contrast, Cx46 isolated from osteoblastic cells shows only a single band, in other words, it does not appear to be phosphorylated. Since most connexin phosphorylation occurs at the cell surface [21,22,26], this may reflect the lack of Cx46 assembly into gap junction channels. However, Cx46 phosphorylation in a secretory compartment may actually be a prerequisite for transport to the plasma membrane and/or gap junction assembly. Consistent with this notion, Cx43 phosphorylation has been found to occur at the Golgi apparatus [27].

Since the TGN has been implicated as the site for Cx43 assembly into gap junction hemichannels, we examined the state of Cx46 assembly in ROS cells, Hela/Cx46 cells, and rat lens. Connexins were solubilized in 1% Triton X-100 at 4°C as described above, and then monomers were resolved from oligomers by sucrose gradient velocity sedimentation [6,8,9,18]. As summarized in table I, all of the Cx46 associated with ROS cells was found in to be in a monomeric form. In contrast, 40-60% of the Cx46 solubilized from either lens or Hela/Cx46 cells was assembled into oligomers. This indicates that Cx46 oligomers are stable in 1% Triton X-100 at 4°C. Also, since Hela/Cx46 cells assembled Cx46 into oligomers, Cx50 is not required for oligomerization. However, Cx50 may help stabilize Cx46 oligomers, since lens showed nearly 50% more Cx46 in the oligomeric form than Hela/Cx46 cells.

**Table I**

| *Cell type* | *Cx* | *Post-translational processing?* | *TX-100 ppt fraction?* | *% oligomer* |
|---|---|---|---|---|
| ROS | Cx43 | Y | Y | 21% |
| ROS | Cx46 | N | N | 0% |
| Hela/Cx46 | Cx46 | Y | Y | 38% |
| lens | Cx46 | Y | Y | 60% |

**Summary of biochemical data on Cx43 and Cx46.** ROS, Hela/Cx46 or lens cells were analyzed as described in the text. The appearance of multiple bands detectable by immunoblot was taken to be evidence of post-translational modification (see figure 2). The percent oligomer was obtained from sucrose gradient velocity sedimentation analysis of the Triton X-100 soluble pool for a given connexin [9].

## 3. Future directions

Cx46 associated with cultured osteoblasts was in a monomeric form. Since Cx46 monomers are present in the TGN, this means that they were neither retained in the ER nor sorted to an early degradative pathway as a result of protein-folding proofreading mechanisms. This, combined with our observation that monensin inhibits Cx43 oligomerization [9] suggests that hemichannel assembly occurs at an aspect of the TGN, at least for these alpha connexins. Whether this is the case for the other connexins remains to be determined. Also, the construction of Cx43/Cx46 chimeras should enable determination of regions of these connexin molecules important for transport and assembly.

Why do osteoblasts express an intracellular pool of Cx46? An appealing idea is that this represents a stimulatable secretory compartment. The best characterized example of this is pre-term myometrium, which has an intracellular pool of Cx43 that is rapidly mobilized upon the onset of labor to enable electrical coupling [28-30]. A major goal is to find culture conditions that create an environment where Cx46 is assembled into gap junctions by osteoblasts. Not only should this shed light on biological roles for Cx46 in osteoblasts, but culture systems showing differential expression of Cx46 in the same cell type would also facilitate studies on mechanisms controlling connexin assembly.

## References


[1] Nicholson, S.M. and Bruzzone, R. Gap junctions: Getting the message through, *Current Biology*, 7 (1997) R340-R344.

[2] Bruzzone, R., White, T.W. and Goodenough, D.A. The cellular internet - on-line with connexins, *Bioessays*, 18 (1996) 709-718.

[3] Goodenough, D.A., Goliger, J.A. and Paul, D.L. Connexins, connexons and intercellular communication, *Ann. Rev. Biochem.* 65 (1996) 475-502.

[4] Kumar, N.M. and Gilula, N.B. The gap junction communication channel, *Cell*, 84 (1996) 381-388.

[5] Sosinsky, G.E. Molecular organization of gap junction membrane channels., *Journal of Bioenergetics & Biomembranes*, 28 (1996) 297-309.

[6] Falk, M.M., Buehler, L.K., Kumar, N.M. and Gilula, N.B. Cell-free synthesis and assembly of connexins into functional gap junction membrane channels, *EMBO J.* 16 (1997) 2703-2716.

[7] Kumar, N.M., Friend, D.S. and Gilula, N.B. Synthesis and assembly of human β1 gap junctions in BHK cells by DNA transfection with the human β1 cDNA, *J. Cell Sci.* 108 (1995) 3725-3734.

[8] Musil, L.S. and Goodenough, D.A. Multisubunit assembly of an integral plasma membrane channel protein, gap junction connexin43, occurs after exit from the ER, *Cell*, 74 (1993) 1065-1077.

[9] Koval, M., Harley, J.E., Hick, E. and Steinberg, T.H. Connexin46 is retained as monomers in a trans-Golgi compartment of osteoblastic cells, *J. Cell Biol.* 137 (1997) 847-857.
[10] Donaldson, P.J., Dong, Y., Roos, M., Green, C., Goodenough, D.A. and Kistler, J. Changes in lens connexin expression lead to increased gap junctional voltage dependence and conductance, *Amer. J. Phys.* 269 (1995) C590-600.
[11] Guerrier, A., Fonlupt, P., Morand, I., Rabilloud, R., Audebet, C., Krutovskikh, V., Gros, D., Rousset, B. and Munari-Silem, Y. Gap junctions and cell polarity: connexin32 and connexin43 expressed in polarized thyroid epithelial cells assemble into separate gap junctions, which are located in distinct regions of the lateral plasma membrane domain, *J. Cell Sci.* 108 (1995) 2609-2617.
[12] Risek, B., Klier, F.G., Phillips, A., Hahn, D.W. and Gilula, N.B. Gap junction regulation in the uterus and ovaries of immature rats by estrogen and progesterone, *J. Cell Sci.* 108 (1995) 1017-1032.
[13] Elfgang, C., Eckert, R., Lichtenberg-Frate, H., Butterweck, A., Traub, O., Klein, R.A., Hulser, D.F. and Willecke, K. Specific permeability and selective formation of gap junction channels in connexin-transfected HeLa cells, *J. Cell Biol.* 129 (1995) 805-817.
[14] White, T.W., Bruzzone, R., Wolfram, S., Paul, D.L. and Goodenough, D.A. Selective interactions among the multiple connexin proteins expressed in the vertebrate lens: the second extracellular domain is a determinant of compatibility between connexins, *J. Cell Biol.* 125 (1994) 879-892.
[15] Koval, M., Geist, S.T., Westphale, E.M., Kemendy, A.E., Civitelli, R., Beyer, E.C. and Steinberg, T.H. Transfected connexin45 alters gap junction permeability in cells expressing endogenous connexin43, *J. Cell Biol.* 130 (1995) 987-995.
[16] Musil, L.S., Beyer, E.C. and Goodenough, D.A. Expression of the gap junction protein connexin43 in embryonic chick lens: molecular cloning, ultrastructural localization, and post-translational phosphorylation, *J. Membrane Biol.* 116 (1990) 163-175.
[17] Konig, N. and Zampighi, G.A. Purification of bovine lens cell-to-cell channels composed of Connexin44 and Connexin50, *J. Cell Sci.* 108 (1995) 3091-3098.
[18] Jiang, J.X. and Goodenough, D.A. Heteromeric connexons in lens gap junction channels, *Proc. Nat. Acad. Sci. USA*, 93 (1996) 1287-1291.
[19] Steinberg, T.H., Civitelli, R., Geist, S.T., Robertson, A.J., Hick, E., Veenstra, R.D., Wang, H.Z., Warlow, P.M., Westphale, E.M., Laing, J.G. and Beyer, E.C. Connexin43 and connexin45 form gap junctions with different molecular permeabilities in osteoblastic cells, *EMBO J.* 13 (1994) 744-750.
[20] Civitelli, R., Beyer, E.C., Warlow, P.M., Robertson, A.J., Geist, S.T. and Steinberg, T.H. Connexin43 mediates direct intercellular communication in human osteoblastic cell networks, *J. Clin. Invest.* 91 (1993) 1888-1896.
[21] Musil, L.S. and Goodenough, D.A. Biochemical analysis of connexin43 intracellular transport, phosphorylation, and assembly into gap junctional plaques, *J. Cell Biol.* 115 (1991) 1357-1374.
[22] Musil, L.S., Cunningham, B.A., Edelman, G.M and Goodenough, D.A. Differential phosphorylation of the gap junction protein connexin43 in junctional communication-competent and -deficient cell lines, *J. Cell Biol.* 111 (1990) Pt 1):2077-88.
[23] Paul, D.L., Ebihara, L., Takemoto, L.J., Swenson, K.I. and Goodenough, D.A. Connexin46, a novel lens gap junction protein, induces voltage-gated currents in nonjunctional plasma membrane of Xenopus oocytes, *J. Cell Biol.* 115 (1991) 1077-1089.
[24] Gupta, V.K., Berthoud, V.M., Atal, N., Jarillo, J.A., Barrio, L.C. and Beyer, E.C. Bovine connexin44, a lens gap junction protein: molecular cloning, immunologic characterization, and functional expression, *Investigative Ophthalmology & Visual Science*, 35 (1994) 3747-3758.
[25] Jiang, J.X., Paul, D.L. and Goodenough, D.A. Posttranslational phosphorylation of lens fiber connexin46: a slow occurrence, *Investigative Ophthalmology & Visual Science*, 34 (1993) 3558-3565.
[26] Lampe, P.D. Analyzing phorbol ester effects on gap junctional communication: a dramatic inhibition of assembly, *J. Cell Biol.* 127 (1994) 1895-905.
[27] Laird, D.W., Castillo, M. and Kasprzak, L. Gap junction turnover, intracellular trafficking, and phosphorylation of connexin43 in brefeldin A-treated rat mammary tumor cells, *J. Cell Biol.* 131 (1995) 1193-1203.
[28] Hendrix, E.M., Myatt, L., Sellers, S., Russell, P.T. and Larsen, W.J. Steroid hormone regulation of rat myometrial gap junction formation: effects on cx43 levels and trafficking, *Biology of Reproduction*, 52 (1995) 547-560.
[29] Chow, L. and Lye, S.J. Expression of the gap junction protein connexin-43 is increased in the human myometrium toward term and with the onset of labor, *Amer. J. Obst. Gyn.* 170 (1994) 788-795.
[30] Balducci, J., Risek, B., Gilula, N.B., Hand, A., Egan, J.F. and Vintzileos, A.M. Gap junction formation in human myometrium: a key to preterm labor? *Amer. J. Obst. Gyn.* 168 (1993) 1609-1615.

*Gap Junctions*
*R. Werner (Ed.)*
*IOS Press, 1998*



# Targeting Connexin-Aequorin Chimerae to the Gap Junction

P.E.M. Martin[1], C. Castro[2], C.H. George[1], J. Capel[2], A.K. Campbell[1], A. Revilla[2], L.C. Barrio[2] and W.H. Evans[1]

[1]*Department of Medical Biochemistry, University of Wales College of Medicine, Heath Park, Cardiff, CF4 4XN, Wales, United Kingdom*

[2]*Departamento de Investigacion, Servicio de Neurolog'a Experimental, Hospital "Ramon y Cajal", 28034-Madrid, Spain*

**Abstract.** Chimeric proteins comprised of gap junction connexins (Cx) 26, 32 and 43 and aequorin (Aeq), a calcium indicator, were constructed by fusing Aeq to the carboxyl terminus of Cx. These recombinant probes were constructed to monitor the $Ca^{2+}$ environment along the trafficking pathways followed by connexins in cells and in the vicinity of the gap junction. The ability of the aequorin and connexin components in the chimeras to function was analysed. With the exception of Cx26-Aeq, where activity was negligible, the chemiluminescent activity of the aequorin component of Cx32-Aeq and of Cx43-Aeq synthesised either *in vitro* or when expressed in transfected COS-1 cells was comparable to that of wild type Aeq. Immunofluorescent staining of COS-1 cells transfected with cDNA to the chimeras showed targeting to the plasma membrane and chimeric proteins of the correct size were expressed. Intercellular channel formation by Cx-Aeq chimeras alone and in combination with wild-type connexins was studied by dye transfer in stably transfected HeLa cell lines and by electrophysiological analysis in paired *Xenopus* oocytes. Functional gap junctions in stable HeLa cell lines expressing Cx43-Aeq was demonstrated by Lucifer yellow transfer. Oocyte pairs expressing Cx43-Aeq were electrically coupled but Cx26-Aeq and Cx32-Aeq oocyte pairs showed no detectable levels of coupling. Oocytes expressing Cx43-Aeq also formed heterotypic junctions when paired with cells expressing wild-type Cx43. The formation of heteromeric channels comprised of chimeric Cx32-Aeq and Cx43-Aeq and their respective parent wild-type subunits was inferred from the novel voltage gating properties of the junctional conductance. Therefore the formation of chimeric cell-to-cell channels was restricted to homomeric channels constructed of Cx43-Aeq, to heterotypic Cx43 and Cx43-Aeq channels, and to heteromeric combinations of Cx32-Aeq or Cx43-Aeq subunits paired to wild-type connexins. The results show that: a). the preservation of function of each of component of the chimeric protein is dependent on connexin subtype, in particular the length of the carboxyl terminal domains of Cx26, Cx32 and Cx43; b). Cx43-Aeq was successfully targeted to the gap junction and c). that heteromeric gap junction hemi- channels can form.

## 1. Introduction

Gap junction channels provide a direct pathway connecting the cytoplasm's of contacting cells. They allow the movement of ions and small molecules <1kD between neighbouring cells. This direct interchange of molecular information is important in many key biological events. e.g. tissue growth and differentiation [1] and co-ordination of cardiac and smooth muscle cells [2].

Calcium plays a central role in cellular communication. It is involved in channel gating where it induces conformational change of isolated gap junctions from open to closed states [3]. It is important to access the calcium levels in the vicinity of the gap junctions, at the plasma membrane and along the secretory pathway to ascertain the role of $Ca^{2+}$ in channel generation and operation as well as the control of signalling through the gap junction. To this end we have used the calcium sensitive photoprotein Aequorin. The use of aequorin to demonstrate the importance of free $Ca^{2+}$ in the gating of gap junction channels was first illustrated in 1975 by Rose and Loewenstein [4]. Recently, many recombinant chimeric proteins expressing aequorin, including proteins that target to various cellular organelles, have been used to monitor the $Ca^{2+}$ environment of e.g. endoplasmic reticulum [5] and nucleus [6].

Therefore, we constructed a series of chimeric proteins expressing aequorin fused to the carboxyl tail of Cx43, Cx32 and Cx26 and analysed their expression and functionality *in vitro* and *in vivo*. These chimeric proteins are being used to study the $Ca^{2+}$ environment conditioning gap junction biogenesis.

## 2. Results

### 2.1. Construction of Connexin-Aequorin chimera

The carboxyl terminus of Cx26, Cx32 and Cx43 was fused in frame to the amino terminus of aequorin by a 2 step PCR procedure [5]. The PCR products were cloned into the TA cloning plasmid PCR3 (Invitrogen) enabling analysis *in vitro* and *in vivo* by means of the T7 and cytomegalovirus immediate early (CMV) promoters respectively. The constructs were fully sequenced to ensure no mutations were present.

### 2.2. *In vitro* expression and aequorin activity

Expression of aequorin was determined *in vitro* by coupled *in vitro* transcription and translation assays (Promega) followed by reconstitution of aequorin with coelenterazine [5,6]. Chimeric protein of the predicted size were synthesised (Cx26-Aeq 47 kD, Cx32-Aeq 53 kD and Cx43-Aeq 64 kD). Aequorin activity from Cx43-Aeq and Cx32-Aeq produced levels of chemiluminescence comparable to Wt-Aeq (Table 1).

**Table 1: Analysis of the effect of connexin on aequorin activity**

| Construct | % Aequorin activity |
|---|---|
| Cx26-Aeq | 1.31 ± 0.08 |
| Cx32-Aeq | 80 ± 10.5 |
| Cx43-Aeq | 100 ± 4.51 |
| Cx26-Aeq (-Cx26) | 62± 1.53 |
| Cx32-Aeq (-Cx32) | 78 ± 6.43 |
| Aeq-WT | 100 |

Total aequorin activity was measured as chemiluminescent counts. Results are the mean of triplicate experiments ± SEM and are expressed as a % aequorin activity compared to Aeq-WT activity.

However, levels of aequorin activity from Cx26-Aeq were markedly reduced. To ascertain if Cx26 accounted for this diminished aequorin activity, aequorin was PCR amplified from Cx26-Aeq and Cx32-Aeq plasmids using a forward primer containing the T7 promoter and the first 15 bp of the Aeq gene. Aequorin expression from the product was analysed following translation in the TNT system.

These experiments indicated that the addition of Cx26 to the amino terminus of aequorin inhibits its activity *in vitro*.

### 2.3 *In vivo* expression and aequorin activity

To study the expression of the chimeric proteins *in vivo* the constructs were transfected into COS-1 cells. Forty-eight hours post transfection protein expression was analysed by Western blot analysis using an anti-aequorin antibody (Figure 1). Protein of the predicted size was expressed in the transfected cells. Measurement of Aequorin activity from the cell extracts [5, 6] showed that Cx43-Aeq and Cx32-Aeq had comparable chemiluminescent activity to WT-Aeq. However, despite similar expression levels being achieved by the cells expressing Cx26-Aeq there was a 90% reduction in the level of aequorin activity similar to that observed *in vitro* (data not shown). These results show that connexins differ in their ability to retain the activity of the reporter group with Cx43-Aeq>Cx32-Aeq>Cx26-Aeq.

Immunolocalisation studies in COS-1 cells showed that Cx43-Aeq and Cx32-Aeq both target to the plasma membrane (Figure 2). As expected partner colocalisation with an anti-aequorin antibody was also evident. Although staining was evident at the plamsa

**Figure 1: Western Blot analysis of Chimeric proteins expressed in COS-1 cells**

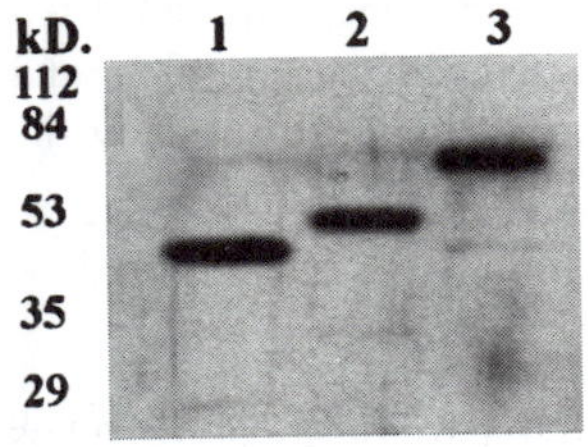


Blots were probed with antibodies to aequorin. Lane 1: Cx26-Aeq, lane 2: Cx32-Aeq, lane 3: Cx43-Aeq.

**Figure 2: Immunolocalisation of Cx32-Aeq and Cx43-Aeq in transiently transfected COS-1 cells.**

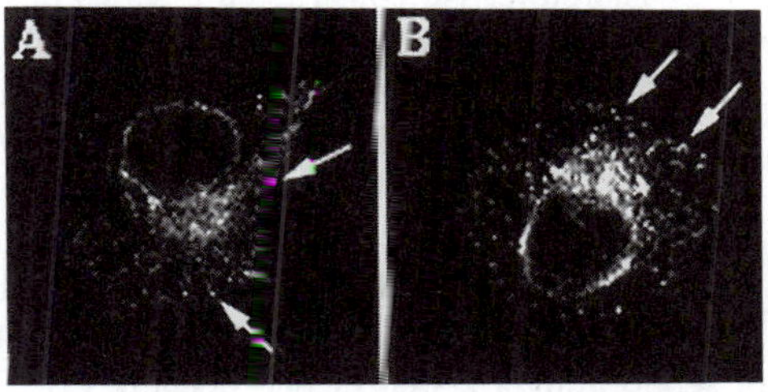


A] Cx32-Aeq, B] Cx43-Aeq stained with monoclonal antibodies against Cx32 or Cx43 respectively.

membrane, the chimeric proteins were also located in intercellular locations, especially the intermediate/Golgi compartment.

### 2.4 Functional analysis of chimeric Cx-Aeq in stable cell lines.

Stable HeLa cell lines expressing the chimeric proteins were established. The ability to form functional cell-cell channels was assessed by Lucifer yellow transfer [8] and aequorin activity was monitored (Table 2).

**Table 2: Dye transfer and aequorin activity in HeLa cells.**

| Cell type | % cells coupled | Aequorin Activity |
|---|---|---|
| Wild type | 1.8 (55) | 43 ± 3 |
| Cx43-Aeq [D7] | 74 (69) | 69803± 2782 |
| Cx43-Aeq [E9] | 59 (68) | 12792± 2072 |

The results are expressed as the % cells transferring dye to 10 or more neighbouring cells. aequorin activity is expressed as relative photon counts obtained from $4 \times 10^5$ cells ± SEM. (n): total number of cells injected

In cell line D7, showing high aequorin activity, 74% of cells transferred Lucifer yellow to more than 10 neighbouring cells and for cell line E9, with low aequorin activity, 60% of cells transferred the dye to the same number of cells. In non-transfected HeLa cells <2% were able to transfer dye. This shows that the ability of the cells to form functional gap junctions and the level of aequorin activity are closely linked.

### 2.5 Electrophysiological properties of chimeric connexons in paired oocytes.

The electrophysiological characteristics of the intercellular communication channels formed by the chimeric proteins were analysed in the paired *Xenopus* oocyte system [9]. The ability of the chimeric proteins to form functional channels alone (i.e. homomeric) or in combination with wtCx (i.e. heteromeric and heterotypic) was analysed.

#### 2.5.1 Homomeric/heterotypic channel formation.

The formation of homotypic channels was assessed by injection of each oocyte in a pair with either wtCx or Cx-Aeq and with an antisense oligonucleotide to abrogate endogenous expression of Cx38 by oocytes [9]. Cx43-Aeq forms homomeric channels at reduced efficiency of Cx43 (Table 3), whereas Cx32-Aeq and Cx26-Aeq have lost their channel forming ability. The ability to form homomeric /heterotypic channels was assessed by injection of one oocyte with wtCx and the other with Cx-Aeq. Cx43-Aeq coupled more efficiently with wtCx43 than with Cx43-Aeq (Table 3) suggesting that heterotypic channel formation is favoured by this chimeric protein. To characterise further the intercellular channels comprised of Cx43-Aeq subunits their voltage gating properties were compared to those of wtCx43 (Figure 4). Cx43 and Cx43-Aeq have a common maximal junctional conductance (gj) at Vj=0. Similarly to wtCx43, the gj

**Table 3: Junctional conductance developed by homotypic and heterotypic Cx43 oocyte pairs**

| Junction type | Junctional conductance (μS) |
|---|---|
| Cx43 : Cx43 | 35 ± 8.4 (6) |
| Cx43-Aeq: Cx43-Aeq | 0.34 ± 0.3 (9) |
| Cx43 : Cx43-Aeq | 1 ± 0.6 (8) |

Typical results are shown. Conductance values are in μS of the mean ± SD of the number of pairs analysed (n).

**Figure 4: Voltage gating properties of Cx43-Aeq channels**

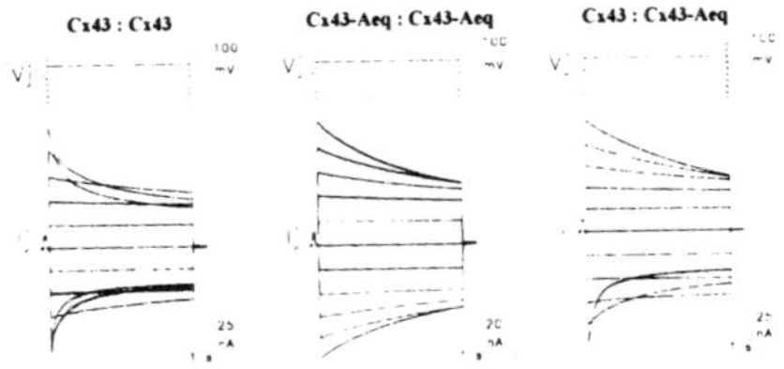


Sample records of junctional currents (Ij) induced by a transjunctional voltage (Vj) of 5 second steps were applied in increments of 20mV up to ± 100mV from a common holding potential of 40mV.

reduction of Cx43-Aeq channels is symmetrical in response to positive and negative Vj. However, Cx43 and Cx43-Aeq display different kinetic properties with Cx43-Aeq showing markedly slower time courses. When the voltage gating characteristics of the Cx43:Cx43-Aeq junctions were analysed (Figure 4) it was evident that the junctions responded asymmetrically to transjunctional voltage suggesting there are two independent Vj gates. Cx43-Aeq has the same polarity of closing as wtCx43 but it responds more slowly to transjunctional voltage. In summary, the Vj dependence of these heterotypic junctions largely reflects the gating properties of each hemichannel in their homomeric junctions.

### 2.5.2 Heteromeric/heterotypic channel formation.

The ability to form heteromeric channels was assessed by injecting one oocyte with both wtCx and Cx-Aeq cRNA and the other with both wtCx and Cx-Aeq or with either wtCx or Cx-Aeq cRNA. Cx32/Cx32Aeq expressing pairs developed lower levels of coupling than that obtained with wtCx (Table 4) and this reduction was less in pairs where only one oocyte expressed both cRNAs. This suggests that channel forming ability of Cx32-Aeq is rescued by oligomerisation of chimeric subunits with Cx32 and this conclusion is further supported when the gating properties of the junctions formed by Cx32/Cx32-Aeq and Cx32 pairs are analysed (Figure 5). The junctions formed react asymmetrically to Vj steps of opposite polarity and induce currents that are less voltage sensitive and display slower kinetics than homomeric Cx32 hemichannels. This asymmetric behaviour supports the idea that there is heterogeneity in Cx composition of the hemichannels and that co-oligomerisation to generate heteromeric Cx32/Cx32-Aeq hemichannels can occur.

## 3. Discussion

In the assembly of chimeric proteins into hemi-channels and gap junctions it is necessary that each partner protein retains a degree of biological activity. The chimeric proteins should operate as near to normal as possible, allowing the formation of functional gap junction channels at the plasma membrane together with the maintenance of the chemiluminescent properties of aequorin. The present work shows that the ability to meet these criteria differed among connexins, with Cx43>Cx32>Cx26. This suggested that the length of the carboxyl tail (118, 78 and 18 amino acids respectively) plays a key role in maintaining functionality of the chimeric proteins.

Cx43-Aeq is targeted to the plasma membrane where it forms homomeric channels as shown by Lucifer yellow transfer in stable HeLa cell lines and by electrophysiological

**Table 4: Junctional conductance developed by heteromeric Cx32 oocyte pairs.**

| Junction type | Conductance (μS) |
|---|---|
| Cx32-Cx32 | 33 ± 10.7 (6) |
| Cx32/Cx32A:Cx32/ Cx32A | 8.4 ± 9.4 (8) |
| Cx32/Cx32A:Cx32 | 36.2 ± 16.6 (8) |

The results are one of three experiments. Conductance values are μS of the mean ± sd of the number of pairs indicated (n).

**Figure 5: Voltage gating characteristics of Cx32-Aeq channels**

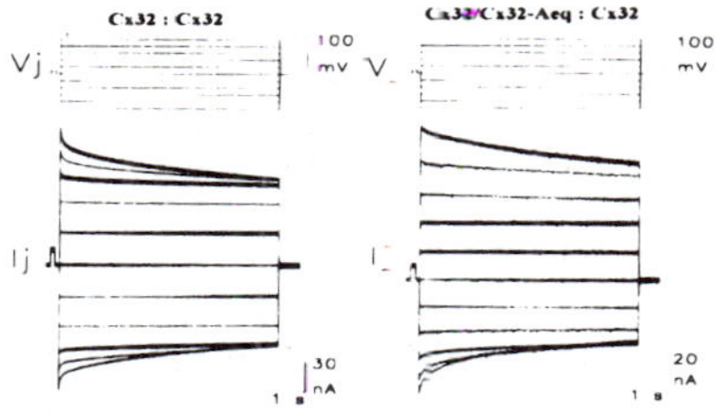


The conditions are as outlined in Figure 4.

measurements in paired oocytes. However, the chimeric connexons induced low levels of macroscopic conductance relative to wt Cx43. Cx32-Aeq was unable to form homotypic cell-cell communication channels in oocytes, but it is able to incorporate with wtCx32 to give electrophysiologically distinguishable heteromeric channels and has a dominant-negative effect on wtCx32 coupling. This suggests that co-oligomerisation with a competent Cx can rescue the complete loss of function of Cx32-Aeq The ability to form heterotypic channels may be crucially dependent on the stoichiometry of Cx32-Aeq wtCx32 subunits with the lowest ratio of chimeric protein inducing highest levels of coupling. Cx26-Aeq was unable to incorporate into functional gap junctional channels. Furthermore, while Cx43-Aeq and Cx32-Aeq showed aequorin levels similar to native aequorin Cx26-Aeq levels were markedly reduced.

Gap junction channels are gated by voltage, intracellular $Ca^{2+}$ and pH. Different domains of connexins have been implicated in regulating the gating mechanism. e.g. M2 [10], and the cytoplasmic loop [11] in voltage gating In contrast, the cytoplasmic carboxyl tail of Cx43 has been suggested not to play a major role in transjunctional voltage gating [12]. Our results show that attaching a large molecule of similar overall size to the carboxyl tail of Cx43 and Cx32 alters at least the voltage dependent gating mechanism as was shown by the novel gating properties displayed by Cx43-Aeq homotypic and Cx32-Aeq heteromeric channels.

In summary, the ability of chimeric connexons to produce functional gap junction channels and to give optimal levels of aequorin activity is governed by the nature of the Cx. We show that by attaching the reporter group to Cx43, containing the longest carboxyl tail, the positioning of the reporter group away from the plasma membrane enables correct processing of both partners of the chimeric protein. The constructs characterised in this work are useful for monitoring the $Ca^{2+}$ environment along the trafficking pathway followed by connexins.

## References

1. J.L. Brissette *et al.* (1994). *Proc. Natl. Acad. Sci* **91**, 6453-6547.
2. G.J Christ. (1996). *Circ. Res.* **79** , 631-646.
3. P. N. T Unwin and P. D. Ennis, (1984). *Nature* **307**, 609-613
4. B. Rose and W.R Loewenstein. (1975). *Nature* **254,** 250-252.
5. J. M.Kendall.*et al*, (1992). *Biochem. Biophys. Res. Comm.* **189** , 1008-1016.
6. M. N Badminton, *et al* (1995) *Expt Cell Res.* **216**, 236-243.
7. C. Elfgang,. *et al*, (1995). *J. Cell. Biol.* **129,** 805-817.
8. G.Dahl *et al.* (1987). *Science* **236,** 1290-1293.
9. L. C Barrio *et al*,. (1991). *Proc. Natl. Acad. Sci.* **88,** 8410-8414.
10. T. M Suchyna *et al*,. (1993). *Nature* **365, 847-849.**
11. X. Wang *et al*, (1996). *Pflugers Archiv European J.Phys.* **431**, 844-852.
12. B. Dunham *et al*,. (1992). *Circ. Res.* **70**, 1233-1243.

This work was supported by a MRC program grant (grant no: G-9305117) to W.H.E and by a FIS grant (no 95/0643) to L.C.B.

Gap Junctions
R. Werner (Ed.)
IOS Press, 1998

# Cx43 Gap Junction Proteolysis Involves Both the Lysosome and the Proteasome

James G. LAING, Peter N. TADROS and Eric C. BEYER
*Departments of Pediatrics, Cell Biology and Physiology, and Medicine*
*Washington University School of Medicine, St. Louis Mo. 63110*

Abstract: Intercellular communication may be modulated by the rather rapid turnover and degradation of gap junction proteins, since many connexins have half-lives of 1-3 hours. We have recently shown that Cx43 proteolysis involves ubiquitination and both the lysosome and the proteasome. In the present study, we examined the sites of Cx43 gap junction proteolysis in BWEM cells. Immunofluorescence studies showed that connexin43 accumulated at the cell surface in lactacystin treated cells, but in vesicles in cells treated with leupeptin. After treatment of cells with Brefeldin A to prevent transport of newly synthesized connexin43 polypeptides to the cell surface, immunoreactive connexin43 rapidly disappeared from appositional membranes. This loss of connexin43 staining was inhibited by co-treatment with the proteasomal inhibitor, lactacystin, or with the lysosomal inhibitor, leupeptin. This indicates that the proteasome and the lysosome are involved in the degradation of the Cx43 gap junction at the cell surface.

## 1. Introduction

In contrast to many other membrane proteins, most connexins are relatively dynamic. Pulse-chase experiments have shown that connexin43 (Cx43) and other gap junction proteins have half-lives of 1-3 hours in cultured cells [1, 2, 3]. Cells degrade proteins through two major systems, the proteasome and the lysosome. The proteasome is involved in the degradation of most cytosolic and nuclear proteins and some membrane proteins, and it removes misfolded or misaggregated proteins in the endoplasmic reticulum [4,5]. The lysosome degrades membrane proteins and extracellular materials that enter the cell via endocytosis [6]. These different pathways of proteolysis can be determined by their sensitivity to different inhibitors. The most specific inhibitor of the proteasome is the fungal metabolite, lactacystin [7]. Lysosomal proteolysis can be inhibited by cysteine protease inhibitors, including leupeptin [8]

## 2. Cx43 degradation is ubiquitin-dependent

Our initial studies of gap junction proteolysis showed the major involvement of the ubiquitin pathway in Cx43 degradation [9]. These experiments were performed primarily in E36 Chinese hamster ovary cells and a temperature sensitive mutant (ts20) which contains a thermolabile ubiquitin activating enzyme, E1. When grown at the restrictive temperature (which inactivated ubiquitination in the ts20 cells), immunoblotting and immunofluorescence experiments showed Cx43 accumulation in the ts20 cells as compared to the parental E36 cells [9]. Pulse-chase experiments in ts20 cells showed that Cx43 turned over more slowly at the restrictive temperature than at the permissive temperature [9,10].

## 3. Cx43 degradation involves both the lysosome and the proteasome

To determine the involvement of different proteases in Cx43 degradation, E36 and BWEM cells were treated with either proteasomal or lysosomal inhibitors; immunoblots showed Cx43 accumulation after inhibition of either pathway [9, 11]. Pulse-chase studies indicated that Cx43 degradation was slowed in the treated cells. In BWEM cells, proteasomal and lysosomal inhibitors had similar effects on Cx43 abundance; in contrast, in E36 cells, more Cx43 was found in cells treated with proteasomal inhibitors than those treated with lysosomal inhibitors [9]. This suggests a difference in the relative importance of these proteolytic pathways in different cell lines.

## 4. The proteasome and the lysosome both degrade Cx43 gap junctions

To examine cellular sites of Cx43 degradation, BWEM cells were treated with protease inhibitors, fixed, permeabilized, and Cx43 localization was examined by indirect immunofluorescence ([11] and Fig. 1). Cells treated with either lactacystin (Fig. 1B) or leupeptin (Fig. 1C) showed greater Cx43 staining than untreated control cells (Fig. 1A). Lactacystin-treated cells had more Cx43 at appositional membranes while leupeptin treatment increased Cx43 staining at many points within the cytoplasm.

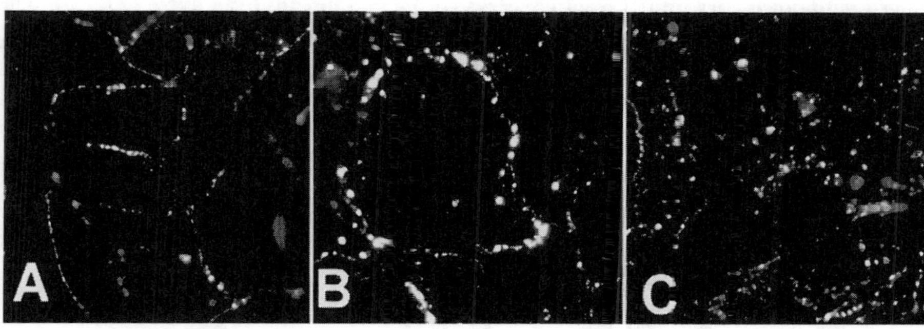


Fig 1. Immunofluorescent determination of the sites of Cx43 degradation. Cx43 was localized in cultures of BWEM cells that were untreated (A) or treated for 3 h with 20 μM lactacystin (B) or 100 μM leupeptin (C).

To selectively examine degradation of Cx43 contained within cell surface gap junctions, we treated BWEM cells with the drug Brefeldin A (BFA) which disrupts delivery of newly synthesized proteins to the cell surface. As has been shown in other cell lines [12,13], a 3 hour treatment with 2 μg/ml BFA markedly reduced the appositional membrane staining for Cx43 ([11] and Fig. 2A, B). However, simultaneous treatment of cells with BFA and the protease inhibitors lactacystin or leupeptin prevented much of the loss of immunoreactive Cx43 (Fig. 2 C,D). Cells treated with BFA and lactacystin exhibited staining at both the plasma membrane and in a cytoplasmic compartment (Fig. 2C). Cells treated with BFA and leupeptin showed a somewhat different staining pattern with Cx43 immunofluorescence at appositional membranes and in a cytoplasmic vesicular compartment (Fig. 2D).

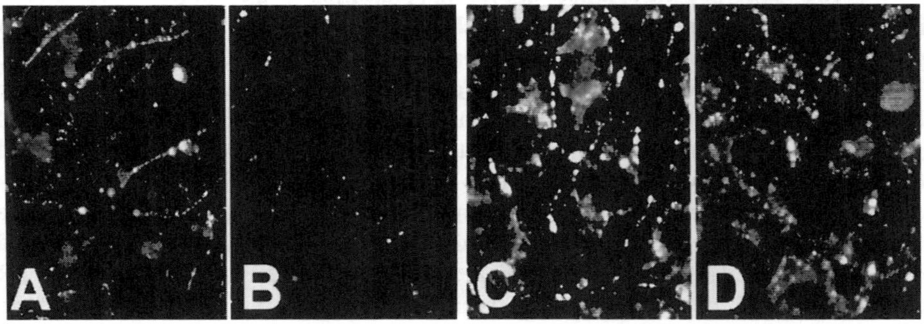


Figure 2. Immunofluorescent analysis of Cx43 gap junction degradation in cells treated with BFA and protease inhibitors. (A) untreated control BWEM cells, (B) BWEM cells treated with 2 $\mu$g/ml BFA for 3 hours, (C) BWEM cells simultaneously treated with BFA and 20 $\mu$M lactacystin, and (D) BWEM cells co-treated with 2 $\mu$g/ml BFA and 100 $\mu$M leupeptin were analyzed by immunofluorescent staining with antibodies directed against Cx43.

## 5. Pathways of Cx43 degradation

We have incorporated our data regarding possible sites of Cx43 degradation and some results of others into current concepts of the gap junction life cycle (Fig. 3). The biosynthesis of Cx43 has been recently reviewed [9] and can be briefly summarized. Connexin polypeptides are initially translated and inserted into the plasma membrane in the endoplasmic reticulum. Connexins oligomerize to form hemichannels in either the ER [14] or the Golgi [13]. Hemichannels are transported to the cell surface where they pair with hemichannels from a neighboring cell to form gap junction channels. These channels cluster in the plasma membrane to form gap junction plaques.

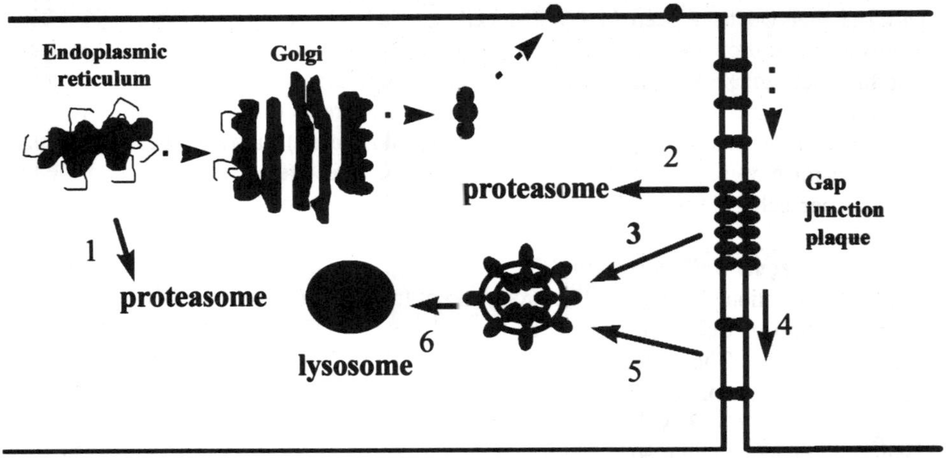


Figure 3. A model for the life cycle of Cx43 in cultured cells. Potential sites of proteolysis are numbered. Four different pathways of degradation of Cx43 would be consistent with parts of our data or previous studies (Proofreading, 1; Proteasomal proteolysis 2; Endocytosis/lysosomal degradation, 3→6; and Sequential proteasomal/lysosomal degradation, 4→5→6) and are discussed more fully in the text.

There likely are multiple sites of connexin proteolysis in the cell (numbered arrows in Fig. 3). There are four potential mechanisms for complete degradation of Cx43 polypeptides that would be consistent with aspects of our data and previous studies: A. "Proofreading" (1); B. Complete proteasomal degradation of a gap junction plaque (2); C. Endocytosis (3) and lysosomal degradation of a gap junction plaque (6); or D. partial proteasomal degradation (4) of a gap junction plaque followed by endocytosis (5) and lysosomal degradation (6). We will discuss each of these possibilities in turn.

***A. Proofreading (1).*** The proteasome has recently been implicated in serving a "quality control" function, facilitating the degradation of misfolded or mis-aggregated proteins from the ER [4, 5]. It would be reasonable to hypothesize a component of ER degradation for Cx43, since newly synthesized polypeptides must oligomerize to form hemichannels either in the ER [14] or the Golgi apparatus [13]. Our data indicate that some Cx43 accumulates in a cytoplasmic compartment (possibly ER/Golgi) in BWEM cells co-treated with BFA and the proteasomal inhibitor lactacystin (Fig. 2C). However, we do not know how much Cx43 proofreading occurs in untreated cells.

***B. Proteasomal degradation (2).*** Our recent data showing Cx43 accumulation at the cell surface even when outward trafficking is blocked ([11] and Fig. 3) demonstrate that the proteasome digests gap junctions at the plasma membrane. It is possible that the proteasome completely degrades Cx43 polypeptides derived from gap junction plaques. However, we are uncomfortable with the concept of exclusive involvement of the proteasome, since the plasma membrane would provide a barrier to degradation of transmembrane and extracellular domains, and since (unlike the ER) there is not an apparent mechanism to thread a protein out of the plasma membrane.

***C. Endocytosis/lysosomal degradation (3→6).*** A number of electron microscopy studies have visualized annular gap junctions or "gap junction-like" structures within vesicles (reviewed in [15]) suggesting that entire gap junctions might be internalized and then digested within a lysosome. This mechanism would be consistent with the staining patterns that we have seen in BWEM cells treated with leupeptin (Fig. 1C). This mechanism could operate in parallel with proteasomal degradation. However, we doubt that this is an exclusive or major pathway of gap junction degradation in all cells. Annular gap junctions are best visualized in cells that have precipitously degraded their gap junctions in response to a physiologic (e.g. granulosa cells) or traumatic (e.g. freshly isolated cardiac myocytes) event [16, 17]. These structures are not so easily identified in many other cells, despite what might be expected based on the short half life of the protein.

***D. Sequential proteasomal/lysosomal degradation (4→5→6).*** In a sequential model, gap junction plaques could be broken up at the plasma membrane (4), followed by internalization of dispersed connexons (by endocytosis, 5) and degradation within the lysosome (6). The proteasome might be involved in this initial partial proteolysis/dispersion event, since proteasomal inhibition leads to accumulation of immunoreactive Cx43 within large spots at appositional membranes (Fig. 2B). This sequential model is supported by the recent data of Fujimoto *et al.* [18]. They used freeze fracture immuno-electron microscopy to monitor the disappearance of gap junctions in the regenerating rat liver after partial hepatectomy. Their studies indicated that Cx32 gap junctions became smaller prior to their disappearance. These data suggest that a cytoplasmic protease catalyzes an initial step in the proteolysis of hepatic gap junctions. However, there are not yet any data to evaluate the role of the proteasome and the lysosome in this system.

Some of the data regarding degradation of cell surface gap junctions are most consistent with parallel operation of the proteasome and the endosome/lysosome (routes B,C), since the amount of Cx43 that accumulates differs with lysosomal and proteasomal inhibitors. However, we favor the sequential model as the most common mechanism of degradation of Cx43 gap junctions, since it incorporates both the proteasome and the lysosome and since it minimizes problems of protein topology and access of proteases across membranes.

These proteolytic processes are not mutually exclusive; some may operate simultaneously in the same cell. For example, cells would need both proof-reading of misformed connexins and degradation of cell surface gap junction plaques. Some pathways might be more prominent in certain cell types or conditions than in others as in comparing the abrupt degradation of all gap junctions in a cell with its steady-state turnover.

## References

[1] B.J. Darrow *et al.*, Expression of multiple connexins in cultured neonatal rat ventricular myocytes, *Circ. Res.* **76** (1995) 381-387.

[2] D.W. Laird *et al.*, Turnover and phosphorylation dynamics of connexin43 gap junction protein in cultured cardiac myocytes, *Biochem. J.* **273** (1991) 67-72.

[3] L.S. Musil *et al.*, Expression of the gap junction protein connexin43 in embryonic chick lens: molecular cloning, ultrastructural localization, and post-translational phosphorylation, *J. Membr. Biol.* **116** (1990) 163-175.

[4] J.L. Brodsky and A.A. McCracken, ER-associated and proteasome-mediated protein degradation: how two topologically restricted events came together. *Trends in Cell Biol.* **7** (1997) 151-155.

[5] R.R. Kopito, ER quality control: the cytoplasmic connection, *Cell* **88** (1997) 427-430.

[6] C.L. Ward *et al.*, Degradation of CFTR by the ubiquitin-proteasome pathway, *Cell* **83** (1995) 121-127.

[7] G. Fenteany *et al.*, Inhibition of proteasome activities and subunit-specific amino terminal modification by lactacystin. *Science* **268** (1995) 726-731.

[8] J. Bonfacino and R.D. Klausner, Degradation of proteins in the endoplasmic reticulum, in: "Cellular proteolytic systems", A.J. Ciechanover et al., Wiley-Liss, New York, (1994) pp. 163-180.

[9] J.G. Laing and E.C. Beyer, The gap junction protein connexin43 is degraded via the ubiquitin proteasome pathway, *J. Biol. Chem.* **270** (1995) 26399-26403.

[10] T.P. McGee *et al.*, Degradation of 3-hydroxy-3-methylglutaryl-CoA reductase in endoplasmic reticulum membranes is accelerated as a result of increased susceptibility to proteolysis. *J. Biol. Chem.* **271** (1996) 25630-25638.

[11] J.G. Laing *et al.*, Degradation of connexin43 gap junctions involves both the proteasome and the lysosome. *Exp. Cell Res.* in press.

[12] D.W. Laird *et al.*, Gap junction turnover, intracellular trafficking, and phosphorylation of connexin43 in Brefeldin A-treated rat mammary tumor cells, *J. Cell Biol.* **131** (1995) 1193-1203.

[13] L.S. Musil and D.A. Goodenough, Multisubunit assembly of an integral plasma membrane channel protein, gap junction connexin43, occurs after exit from the ER, *Cell* **74** (1993) 1065-1077 .

[14] M.M. Falk *et al.*, Membrane insertion of gap junction connexins: polytopic channel forming membrane proteins, *J. Cell Biol.* **127** (1994) 343-355.

[15] D.W. Laird, The life cycle of a connexin: gap junction formation, removal, and degradation. *J. Bioen. & Biomem.* **28** (1996) 311-318.

[16] W. J. Larsen and H.-N. Tung, Origin and fate of cytoplasmic gap junctions in rabbit granulosa cells. *Tissue Cell* **10** (1978) 585-598.

[17] N. J. Severs *et al.*, Fate of gap junctions in isolated adult mammalian cardiomyocytes. *Circ. Res.* **65** (1989) 22-42.

[18] K. Fujimoto *et al.*, Dynamics of connexins, E-cadherin and α-catenin on cell membranes during gap junction formation. *J. Cell Sci.* **110** (1997) 311-322.

*Gap Junctions*
*R. Werner (Ed.)*
*IOS Press, 1998*



# Functional Stabilization of Connexins by Inhibitors of Protein Synthesis

Linda S. Musil and Lori M. Roberts
*Vollum Institute, Oregon Health Sciences University, Portland, Oregon*

**Abstract**. The metabolic half-life of the "average" resident plasma membrane protein has been estimated to exceed 50 hours (1). In contrast, numerous studies have documented that connexins (with the exception of those expressed in mature lens fiber cells) have a $t_{1/2}$ of only 1.5-5 hours (2-4). In the case of connexin43 (Cx43), communication-competent, gap junction-rich cell types (e.g., NRK fibroblasts) as well as communication-deficient cells with few or no detectable gap junctions (L929; S180 cells) turn over Cx43 with equally rapid kinetics, suggesting that connexins are not stabilized by assembly into junctional plaques [2]. Although highly unusual for integral plasma membrane proteins, rapid degradation kinetics are a common characteristic of cytosolic and nuclear molecules (including cyclins and many transcription factors) that are involved in signal transduction. For several such regulatory molecules, a decrease in the rate of degradation is a physiologically important mechanism whereby their signalling function is prolonged or enhanced [5]. Given that connexins are regulatory molecules specialized for intercellular signalling, we were interested in determining whether the turnover rate of connexins could also be reduced and whether such metabolic stabilization would be accompanied by an upregulation of gap junction assembly and/or function. We show here that inhibitors of protein synthesis increase the metabolic stability of connexins in a variety of cells and enhance gap junction assembly and/or function in a cell type-specific manner. Regulation of the rate of connexin turnover may be a means of post-translationally altering gap junctional intercellular communication distinct from either modulation of assembly or gating of the junctional channel.

## 1. Results

Our strategy was based on reports demonstrating that the B$_1$ integrin subunit, an integral plasma membrane protein involved in cell-substrate signalling, has a relatively rapid turnover rate that is substantially reduced in the presence of the protein synthesis inhibitor cycloheximide by an uncharacterized mechanism [6][7]. To test whether inhibitors of protein synthesis also increase the metabolic stability of connexins, CHO cells were pulse-labeled with [$^{35}$S]methionine for 30 minutes and then chased for 7 hours in the presence of 50 uM cycloheximide (CHX), conditions under which protein synthesis is blocked but cell viability was not significantly affected. When compared to untreated controls in the same experiment, CHX consistently enhanced the amount of [$^{35}$S]methionine-labeled Cx43 that survived the chase period between 5- and 10- fold (Fig. 1A). The same concentration of CHX also increased (2-8 fold) the recovery of pulse-labeled Cx43 in other gap junctional communication-competent (NRK and S180L fibroblasts; primary lens epithelial cells) as well as in communication-deficient (L929; S180) cell lines. The ability of CHX to decrease the rate of degradation of Cx43 was time- and concentration- dependent, reversible, and could be mimicked by synthesis-blocking concentrations of the unrelated elongation inhibitors puromycin and anisomycin, suggesting that inhibition of protein synthesis <u>per se</u> rather than some other property of CHX is responsible for this effect. Notably, CHX also comparably increased the survival (~6-fold) of pulse-labeled connexin32 stably expressed in PC12 cell transfectants, demonstrating that this phenomenon is not confined to a single connexin type or to a particular immunoprecipitating antibody.

To be of functional significance, the CHX-stabilized connexins must assemble into active gap junctional channels. To address this issue, we used morphological (immunofluorescence microscopy) and functional (scrape-load dye transfer) techniques to examine the effect of CHX on Cx43 in CHO cells. Although CHOs are considered communication-competent, they possess relatively few Cx43-containing gap junctions (Fig. 2A) and show correspondingly modest levels of intercellular transfer of Lucifer Yellow under our basal conditions (Fig. 2E). These features belie the fact that CHO cells (especially when fed with fresh culture medium) synthesize substantial amounts of Cx43 which is assembled into connexons to an extent comparable to that observed in gap junction-rich and highly communication-competent NRK cells [8]. CHO cells therefore appear to be much less efficient than NRK cells in the subsequent stages of gap junctional plaque formation. Strikingly, exposure of CHO cells to 50 uM CHX for 7 hours significantly enhanced the level of gap junctional staining (Fig. 2B) and intercellular dye coupling (Fig. 2F) over untreated control cells (Fig. 2A and E) despite the fact that Cx43 synthesis was almost completely blocked. Brefeldin A is a fungal antibiotic that prevents the export of newly synthesized proteins, including Cx43, to the cell surface by blocking vesicular traffic from the endoplasmic reticulum and the Golgi stack [8]. CHX was also able to enhance (relative to cells treated with brefeldin A alone) gap junction formation and function in the presence of brefeldin A, although the level of junctions detectable under these conditions was less than in cells exposed to CHX alone (not shown). These data indicate that some (but not all) of the connexin molecules recruited for the formation of gap junctions in the presence of CHX originate from cellular compartments distal to the Golgi stack, possibly the plasma membrane itself.

The effect of CHX on gap junction formation and function varied between different cell types and correlated with the efficiency with which cells assembled gap junctions under basal conditions. All junctional communication-competent cells examined had more immunofluorescently detectable gap junctions after a 7 hour treatment with brefeldin A plus CHX than if incubated with brefeldin A alone, similar to what was observed in CHO cells. However, in the absence of brefeldin A, cells in which a relatively high percentage of pulse-labeled Cx43 becomes assembled into junctional plaques (e.g., NRK cells) possessed an equal or lesser number of gap junctions when exposed to CHX for 6-12 hours than if protein synthesis was allowed to continue. S180L fibroblasts, which assemble Cx43 into plaques

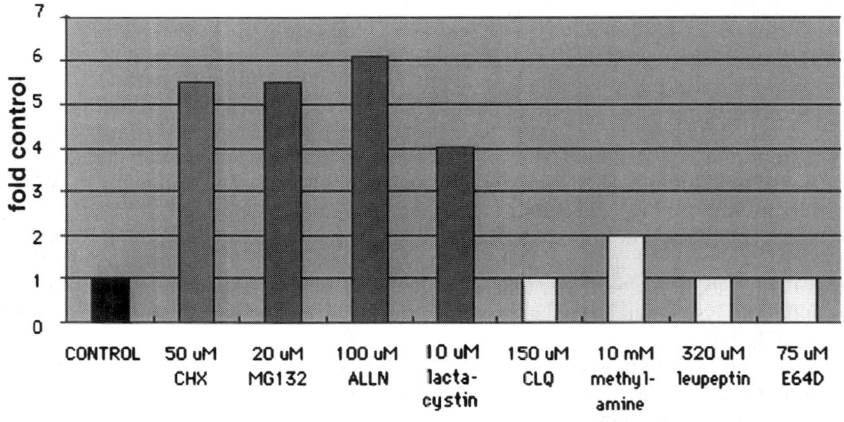


Figure 1. Effect of inhibitors of protein synthesis (CHX), the proteasome (MG132; ALLN; lactacystin), or lysosomal degradation (chloroquine; methylamine; leupeptin; E64D) on the metabolic stability of Cx43 in CHO cells. Confluent monolayers of CHO cells were pulse-labeled with [$^{35}$S]-methionine for 30 min and then chased for 7 h in the presence of excess unlabeled methionine and the indicated inhibitor. Cx43 was immunoprecipitated from SDS-boiled lysates of the labeled cells, analyzed on SDS-PAGE, and the amount of [$^{35}$S]met-Cx43 recovered quantitated on a phosphoimager and normalized to that obtained from untreated controls, in which ~5-9% of the pulse-labeled Cx43 remained after the 7 h chase. All cells were fed with fresh 10% fetal calf serum-containing medium 5 h before labeling to enhance the expression of Cx43.

less efficiently than NRK cells but more so than CHOs, showed more immunodetectable gap junctional plaques when incubated for 7 hours with 50 uM CHX than in its absence, but the accompanying increase in intercellular Lucifer Yellow transfer was not as striking as in CHX-treated CHO cells. These findings can be explained if the overall amount of gap junctions present in a cell is a function of both the total number of connexin molecules and the efficiency with which those molecules are incorporated into junctional plaques. In CHOs, the CHX-induced enhancement in the efficiency of gap junction assembly is enough to result in an increase in the total number of gap junctions despite the absence of new connexin production. In contrast, in NRKs and other cells in which gap junction assembly is relatively efficient in the absence of CHX, the level of gap junction formation is dictated predominantly by the amount of available connexin precursors such that inhibition of connexin synthesis eventually results in fewer gap junctions as pre-existing junctions are degraded and cannot be replaced. In such cells a CHX-induced increase in gap junction formation is detectable only under conditions (such as brefeldin A exposure) in which nascent connexin molecules are trapped within the ER and Golgi and therefore only post-Golgi Cx43 pools can participate in gap junction formation. Interestingly, protein synthesis inhibitors did not (either with or without brefeldin A) promote gap junction formation in cells with no or very low levels of detectable gap junctions such as S180 or L929 cells. Taken together, these findings suggest that inhibitors of protein synthesis can increase the efficiency with which communication-competent cells assemble gap junctional plaques but cannot change the phenotype of communication-deficient cells.

How do protein synthesis inhibitors increase the metabolic stability of connexins? To date, two mechanisms have been implicated in the turnover of connexins. Morphological data from several systems have indicated that lysosomes are involved in the proteolysis of gap junctional plaques, in keeping with the predominant role of this organelle in the degradation of integral plasma membrane proteins [9]. In addition, Laing and Beyer [10] have reported that the $t_{1/2}$ of Cx43 is increased by inhibitors of the proteasome, the multicatalytic protease complex responsible for the majority of ubiquitin-dependent degradation of short-lived cytosolic and nuclear proteins. Although it has not yet been demonstrated that connexins are direct substrates for proteasomal degradation, evidence has accumulated that the proteasome may play a role in the turnover of certain other integral membrane proteins [11][12]. To investigate whether CHX might be acting by interfering with either lysosomal or proteasomal degradation, we compared the effects on Cx43 of inhibitors of these pathways with those obtained with CHX (Fig. 1). Addition of any of four established inhibitors of lysosomal proteolysis (leupeptin, chloroquine, methylamine, E64D) to the chase medium of pulse-labeled CHO cells was unable to substantially increase the recovery of [$^{35}$S]methionine-Cx43 after a 7 hour chase when compared to untreated controls. In contrast, degradation of Cx43 in these cells was reproducibly reduced by inhibitors of proteasomal degradation, including the most specific blocker of the proteasome available, lactacystin [13]. These results are consistent with the findings of Laing and Beyer, who reported that Cx43 in CHOs is metabolically stabilized to a much greater extent by the proteasome inhibitor ALLN that by the lysosomal blocker primaquine [10]. When examined by immunofluorescence microscopy (Fig. 2C) and scrape-load dye transfer (Fig. 2G), we observed that a 7 hour treatment of CHO cells with any of the three proteasome inhibitors tested increased gap junctional staining and intercellular transfer of Lucifer Yellow to an extent equal to, or even greater than, that observed in the presence of CHX. In contrast, blocking lysosomal degradation with either chloroquine or leupeptin had no such effect (Fig. 2D, 2H).

Subsequent pulse-chase experiments demonstrated that proteasome inhibitors enhanced the metabolic stability of Cx43 and Cx32 in all cell types examined (not shown). [$^{35}$S]-methionine-labeled connexin degradation in some of these cells (NRK, S180L, and S180 fibroblasts) was (unlike in CHOs) also decreased by blockers of lysosomal proteolysis, often to an extent comparable to that obtained with CHX. Regardless of their effect on connexin turnover, however, lysosomal inhibitors were unable to substantially enhance either connexin localization in gap junctional plaques (as assessed by immunofluorescence microscopy) or gap junction-mediated intercellular communication (as determined by scrape-loaded Lucifer Yellow transfer) in any cell type examined, suggesting that undegraded connexin molecules accumulated intracellularly instead of participating in gap junction formation.

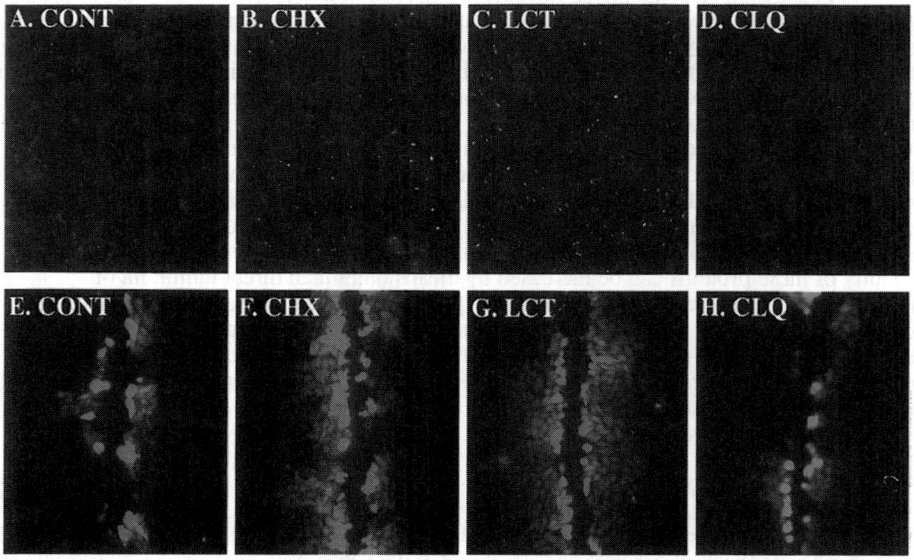


Figure 2. Effect of inhibitors of protein synthesis (50 uM cycloheximide), the proteasome (10 uM lactacystin), or lysosomal degradation (150 uM chloroquine) on the localization and function of Cx43 in CHO cells. Confluent monolayers of CHO cells were exposed to the indicated inhibitor for 7 h prior to: (A-D) immunofluorescence staining of Cx43 using a monospecific, affinity-purified anti-Cx43 antibody preparation, or (E-H) scrape-load analysis of intercellular transfer of Lucifer Yellow. For E-H, junction-impermeant rhodamine dextran served as a control for cell damage and remained confined to cells directly bordering the scrape. Cells were photographed 5 min after dye loading. A and E: untreated controls. All cells were fed with fresh 10% fetal calf serum-containing medium 5 h before further treatment to enhance Cx43 expression.

## 2. Discussion

In conclusion, the pulse-chase studies demonstrate that inhibitors of protein synthesis decrease the rate of degradation of newly synthesized connexin in a variety of cell types. In gap junctional communication-competent, but not in communication-deficient cells, these inhibitors can also increase the fraction of connexin molecules on the cell surface that are localized in functional gap junctional plaques as assessed by immunofluorescence microscopy and intercellular dye coupling. It seems quite likely, although has not yet been experimentally proven, that these two effects are interrelated; that is, that protein synthesis inhibitors upregulate gap junctions at least in part by increasing the period of time in which a connexin molecule can become assembled into, and function as, a gap junctional plaque before being degraded. From our current data we cannot, however, rule out that these agents might also be affecting connexins via mechanisms not directly related to inhibition of protein synthesis or to connexin degradation. The fact that inhibitors of the proteasome, but not lysosomal blockers, have an effect on connexin localization, metabolic stability, and function that is very similar to that obtained with CHX suggests that CHX may be acting by inhibiting the turnover of connexin molecules by the proteasome. Although CHX is not itself a general inhibitor of proteasomal function, protein synthesis inhibitors have been reported to reduce the degradation of several putative substrates of this pathway, including estrogen receptors [14] and p53 [15]. A possible explanation for this effect arises from the fact that proteins often require some type of modification, such as phosphorylation or a change in protein-protein association, before they can be ubiquitinated as a prerequisite for proteasomal degradation [5]. It is conceivable that CHX somehow interferes with such a modification, perhaps by reducing the levels of a short half-life protein required for the tagging of connexins for proteasomal destruction.

A central issue is the physiological significance of the ability of inhibitors of protein synthesis to upregulate gap junctions. Although not a common event, protein synthesis is greatly reduced in vivo in some pathophysiological situations such as starvation, viral infection, or exposure to certain (e.g., ricin; diphtheria ) toxins. It is conceivable that under such conditions connexins become metabolically stabilized as a means to maintain or even increase gap junction function. Alternatively or in addition, protein synthesis inhibitors could affect a step in connexin processing that is regulated in vivo quite differently. For instance, CHX may reduce the amount of a fast turnover protein required to target connexins for proteasomal degradation, whereas physiologically the function of this protein may be regulated by a post-translational modification such as phosphorylation. Regardless of the role of protein synthesis inhibition in connexin biology, our studies demonstrate that the stability of these proteins can be increased by means other than direct inhibitors of proteolysis and that such stabilization can (depending on the cell type) be associated with increased gap junction assembly and function. A major goal of future studies will be to determine whether the pathway activated experimentally by inhibition of protein synthesis or of proteasome-mediated degradation is used to upregulate gap junctions in response to other, more generally encountered stimuli.

## References

[1] J.F. Hare and K. Taylor. 1991. Mechanisms of plasma membrane protein degradation: recycling proteins are degraded more rapidly than those confined to the cell surface. Proc. Natl. Acad. Sci. **88**: 5902-5906.

[2] L.S. Musil, B.A. Cunningham, G.M. Edelman, and D.A. Goodenough. 1990. Differential phosphorylation of the gap junction protein connexin43 in junctional communication-competent and -deficient cell lines. J. Cell Biol. **111**:2077-2088.

[3] O. Traub, J. Look, R. Dermietzel, F. Brummer, D. Hulser, and K. Willecke. 1989. Comparative characterization of the 21-kD and 26-kD gap junction proteins in murine liver and cultured hepatocytes. J. Cell Biol. **108**:1039-1051.

[4] R. F. Fallon and D.A. Goodenough. 1981. Five-hour half-life of mouse liver gap-junction protein. J Cell Biol. **90**:521-526.

[5] H.L. Paul and P.A. Baeuerle 1996. Control of gene expression by proteolysis. Curr. Opin. Cell Biol. **8**:340-347.

[6] S. Vekeman, M. Jaspers and J.-J. Cassiman. 1993. Inhibition of the degradation of the precursor and of the mature B1 integrin subunit by different protein synthesis inhibitors and by ATP depletion. FEBS Lett. **327**:207-212.

[7] N.A. Hotchin, A. Gandarillas, and F.M. Watt. 1995. Regulation of cell surface B1 integrin levels during keratinocyte terminal differentiation. J. Cell Biol. **128**:1209-1219.

[8] L.S. Musil and D.A. Goodenough. 1993. Multisubunit assembly of an integral plasma membrane channel protein, gap junction connexin43, occurs after exit from the ER. Cell **74**:1065-1077.

[9] W.J. Larsen and H.N. Tung. 1978. Origin and fate of cytoplasmic gap junction vesicles in rabbit granulosa cells. Tissue Cell. **10**:585-598.

[10] J.G. Laing and E.C. Beyer. 1995. The gap junction protein connexin43 is degraded via the ubiquitin proteasome pathway. J. Biol. Chem. **270**:26399-26403.

[11] C. L. Ward, S. Omura, and R.R. Kopito. 1995. Degradation of CFTR by the ubiquitin-proteasome pathway. Cell **83**:121-127.

[12] S. Mori, K. Tanaka, S. Omura, and Y. Saito. 1995. Degradation process of ligand-stimulated platelet-derived growth factor B-receptor involves ubiquitin-proteasome proteolytic pathway. J. Biol. Chem. **270**:29447-29452.

[13] G.Fenteany, R.F. Standaert, W.S. Lane, S. Choi, E.J. Corey, and S.L. Schreiber. 1995. Inhibition of proteasome activities and subunit-specific amino-terminal threonine modification by lactacystin. Science **268**:726-731.

[14] M. Borras, I. Laios, A. El Khissiin, H-S Seo, F. Lempereur, N. Legros, and G. Leclercq. 1996. Estrogenic and antiestrogenic regulation of the half-life of covalently labeled estrogen receptor in MCF-7 breast cancer cells. J. Steriod Biochem. Molec. Biol. **57**: 3/4:203-213.

[15] N.C. Reich, M. Oren, and A.J. Levine. 1983. Two distinct mechanisms regulate the levels of a cellular tumor antigen, p53. Mol. Cell. Biol. **3**:2143-2150.

# Section 5:

# Biosynthesis of Gap Junction Channels

*Gap Junctions*
*R. Werner (Ed.)*
*IOS Press, 1998*



# Activation of Cx46 Hemichannels Increases the Rate of Gap Junction Formation in the Xenopus Paired-Oocyte Expression System

Derek L. Beahm and James E. Hall
*Department of Physiology and Biophysics, UC Irvine, Irvine, CA 92697-4560, USA*

**We are using the Xenopus paired-oocyte expression system to examine the relationship between connexon hemichannel activity and the rates of gap junction formation. To quantitate the precursor pool, we are exploiting the ability of cx46 to form functional ion conducting channels in the nonjunctional plasma membrane of the oocytes. The activity of cx46 hemichannels is under experimental control, thereby allowing us to examine the role of hemichannel gating in the process of gap junction formation. During the course of these experiments, we discovered that activation of cx46 hemichannels dramatically increases the rate of gap junction formation through an unknown mechanism. The rate of increase decays with single exponential kinetics to a junctional conductance level that is linearly correlated with the availability of hemichannels as measured by nonjunctional membrane conductance. We are currently investigating molecular mechanisms which could explain the observed phenomenon.**

## 1. Introduction... Gap Junction Formation Revisited

The extent of gap junctional communication between two cells is a function of the number, conductance, and open probability of individual channels. Hence, all the stages leading to the functional expression of a gap junction channel are the first potentially important control points for the regulation of intercellular communication. Unfortunately, the molecular mechanisms and intermediates involved in gap junction channel assembly are not well understood. We do know that gap junction channels assemble from precursors present in the plasma membrane [1-5]. Progress in identifying the protein trafficking pathways of cx43 demonstrates that oligomerization occurs prior to insertion into the plasma membrane [6,7]. The discovery of connexins that form conductive hemichannels in non-junctional membane also suggests that at least a fraction of plasma membrane-resident connexin polypeptide is in the form of hexameric assemblies [8-13]. So although direct proof is still unavailable, there is increasing evidence that hemichannels represent the immediate precursor from which gap junction channels form. Our goal is to understand if and how hemichannels assemble into gap junction channels, and to identify factors that directly regulate this process.

Even without definitive knowledge of the gap junction precursor, several factors influencing gap junction formation have been identified. Sequence-specific factors include regions in the two extracellular domains of the connexin polypeptide, which determine whether or not gap junctions can form at all [4,14,15]. In addition, there are external factors which influence the rate of gap junction formation. Of particular interest is the role of divalent cations in the process of gap junction formation. In the paired-oocyte system, where initial cell contact is controlled manually, the rate of cx32 homotypic gap junction formation is inversely related to the concentration of divalent cations in the external media

[5]. Divalent cations are also responsible for blocking hemichannel currents, so it would be interesting to determine if their effect on rates of gap junction formation are linked to their effects on hemichannel conductance.

Our inability to quantitatively assay the non-junctional precursor pool is a principal obstacle in understanding the mechanism and kinetics of gap junction channel assembly. We are currently exploiting the hemichannel forming properties of cx46 to examine the relationship between the hemichannel pool and gap junction formation, including how factors influencing hemichannel activity can affect rates of gap junction channel assembly. Our general technique is to express cx46 in oocytes, allow the accumulation of precursors to build up, pair the oocytes, and measure hemichannel and gap junctional conductances over time while subjecting the oocyte pair to conditions that are known to influence either hemichannel activity or gap junction formation.

Initially, experiments were designed to quantitate the relationship between cx46 hemichannel abundance and the rate of cx46 homotypic gap junction formation. The hemichannel pool size was periodically assayed by the magnitude of tail currents resulting at the end of a depolarizing voltage step and could subsequently be correlated with the actual number of hemichannels determined by freeze-fracture analysis (collaboration w/ Guido Zampighi). However, it became immediately clear that the pulse protocol originally used to quantitate hemichannel abundance dramatically influenced the rate of gap junction formation in a reproducible manner. This paper describes how hemichannel activity can influence the rate of gap junction formation and represents the identification of a new control point for the potential regulation of intercellular communication.

## 2. Cx46 Hemichannel Activation Increases the Rate of Gap Junction Formation

The magnitude of cx46 hemichannel current is voltage, time, and calcium dependent [16]. Under normal physiological conditions of high $[Ca^{++}]$, the channels open upon membrane depolarizations greater than -10mV. Reducing external calcium shifts the steady state activation curve towards more negative potentials and results in a small percentage of hemichannels being open at the holding potentials of -30mV to -40mV used in the following experiments.

Figure 1 illustrates the typical time course of cx46 gap junctional conductance increase in oocyte pairs before and after hemichannel activation by voltage (1A) or reduced external calcium concentration (1B) The rate of gap junction formation increases immediately after hemichannel activation and decays slowly back to its baseline rate. The kinetics follow a single exponential growth curve of the form $G(t)=G_{max}(1-\exp(-t/\tau))$. Subsequent voltage activation of hemichannels during the growth phase usually results in only a minor perturbation away from the modelled fit; however, in oocyte pairs expressing high levels of cx46, a second growth curve can be generated by reducing external calcium (data not shown).

To rule out non-specific effects of depolarization pulses, oocyte pairs are routinely

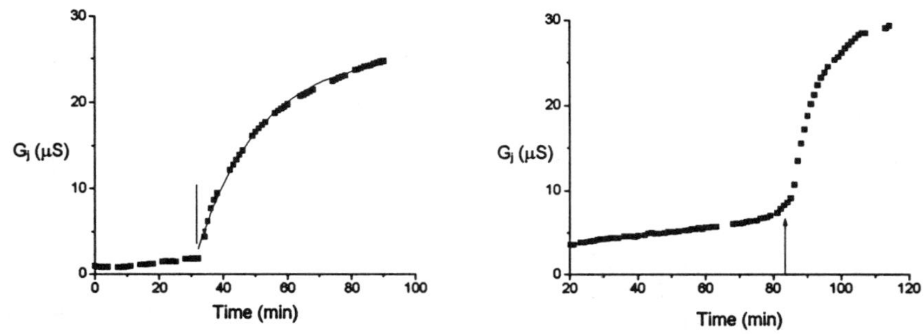


Figure 1. Activation of cx46 hemichannels by a +20mV pulse for 30 seconds (1A, arrow) or by reducing [Ca++] from 1.0mM to 0.5mM (1b arrow) results in a rapid increase in the rate of gap junction formation that decays with single exponential kinetics (solid line in 1A, see text) Oocyte pairs are injected with cx46 cRNA 24 hours prior to pairing and dual-TEVC is applied 2hrs after pairing.

subjected to large and/or long depolarizations in the presence of 1mM cobalt, which blocks hemichannels, at the beginning of each experiment. Under these conditions, depolarizations never produced a change in the rate of gap junction formation. Gap junction formation rates are also insensitive to reductions in external calcium if cobalt is maintained in the media. No junctional currents were seen in non-injected oocytes pairs or pairs in which only one of the oocytes expressed cx46.

## 3. Hemichannel Currents Must Reach a Threshold Level to Stimulate Gap Junction Channel Formation

There appears to be a minimum threshold level of hemichannel activation required to stimulate an increase in the rate of gap junction formation. The effect of increasing depolarization times, resulting in larger hemichannel whole cell currents, is shown in Figure 2A. Alternatively, a threshold level of hemichannel activation can also be observed by increasing the magnitude rather than the duration of a depolarization pulse (data not shown). For oocyte pairs expressing lower levels of cx46, as in Figure 2B, voltage activation of hemichannels does not result in large enough currents to stimulate gap junction channel formation unless the external calcium concentration is reduced prior to activation.

It appears that the threshold level of cx46 hemichannel current is similar for all oocyte pairs tested, regardless of the expression levels. Hence, the size of the hemichannel pool determines the type of stimulus required to initiate an increase in gap junction formation. In oocyte pairs expressing large amounts of cx46, threshold hemichannel currents are easily achieved by voltage activation in 1mM calcium or simply reducing calcium to 0.5mM, as seen in Figure 1A,B and Figure 2A. In lower expressing pairs, voltage activation in 0.5mM calcium, reducing calcium to 0.1mM, or voltage activation in 0.1mM calcium is required to cause an increase in the rate of gap junction formation, as seen in Figure 2B.

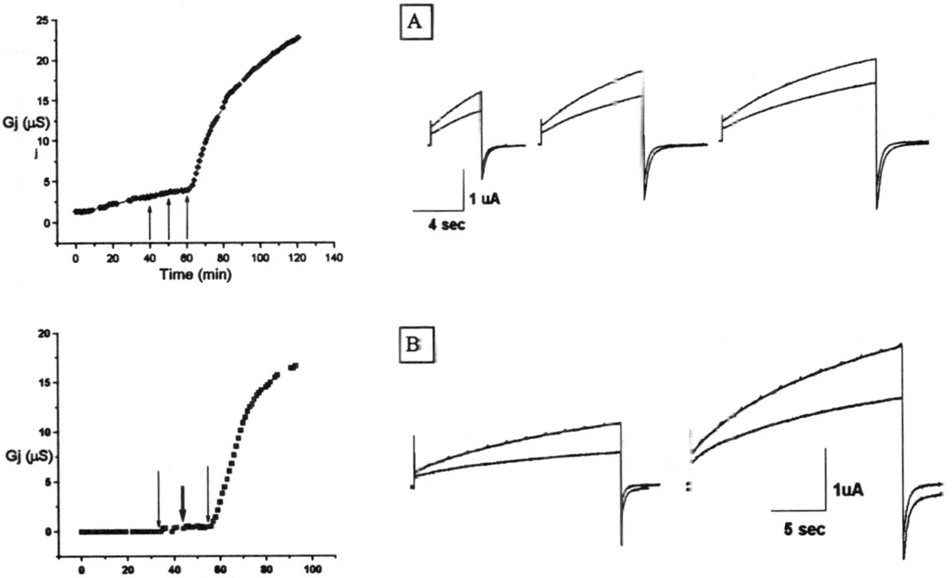


Figure 2. The rate of gap junction formation increases after a threshold level of hemichannel activation. A: Paired-oocytes were subjected to a +20mV depolarization (arrows) of increasing duration followed by a period of junctional current measurements. B: Paired-oocytes were subjected to the same +40mV, 20 second pulse before and after (thin arrows) the external calcium was reduced (thick arrow) from 1.0mM to 0.5mM. The raw current traces from both oocytes are superimposed for each event and shown on the right.

## 4. What Regulates the Extent and Duration of Growth in Junctional Conductance Following Hemichannel Activation?

Several factors may contribute to the kinetics of stimulated gap junction formation. Note that in Figure 2, the net increase in junctional conductance is different for the two oocyte pairs. Although the different pulse protocols used to stimulate gap junction formation in these examples did not allow quantitative comparison of cx46 expression levels, one can deduce from the hemichannel current traces obtained in 1mM $Ca^{++}$ that the pair which demonstrates a higher level of cx46 expression underwent a greater change in junctional conductance in response to hemichannel activation. In other experiments, whole-cell hemichannel conductance was measured under identical conditions by tail current analysis and was found, by linear regression (n=12, R=0.78), to be correlated with the magnitude of stimulated junctional conductance increase over a wide range (5-50uS). Other parameters that may affect the kinetics include the extent of hemichannel activation beyond the threshold level, the insertion rate of new hemichannels, and changes in the holding potential.

## 5. Conclusion

In conclusion, we have found that activation of cx46 hemichannels in the paired-oocyte expression system leads to a dramatic increase in the rate of gap junction channel formation. Hence, the extent of intercellular communication can be regulated at the level of hemichannel activity. The initial rate increase is greater than 10-fold that of the normal baseline rate and decays back to baseline with single exponential kinetics over a 10-45 minute period. There appears to be a threshold level of hemichannel current required to stimulate channel assembly, and this threshold is independent of cx46 expression levels. The total change in junctional conductance, however, is proportional to the size of the hemichannel pool.

The mechanism responsible for this effect is currently being investigated and may prove physiologically important in systems endogenously expressing conductive hemichannels. Possible mechanisms include an increase in cell contact area, an increase in insertion rate of new hemichannels, removal of steric hindrance [17], or an alteration in hemichannel affinity. A potential role for functional cx46 hemichannels in the lens is to increase cell contact area, thereby reducing extracellular volume which is a requirement to maintain clarity of the lens tissue.

### References

[1] Dahl G, Werner R, Levine E: Paired Oocytes: An Expression System For Cell-Cell Channels, in Anonymous*Gap Junctions*. 1988, pp 183-197

[2] Swenson KI, Jordan JR, Beyer EC, Paul DL: Formation of gap junctions by expression of connexins in *Xenopus* oocyte pairs. *Cell* 1989;57:145-155

[3] Werner R, Levine E, Rabadan-Diehl C, Dahl G: Formation of hybrid cell-cell channels. *Proc Natl Acad Sci USA* 1989;86:5380-5384

[4] Dahl G, Levine E, Rabadan-Diehl C, Werner R: Cell/cell channel formation involves disulfide exchange. *Eur J Biochem* 1991;197:141-144

[5] Dahl G, Werner R, Levine E, Rabadan-Diehl C: Mutational analysis of gap junction formation. *Biophys J* 1992;62:172-80; discussion 180-2

[6] Musil LS, Goodenough DA: Multisubunit assembly of an integral plasma membrane channel protein, gap junction connexin43, occurs after exit from the ER. *Cell* 1993;74:1065-1077

[7] Goodenough DA, Musil LS: Gap junctions and tissue business: problems and strategies for developing specific functional reagents. *J Cell Sci Suppl* 1993;17:133-138

[8] DeVries SH, Schwartz EA: Hemi-gap junctions in solitary horizontal cells: lucifer yellow permeability and single channel currents. *International Meeting on gap junctions,Asilomar 1991* 1991;Abstract book:91(abstract)

[9] DeVries SH, Schwartz EA: Hemi-gap-junction channels in solitary horizontal cells of the catfish retina. *J Physiol (Lond)* 1992;445:201-230

[10] White TW, Goodenough DA, Paul DL: A novel connexin expressed in rodent and chick lens. *J Cell Biol* 1991;115:189(abstract)

[11] Ebihara L: *Xenopus* connexin38 forms hemi-gap-junctional channels in the nonjunctional plasma membrane of *Xenopus* oocytes. *Biophys J* 1996;71:742-748

[12] Ebihara L, Berthoud VM, Beyer EC: Distinct behavior of connexin56 and connexin46 gap junctional channels can be predicted from the behavior of their hemi-gap-junctional channels. *Biophys J* 1995;68:1796-1803

[13] Li H, Liu TF, Lazrak A, Peracchia C, Goldberg GS, Lampe PD, Johnson RG: Properties and regulation of gap junctional hemichannels in the plasma membranes of cultured cells. *J Cell Biol* 1996;134:1019-1030

[14] Warner A, Clements DK, Parikh S, Evans WH, DeHaan RL: Specific motifs in the external loops of connexin proteins can determine gap junction formation between chick heart myocytes. *J Physiol (Lond)* 1995;488:721-728

[15] White TW, Bruzzone R, Wolfram S, Paul DL, Goodenough DA: Selective interactions among the multiple connexin proteins expressed in the vertebrate lens: the second extracellular domain is a determinant of compatibility between connexins. *J Cell Biol* 1994;125:879-892

[16] Ebihara L, Steiner E: Properties of a nonjunctional current expressed from a rat connexin46 cDNA in *Xenopus* oocytes. *J Gen Physiol* 1993 102:59-74

[17] Levine E, Werner R, Dahl G: *Am J Physiol* 1991;261:C1025

Gap Junctions
R. Werner (Ed.)
IOS Press, 1998

# Biogenesis of liver gap junctions

Juan A. Díez, Shoab Ahmad and W. Howard Evans
*University of Wales College of Medicine, Department of Medical Biochemistry, Heath Park, Cardiff CF4 4XN, Wales United Kingdom*

**Abstract.** Connexins, the channel-forming gap junction proteins, oligomerise en route to the plasma membrane to form hexameric hemichannels (connexons) that dock with connexons in contacting cells to generate intercellular communication channels. The synthesis of connexins and the subcellular location of their assembly into connexons as well as their homomeric or heteromeric nature in liver were examined. We have used guinea pig liver where connexin 26 (Cx26) and connexin 32 (Cx32) are expressed in an almost equal amount (60% and 40% respectively). To study the biogenesis and trafficking of Cx26 and Cx32, different hepatocyte compartments were isolated by subcellular fractionation and the relative amount of the two connexins estimated by SDS-PAGE and immunoblots using anti-peptide antibodies. The oligomeric state of connexins in the subcellular fractions was assayed by rate-zonal centrifugation in linear sucrose gradients. Finally, the homomeric or heteromeric nature of Cx26 and Cx32 in connexons was analysed by chemical cross-linking and immunoprecipitation techniques.

Although both connexins were present in equal amounts in liver gap junctions and plasma membranes, Cx26 could only be detected in ER and Golgi using a high titre antibody. Cx32 is present mainly as a monomer in the ER and as a hexamer in Golgi and plasma membranes, indicating that its oligomerisation occurs at the ER-Golgi intermediate compartment (ERGIC) but not later than the Golgi in the trafficking pathway. Cross-linking experiments showed that Cx26 and Cx32 form homomeric hexamers. However, using solubilised gap junctions both connexins were co-immunoprecipitated in intact connexons suggesting the presence in guinea pig hepatocytes of heteromeric connexons, although one connexin would be predominant.

A cell-free analysis approach shows that Cx26, unlike Cx32, can be inserted into membranes post-translationally. Our interpretation of these results is that Cx26 and Cx32 connexons may follow a different biogenetic intracellular pathway to the gap junction.

## 1. Introduction

Cells communicate directly in organised tissues through gap junctions that co-ordinate integrative behaviour by allowing the movement of small molecules between adjacent cells. Each cooperating cell contributes to each channel a connexon (hemichannel) formed by six subunits of the gap junction protein, connexin. Connexins belong to a 13 member family comprised of distinct integral membrane proteins, each encoded by a separate gene [1]. Detailed studies of tissues expressing more than one connexin show that they are often located in the same gap junction plaque [2]. However, different connexins seem to have a different distribution in the same cell [3]. The expression of two different connexins by the same cell raises the possibility that heteromeric connexons are assembled, as reported in insect cells over-expressing recombinant Cx26 and Cx32 [4] or in bovine lens fiber [5]; however, connexon homomericity has been reported in mouse liver gap junctions [6]. How and where connexins oligomerise en route to the plasma membrane in liver is still unclear. Using guinea pig liver as a model, we have studied the biogenesis and oligomeric nature of gap junctions, and used a cell-free system in which connexins and connexon hemichannels assemble.

## 2. Results

### 2.1. Cx26 and Cx32 are present in the secretory pathway in guinea pig liver

Subcellular fractions originating from various points in the trafficking pathway were isolated from guinea pig liver. Cx26 and Cx32 in the fractions were resolved by SDS-PAGE and detected by Western blotting using specific anti-peptide antibodies [7]. Figure 1 shows that guinea pig liver gap junctions express approximately equal amounts of Cx26 and Cx32; a similar ratio is found in plasma membranes. Cx32 was present in ER, intermediate compartment (IC), Golgi, sinusoidal and lateral plasma membrane. However, although Cx26 was also detected in similar amounts in plasma membranes (PM) and gap junctions, the low amount in ER, IC and Golgi fractions could only be detected by using a higher titre antibody.

### 2.2. Connexon composition

In order to study the connexon composition in guinea pig liver and determine their oligomeric nature we used two different experimental approaches: chemical cross-linking of connexons dissociated from intact gap junctions and immunoprecipitation.

#### 2.2.1. Chemical cross-linking

Gap junction plaques were solubilised in dodecyl maltoside (a mild detergent that does not disassemble connexons) and the solubilised connexons were cross-linked using dimethyl suberimidate (DMS), a homobifunctional reagent which reacts with primary amines. Reaction products were analyzed by SDS-PAGE and Western blotting (Figure 2).

Hexameric cross-linked connexins were clearly demonstrated for the first time using this technique. Also, the oligomeric products of the cross-linking of connexons are exclusively formed by Cx26 or Cx32, since they do not react with antibodies against both connexins simultaneously. These results points to a homomeric composition with respect to Cx26 and Cx32 in guinea pig liver connexons.

#### 2.2.2. Immunoprecipitation of isolated connexons

Gap junction plaques were solubilised in dodecyl maltoside or SDS to yield connexons or connexins respectively. After immunoprecipitation using antibodies specific to Cx26 or Cx32 (Figure 3) both connexins co-immunoprecipitated when connexons were solubilised in dodecyl maltoside whereas only the specific Cx26 (but not Cx32) antibody precipitated

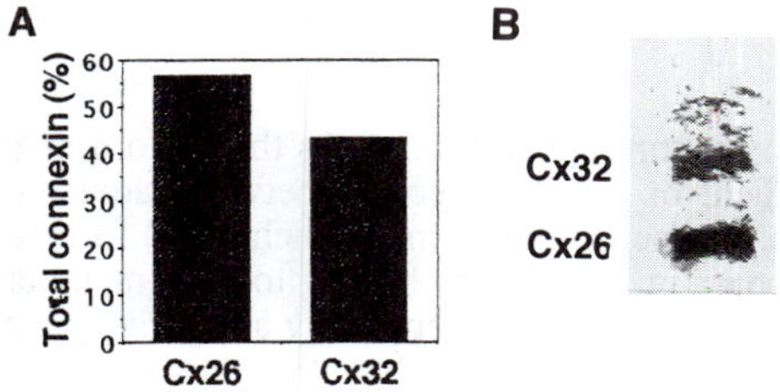


**Figure 1. Connexin composition in guinea pig liver gap junctions**

Purified guinea pig liver gap junctions were analysed by SDS-PAGE and stained with Coomassie blue (panel B). The intensity of the bands corresponding to Cx26 and Cx32 was measured by densitometry (panel A).

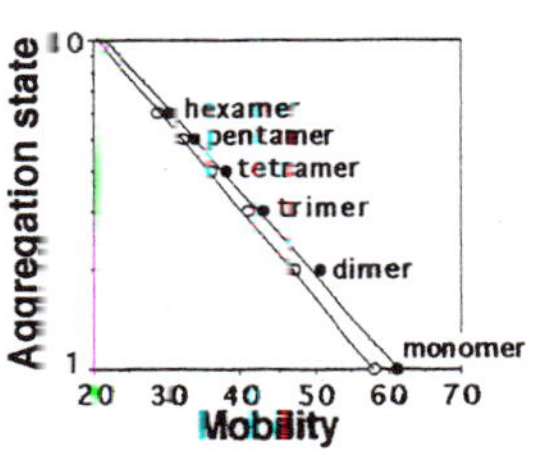


**Figure 2. Cross-linking of connexins isolated from guinea pig liver gap junctions.**

Purified gap junctions were solubilised in dodecyl maltoside and cross-linked with dimethyl suberimidate. Cross-linked products were processed by SDS-PAGE followed by immunoblotting with antibodies against Cx26 or Cx32. The graph represents the mobility in the gel of the different oligomers of Cx26 (closed circles) and Cx32 (open circles).

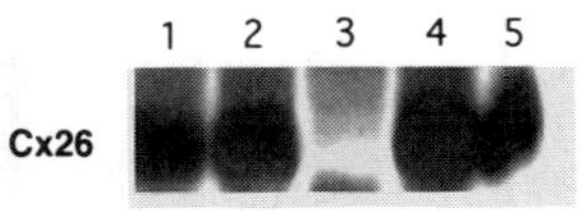


**Figure 3. Immunoprecipitation of connexins from intact connexons or connexins.**

Cx26 was immunoprecipitated using antibodies against Cx32 (lanes 1 and 3) or Cx26 (lanes 2 and 4) from purified gap junctions solubilised in dodecyl maltoside (lanes 1 and 2) or in SDS (lanes 3 and 4). Lane 5 is the control before immunoprecipitation.

Cx26 when connexons were fully dissociated into connexins in SDS. Thus, in contrast with the results obtained in the cross-linking experiments, the immunoprecipitation approach demonstrates that Cx26 and Cx32 reside in the same connexon.

## 2.3. Oligomerisation of connexins in liver occurs between the ER and Golgi

To elucidate the location on the secretory pathway where oligomerisation of connexins occurs, we studied the oligomerisation state of Cx26 and Cx32 in the different subcellular compartments isolated from liver. The subcellular fractions, solubilised in dodecyl maltoside, were centrifuged in linear sucrose gradients to separate, according to their different sedimentation coefficients, connexins from connexons (Figure 4). The ER contained predominantly connexins migrating in the upper part of the linear sucrose gradient. In contrast, Golgi contained mainly connexons that sedimented in the lower part of the gradient in a similar position to that adopted by connexons extracted by dodecyl maltoside from purified gap junctions. In the ERGIC compartment both connexins and connexons were observed. This result demonstrates that oligomerisation of Cx26 and Cx32 in liver is likely to occur between the ER and the Golgi, problably in ERGIC.

We also employed a cell-free system to study the synthesis and oligomerisation of Cx26 and Cx32 (Figure 5), using methodologies previously reported [8]. Both connexins were inserted into membranes (ER derived canine pancreatic microsomes) co-translationally, but Cx26 was also inserted post-translationally, this being an unusual characteristic of mammalian proteins [9]. After translation in this cell-free system, approximately 15% of connexins (Cx26 or Cx32) were in oligomeric form, as deduced by their migration position by sedimentation analysis in linear sucrose gradients. Thus, oligomerisation of connexins may commence early in the trafficking pathway, probably in the ER. Supplementation of the *in vitro* translation system with Golgi membranes produced a four-fold increase in the amount of oligomerised Cx32 but only a marginal increase was recorded with Cx26. This result indicates that the oligomerisation properties of Cx26 and Cx32 are different, especially with respect to the effects of addition of Golgi membrane on the efficiency of the assembly process.

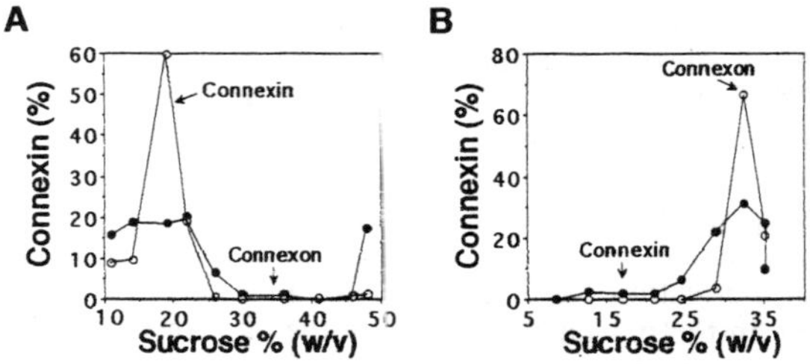


**Figure 4. Oligomerisation assay of guinea pig liver ER and Golgi.**

Purified ER and Golgi fractions were solubilised in dodecyl maltoside and centrifuged in linear sucrose gradients. Fractions collected from the gradients were analysed by SDS-PAGE and immunoblotted to detect the position of connexons or connexins. Graphs represent the densitometric values of Cx26 (closed circles) and Cx32 (open circles) in the different fractions: ER (A) and Golgi (B).

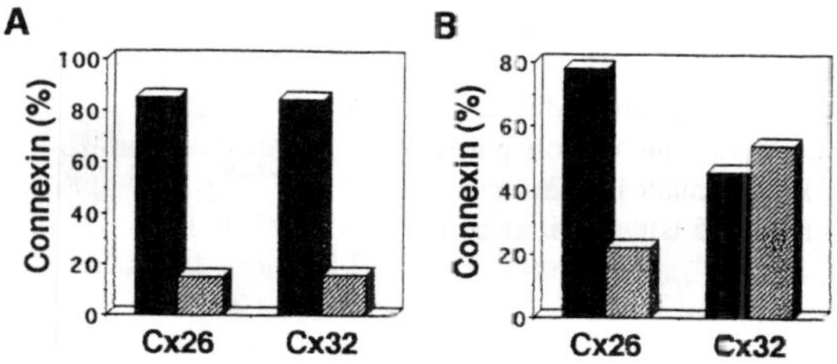


**Figure 5. Oligomerisation assay of *in vitro* synthesised Cx26 and Cx32.**
Cx26 and Cx32 were synthesised independently in a cell-free system in the presence of canine pancreatic microsomes (panel A). In panel B, the medium was supplemented with purified liver Golgi membranes. The products (connexins/connexons) were separated in a linear sucrose gradient and fractions collected. The graphs represent the proportion of oligomeric form (black bars) and monomeric form (grey bars) of connexins separated in the sucrose gradients.

## 3. Discussion

To elucidate the intracellular trafficking pathway followed by Cx26 and Cx32 prior to gap junction formation we isolated different intracellular compartments from guinea pig liver and analysed the oligomeric states of the two connexins. Both connexins were detected in ER, IC, Golgi, sinusoidal plasma membrane and lateral plasma membrane, confirming the route map of Cx32 in rat liver [7]. However a substantially lower amount of Cx26 relative to Cx32 was present in the ER, IC and Golgi stations of the biogenetic pathway although in the gap junction plaques and plasma membrane approximately equal amounts of the two connexins were found.

An immunoprecipitation approach showed that connexons in guinea pig liver were heteromeric, with both connexins (Cx26 and Cx32) present in the same connexon. In contrast, cross-linking of connexons showed no evidence for heteromericity suggesting that the constituent connexons are homomeric. The apparent contradiction between these two results, with homomeric connexons demonstrated by cross-linking and heteromeric connexons demonstrated by immunoprecipitation, can be explained if some connexons are homomeric while others are heteromeric; heteromeric connexons would be composed mainly of one connexin, perhaps in a ratio 5:1, and after cross-linking this arrangement would produce mostly homomeric oligomers, despite the overall heteromeric composition. Thus, in cells expressing more than one connexin there would be three putative connexon configurations: homomeric, heteromeric and a mixture of both.

Oligomerisation of connexins results in the formation of hemichannels (or connexons) that dock with their partners in the plasma membrane of the adjacent cell. In which organelle en route to the plasma membrane this oligomerisation occurs has not been established in liver. Cx43 was reported to oligomerise in the TGN in cultured Chinese hamster ovary (CHO) cells [10]. Here we show that in liver gap junctions oligomerisation of Cx26 and Cx32 is an early event that is initiated at the ER-Golgi boundary, and is completed in the Golgi apparatus.

Analysis of the synthesis and oligomerisation of Cx26 and Cx32 using cell-free system demonstrates that oligomerisation can occur, at least *in vitro*, in the ER. However, there was a dramatic increase in the presence of Golgi in the case of Cx32. Again, Cx26 shows different and unexpected behaviour, since its *in vitro* oligomerisation did not increase in the presence of Golgi to the same degree as that of Cx32. Since Cx26 can insert into the membrane post-translationally, this behaviour provides a mechanism in hepatocytes to separate Cx26 and Cx32 and limit mixing in the same connexon. These results can explain the minority presence of Cx26 in the classical trafficking pathway, despite the 60% represented in the gap junction plaques. They support the possible existence of an alternative trafficking pathway for Cx26 in liver, bypassing the Golgi apparatus and inserting this protein, possibly directly, into the plasma membrane (Figure 6).

Cx26 is unique among the connexin family: it has the shortest carboxyl tail, is not phosphorylated (with oligomerisation being the main postranslational modification) and is practically always co-expressed with another connexin. A rapid specific appearance of Cx26 in the plasma membrane after perfusion of female rat liver without an increase in total protein or mRNA levels has been reported [11]. Such a result would be unlikely if both connexins, Cx26 and Cx32, traffick together (even mixed in the same connexon) along the secretory pathway. The alternative pathway for Cx26 trafficking in liver now proposed could be a general pathway for targeting to the plasma membrane integral membrane-type proteins not containing conventional signal peptide sequences and that do not require ER/Golgi postranslational modification. Connexins lack a signal peptide sequence and Cx26 is not phosphorylated. Basic fibroblast growth factor (FGF-2) is another example of a protein exported from cells via a non-classical signal-peptide secretion [12]. These multiple considerations taken together with the unusual postranslational insertion properties of Cx26, further support the possiblility that this connexin follows an alternative non-classical trafficking pathway to the gap junction.

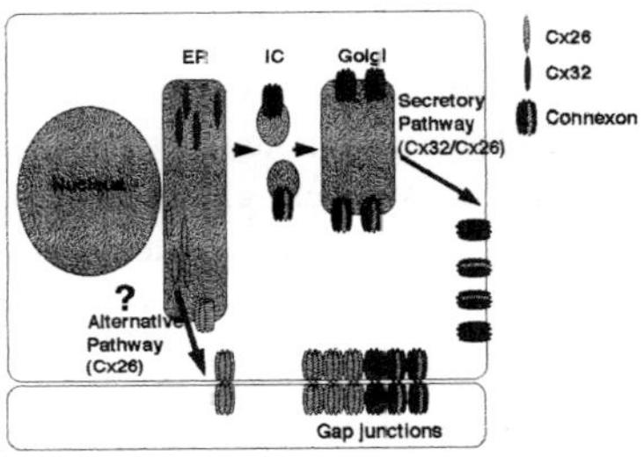


**Figure 6. Trafficking pathways for Cx26 and Cx32 in liver.** The scheme represents the "classical" trafficking pathway in a hepatocyte followed by Cx32 (ER-IC-Golgi-plasma membrane) and a possible "alternative" pathway followed by Cx26 (bypassing Golgi).

## Acknowledgement

J.A.D. thanks the Basque Goverment and the Welsh office, and S.A. the Commonwealth Fellowship Scheme for support. This work was suported by a MRC program grant (grant No: G-9305117) to W.H.E.

## References

1. K. Willecke *et al.* (1991). *Eur. J. Cell. Biol.* **56**, 1-7.
2. M. Bennett *et al.* (1991). *Neuron* **6**, 305-320.
3. O. Traub *et al.* (1989). *J. Cell. Biol.* **108**, 1039-1051.
4. K. A. Stauffer (1995). *J. Biol. Chem.* **270**, 6768-6772.
5. J. X. Jiang and D. Goodenough (1996). *Proc. Natl. Acad. Sci. USA* **93**, 1287-1291.
6. G. Sosinsky (1995). *Proc. Natl. Acad. Sci. USA* **92**, 9210-9214.
7. S. Rahman *et al* (1993). *J. Biol. Chem.* **268** (2), 1260-1265.
8. M. M. Falk *et al* (1997). *EMBO J.* **16**, 2703-2716.
9. J. Zhang *et al* (1996). *Mol. Biol. Cell* **7**, 471-482.
10. L. S. Musil and D. Goodenough (1993). *Cell* **74**, 1065-1077.
11. T. Kojima *et al* (1994). *J. Cell. Sci.* **107**, 3579-3590.
12. R. Z. Florkiewicz *et al* (1995). *J. Cell. Physiol* **162**, 388-399.

*Gap Junctions*
*R. Werner (Ed.)*
*IOS Press, 1998*


# Cell-free Expression of Functional Gap Junction Channels

Matthias M. Falk, Lukas K. Buehler, Nalin M. Kumar, and Norton B. Gilula*
*Department of Cell Biology, The Scripps Research Institute, La Jolla, CA 92037, USA*

**Abstract.** We have synthesized different connexin isotypes in standard cell-free translation/translocation systems to investigate and understand the synthesis and assembly of connexins into gap junction channels. A combined biochemical and biophysical analysis indicated that connexins, integrated into rough endoplasmic reticulum derived microsomal membranes, assembled specifically into homo-oligomeric structures with properties consistent with the properties of hexameric gap junction connexons. Furthermore, co-expressed $\alpha_1$ (Cx43) and $\beta_1$ (Cx32) did not assemble into hetero-oligomers indicating that the interaction between different connexin isotypes is selective. Molecular characterization of signals that regulate the interaction of connexin subunits suggests that specific "assembly" as well as "selectivity" signals are located within the connexin polypeptide sequence.

## 1. Cell-free Synthesis of Connexins

More than a dozen different connexin (Cx) isotypes have been cloned and sequenced from rodents (reviewed in [10]). Although the Cxs show a specific tissue distribution, several cell types in different organs have been characterized that express more than one Cx isotype [2, 12, 14]. Furthermore, immunohistochemical analyses have shown that the different Cxs can be localized within the same gap junction (GJ) plaque [12, 15]. Finally, experimental data obtained *in vitro* and *in vivo* suggest that besides homo-oligomeric GJ channels, hetero-oligomeric channels could exist as well [7, 9, 16, 17].

To investigate the synthesis and maturation of the Cx polypeptides and their assembly into GJ channels, we have expressed different Cx isotypes in standard cell-free translation/membrane translocation systems [1]. This system consists of a translation competent cell lysate that contains all components required for protein synthesis, endoplasmic reticulum derived membrane vesicles (microsomes) that contain all components required for membrane protein translocation, a radiolabeled amino acid for protein detection, and a synthetic RNA encoding the protein of interest. During translation integral membrane proteins are incorporated into the microsomes identical to the steps involved in membrane-protein synthesis *in vivo*. Newly synthesized proteins are analyzed by SDS-PAGE and visualized by autoradiography.

The microsomes used as acceptor membranes have an average diameter of 200-300 nm and are covered with multiple ribosomes, indicating that each membrane vesicle has many protein insertion sites (Fig. 1A). This is a prerequisite for studying protein oligomerization, since several polypeptides have to insert into the same microsomal vesicle. Under the conditions applied complete GJ channels - consisting of two paired connexons - were not assembled in the system in a detectable amount. This, however, is a great advantage for the analysis of the connexon-composition, since a potential hetero-oligomer can not be derived from a hetero-typic interaction of two different connexons.

---

* This work was supported by grants from the National Institutes of Health (GM 37904), the Lucille P. Markey Charitable Trust (N.B.G.) and a Deutsche Forschungsgemeinschaft fellowship Fa 261/1-1 to M.M.F.

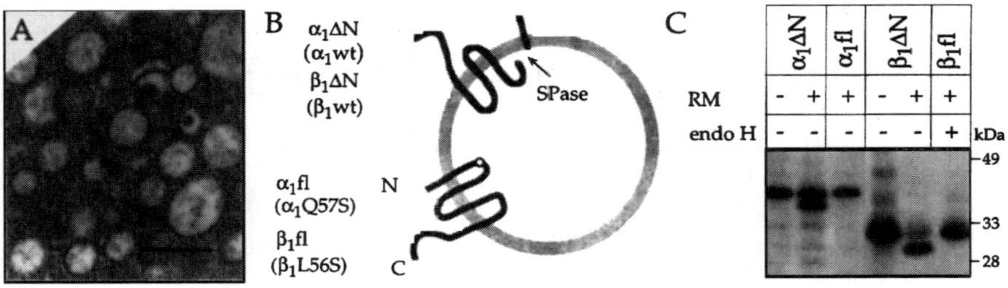


**Fig. 1** Cell-free synthesis of Cx polypeptides. (**A**) Electron micrograph of negatively stained rough endoplasmic reticulum derived membrane vesicles (microsomes). Scale bar = 500 nm. (Courtesy of F. George Klier). (**B**) Schematic representation of Cx polypeptides integrated into microsomes. (**C**) Translation products of $\alpha_1$ and $\beta_1$ Cx obtained in the absence (-) and in the presence (+) of microsomes (RM).

We analyzed GJ connexon assembly by expressing the two most abundant Cx isotypes $\alpha_1$ (Cx43) and $\beta_1$ (Cx32). As previously described [6], Cxs can efficiently be synthesized in cell-free translation systems. N-glycosylation consensus site tagging showed that upon microsomal membrane integration the Cxs adopt their natural transmembrane organization with four domains traversing the membrane bilayer, and N- and C-termini located outside the microsomal vesicles (in the cytoplasm) (Fig. 1B, and [Falk and Gilula, submitted]. However, to avoid an aberrant cleavage of the wild type Cx polypeptides by signal peptidase which removes the N-terminal domain and the first transmembrane domain (TM1) immediately after membrane integration [5] ($\alpha_1\Delta$N and $\beta_1\Delta$N in Fig. 1B, C), a single amino acid exchange in the first extracellular loop region (E1) was required. This amino acid exchange (Q57->S in $\alpha_1$ Cx, and L56->S in $\beta_1$ Cx) completely abolished cleavage and allowed the synthesis of appropriately membrane integrated, full length Cxs ($\alpha_1$fl and $\beta_1$fl in Fig. 1B, C).

## 2. Homo-oligomeric Connexon Assembly

To determine whether the Cx polypeptides oligomerize within the microsomes, sucrose gradient sedimentation analysis was applied. After solubilization of the microsomal membranes in detergent, the translation products separated into two distinct peaks, a wide 5S particle containing peak and a narrow 9S particle containing peak (Fig. 2A). Previous analysis of GJ channels assembled in tissue and in cultured cells indicated that the 5S particles correspond to unassembled Cx polypeptides, while the 9S particles correspond to assembled, hexameric connexons [4, 8, 11]. Integration of the area of the 9S peak showed that between 17% and 32% of the Cx polypeptides could be recovered as assembled connexons, dependent on the type of detergent used for the solubilization of the membrane vesicles (Table I in [5]).

Successful connexon assembly within microsomes was further indicated by results obtained by native polyacrylamide gel electrophoresis (PAGE). This technique allows detection of protein-protein interactions in the absence of chemical modifications. A 9S particle containing fraction, a 5S particle containing fraction, and a fraction between the two peaks taken from the sucrose gradient were analyzed. The 9S particles showed a drastically reduced mobility when compared to the mobility of the 5S particles (Fig. 2B, lanes 1, 3). No sharp protein band was detected in the control fraction (Fig. 2B, lane 2).

To determine whether the connexons assembled in the cell-free system were functional, microsomes containing cell-free expressed Cxs were fused with planar lipid bilayers and single channel activities were recorded. Distinct channel activities of 170 pS were recorded from fused microsomes containing $\beta_1$ Cx32, and channels with an average conductance of 52.4 pS were recorded from the fused microsomes containing $\beta_2$ Cx26 (Table II in [5]). The recorded channels were similar in behavior and conductance to channels previously recorded from these connexons expressed and purified from *in vivo* sources [13], or baculovirus infected insect cells [3], respectively. Control microsomes without cell-free expressed Cxs had some endogenous channel activities. However, these channels were clearly different

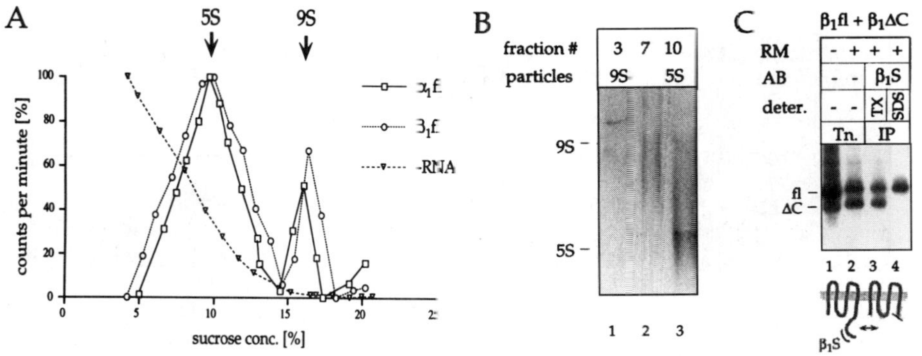


**Fig. 2.** Assembly of homo-oligomeric connexons in a cell-free expression system. **(A)** Hydrodynamic analysis of full-length $\alpha_1$ and $\beta_1$ Cx translation products. 9S particles represent assembled connexons, while 5S particles represent unassembled Cx polypeptides. **(B)** Native PAGE analysis of full length $\beta_1$ Cx derived from selected gradient fractions. **(C)** Immunoprecipitation analysis of co-translated full length (fl) and C-terminal truncated ($\Delta$C) $\beta_1$ Cx with a monoclonal antibody directed against the extreme C-terminal amino acid residues of $\beta_1$ Cx, after solubilization of the microsomes in 1% TX-100 (TX), or 1% SDS, respectively.

from the channels recorded from Cx containing microsomes (Table II in [5]).

Interaction of the Cx polypeptides within the oligomers was directly shown by co-immunoprecipitation (Fig. 2C). To be able to show an interaction between identical subunits within the homomeric oligomers, C-terminal truncated $\beta_1$ Cx cDNAs were generated and translated together with full-length $\beta_1$ Cx. A monoclonal antibody to $\beta_1$ Cx ($\beta_1$S, described in [5]) recognizing the extreme C-terminal tail was then used for immunoprecipitation after lysis of the microsomes in 1% TX-100. Co-immunoprecipitation of the truncated Cx polypeptides indicated that they paired and assembled with the full length Cx polypeptides (Fig. 2C, lane 3). When microsomes were lysed in 1% SDS which has been shown to disassemble Cx oligomers as well, no co-immunoprecipitation of the truncated Cx polypeptides occurred (Fig. 2C, lane 4). Control experiments showed that the Cx assembly in the cell-free expression system was a highly specific interaction that was restricted to appropriately membrane integrated Cx polypeptides. Other membrane proteins translated together with Cx polypep-

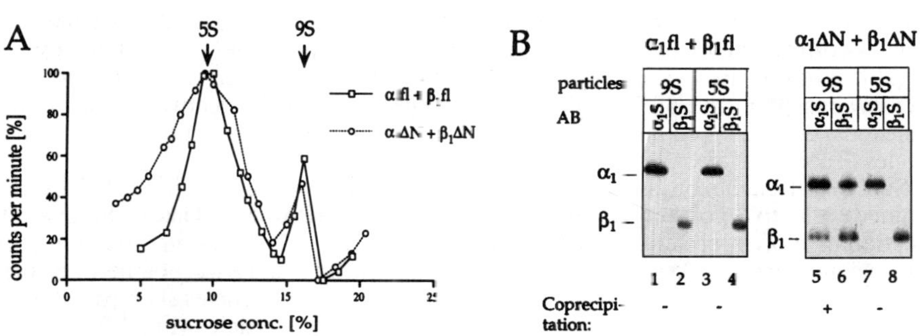


**Fig. 3.** Selective assembly of hetero-oligomeric connexons. **(A)** Hydrodynamic analysis of translation products of co-translated full length (fl), and N-terminal truncated ($\Delta$N) $\alpha_1$ and $\beta_1$ Cx, respectively. The N-terminal domain and the TM1 domain is removed from the Cx polypeptides immediately after their membrane integration by signal peptidase. Cx polypeptide truncation did not interfere with correct membrane insertion, or with oligomerization of the truncated Cxs into "connexon core structures" (9S particles). **(B)** Analysis of the composition of the 9S particles assembled in the co-translation reactions shown in (A) by immunoprecipitation using $\alpha_1$ ($\alpha_1$S) and $\beta_1$ ($\beta_1$S) specific monoclonal antibodies. As a control, 5S particles (unassembled Cx polypeptides) were immunoprecipitated as well. Co-immunoprecipitation indicates co-assembly of the polypeptides.

tides did not co-assemble with the Cxs, and hence, were not co-precipitated (Fig. 6 in [5]).

Taken together, these results indicate that functional homo-oligomeric GJ connexons can be assembled in microsomal vesicles following cell-free expression of Cx polypeptides.

## 3. Hetero-oligomeric Connexon Assembly

To further investigate whether Cxs also assemble into hetero-oligomeric connexons, full length $\alpha_1$ Cx ($\alpha_1$fl), and full length $\beta_1$ Cx ($\beta_1$fl) were translated together in the presence of microsomes. After vesicle lysis translation products were analyzed by hydrodynamic analysis, as described above (Fig. 3A). The composition of the obtained 9S particles was analyzed by immuprecipitation using $\alpha_1$ ($\alpha_1$S) and $\beta_1$ ($\beta_1$S) specific monoclonal antibodies (described in [5]). No co-precipitation of the other Cx isotype was observed under these conditions (Fig. 3B, lanes 1,2) indicating that the connexons (9S particles) were homo-oligomers either composed of $\alpha_1$, or $\beta_1$ Cx, respectively.

This result indicates that different Cx isotypes do not assemble randomly with each other, but interact selectively, allowing only the assembly of homo-oligomeric and possibly certain types of hetero-oligomeric connexons.

## 4. Characterization of Connexin Selectivity and Assembly Signals

To characterize potential signals within the Cx polypeptide sequence that regulate a selective interaction of Cxs, and thus would lead to the assembly of hetero-oligomeric connexons, N-terminal truncated $\alpha_1$ Cx ($\alpha_1\Delta$N), and $\beta_1$ Cx ($\beta_1\Delta$N) - with N-terminal domain and TM1 post-integrationally removed by signal peptidase as described above - were cotranslated, and translation products were analyzed by hydrodynamic analysis. 9S particles (connexon core structures) were still assembled from the truncated Cx polypeptides (Fig. 3A). Immunoprecipitation analysis, however, showed that the obtained connexon core structures were composed of both Cx isotypes (Fig. 3B, lanes 5, 6). This result indicates that the N-terminal truncated Cx polypeptides had lost their selectivity that was displayed by the full-length $\alpha_1$ and $\beta_1$ Cx proteins. Further analysis by immunoprecipitation technology indicated that selectivity was already lost when one of the two cotranslated Cx isotypes was N-terminal truncated (Fig. 5C in [5]). The loss in selectivity suggests that a distinct signal located in the N-terminal portion (N-terminus, TM1, E1) of the Cx polypeptide sequence either completely prevents hetero-oligomeric connexon assembly, or more likely, regulates the selective assembly of certain Cx isotypes into distinct hetero-oligomeric connexons.

To characterize an additional potential signal that in general allows Cx polypeptides to recognize each other, and thus leads to the assembly of the hexameric connexon, a set of $\beta_1$ Cx proteins progressively truncated from the C-terminus was constructed (Fig. 4A). The

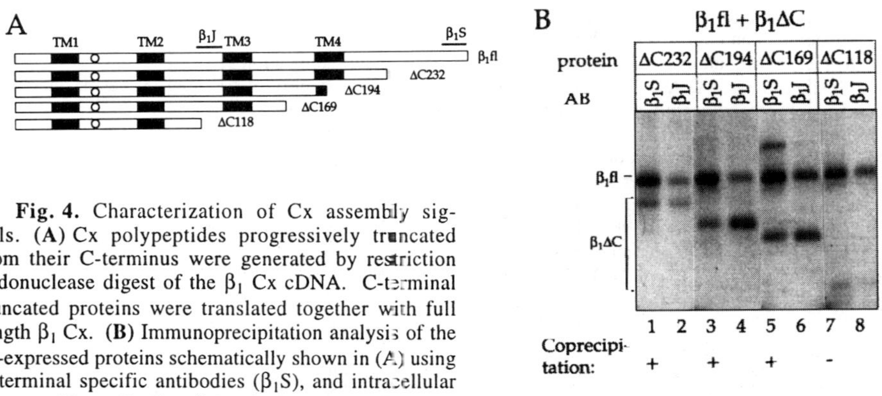


**Fig. 4.** Characterization of Cx assembly signals. **(A)** Cx polypeptides progressively truncated from their C-terminus were generated by restriction endonuclease digest of the $\beta_1$ Cx cDNA. C-terminal truncated proteins were translated together with full length $\beta_1$ Cx. **(B)** Immunoprecipitation analysis of the co-expressed proteins schematically shown in (A) using C-terminal specific antibodies ($\beta_1$S), and intracellular loop specific antibodies ($\beta_1$J). Co-immunoprecipitation indicates co-assembly of the proteins.

truncated Cx proteins were then co-translated with full length $\beta_1$ Cx and a potential interaction was detected by immunoprecipitation analysis using $\beta_1$ Cx antibodies which were specific for the C-terminus ($\beta_1$S), or the intracellular loop ($\beta_1$J, described in [5]), respectively (Fig. 4B). While truncated $\beta_1$ Cx proteins extending from the N-terminus past TM3 (ΔC232, ΔC194, ΔC169) coprecipitated and assembled with full length $\beta_1$ Cx protein (Fig. 4B, lanes 1 to 6), no assembly was observed when the truncated proteins extended only from the N-terminus past TM2 (ΔC118; Fig. 4B, lanes 7, 8). A similar loss in assembly was observed when the truncated $\beta_1$ Cx proteins were translated together with N-terminal truncated $\alpha_1$ Cx polypeptides (not shown). Furthermore, no assembly was observed between the C-terminal truncated $\beta_1$ Cx polypeptides and full length $\alpha_1$ Cx polypeptides, as previously described for full length $\alpha_1$, and full length $\beta_1$ Cxs (not shown).

Taken together, these results suggest that an "assembly" signal regulating basic Cx subunit recognition is located in the C-terminal portion (e. g. TM3) of the Cx polypeptides, while a "selectivity" signal regulating potential assembly of hetero-oligomeric connexons is located in the N-terminal portion of the Cx polypeptide sequence. All our conclusions are in agreement with the high resolution structure of GJ channels [18]. The structural data indicate that the individual Cx polypeptides within the connexon interact with one or two TM spanning domains of the adjacent Cx polypeptides. It is, therefore, likely that signals regulating Cx assembly are located within the TM domains, and/or immediately adjacent sequences. Further analysis of the Cx assembly will provide an extended understanding of the potential hetero-oligomeric connexon assembly, and a more detailed structural picture of the signals that regulate the assembly of Cx polypeptides.

## References


[1] Blobel, G., and B. Dobberstein. 1975. Transfer of proteins across membranes. II. Reconstitution of functional rough microsomes from heterologous components. *J. Cell Biol.* 67:852-862.

[2] Brissette, J.L., N.M. Kumar, N.B. Gilula, J.E. Hall, and G.P. Dotto. 1994. Switch in gap junction protein expression is associated with selective changes in junctional permeability during keratinocyte differentiation. *Proc. Natl. Acad. Sci. USA* 91:6453-6457.

[3] Buehler, L.K., K.A. Stauffer, N.B. Gilula, and N.M. Kumar. 1995. Single channel behavior of recombinant $\beta_2$ gap junction connexons reconstituted into planar lipid bilayers. *Biophys. J.* 68:1767-1775.

[4] Cascio, M., N.M. Kumar, R. Safarik, and N.B. Gilula. 1995. Physical characterization of gap junction membrane connexons (hemi-channels) isolated from rat liver. *J. Biol. Chem.* 270:18643-18648.

[5] Falk, M.M., L.K. Buehler, N.M. Kumar, and N.B. Gilula. 1997. Cell-free synthesis and assembly of connexins into functional gap junction membrane channels. *EMBO J.* 16:2703-2716.

[6] Falk, M.M., N.M. Kumar, and N.B. Gilula. 1994. Membrane insertion of gap junction connexins: polytopic channel forming membrane proteins. *J. Cell Biol.* 127:343-355.

[7] Jiang, J.X., and D.A. Goodenough. 1996 Heteromeric connexons in lens gap junction channels. *Proc. Natl. Acad. Sci. USA*. 93:1287-1291.

[8] Kistler, J., K. Goldie, P. Donaldson, and A. Engel. 1994. Reconstitution of native-type noncrystalline lens fiber gap junctions from isolated hemichannels. *J. Cell Biol.* 126:1047-1058.

[9] Konig, N. and G.A. Zampighi. 1995. Purification of bovine cell-to-cell channels composed of connexin44 and connexin50. *J. Cell Sci.* 108:3091-3098.

[10] Kumar, N.M., and N.B. Gilula. 1996. The gap junction communication channel. *Cell.* 84:381-388.

[11] Musil, L.S., and D.A. Goodenough. 1993. Multisubunit assembly of an integral plasma membrane channel protein, gap junction connexin43, occurs after exit from the ER. *Cell* 74:1065-1077.

[12] Nicholson, B., R. Dermietzel, D. Teplow, D. Traub, K. Willecke, and J.-P. Revel. 1987. Two homologous protein components of hepatic gap junctions. *Nature* 329:732-734.

[13] Rhee, S.K., C.G. Bevans, and A.L. Harris. 1996. Channel-forming activity of immunoaffinity-purified connexin32 in single phospholipid membranes. *Biochemistry* 35:9212-9223.

[14] Risek, B., F.G. Klier, A. Phillips, D.W. Hahn, and N.B. Gilula. 1995. Gap junction regulation in the uterus and ovaries of immature rats by estrogen and progesterone. *J. Cell Sci.* 108:1017-1032.

[15] Risek, B., F.G. Klier, and N.B. Gilula. 1994. Developmental regulation and structural organization of connexins in epidermal gap junctions. *Developmental Biol.* 164:183-196.

[16] Sosinsky, G. 1995. Mixing of connexins in gap junction membrane channels. *Proc. Natl. Acad. Sci. USA*. 92:9210-9214.

[17] Stauffer, K.A. 1995. The gap junction proteins $\beta_1$-connexin (connexin-32) and $\beta_2$-connexin (connexin-26) can form heteromeric hemichannels. *J. Biol. Chem.* 270:6768-6772.

[18] Unger, V.M., N.M. Kumar, N.B. Gilula and M. Yeager. 1997. Projection structure of a gap junction membrane channel at 7 Å resolution. *Nature Structural Biol.* 4:39-43.

Gap Junctions
R. Werner (Ed.)
IOS Press, 1998

# Reporting the Calcium Environment in Trafficking Pathways Leading to Gap Junction Biogenesis Using Connexin-Aequorin Chimerae

Christopher H. George, Jonathan M. Kendall, Anthony K. Campbell, Patricia E.M. Martin, W. Howard Evans

*Department of Medical Biochemistry, University of Wales College of Medicine, Cardiff, Wales, United Kingdom, CF4 4XN.*

**Abstract.** Serial biogenetic processes leading to the formation of gap junction intercellular communication channels (GJIC) are calcium-dependent. In order to investigate the role of calcium in the intracellular processing of connexons and at gap junctions we have constructed full length connexin-aequorin chimerae (Cx-Aeq) which have been expressed in Cos cells. Fusion of the aequorin moiety to the carboxyl terminus of Cx32 and Cx43 ensures aequorin retains full activity. Immunofluorescent staining of Cos cells expressing chimeric Cx32-Aeq or Cx43-Aeq using an anti-aequorin antibody and antibodies to specific sub-cellular markers along the secretory pathway indicate that the chimeric proteins are predominantly located at the endoplasmic reticulum/ Golgi (ER/Golgi) interface. The calcium dependent light emission of aequorin reports microdomains of cytoplasmic calcium along trafficking pathways. Using a photon counting camera to measure chemiluminescent activity of the chimerae in living Cos cells, we show that the calcium environment surrounding Cx43-Aeq is significantly higher than that reported by Cx32-Aeq indicating that the microdomains surrounding the ER/Golgi have different calcium concentrations. Both chimerae detect a calcium level higher than that of the bulk cytoplasm reported by a cytoplasmically located aequorin. When the cells are exposed to brefeldin A or nocodazole, the intermediate compartment and Golgi complex are redistributed to different intracellular regions which alters the calcium environment surrounding the chimerae. The chimerae have also been used to measure calcium at the plasma membrane, which we show to be dramatically higher than that of the bulk cytoplasm. By manipulating the trafficking of connexin-aequorin chimerae, we are able to explore the roles of heterogeneous cytoplasmic calcium domains in the biogenesis of gap junctions.

## 1. Introduction

The oligomerisation of individual connexins (Cx) into a hexameric connexon half channel and its subsequent trafficking through the cell are key events in the establishment of gap junction intercellular communication (GJIC). Many of the processes characterising the classical protein trafficking pathway in cells are calcium-dependent [1,2] although the precise role of calcium in the oligomerisation of connexins is unknown. At the plasma membrane, calcium is implicated in the gating of the gap junction channel [3] and it has been shown that elevated, i.e. micromolar levels of cytoplasmic calcium in the vicinity of the plasma membrane results in the closing of the gap junction pore [4]. The mechanisms of calcium induced pore closure are poorly understood but the effects are likely to be via such intermediaries as calmodulin [5,6]. We have characterised two chimerae in which the

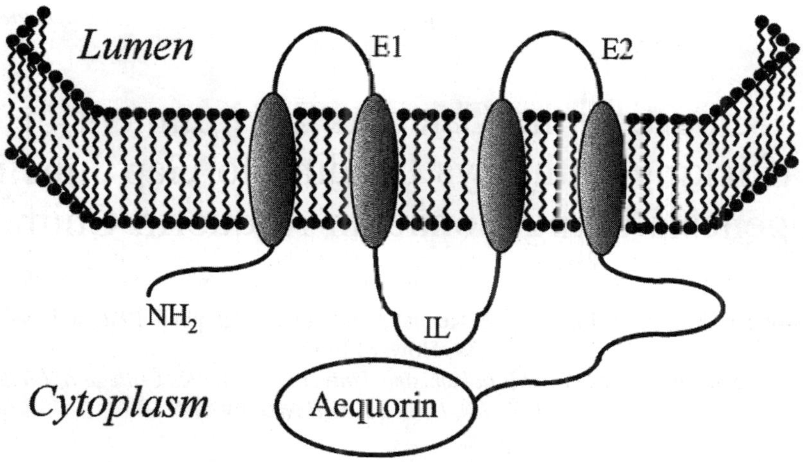


Figure 1. Diagrammatic representation of the topology of connexin-aequorin chimerae in the ER membrane. E1 and E2 are extracellular loops 1 and 2 respectively, IL is the intracellular cytoplasmic loop.

calcium-dependent photoprotein, aequorin [7], has been fused to the carboxyl terminus of Cx32 (Cx32-Aeq) and Cx43 (Cx43-Aeq) (Fig.1).

These proteins are assembled into connexons that form functional gap junctions (Martin et al., submitted). Here, we demonstrate that expression of these chimerae in Cos cells can be used to measure levels of free cytoplasmic calcium along the protein trafficking pathway and also at the cell surface following integration of the chimerae into the plasma membrane and the gap junction.

Aequorin has been targeted to different intracellular organelles including the ER [8], nucleus [9] and mitochondria [10]. The calcium environments reported by the photoprotein reveal a heterogeneity between organelles. Not only does this heterogeneity exist between organelles, but recent studies demonstrate functionally separate calcium stores within an organelle [11, 12], and dramatically elevated levels of calcium beneath the plasma membrane [13]. These results together with the limited diffusion of calcium in the cytoplasm [14], suggests that heterogeneous microenvironments of calcium exist within the cytoplasm. To investigate the functional significance of such these calcium microenvironments, we are using the Cx-Aeq chimerae to measure free calcium at the mouth of the gap junction channel and the influence of microdomains of cytoplasmic calcium in gap junction biogenesis.

## 2. Strategy

From the cDNA encoding Cx32, Cx43 and aequorin, chimeras were generated by a two step PCR procedure [15]. Following expression of Cx32-Aeq or Cx43-Aeq in Cos cells the intracellular location of these proteins was determined immunocytochemically with antibodies to aequorin in combination with organelle specific markers that identify the intermediate compartment (ERGIC) and the trans-Golgi network (TGN) [16,17,18].

To investigate the effect of brefeldin A and nocodazole on the intracellular trafficking of the connexon, Cos cells expressing Cx32-Aeq or Cx43-Aeq were incubated with brefeldin A or nocodazole as previously described [19].

To measure intracellular calcium, aequorin was reactivated in Cos cells expressing recombinant chimerae by incubation with coelentrazine [16]. To measure sub- plasma membrane calcium, reactivation with coelentrazine was carried out in the presence of 1mM

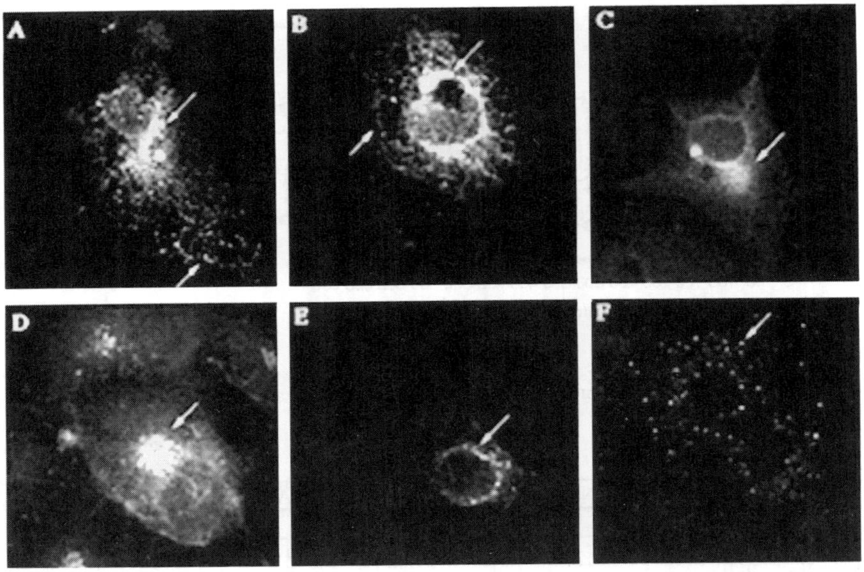

Figure 2. Immunostaining of Cx32-Aeq (A) and Cx43-Aeq (B) expressed in Cos cells. Immunolocalisation of ERGIC (C) and TGN (D) reveals overlap between the compartments in wild-type Cos cells. Redistribution of chimeric protein occurs in Cos cells expressing Cx43-Aeq following treatment with 5μg/ml brefeldin A (E) or 20μg/ml nocodazole (F). Brefeldin A and nocodazole cause a similar redistribution of Cx32-Aeq within the cell (data not shown).

EGTA [13]. Chemiluminescence was measured by a photon counting camera via direct contact with a fibre optic [9]. Measurement of basal cytoplasmic calcium levels was as described [9] and rate constants were converted into calcium concentrations using equations generated by calibration of each of the chimerae. A cytoplasmically located aequorin [16] was used as a control in all experiments in which calcium was measured.

## 3. Results and Discussion

Calibration of the aequorin moiety of each chimerae as described [20] demonstrated that the addition of the Cx32 or Cx43 to the amino terminus of aequorin did not significantly affect the calcium dependent light emission *in vitro* when compared to unmodified recombinant aequorin.

A comparison of immunofluorescent staining of Cos cells expressing Cx32-Aeq or Cx43-Aeq (Figure 2(A) and (B) respectively) with staining of the ERGIC and Golgi (Figure 2(C) and (D) respectively) reveals the chimerae are located in the ER/Golgi regions of the cell aswell as at the plasma membrane. The cytoplasmic calcium environment reported by Cx43-Aeq is significantly higher than that detected by Cx32-Aeq and both proteins are located in domains of elevated calcium when compared to the bulk cytosol (Figure 3).This suggests that the chimerae are assembled into connexons that are located in separate regions of the ER/Golgi according to the connexin isoform used. Subcellular fractionation of these cells confirms by independent technology that Cx32-Aeq and Cx43-Aeq are found in different functional axes of the Golgi which can not be resolved by immunocytochemical techniques (data not shown).

Redistribution of the chimeric proteins to a perinuclear region (Figure 2(E)) occurs following treatment of the cells with brefeldin A, and this is accompanied by a marked decrease in the calcium level surrounding the proteins (Table 1).

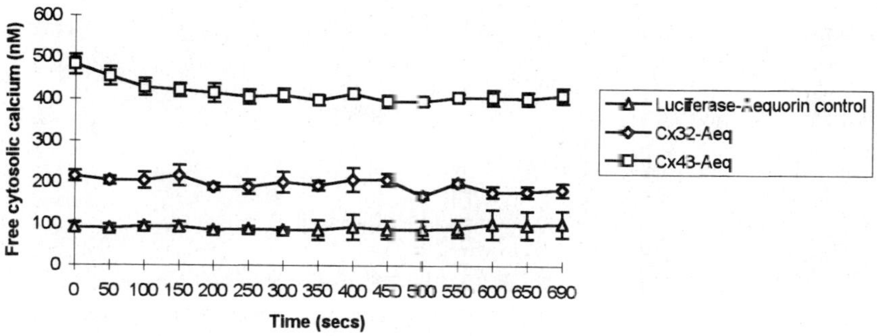


Figure 3. Concentrations of free cytoplasmic calcium reported by Cx32-Aeq, Cx43-Aeq and a cytoplasmically located aequorin expressed in live Cos cells. Error bars are ± SEM.

| | No treatment | Brefeldin A | Nocodazole |
|---|---|---|---|
| Cx32-Aeq | 195 ± 16 | 144 ± 8 | 205 ± 39 |
| Cx43-Aeq | 414 ± 11 | 171 ± 84 | 434 ± 71 |
| Luciferase-Aequorin | 91 ± 19 | 94 ± 4 | 102 ± 5 |

Table 1. Effect of brefeldin A and nocodazole on free cytoplasmic calcium levels detected by Cx32-Aeq and Cx43-Aeq expressed in live Cos cells. All values are given as nanomolar concentrations of calcium ± SEM.

This result indicates that an intact Golgi apparatus maintains separate 'pools' of connexons en route to the plasma membrane and we are currently investigating the functional significance of these discrete stores of connexons prior to connexon integration into gap junctions. The dispersal of the chimeric proteins following incubation with nocodazole (Figure 2(F)) was observed only in a few cells and, unlike brefeldin A, nocodazole treatment of cells does not produce a significant alteration in calcium environment surrounding Cx32-Aeq or Cx43-Aeq (Table 1). We have shown that neither brefeldin A nor nocodazole had any significant effect on the levels of calcium in the bulk cytosol as determined by the cytoplasmically located luciferase-aequorin control (Table 1) and that the overall structure of the ER is unaffected by treatment with brefeldin A or nocodazole (data not shown). Therefore the decrease in calcium levels reported by the chimeric proteins following incubation with brefeldin A is unlikely to be due to increased sequestration of calcium from the cytoplasm or fragmentation of the ER.

Measurement of cytoplasmic levels of calcium by this method [9] does not allow sub-plasma membrane levels of calcium to be determined. Using a protocol in which cells expressing chimerae were reactivated with coelentrazine in 1mM EGTA, we are able to maintain the functionality of plasma membrane associated chimeric protein. When the cells were exposed to media containing physiological levels of calcium (1.3mM), both Cx32-Aeq and Cx43-Aeq report a sub-plasma membrane calcium environment up to *twenty* times higher than detected in the bulk cytosol of resting cells (~ 2μM). This result supports the view that a rim of high calcium levels are present beneath the plasma membrane [13] and that the cytoplasm is capable of supporting domains of varying concentrations of calcium [21]. The role of such calcium domains in the biogenesis and function of gap junctions remains to be investigated.

The results demonstrate the utility of connexin-aequorin chimeras in analysing the diverse calcium environments that exist along the protein trafficking pathway and at the plasma membrane. We are now able to investigate the effects of physiological stimuli on these calcium domains and the subsequent effects on GJIC using these chimeric probes.

**Acknowledgements.** C.H.G. is in receipt of an M.R.C. studentship. This work was supported by the M.R.C.

**References**

[1] J.E. Rothman and F.T. Wieland (1996) Science **272** 227-234
[2] J.F. Sambrook (1990) Cell **61** 197-199
[3] P.N.T. Unwin and P.D. Ennis (1989) Journal of Cell Biology **97** 1459-1466
[4] B. Rose and W.R. Loewenstein (1975) Nature **254** 250-252
[5] C. Peracchia *et al.*, (1996) European Journal of Physiology **431** 379-387
[6] K. Torok *et al.*, (1997) Biochemical Journal **326** 479-483
[7] O. Shimomura and F.H. Johnson (1969) Biochemistry **8** 3991-3997
[8] J.M. Kendall *et al.*, (1992) Biochemical and Biophysical Research Communications **189** 1008-1016
[9] M.N. Badminton *et al.*, (1996) Journal of Biological Chemistry **271** 31210-31214
[10] R. Rizzuto *et al.*, (1992) Nature **358** 325-328
[11] D. Button and A. Eidsath (1996) Molecular Biology of the Cell **7** 419-434
[12] V.A. Golovina and M.P. Blaustein (1997) Science **275** 1643-1648
[13] R. Marsault *et al.*, (1997) EMBO Journal **16** 1575-1581
[14] N.L. Allbritton *et al.*, (1992) Science **258** 1812-1815
[15] G. Sala-Newby and A.K. Campbell (1991) Biochemical Journal **279** 727-732
[16] J.M. Kendall *et al.*, (1996) Cell Calcium **19** 125-1325
[17] B.L. Tang *et al.*, (1995) European Journal of Cell Biology **68** 398-410
[18] M. Girotti and G. Banting (1996) Journal of Cell Science **109** 2915-2926
[19] N.B. Cole *et al.*, (1996) Molecular Biology of the Cell **7** 631-650
[20] R. Tsien and T. Pozzan (1989) Methods in Enzymology **172** 230-262
[21] J. Meldolesi *et al.*, (1990) Biochimica et Biophysica Acta **1055** 130-140

*Gap Junctions*
*R. Werner (Ed.)*
*IOS Press, 1998*



# Growth Factors Alter Connexin Expression but Not Junctional Function in Endothelial Cells, LDL has no Effect

Janis M. Burt, Anna Fletcher and Mary Olsen
*Department of Physiology, University of Arizona, Tucson*

In the vascular wall, endothelial cells (ECs) are typically growth arrested; however, they proliferate in response to various pathophysiologic conditions. Whereas in the normal vessel Cxs 40, 37 and 43 are expressed in ECs, in regions of pathology and in cell culture Cx43 expression is elevated and Cx40 and 37 expression are reduced. This change in connexin expression could be a part of the proliferative response of these cells to pathophysiologic conditions, including standard cell culture. This possibility was evaluated in the current study in rat microvascular coronary endothelial cells (CECs) and in aortic endothelial cells (AECs). ECs proliferated in DMEM supplemented with 15% FBS, heparin and EC growth factor. Proliferation was blocked when the FBS content of the medium was reduced to 1%. Neither native or mildly oxidized LDL (ox-LDL) stimulated proliferation of ECs. Growth arrested AECs exhibited an ~5 and 16 fold reduction in Cx43 and Cx40 mRNA expression, respectively. In contrast, growth arrested CECs exhibited an ~ 2 fold increase in both Cx43 and Cx40 expression. The Cx43:Cx40 mRNA ratio in proliferating AECs and CECs was ~ 5:1 and 3.6:1, respectively; these ratios in growth arrested AECs and CECs were 17:1 and 3.6:1, respectively. Consistent with ~5 fold higher Cx43 levels in AECs, these cells were better coupled than the CECs. Lucifer Yellow diffused to ~50 neighbors in the AECs vs. 9 neighbors in the CECs. However, despite large differences in connexin levels in the growth arrested vs. proliferating AECs and CECs, differences in dye coupling could not be detected. LDL and ox-LDL also had no effect on dye coupling.

## 1. Introduction

Available data indicate that gap junctions contribute to regulation of blood flow by communicating the effects of localized stimuli to distant upstream regulatory sites [1-4]. Gap junctions are also likely to contribute to homeostasis of the endothelium and consequently of the vessel wall, and to angiogenesis. Definitive proof of these roles for gap junctions in vessel function (and related dysfunction) is lacking, in large part because i) there are no specific inhibitors of gap junction channel function and ii) quantitative measures of junctional function in endothelial cells of intact vessels has not been possible. To better define the roles of gap junctions in vascular function and pathophysiology, numerous investigators have used cultured endothelial cells [5-7]. However, whereas connexins 40 and 37 predominate in the endothelium of the normal vessel wall, in culture (and in regions of vessel pathology) Cx43 predominates [7]. The mechanisms underlying these alterations of connexin expression remain obscure. More importantly, the physiologic consequences of altered expression are unknown.

In the current study we determine whether growth arrest via growth factor-reduction influences connexin expression or gap junction function in endothelial cells from large and small vessels. We also determine whether LDL or ox-LDL influence growth or gap junction function in these endothelial cells.

## 2. Materials and Methods

CECs were isolated as follows. Rat hearts were minced finely, digested in collagenase, and the cells separated from debris on a 45% Percoll gradient. Cells were plated on gelatin coated dishes and maintained at 37°C in a 5% $CO_2$, humidified incubator. For AECs, following removal of the adventitia from an isolated thoracic aorta, the distal end was tied and the vessel filled to distention with collagenase. After 20 min. digestion, the enzyme/cell mixture was collected. Following a second digestion period, cells from both incubation periods were combined, pelleted by centrifugation and plated on gelatin coated tissue culture dishes in growth medium. Growth medium contained DMEM (low glucose) with glutamine, 15% heat inactivated fetal bovine serum (FBS), heparin, crude endothelial cell growth factor (ECGF), Pennicillin and Streptomycin. ECGF was prepared from bovine hypothalamus. Media for growth arrest experiments were made with lipoprotein deficient FBS (LPD-FBS). Lipoproteins were removed by density centrifugation. Cells were passaged with trypsin-EDTA and replated at a 1:3 dilution. All experiments were done on confluent cells from passages 2 or 3.

Proliferation was assessed by quantifying absorbance (590nm) following fixation of the cells in 1% glutaraldehyde, staining with 0.1% crystal violet, and liberation of stain from the cells in 10% acetic acid [8]. Absorbance and cell number were linearly related (not shown).

LDL was prepared from human blood by density centrifugation (KBr). Blood was protected from oxidation throughout the isolation procedure by EDTA; sodium azide, PMSF, and aprotinin were also present once plasma had been isolated. The final LDL component (1.06-1.63 g/ml density fraction) was dialyzed against saline/ EDTA for 24hr at 4°C and the protein content of the sample assayed. LDL was stored under $N_2$ in the refrigerator. Following removal of the EDTA by dialysis, LDL was oxidized for 24h at 37°C by addition of sufficient $CuSO_4$ and PBS to bring the Cu concentration to 6 μM and the final protein concentration to 4mg/ml. Oxidation was terminated by addition of excess EDTA and the degree of oxidation quantified as the change in mobility of the electrophoresed LDL sample. The ox-LDL used in the current study exhibited a 5-15% increase in mobility.

Extent of cell-cell coupling was determined following microinjection of Lucifer Yellow into individual endothelial cells, by determining, one minute later, the number of neighboring cells containing dye.

Total RNA was isolated using standard guanidinium thiocyanate procedures. RNA for Northern blot evaluation was denatured in formaldehyde/formamide at 55°C and electrophoretically separated on 1% agarose gels containing 0.4M formaldehyde. RNA was transferred passively to Zeta probe GT membrane and hybridized with radiolabelled cDNA probes (random prime label) at high stringency for Cx43, Cx40, Cx37 and GAPDH. Band density was determined by PhorphorImager using the manufacturer's Densitometry Technology program. Connexin data were normalized against GAPDH data.

## 3. Results

To assess the effects of growth arrest on connexin expression and gap junction function conditions that resulted in growth arrest had to be identified. Shown in Figure 1 are the results of an experiment in which the effects of serum concentration on the proliferation of CECs and AECs as a function of time was determined. Note that proliferation of both cell

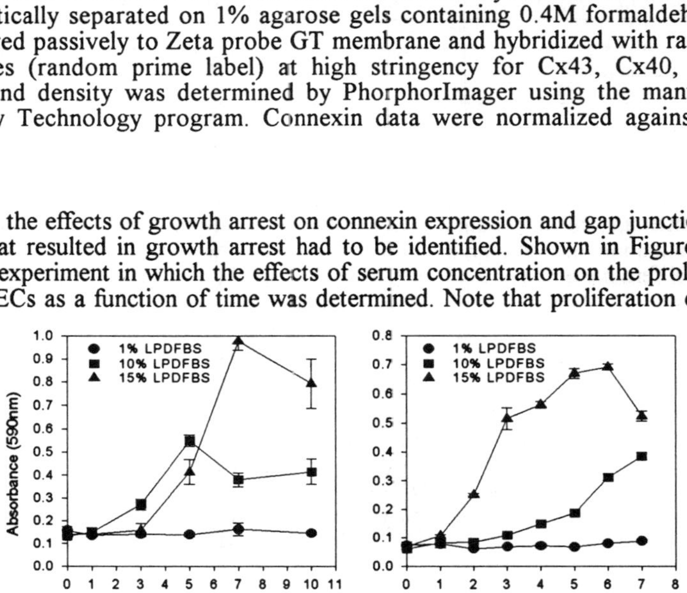


Figure 1 - CECs(left) and AECs (right) failed to proliferate in 1% LPDFBS, and proliferated at a reduced rate in 10% LPDFBS without ECGF or heparin.

types was blocked when serum concentration was reduced to 1%. To determine whether LDL could stimulate growth arrested ECs to proliferate, either native LDL or mildly oxidized LDL were added to the culture medium and proliferation monitored as a function of time. The results are illustrated in Figure 2. Note that neither form of LDL stimulated proliferation of the cells (CECs data not shown).

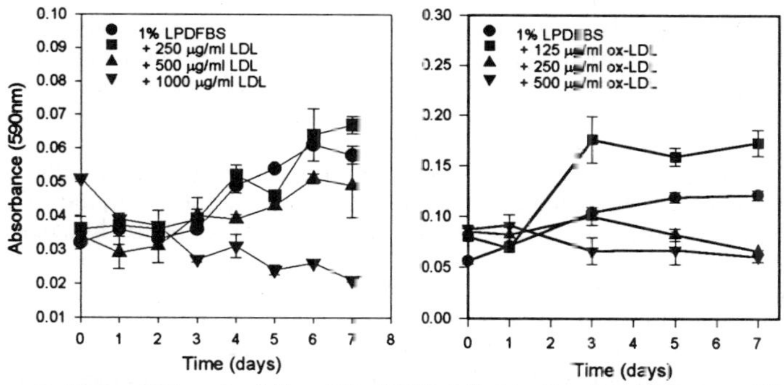


Figure 2 - Native LDL and mildly oxidized LDL failed to stimulate AECs to proliferate.

To determine whether connexin expression was affected by the presence of growth factors, total RNA was isolated from confluent ECs that were in 15% FBS- or 1% FBS-medium. The results of two experiments are presented in Table 1. In AECs, growth arrest resulted in an ~5 and 16 fold **reduction** in Cx43 and Cx40 mRNA, respectively, whereas in CECs growth arrest resulted in an ~ 2 fold **increase** in both connexins. In AECs, the Cx43:Cx40 mRNA ratio was 5:1 in proliferating cells 17 1 in growth arrested cells. In CECs, this ratio was ~ 3.6:1 in both proliferating and growth arrested cells. Cx43 mRNA was 4.3 fold more abundant in the proliferating AECs than in proliferating CECs.

Table 1 - Influence of serum concentration on connexin mRNA levels* in CECs and AECs.

| | CECs | | | AECs | | |
|---|---|---|---|---|---|---|
| | 1% FBS | 10% FBS | Control | 1% FBS | 10% FBS | Control |
| Cx40 | 1.88, 1.81 | 1.09, 0.70 | 1.01 | 0.17, 0.102 | 0.293, 1.05 | 2.4, 3.51 |
| Cx43 | 7.43, 5.24 | 7.39, 2.32 | 4.37, 2.58 | 1.87, 4.58 | 5.87, 1.48 | 12.9, 17.2 |

* data expressed in arbitrary units determined by densitometry

To assess the effects of growth arrest and putative modulators of growth on gap junction function, number of coupled neighbors was determined one minute after Lucifer yellow injection into individual ECs. Proliferating AECs were coupled to more neighbors than the proliferating CECs, 50 ± 1 (n=75) vs. 9.4 ± 0.4 (n=151) neighbors, respectively,

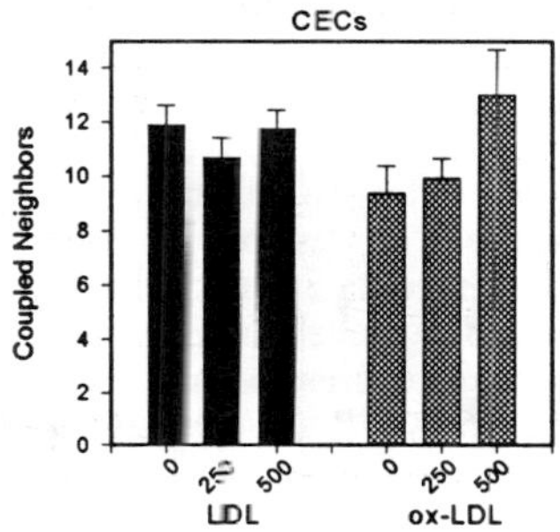


Figure 3 - Extent of dye coupling (Lucifer Yellow) between growth arrested CECs was not affected by native or mildly oxidized LDL. Comparable data for AECs (not shown) also revealed no LDL-induced change in coupling.

consistent with the ~5fold greater Cx43 content of the AECs. However, despite significant changes in Cx mRNA levels in growth arrested vs. proliferating ECs, no differences in Lucifer Yellow dye coupling could be detected (Fig. 3). Further, neither LDL or ox-LDL had any effect on coupling, even after 24h exposure.

## 4. Discussion

We hypothesized that removal of growth factors would restore connexin expression to the pattern observed in normal vessels *in vivo*. For the AECs this would mean that Cx43 expression would be reduced and Cxs 40 and 37 increased [7, 9, 10]. Our data clearly demonstrate that reduced levels of growth factors (heparin, ECGF, and serum) cause **down-regulation** of both Cxs 43 and 40 in AECs, but fail to restore expression to the *in vivo* pattern. Our results further support those of Larson *et al.* [5] who demonstrated that Cx43 expression in bovine AECs is related to their proliferative activity. However, our data also indicate that at confluence, when proliferation is arrested, the continued presence of growth factors blocks the complete down-regulation of Cx43 that otherwise would occur if the growth factors were absent.

Connexin expression in the normal rat microvascular coronary artery has not been reported, and consequently we can't evaluate whether expression is altered by culture. However, our data indicate that growth factor removal causes significant **up-regulation** of Cx43 and Cx40. This result is exactly opposite to that of Pepper *et al.* [11], who observed up-regulation of Cx43 in microvascular adrenal ECs in response to growth factor addition. The basis for the differences are not known, but it is interesting that our cells were much better coupled under all conditions than theirs.

Despite significant changes in Cx mRNA expression in growth arrested vs. proliferating cells, we could not detect any change in the extent of dye-coupling between either the AECs or the CECs. The maintenance of coupling in the face of altered mRNA levels indicates that post-transcriptional regulation of coupling must occur in parallel with regulation of mRNA level (via transcription or mRNA half-life mechanisms). This conclusion agrees with that of Larson *et al.* [5].

Meyer *et al* [12] reported that LDL (and Apo-B) enhanced the assembly of gap junctions and thereby increased functional coupling in Novikoff hepatoma cells. In our experiments we could detect no effect of nLDL on intercellular communication between either type of endothelial cell. The failure to observe any increase in functional coupling in the endothelial cells could indicate that the pathways activated by LDL in the hepatoma cells were already maximally activated by alternate means in the cultured endothelial cells. Alternatively, the discrepancy could reflect cell-specific (hepatoma vs. endothelial) differences in regulation of junction assembly/ function.

Finally, although oxidized LDL is mitogenic in vascular smooth muscle [13-15] it does not appear to be so in cultured ECs. For smooth muscle recent evidence from Chai *et al.* [15] suggests that oxidized LDL is indirectly mitogenic, stimulating release of growth factors which subsequently stimulate proliferation. Our data indicate that LDL and ox-LDL are not mitogenic in ECs by a direct or indirect mechanism. Since significant endothelial cell proliferation does not occur in atherogenesis, this result is not surprising, and indicates that the pathways activated by ox-LDL in smooth muscle cells are either not present in the ECs or are not activated by ox-LDL in the ECs.

In summary, connexin expression differs significantly in growth arrested vs. proliferating AECs and CECs. Despite this change in expression, no differences in gap junction mediated dye coupling could be detected. The data suggest that there is no simple relationship between mRNA and functional coupling levels in ECs. Further the data indicate that the mechanisms for regulation of connexin expression and function differ significantly in large (aorta) vs. small (microvascular coronary) ECs. Such differences could readily contribute to the physiological and pathophysiological differences of these vessel types.

## References


[1] S. S. Segal and B. R. Duling, Flow control among microvessels coordinated by intercellular conduction. *Science*, vol. 234, pp. 868-870, 1986.

[2] S. S. Segal, D. N. Damon, and B. R. Duling, Propagation of vasomotor responses coordinates arteriolar resistances. *Am. J. Physiol.* vol. 256, pp. H832-H837, 1989.

[3] D. T. Kurjiaka and S. S. Segal, Interaction between conducted vasodilation and sympathetic nerve activation in arterioles of hamster striated muscle. *Circ. Res.* vol. 76, pp. 885-891, 1995.

[4] M. D. S. Frame and I. H. Sarelius, L-arginine-induced conducted signals alter upstream arteriolar responsivity to L-arginine. *Circ. Res.* vol. 77, pp. 695-701, 1995.

[5] D. M. Larson, M. J. Wrobleski, G. D. V. Sagar, E. M. Westphale, and E. C. Beyer, Differential regulation of connexin43 and connexin37 in endothelial cells by cell density, growth, and tgf-beta-1. *American Journal of Physiology - Cell Physiology*, vol. 41, pp. C 405-C 415, 1997.

[6] H. V. M. Van Rijen, M. J. A. van Kempen, L. J. S. Analbers, M. B. Rook, A. C. G. Van Ginneken, D. Gros, and H. J. Jongsma, Gap junctions in human umbilical cord endothelial cells contain multiple connexins. *American Journal of Physiology - Cell Physiology*, vol. 41, pp. C 117-C 130, 1997.

[7] J. E. Gabriels and D. L. Paul, Differential expression of three gap junction proteins, Cx40, Cx37, and Cx43, *in vivo* and *in vitro*, in normal and wounded aortic endothelial cells of the rat. *Gap Junctions*, vol. 5, pp. 301995.(Abstract)

[8] R. J. Gillies, N. Didier, and M. Denton, Determination of cell number in monolayer cultures. *Anal. Biochem.* vol. 159, pp. 109-113, 1986.

[9] R. Bruzzone, J.-A. Haefliger, R. L. Gimlich, and D. L. Paul, Connexin40, a component of gap junctions in vascular endothelium, is restricted in its ability to interact with other connexins. *Mol. Biol. Cell*, vol. 4, pp. 7-20, 1993.

[10] K. E. Reed, E. M. Westphale, D. M. Larson, H.-Z. Wang, R. D. Veenstra, and E. C. Beyer, Molecular cloning and functional expression of human connexin37, an endothelial cell gap junction protein. *J.Clin.Invest.* vol. 91, pp. 997-1004, 1993.

[11] M. S. Pepper and P. Meda, Basic fibroblast growth factor increases junctional communication and connexin 43 expression in microvascular endothelial cells. *Journal of Cellular Physiology*, vol. 153, pp. 196-205, 1992.

[12] R. A. Meyer, P. D. Lampe, B. Malewicz, W. J. Baumann, and R. G. Johnson, Enhanced gap junction formation with LDL and apolipoprotein. *Exp. Cell Res.* vol. 196, pp. 72-81, 1991.

[13] A. Sachinids, Y. Ko, A. Wieczorek, B. Weisser, R. Lochner, W. Vetter, and H. Vetter, Lipoproteins induce expression of the early growth response gene-1 in vascular smooth muscle cells from rat. *Biochem. Biophys. Res. Comm.* vol. 192 pp. 794-799, 1993.

[14] S. Harris-Hooker, G. L. Sanford, V. Montgomery, R. Rivers, and N. Emmett, Influence of low density lipoproteins on vascular smooth muscle cell growth and motility: modulation by extracellular matrix. *Cell Biol. Internatl. Rep.* vol. 16, pp. 433-450, 1992.

[15] Y. C. Chai, P. H. Howe, P. E. DiCorleto, and G. M. Chisolm, Oxidized low density lipoprotein and lysophosphatidylcholine stimulate cell cycle entry in vascular smooth muscle cells -Evidence for release of fibroblast growth factor-2. *J. Biol. Chem.* vol. 271, pp. 17791-17797, 1996.

# Section 6:

# Function of Gap Junctions in Different Tissues

*Gap Junctions*
*R. Werner (Ed.)*
*IOS Press, 1998*



# Search for the gap junction-permeant signals that potentiate secretion of β-cells in contact

Domenico BOSCO and Paolo MEDA
*Department of Morphology, University of Geneva, CH-1211 Geneva 4, Switzerland*

**Abstract.** Glucose stimulation of insulin release is potentiated by contact between pancreatic β-cells. We have found that 1) this potentiation is not observed under conditions that block gap junction channels, in the presence of subthreshold concentrations of glucose and following stimulation by secretagogues that bypass the early steps of glucose metabolism ; 2) glucose-starved β-cells which do not respond to the sugar can be recruited to secrete after contact with control cells ; 3) contact with cells rich in glucose-6-phosphate induces some β-cells to secrete in the presence of non stimulatory levels of glucose. The data are consistent with the view that the glucose-induced potentiation of insulin release is dependent on junctional coupling, possibly via the intercellular exchange of glycolytic metabolites.

## 1. Introduction

Insulin secretion is a multicellular event that implies the coordinated activity of numerous β-cells within pancreatic islets. These cells achieve such a coordination by communicating via hormones, neuromediators and other signals that transit in the extracellular fluid, as well as via ions and molecules that pass Cx43 channels [1,2]. The influence of the latter mechanism is indicated by the following observations : single β-cells show secretory defects that are partly corrected by cell contact; pharmacological blockade of gap junction channels prevents this correction; sustained stimulation of insulin release increases β-cell coupling and Cx43 expression, while acute stimulation enhances the electrical synchronization of β-cells and their $Ca^{2+}$ oscillations [1] ; conditions that inhibit insulin release decrease or abolish β-cell coupling; cell lines which show abnormal insulin secretion do not express connexins and feature improved insulin biosynthesis and release after stable transfection for Cx43, provided this protein is expressed as in native pancreatic islets; overexpression of Cx43 affects the growth and secretion of tumoral insulin-producing cells, resulting in alterations of circulating levels of glucose and insulin; the expression of an ectopic connexin alters the glucose responsiveness of pancreatic β-cells of transgenic mice [1-5].

These observations indicate that adequate expression of Cx43 and junctional communication is required for proper control of insulin biosynthesis and secretion, in vitro as well as in vivo. However, the data do not provide information about the junction-permeant signals that are responsible for the improved functioning of coupled β-cells, nor about the reason(s) why these cells need to be coupled. To address these issues, we have used an immunological assay which allows for the parallel assessment of junctional coupling and insulin secretion of individual β-cells [6-8].

## 2. Glucose-induced insulin secretion is potentiated between β-cells by a mechanism which involves junctional coupling

By comparing the behaviour of single and paired β-cells under basal and stimulatory conditions, we have found that stimulation by glucose markedly increases the proportion of cells contributing to insulin output, while slightly enhancing the amounts of hormone released by each secreting cell [6-8]. However, whereas in the presence of glucose concentrations below 5.7 mM, single β-cells and β-cell pairs behaved similarly, in the presence of higher concentrations of the sugar β-cell pairs secreted about twice more frequently than single β-cells and also released, on the average, twice as much insulin [7]. From these changes, we calculated that, under conditions of maximal stimulation, pairs of β-cells secreted four times more insulin than single β-cells, i.e. that, on the average, each cell functioned twice as much when it was paired to another cell than when it was single.

Knowing that about 60 % of single β-cells do not release detectable amounts of insulin under maximal stimulatory conditions [6-8] and assuming that cell adhesion occurs at random between active and quiescent cells, only 16 % of the pairs would be expected to comprise two active β-cells and, thus, to release more insulin than single cells. This figure is significantly smaller than that (about 70%) of the pairs which actually show some potentiation of insulin release. The latter proportion fits nicely with that which can be computed assuming that contact actually activated all quiescent cells (60% of the population) that had encountered an active companion cell (40 % of the population). Equation $[0.4 \times 0.4 + 2 \times (0.4 \times 0.6)]$ predicts that this proportion should be 64 %.

Experiments further showed that the contact-dependent potentiation of insulin release was established within 10 min of contact, was not observed with pairs exposed to glucose concentrations $\leq 5.7$ mM and, in pairs stimulated by higher glucose levels, was abolished under conditions that elicited uncoupling without altering the secretion of single cells. Under these conditions, paired β-cells secreted, on the average, as single cells, even though they had preserved a close appositional contact with a companion cell. Thus, potentiation of secretion cannot be simply accounted for by the changes in cell adhesion or in the polarity of membrane and cytoplasm which obligatorily accompany cell contact and may depend on junctional coupling.

## 3. Potentiation of insulin secretion is not seen with all secretagogues

Glucose stimulates β-cells via a multistep mechanism which involves the metabolic processing of the sugar first through the glycolytic pathway and, then, through the citrate cycle. The two pathways concur to increase the cytoplasmic ratio of ATP/ADP concentrations which, in turn, elicits a cascade of electrophysiological and phosphorylation changes that eventually lead to the exocytosis of secretory granules [9].

Therefore, in the search of signal(s) potentiating insulin release, we investigated β-cell pairs exposed either to non metabolizable esters of succinic and glutamic acid or to D-glyceraldehyde. While the former molecules directly stimulate the citrate cycle [10], the latter activates the distal portion of the glycolytic pathway, bypassing the first steps of glucose metabolism [11]. We have found that none of these compounds potentiates the insulin release of β-cells in contact, in spite of the fact that all of them stimulated the secretion of single cells to the same extent (at the concentrations tested) observed in the presence of maximal stimulatory concentrations of glucose. These experiments suggest that the gap junction-permeant signal(s) that potentiate insulin release in the presence of suprathreshold concentrations of glucose may be metabolites generated in the glycolytic pathway, before the splitting of glucose into two molecules of triose-phosphate.

## 4. Potentiation of insulin secretion may be mediated by cell-to-cell transfer of glucose-6-phosphate

Among the candidate metabolites, we have considered glucose-6-phosphate, since this molecule plays a crucial role in the chain of events linking the stimulation of a β-cell to the secretion of insulin [9], crosses connexin channels and may coordinate the metabolic activity of β-cells in contact [12].

By comparing the secretion of β-cells after exposure to glucose levels expected to increase or to lower the intracellular concentration of glucose-6-phosphate, we have found that, when acutely challenged by a maximal stimulatory concentration of glucose, the former cells significantly increase their insulin release, whereas the latter cells do not. Mixing secreting and quiescent β-cells to favour the formation of pairs, we have found that in the presence of high glucose concentrations, the insulin release of pairs formed by either two secreting cells or by a secreting and a non secreting cell was four times larger than that of single cells. Thus, in these cases, each paired cell secreted, on the average, twice as much as a single cell. Since such a potentiation was not observed after contact between two non secreting β-cells, the data imply that quiescent cells were efficiently recruited for secretion by the normally responsive ones, via a mechanism other than cell-to-cell adhesion.

To assess whether such a mechanism could involve the junctional transfer of glucose-6-phosphate from a secreting into a quiescent cell, we studied a clone of RIN2A cells that had been stably transfected for Cx43 [13]. In contrast to normal β-cells, RIN cells show fairly low levels of stored insulin and reversed ratios of the GLUT2 and GLUT1 isoforms of glucose transporters and of the glucokinase and hexokinase isoforms of glucose phosphorylating enzymes [14-16]. As a result of these changes, RIN cells contain significantly higher levels of glucose-6-phosphate than primary β-cells and respond to levels of glucose that are too low to stimulate native β-cells. However, in contrast to these cells, RIN cells cannot increase their insulin output in response to a further elevation in the concentration of the sugar, in a range thought to be of physiological significance. Preliminary experiments under conditions which do not allow for a detectable release of insulin from either single RIN or β-cells and which do not potentiate secretion of pairs formed by either one of the two cell types, have revealed some insulin release from heterologous pairs formed by one RIN and one β-cell. Since this release cannot be accounted for by the secretion of individual RIN cells (the amount of insulin released by these cells is below the detection limit of our assay), this finding implies that cell contact was sufficient, at least in some cases, to stimulate β-cells in the presence of substimulatory levels of glucose. This unusual response may be mediated by the transfer of some signal from the glucose-6-phosphate-rich RIN cells into β-cells.

## 5. β-cells require junctional coupling because they are intrinsically heterogeneous

The reason(s) why junctional coupling is beneficial for insulin secretion remain(s) to be elucidated. As compared to other forms of cell-to-cell communication, coupling achieves a rapid intercellular equilibration of electrochemical gradients of ions and molecules that may be crucial for secretion. This equilibration, in turn, could ensure the recruitment and coordination of a population of cells that, taken individually, would otherwise function asynchronously. For example, cells deficient in some critical metabolite may be secretory quiescent until gaining from coupled neighbors sufficient amounts of the required cytoplasmic molecule to reach the threshold level for activation. Thus, coupling could account for both the recruitment and the activation of quiescent β-cells, which is observed during glucose stimulation [1,2].

A central feature of this model is that individual β-cells should be substantially heterogeneous in terms of metabolic and secretory characteristics. Increasing evidence indicates that this is actually the case. Thus, individual β-cells have been shown to markedly differ in terms of several structural and metabolic parameters, which result in widely different abilities to release insulin and and to biosynthesize proteins, including insulin [1,2,6,7].

## 6. Outlook

Insulin secretion results from the activity of numerous and functionally heterogeneous β-cells whose integration depends on cell-to-cell signaling by nutrients, intrinsic and extrinsic neural inputs, local and circulating hormones, as well as direct cell-to-cell communications. The contribution of each of these mechanisms to the net response of pancreatic islets remains to be fully understood, as does their hierarchic organization. In this context, cell-to-cell coupling has become an obligatory feature, probably because it provides the most direct way for compensating intrinsic metabolic and effector differences of secretory cells, hence permitting to build an appropriate output of insulin, starting from a highly heterogeneous β-cell population. Exchanges of early glycolytic metabolites through connexin channels appear important to ensure proper glucose responsiveness, and to upregulate secretagogue-induced insulin release. Validation of this mechanism awaits direct experimental testing and does not exclude the contribution of critical ion exchanges across connexin channels.

## Acknowledgments

Supported by grants from the Swiss National Science Foundation (32-34086.95), the Juvenile Diabetes Foundation International (197124) and the European Union (BMH4-CT96-1427).

## References

[1] Meda P. (1996) The role of gap junction membrane channels in secretion and hormonal action. J Bioenerg Biomembr 28:369-377.

[2] Meda P (1995) Junctional coupling of pancreatic ß-cells. In: Huizinga JD (ed) Pacemaker activity and intercellular communication. CRC Press, Boca Raton, Florida, pp 275-291.

[3] White TW, Chanson M, Huarte J, Meda P (1996) Generation of transgenic mice expressing Connexin 32 in pancreatic β-cells. Mol Biol Cell 7 suppl 461A.

[4] Vozzi C, Bosco D, Dupont E, Charollais A, Meda P (1997) Hyperinsulinemia-induced hypoglycemia is enhanced by overexpression of connexin 43. Endocrinology 138:2879-2885.

[5] Gylfe E, Grapengiesser E, Hellman B (1991) Propagation of cytoplasmic $Ca^{2+}$ oscillations in clusters of pancreatic β-cells exposed to glucose. Cell Calcium 12:229-240.

[6] Salomon D, Meda P (1986) Heterogeneity and contact-dependent regulation of hormone secretion by individual B cells. Exp Cell Res 162:507-520.

[7] Bosco D, Orci L, Meda P (1989) Homologous but not heterologous contact increases the functioning of individual secretory cells. Exp Cell Res 184:72-80.

[8] Bosco D, Meda P (1992) Actively synthetizing β-cells secrete preferentially during glucose stimulation. Endocrinology 129:3157-3166.

[9] Sweet IR, Najafi H, Li G, Grodberg J, Matschinsky FM (1997) Measurement and modeling of glucose-6-phosphatase in pancreatic islets. Am J Physiol 272:E696-E711.

[10] Bosco D, Meda P, Thorens B, Malaisse WJ (1995) Heterogenous secretion of individual B-cells in response to D-glucose and to non-glucidic nutrient secretagogues. Am J Physiol C611-C618.

[11] Alcazar O, Giné E, Quiu-Yue Z, Tamarit-Rodríguez J (1995) The stimulation of insulin secretion by D-glyceraldehyde correlates with its rate of oxidation in islet cells. Biochem J 310:215-220.

[12] Kohen E, Kohen C, Thorell B, Mintz DH, Rabinovitch A (1979) Intercellular communication in pancreatic islet monolayer cultures: a microfluorometric study. Science 204:862-865.

[13] Vozzi C, Ullrich S, Charollais A, Philippe J, Orci L, Meda P (1995) Adequate connexin-mediated coupling is required for proper insulin production. J Cell Biol 131:1561-1572.

[14] Ferber S, BeltrandelRio H, Johnson JH, Noel RJ, Cassidy LE, Clark S, Becker TC, Hughes SD, Newgard CB (1994) GLUT-2 gene transfer into insulinoma cells confers both low and high affinity glucose-stimulated insulin release. Relationship to glucokinase activity. J Biol Chem 269:11523-11529.

[15] Vischer U, Blondel B, Wollheim CB, Höppner W, Seitz HJ, Iynedjian PB (1987) Hexokinase isoenzymes of RIN-m5F insulinoma cells. Expression of glucokinase gene in insulin-producing cells. Biochem J 241:249-255.

[16] Newgard CB (1994) Cellular engineering and gene therapy strategies for insulin replacement in diabetes. Diabetes 43:341-350.

[17] Pipeleers DG (1992) Heterogeneity in pancreatic ß-cell population. Diabetes 41:777-781.

[18] Stefan Y, Meda P, Neufeld M, Orci L (1987) Stimulation of insulin secretion reveals heterogeneity of pancreatic B-cells in vivo. J Clin Invest 80:175-183.

*Gap Junctions*
*R. Werner (Ed.)*
*IOS Press, 1998*

# On the Role of Gap Junctional Communication in Pancreatic Acinar Cell Function

Marc Chanson[1], Habo J. Jongsma[2], Patrice Mollard[3], Klaus Willecke[4], Susanne Suter[1] and Paolo Meda[5]

[1]*Dept. of Pediatrics, Children Hospital, Geneva, Switzerland,* [2]*Dept. of Medical Physiology and Sports Medicine, Utrecht University, The Netherlands,* [3]*CNRS-INSERM Unit 401, Montpellier, France,* [4]*Institut für Genetik, Universität Bonn, Germany and* [5]*Dept. of Morphology, Medical Faculty, Geneva, Switzerland*

**Abstract.** The role of gap junctional communication in pancreatic acinar cell secretion is not understood. Using confocal digital imaging techniques and patch-clamp electrophysiology, we have found that junctional coupling provides for the intercellular diffusion of signaling molecules able to mobilize $Ca^{2+}$ from internal stores which, in turn, stimulate exocytosis. This mechanism may contribute to the recruitement of pancreatic acinar cells to secrete in response to increasing doses of the $Ca^{2+}$-mobilizing agonist acetylcholine (ACh). Maximal concentrations of ACh induced uncoupling of acinar cells, an effect that restricts the passage of $Ca^{2+}$-signaling molecules and should be expected to attenuate stimulated secretion. However, pancreatic acinar cells from Cx32-deficient mice showed enhanced basal secretion as compared to control (coupled) acinar cells, a finding which is in agreement with the previous observation that enzyme release is increased during heptanol-mediated acinar cell uncoupling. Enhanced basal secretion during uncoupling may therefore represent a compensatory mechanism to sustain enzyme output. Such a mechanism would explain why acinar cell uncoupling induced by $Ca^{2+}$-mobilizing agonists correlates with maximal secretory activity.

## 1. Introduction

The exocrine pancreas is formed by numerous secretory units of acinar cells, referred to as acini. The acini release a mixture of digestive enzymes in response to hormones and neurotransmitters which have specific receptors on acinar cell membranes [1,2]. The observation that individual acini showed morphological and biochemical differences has led to the suggestion that secretion of the exocrine pancreas may not be similarly contributed by each subunit of the gland. In keeping with this idea, substantial heterogeneity in the response to pancreatic secretagogues [3] and individual acinar cells [4,5] has been reported. *In vitro* studies have documented that dispersion of pancreatic acinar cells resulted in a marked reduction of exocrine secretion [6,7]. The finding that re-establishement of cell-cell contacts promote basal and stimulated secretion indicates that additional mechanisms should participate in the regulation of the exocrine pancreas function [5]. In this context, intercellular communication through gap junction channels could play a major role in the integration of signaling among acinar cells.

Pancreatic acinar cells are electrically and chemically coupled through Cx32- and Cx26-formed gap junction channels [8]. This extensive pattern of cell-to-cell communication is not affected by hormones and neurotransmitters with the exception of $Ca^{2+}$-mobilizing agonists, such as acetylcholine (ACh), cholecystokinine and bombesin [5,9-12]. This group of secretagogues induces cell uncoupling at concentrations which stimulate maximally the release of amylase, the most abundant digestive enzyme in murines. This modulation appears to be of physiological importance since an inverse relationship between coupling and secretion could be observed *in vivo* as well [13]. Two important questions have, therefore, to be answered. First, is the extensive acinar cell coupling involved during amylase secretion elicited by low agonist concentrations ? Second, is the uncoupling of acinar cells induced by high concentrations of secretagogues important for their secretory activity ? A response to these questions requires a better understanding of the temporal changes of cell-to-cell communication and second messengers involved in the control of secretion. The study of secretion in communication-deficient acinar cells as observed in genetically engineered mice lacking Cx32, should also provide a valuable experimental model to assess the role of gap junctional coupling in pancreatic secretion.

## 2. Extensive coupling favors recruitment of acinar cells to secrete

ACh stimulates amylase secretion in a dose-dependent manner up to a plateau. It is well established that ACh binds to $G_q$-linked membrane receptors, resulting in the production of diacylglycerol and inositol 1,4,5-trisphosphate ($IP_3$). The activation of protein kinase C by diacylglycerol and the $IP_3$-induced release of cytosolic free $Ca^{2+}$ ($[Ca^{2+}]_i$) from intracellular stores are key events in the regulation of enzyme secretion [14-16]. In addition of triggering exocytosis, a rise in $[Ca^{2+}]_i$ activates $Ca^{2+}$-dependent chloride and cationic channels. Because the activation of these membrane conductances parallels $Ca^{2+}$ mobilization [17-19], the former parameter is often used to indirectly monitor $[Ca^{2+}]_i$ changes (Fig.1).

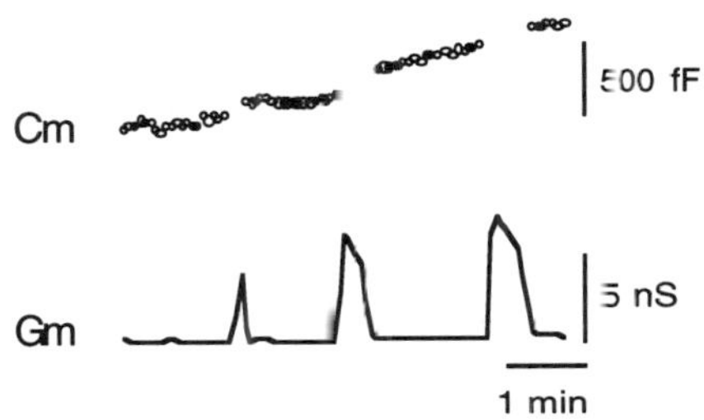


**Figure 1:** Relationship between activation of $Ca^{2+}$-dependent membrane conductances (lower trace, $G_m$) and exocytosis, as determined by measurement of membrane capacitance (upper trace, $C_m$) in a single rat pancreatic acinar cell successively stimulated with 100 nM, 1 µM and 5 µM ACh. $G_m$ and $C_m$ were monitored under perforated patch-clamp conditions.

In a first series of experiments, we compared the $Ca^{2+}$-dependent membrane conductance ($G_m$) changes induced by increasing concentrations of ACh in single and paired rat pancreatic acinar cells. We observed that single acinar cells were markedly heterogenous in terms of ACh-induced $G_m$ responses. At low ACh concentrations (50-100 nM), 67.5% of single cells were unresponsive. Responsive cells exhibited either a single transient or an oscillatory $G_m$ pattern (Fig.2). At ACh concentrations >500 nM, oscillations were replaced by transient responses.

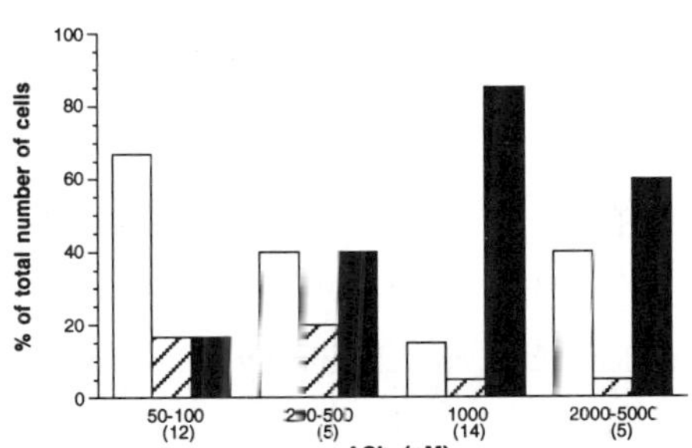


**Figure 2:** Patterns of $Ca^{2+}$-dependent conductance responses in single acinar cells and pairs exposed to increasing ACh concentrations. Single cells either did not respond (open columns), exhibited an oscillatory patterns of $Ca^{2+}$-dependent conductances (hatched columns) or responded transiently (solid columns).

Measurement of electrical coupling in pairs of acinar cells revealed that junctional conductance (42 ± 4 nS, n=37) was not affected during the $Ca^{2+}$ mobilization evoked by ACh concentrations up to 1 µM. Under these conditions, the proportion of unresponsive acinar cells was markedly decreased, representing 33.3% of the population, and an additional pattern of $G_m$ changes consisting of both transient and oscillatory characteristics was observed in 5-25% of cell pairs. Interestingly, this mixed $Ca^{2+}$ response was suppressed in the presence of uncoupling concentration of heptanol (Fig.3).

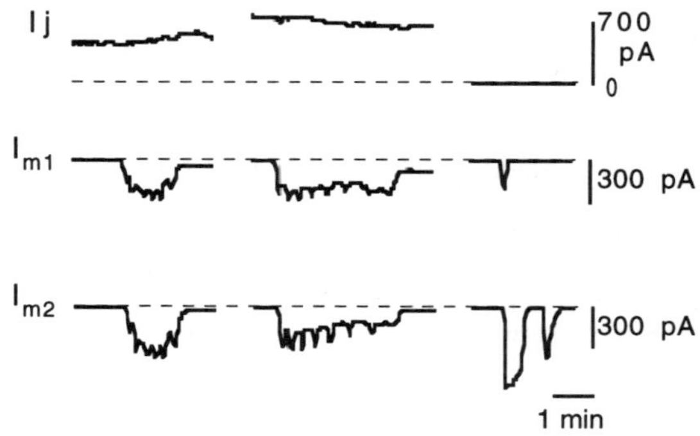


**Figure 3:** Examples of mixed $Ca^{2+}$-dependent current responses to ACh in two coupled acinar cell pairs (left and middle traces) and in one uncoupled acinar cell pair (right traces). $I_{m1}$ and $I_{m2}$ correspond to the membrane currents recorded in each cell of a pair, and $I_j$ to the junctional currents. The apparent simultaneous changes in $I_{m1}$ and $I_{m2}$ reflect a large electrical coupling.

These observations suggest that acinar cells in contact are metabolically coupled in terms of $Ca^{2+}$ mobilization [20-22], either because they release $[Ca^{2+}]_i$ at the same rate from their stores or because the electrochemical gradient of the second messenger was rapidly equilibrated across open gap junction channels. In a second series of experiments, we applied confocal digital imaging techniques on Fluo3-loaded pancreatic acinar cells to determine the temporal changes in $[Ca^{2+}]_i$. Studying individual cells within a cluster, we found that $Ca^{2+}$ mobilization was asynchronous between cells. Although the delay betwen the response of individual cells decreased with increasing agonist concentrations, high resolution of the signal revealed that $[Ca^{2+}]_i$ changes remained asynchronous during the response (Fig.4).

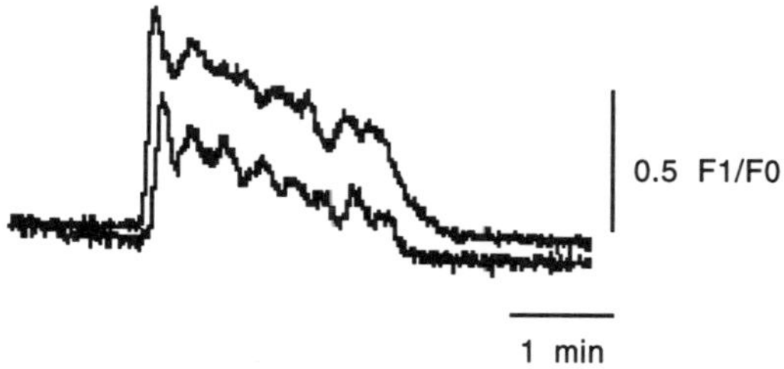


**Figure 4:** Asynchronous $[Ca^{2+}]_i$ changes in response to ACh measured using confocal digital imaging in the two cells of a pair loaded with Fluo3.

This observation is consistent with the view that $Ca^{2+}$-signaling molecules diffuse from one cell to another and not with the possibility that $Ca^{2+}$ is simultaneously released in all coupled cells. This idea was further confirmed by intracellular injection of 10-40 μM $IP_3$ into individual cells of clusters [23].We found that $IP_3$ evoked a rise of $[Ca^{2+}]_i$ in the injected cell and, subsequently, in up to 10 neighboring cells, indicating the intercellular passage of $Ca^{2+}$-signaling molecules. However, perfusion of $IP_3$ did not necessarilly result in $[Ca^{2+}]_i$ changes in acinar cells in direct contact with the stimulated one. These findings indicate that rat pancreatic acinar cells are not similarly sensitive to conditions mobilizing $Ca^{2+}$ from internal stores, as previously suggested by other observations [3-5]. With increasing concentrations of ACh, the rapid production of $IP_3$ in stimulated cells may equilibrate across open gap junction channels, thereby fine-tuning the $Ca^{2+}$ response. The role of other regulatory molecules involved in the receptor-mediated hydrolysis of phosphoinositides cannot, however, be ruled out. Together with the demonstration that the number of responsive acinar cells increased during secretagogue stimulation, our results strenghten the view that gap junctional communication contributes to the mechanism which enhances the recruitement of acinar cells able to mobilize $[Ca^{2+}]_i$ and to secrete.

## 3. Uncoupling increases basal acinar cell secretion

Intercellular communication has been repeatedly shown to decrease during maximal secretory activity of acinar cells, in apparent contradiction with the observation that extensive coupling favors cell recruitement. In an attempt to resolve this potential discrepancy, we investigated the relationship between coupling and secretion using gap junction blockers [3,24,25]. We found that long-chain alcohols, which rapidly uncouple pancreatic acinar cells, increased the basal release of amylase but not that elicited by maximal concentration of either ACh, or its analog carbamylcholine (CCh). This effect appeared directly related to cell uncoupling since long-chain alcohols increased in an almost additive manner the amylase release stimulated by other secretagogues that, alone, do not block cell-to-cell communication [12]. In addition, gap junction blockers did not cause detectable changes in the cytosolic concentrations of $Ca^{2+}$, $H^+$ and cAMP [25,27], nor did they cause activation of protein kinase C [27].

Recently, homozygous Cx32 deficient mice (KO32) have been generated by gene targetting [28]. These animals provides an excellent model to evaluate the physiological role of cell coupling in tissues expressing Cx32, such as pancreatic acinar cells. As expected, acinar cells from KO32 pancreata failed to express Cx32 and showed a drastically decreased exchange of Lucifer Yellow. Preliminary experiments showed that basal amylase release of acini from the knockout mice was increased as compared to that of control C57BL/6 mice. This enhanced basal secretion was similar to that elicited in control acini exposed to uncoupling concentrations of heptanol. These results suggest that the extensive coupling of acinar cells exert a tonic inhibitory effect on amylase secretion. During cell uncoupling, this inhibition is relieved and secretion is enhanced.

In such a system, a reduction in cell coupling, and therefore in the volume of cytoplasm of the interconnected acinar cells, could maintain the intracellular concentration of factors important for secretion at levels high enough to stimulate the secretory machinery. Alternatively, the diffusion of factors that negatively control secretion could be prevented during uncoupling, leading to enhanced secretory activity of acinar cells. Enhanced basal secretion during uncoupling may therefore represent a compensatory mechanism to sustain enzyme output.

## 4. Conclusions

Stimulation by ACh evokes a biphasic increase in $[Ca^{2+}]_i$ mobilization. The first rapid phase reflects $Ca^{2+}$ release from internal stores, whereas the second sustained phase is dependent on $Ca^{2+}$ influx into the cell [15,16,29]. Our results suggest that gap junctional communication fullfils two important roles during pancreatic acinar cell functioning. On the one hand, extensive coupling allows the cell-to-cell diffusion of $Ca^{2+}$-signalling molecules in response to ACh concentrations ranging from 50 nM to 1 μM, favoring the secretory recruitment of acinar cells. On the other hand, uncoupling, as observed in the presence of 1 μM ACh, enhanced the individual responsiveness of cells. In recent experiments, we observed that ACh-induced uncoupling was shortly delayed from the $[Ca^{2+}]_i$ changes, which take place during the $Ca^{2+}$ repletion phase. Such a delayed uncoupling may be required to temporarily separate nearby cells which differ in sensitivity to ACh and/or to allow for efficient reloading of internal $Ca^{2+}$ stores. Closure of gap junction channels is, however, associated with enhanced basal secretion, providing an explanation to the observation that acinar cell uncoupling correlates with maximal secretory activity.

*Supported by grants from the European Union (SC1-CT91-0644 to FHJ and BMH4-CT96-1427 to PM), the Swiss National Science Foundation (32-34086.95 to PM and 32-34090.92 to MC), the Juvenile Diabetes Foundation International (197124 to PM) and the European Science Foundation (travelling grant to MC, PM and HHJ).*

## References

[1] G.A. Scheele, Two dimensional gel analysis of soluble proteins. Characterization of guinea-pig exocrine pancreatic proteins, *J. Biol. Chem.* 250 (1975) 5375-5385.

[2] J.A. Williams and S.R. Hootman, The exocrine pancreas: Biology, Pathobiology and Diseases, In: V.L.W. Go, J.D. Gardner, F.P. Brooks, E. Lebenthal, E.D. Di Magno and G.A. Scheele (eds.), Raven Press, New York, 1986, pp. 123-139.

[3] D. Bosco *et al.*, Visualization of amylase secretion from individual pancreatic acini, *Am. J. Physiol.* 254 (1988) G664-G670.

[4] P. Willems *et al.*, Dose-dependent recruitment of pancreatic acinar cells during receptor-mediated calcium mobilization, *Cell Calcium* 14 (1993) 145-159.

[5] D. Bosco *et al.*, Heterogeneity and contact-dependent regulation of amylase release by individual acinar cells, *J. Cell Physiol.* 160 (1994) 378-388.

[6] A. Amsterdam and J.D. Jamieson, Studies on dispersed pancreatic exocrine cells. II. Functional characteristics of separated cells, *J. Cell Biol.* 63 (1977) 1057-1073.

[7] J.D Gardner and M.J. Jackson, Regulation of amylase release from dispersed pancreatic acinar cells, *J. 7hysiol.* 270 (1977) 439-454.

[8] P. Meda *et al.*, Junctional coupling and secretion of pancreatic acinar cells. In: E.L. Hertzberg and R.G. Johnson (eds.), Gap junctions. Modern Cell Biology 7, Alan R. Liss, Inc., New York, 1988, pp. 353-364.

[9] O.H Petersen and N. Ueda, Pancreatic acinar cells: the role of calcium in stimulus-secretion coupling, *J. Physiol.* 254 (1976) 583-606.

[10] N. Iwatsuki and O.H. Petersen, Acetylcholine-evoked electrical uncoupling and its ionic dependency, *J. Physiol.* 274 (1878) 81-96.

[11] O.H. Petersen and N. Iwatsuki, Hormone receptors in Digestion and Nutrition, In: G. Rosselin, P. Fromageot and S. Bonfils (eds.), Elsevier/North Holland Biomedical Press, Amsterdam, 1979, pp. 191-202.

[12] P. Meda *et al.*, Gap junctional coupling modulates secretion of exocrine pancreas,*Proc. Natl. Acad. Sci. (USA)* 84 (1987) 4901-4904.

[13] M. Chanson *et al.*, Extent and modulation of junctional communication between pancreatic acinar cells in vivo, *Am. J. Physiol.* 261 (1991) G28-G36.

[14] Y. Nishizuka, Studies and perspectives of protein kinas C, *Science* 233 (1986) 305-312.

[15] M.J. Berridge, Inositol trisphosphate and calcium signalling, *Nature* 361 (1993) 315-325.

[16] D.I. Yule *et al.*, Acetylcholine and cholecystokinin induce different patterns of oscillating calcium signals in pancreatic acinar cells, *Cell Calcium* 12 (1991) 145-151.

[17] Y.V. Osipchuk *et al.*, Cytoplasmic $Ca^{2+}$ oscillations evoked by receptor stimulation, G-protein activation, internal application of inositol trisphosphate or $Ca^{2+}$: simultaneous microfluorimetry and $Ca^{2+}$-dependent $Cl^-$ currents recording in single pancratic acinar cells. *EMBO J.* 9 (1990) 697-704.

[18] Y. Maruyama *et al.*, Agonist-induced changes in cell membrane capacitance and conductance in dialysed pancreatic acinar cells of rats, *J. Physiol.* 406 (1993) 299-313.

[19] P. Thorn *et al.*, Local and global cytosolic $Ca^{2+}$ oscillations in exocrine cells evoked by agonists and inositol trisphosphate, *Cell* 74 (1993) 661-668.

[20] C.C. Petersen and O.H. Petersen, Receptor-activated cytoplasmic $Ca^{2+}$ spikes in communicating clusters of pancreatic acinar cells, *FEBS Lett.* 284 (1991) 113-116.

[21] A. Ngezahayo and H.-A. Kolb, Gap junctional conductance tunes phase difference of cholecystokinin evoked calcium oscillations in pairs of pancreatic acinar cells, *Pfluegers Archiv* 422 (1993) 413-415.

[22] P. Loessberg Stauffer *et al.*, Gap junction communication modulates $[Ca^{2+}]_i$ oscillations and enzyme secretion in pancreatic acini, *J. Biol. Chem* 268 (1993) 19769-19775.

[23] D.I. Yule *et al.*, Intercellular calcium waves in rat pancreatic acini: mechanism of transmission, *Am. J. Physiol.* 271 (1994) C1285-C1294.

[24] P. Meda *et al.*, Blockage of cell-to-cell communication within pancreatic acni is associated with increased basal release of amylase, *J. Cell Biol.* 103 (1986) 475-483.

[25] M. Chanson *et al.*, Effects of n-alcohols on junctional coupling and amylase secretion of pancreatic acinar cells, *J. Cell Physiol.* 139 (1989) 147-156.

[26] R. Bruzzone *et al.*, Regulation of pancreatic exocrine function: a role for cell-to-cell communication? *Pancreas* 2 (1987) 262-271.

[27] M. Chanson *et al.*, Increase in pancreatic exocrine secretion during uncoiupling: evidence for a protein kinase C-independent effect, *Exp. Cell Res.* 182 (1989) 349-357.

[28] E. Nelles *et al.*, Defective propagation of signals generated by sympathetic nerve stimulation in the liver of connexin32-deficient mice, *Proc. Natl. Acad. Sci. (USA)* 93 (1996) 9565-9570.

[29] J.W. Putney and G. St. J. Bird, The signal for capacitative calcium entry, *Cell* 75 (1993) 199-201.

*Gap Junctions*
*R. Werner (Ed.)*
*IOS Press, 1998*



# Effects of an Antiarrhythmic Peptide on Intercellular Coupling via Gap Junctions

Stefan Dhein, Michaela Gottwald, Thomas Schaefer, Andreas Müller, Tatjana Tudyka, Kathi Krüsemann, Rajiv Grover
*Institute of Pharmacology, University of Cologne, Gleueler Str.24, D-50931 Cologne, Germany*

**Abstract.** We recently reported on a synthetic antiarrhythmic peptide (AAP10, NH2-GLY-ALA-GLY-HYP-PRO-TYR-CONH2) which was found to be effective against arrhythmia in the late ischemic period in isolated rabbit hearts. This peptide enhanced gap junctional current in pairs of adult guinea pig cardiomyocytes. In this study we wanted to investigate whether AAP10 acts on uncoupled guinea pig papillary muscles. After 30 minutes of equilibration at normoxic conditions the muscles were submitted to hypoxia with glucose free superfusion for 20 minutes with or without pretreatment with 10 nM AAP10. Under these conditions intracellular action potentials were recorded and the delay between stimulus and propagated action potential (stimulus-response interval, SRI) was evaluated. We found no effect of AAP10 under normoxic conditions on SRI or on action potential morphology. Resting membrane potential, amplitude, action potential duration, $dU/dt_{max}$ were not altered. However, while in untreated muscles uncoupling occurred after 12 minutes, this was not the case in muscles treated with AAP10. In additional experiments, we could demonstrate that uncoupling via 50 mM Na-propionate could be antagonized by 10 nM AAP10 without affecting other parameters than SRI. This AAP10 effect could be fully inhibited by 10 µM genistein and 1 µM bisindolylmaleimide I (a specific inhibitor of PKC), but not by 2 µM H8 (a specific PKA blocker) and not by 5 µM genistein. Using $^{14}$C-labeled AAP10 we found that the substance binds to membrane proteins but not to connexin 43. From these results we conclude that AAP10 can enhance intercellular coupling especially in situations with reduced coupling probably via a protein kinase C mediated mechanism.

One of the most intriguing problems in cardiovascular medicine is the prophylaxis of arrhythmia. When administered prophylactically, i.e. in periods of eurhythmic beating, antiarrhythmic drugs often exhibit adverse effects and provoke arrhythmia, which has been termed proarrhythmic risk [1-3]. In some cases this proarrhythmic activity was based on an increase in the dispersion of the action potential duration, which predisposes the heart to re-entrant arrhythmia because of the simultaneous occurrence of long and short action potentials in regions in close vicinity. Such enhancement in dispersion can be caused by reduced intercellular coupling, if there are less low resistance pathways between the cells. Thus, in computer simulations Lesh et al. [4] and Müller & Dhein [5] observed enhanced heterogeneity of action potential duration if coupling was reduced. From these findings we made up the hypothesis, that enhancement of cellular coupling may exert antiarrhythmic effects. Searching for substances which might enhance coupling, we became aware of an article describing a natural antiarrhythmic peptide, isolated from bovine atria, which enhanced the synchronization of spontaneously beating embryonic chick heart cell aggregates [6]. We found that this peptide reduced the dispersion of epicardial action potential duration in isolated rabbit hearts. Therefore, we developed a number of synthetic derivatives among which AAP10 (structure see fig. 1) was found to be most effective in reducing dispersion in isolated hearts. This synthetic antiarrhythmic peptide (AAP10) suppressed arrhythmia in the late ischemic period in isolated rabbit hearts [7, 8]. AAP10 enhances gap junctional coupling in pairs of adult guinea pig ventricular myocytes [9-11]. Since such an action might be of therapeutic interest especially in situations with reduced intercellular coupling, e.g. regional ischemia and acidosis, the goal of this study was to find out, whether this peptide may act preferentially on uncoupled preparations.

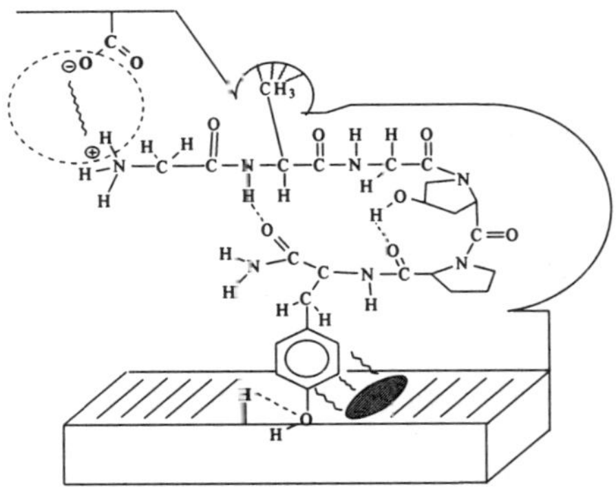

Figure 1: Spatial structure of AAP10 as revealed by two-dimensional NMR analysis. Note the horseshoe-like secondary structure being stabilized by intramolecular H-bonds. The figure also shows how this peptide could fit into a simplified putative receptor binding site as revealed from molecular modeling studies using AAP10-derivatives.

## 1. Methods

AAP10 was synthesized according to the Merrifield technique (Fmoc strategy) and was purified using HPLC (purity >99%). The substance is stable in aqueous solutions at pH 7.0 for about 2 hours and cannot be thawed more than 3 times. AAP10 sticks to glass and some plastic surfaces and has to be used with addition of 0.05% bovine serum albumin. At pH 8 or higher AAP10 is more stable. Guinea pig papillary muscles were placed in an organ bath and superfused with Tyrode solution ($Na^+$ 161.02, $K^+$ 5.36, $Ca^{++}$ 1.8, $Mg^{++}$ 1.05, $Cl^-$ 147.86, $HCO_3^-$ 23.8, $H_2PO_4^-$ 0.42, glucose 11 mM, pH 7.4, equilibrated with 5% $CO_2$ and 95% $O_2$, 37°C) at 7.5 ml/min. After 30 minutes of equilibration, the muscles were submitted to hypoxia and glucose free superfusion for 20 minutes with and without 10 nM AAP10. We recorded intracellular action potentials and evaluated the stimulus-response interval (SRI, delay between stimulus and the propagated action potential). In addition, papillary muscles were uncoupled by 50 mM sodium propionate. After 10 min. 10 nM AAP10 were administered additionally. In order to investigate the mechanism of action of AAP10, we carried out additional series of experiments: (1) in the presence of genistein (5 μM) to inhibit the tyrosine kinases, (2) of 100 μM genistein to inhibit other protein kinases, (3) in the presence of the specific PKC inhibitor bis-indolylmaleimide (1 μM) and (4) in the presence of the specific PKA inhibitor H8 (2 μM). The effect of AAP10 on gap junctional channels was established in freshly isolated pairs of guinea pig ventricular cardiomyocytes using double whole cell patch clamp with two SEC-05 switch clamp amplifiers (npi-electronic, Tamm, FRG) at a switching frequency of 25-30 kHz, 10kHz sampling rate, with two 2-3 MΩ glass-microelectrodes filled with: CsCl 125, NaCl 8, $CaCl_2$ 1, EGTA 10, $Na_2ATP$ 2, MgATP 3, $Na_2GTP$ 0.1, HEPES 10 mM, pH 7.2 with CsOH. Cells were superfused with Tyrode solution. We used the following voltage clamp protocol: both cells were clamped to a holding potential of -40mV. Thereafter, we adjusted the potential of one cell to values between -90 and +10 mV in steps of 10 mV for 200 ms, thus applying transjunctional voltages between -50 and +50 mV. We took the current elicited in the non-pulsed cell as the gap junction current [12, 13]. For further methodological details see [10, 11]. In order to investigate the site of action of the drug, we synthesized radioactive $^{14}$C-AAP10 and infused it (25 μCi) into isolated rabbit hearts. From these hearts Cx43 was purified via differential centrifugation and multistep density gradient centrifugation and finally SDS-PAGE according to the protocols given by [14-16] and we monitored the radioactivity during the isolation using a scintillation counter.

## 2. Results

In papillary muscle we found that AAP10 (10 nM; 1 μM) did not change resting membrane potential (-79.6±2 vs.-79.4±3 mV), $V_{max}$ (140±10 vs. 149±18 V/s), action potential duration ($AD_{90}$: 166±3 vs. 160±5ms) or action potential amplitude (113±2 vs. 108±1 mV). During normoxia there was no effect on stimulus-response interval (SRI) in the presence of 10 nM AAP10. In contrast, during hypoxia AAP10 prevented from the hypoxia-induced uncoupling (fig. 2), while hypoxia-induced depolarization (from -79.6 to -74.2 vs. -79.4 to -74.6mV) and action potential shortening (from 166 to 104 vs. from 160 to 107ms) were unaffected by AAP10. In parallel to uncoupling $V_{max}$ was slightly increased during hypoxia (from 140 to 158 V/s), while it was decreased during hypoxia in the presence of AAP10 (from 149 to 113 V/s), consistent with reduced coupling in hypoxia and an increase in coupling by AAP10. Under normoxic conditions, 1 μM AAP10 also reduced SRI (4.8±0.16 vs. 4.3±0.14 ms, $p<0.05$) without affecting other parameters of the action potential.

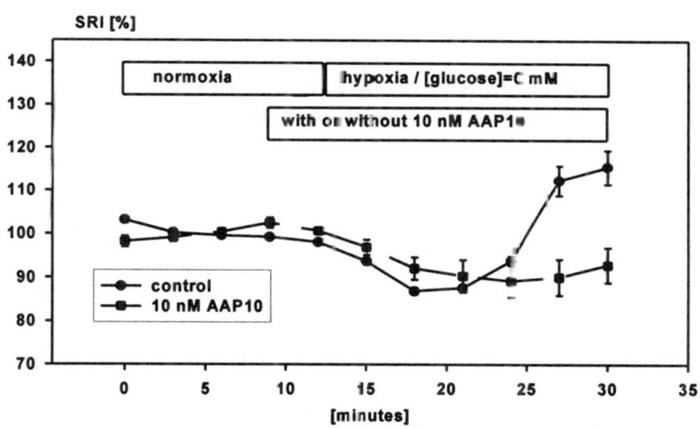


Figure 2: Stimulus-response-interval (SRI) of guinea pig papillary muscles during hypoxia in the absence and presence of 10 nM AAP10, which was infused into the bath after the timepoint as indicated (control: circles, n=14; AAP10 treated muscles: squares; n=6).

Application of sodium propionate to the papillary muscles resulted in a prolongation of SRI without affecting any other action potential parameter. This SRI-prolongation was significantly antagonized by 10 nM AAP10 (fig. 3). This effect of AAP10 was not inhibited by 5 μM genistein, but was antagonized by 100 μM genistein. It was also prevented by 1 μM bis-indolylmaleimide, but was not influenced by H8 (fig. 3). In isolated cell pairs the application of AAP10 (10 nM) resulted in an increase in gap junctional current as elicited by transjunctional voltages ranging from -50 to +50 mV (fig 4). The initial gap junction conductance was 160±25nS and decreased with time by -2.5±2 nS/min. During application of 10 nM AAP10 this reversed to an increase by 1.0±0.7nS/min ($p<0.05$). After wash-out, a decrease in Gj by -1.3±1.0 nS could be observed again ($p<0.05$; n=9).

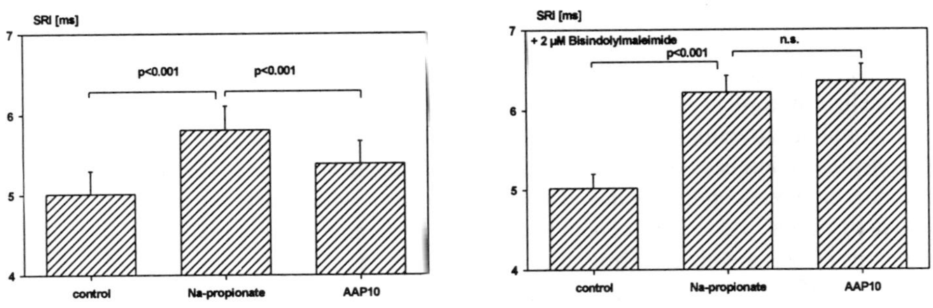


Figure 3: Effect of 10 nM AAP10 on the Na-propionate-induced increase in stimulus-response interval (SRI) in guinea pig papillary muscles (left). The influence of bisindolylmaleimide (2µM, right) on the AAP10 effect is also given (n=6 for each series).

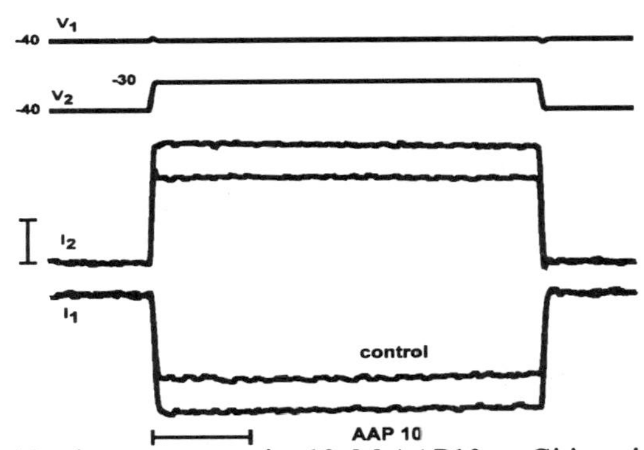


Figure 4: Effect of a 10 min. treatment with 10nM AAP10 on Gj in pairs of adult guinea pig cardiomyocytes. V1, V2, I1, I2 = voltages/currents measured in cell 1 and 2. Voltages and currents represent actual measurements. (Bars: 500 pA, 50 ms; clamp: -40 / -30mV)

Table 1: $^{14}$C-AAP10 radioactivity found in the isolation steps from isolation of membrane proteins from rabbit hearts perfused with 25 µCi $^{14}$C-AAP10.

| Pellet (cpm) | Supernatant (cpm) | Isolation-steps (according to [14, 15]; a detailed protocol with the step numbers is given by [16]) |
|---|---|---|
| 633 | 11962 | S9: after homogenization at pH 8.2, centrifuge: 33.000g, 15 min |
| 157 | 163 | S11: after rehomogenisation of the pellet, centrifuge: 33.000, 15 min |
| 90 | 62 | S16: after resuspension of pellet in 0.6 M KI-buffer: centrifuge: 27.000g, 30 min |
| 71 | 58 | S18: after resuspension of pellet in 0.6 M KI-buffer: centrifuge: 27.000g, 30 min |
| 83 | 39 | S21: after washing in ethanol and resupspension of the pellet in 0.6 M KI centrifuge: 27.000g, 15 min; (corresponds to a preparation of membrane proteins) |
| | | S22-S42: after resuspension of the pellet centrifuge in 35/44.5% sucrose density gradient (65.000g, 60 min, 15°C), repeat with the interface material, centrifuge second interface at 106.000g, 30 min, 15°C, centrifuge the pellet in 31.5/35% sucrose density gradient: 65.000g, 90 min, 15°C, centrifuge the interface material at 106.000g for 30 min, 15°C, wash, --> gap junction pellet: use for SDS-PAGE |
| 0 | Cx43-band | SDS-PAGE |

Biochemical analysis showed that $^{14}$C-AAP10 was bound to membrane proteins, but not to connexin 43 (see tab. 1).

## 3. Discussion

From these results it can be concluded that AAP10 can increase gap junctional current in cardiac cells. In cardiac tissue this action is reflected by a reduced stimulus-response interval without affecting other parameters of the action potential. Under normoxic conditions this action became obvious only with high concentrations of the peptide (1 µM), while under hypoxic conditions this effect could be seen with concentrations as low as 10 nM AAP10. Thus, it can be concluded that AAP10 acts preferentially on uncoupled preparations as became obvious from the difference in action during normoxia and hypoxia. This can explain the antiarrhythmic effect seen with the substance in earlier studies, especially against type Ib ventricular fibrillation in regional ischemia [7, 9, 17]. Furthermore, it is concluded that AAP10 acts via protein kinase C. Since PKC is normally activated via signaling cascades initiated by G-protein coupled membrane receptors, this finding indirectly suggests that the peptide acts via a membrane receptor. This is further supported by the observation, that an enrichment of 14C-AAP10 was found with the membrane proteins. PKC can enhance intercellular electrical coupling in Cx43 transfected SkHep1 cells [18, 19] via favoring a smaller conductance state. According to these studies the single channel conductance is shifted to lower conductance states following PKC activation, thereby reducing the dye permeability of the channel, but since the frequency of this conductance state is enhanced electrical coupling is increased by PKC. Presently, it is uncertain whether PKC directly mediates the effect of AAP10 via a phosphorylation of the channel, or indirectly for example via activation of MAP-kinase. Because of the lack of effect of H8 and 5 µM genistein tyrosine kinase and PKA probably are not involved in AAP10 action. Besides the mechanisms discussed above a recruitment of additional channels could also represent a possible mechanism of action. Future studies are directed to the mechanism of action on the single channel level. For evaluating a possible therapeutic role for substances which increase cardiac gap junctional coupling it is necessary to differentiate between treatment of acute manifest arrhythmia and prophylaxis. Enhancement of coupling can be a prophylactic antiarrhythmic approach in situations with reduced coupling, whereas reduction in coupling can terminate acute arrhythmia. (supported by the DFG and Köln-Fortune)

## References

1 Podrid PJ: Drugs 29, Suppl.4, (1985) 33-44.
2. Echt DS et al.: New Engl J Med 324 (1991) 781-788.
3. Dhein S et al.: Circulation 87 (1993) 617-630.
4. Lesh MD et al., Circ Res 65 (1989) 1426-40.
5. Müller A, Dhein S: Basic Res Cardiol 88 (1993) 11-22.
6. Aonuma S et al., Chem Pharm Bull 28 (1980) 3340-6.
7. Dhein S et al.: Naunyn Schmiedeberg's Arch Pharmacol 350 (1994) 174-184.
8. Dhein S, Tudyka T: Drugs 49 (1995) 851-855.
9. Dhein S et al.: Circulation 92, Suppl., (1995) I-641.
10. Müller A et al.: Naunyn Schmiedeberg's Arch Pharmacol 356 (1997) 76-82.
11. Müller A et al.: Eur J Pharmacol 327 (1997) 65-72.
12. Spray DS et al., J Gen Physiol 77 (1981) 77-90.
13. Weingart R, J Physiol (Lond.) 370 (1986) 267-284.
14. Manjunath CK et al.: Biochem Biophys Res Comm 142 (1987) 228-234.
15. Dhein S et al.: J Clin Exp Cardiol 2 (1997) 51-58.
16. Dhein S: Cardiac gap junctions. Karger Verlag, Basel, 1997 (in press)
17. Dhein S et al., Circulation 94, Suppl., (1996) I-715.
18. Kwak B et al.: Mol Biol Cell 6 (1995) 1707-1719.
19. Takens-Kwak B, Jongsma H: Pflüger's Arch 422 (1992) 198-200.

*Gap Junctions*
*R. Werner (Ed.)*
*IOS Press, 1998*

# Gap junction hemichannels in fiber cells of the mammalian lens

Eckert R., Donaldson P.J., and Kistler J.
*University of Auckland, School of Biological Sciences, Private Bag 92019, Auckland, New Zealand*

**Abstract.** We have identified a current in isolated lens fiber cells which closely resembles the current elicited by connexin46 hemichannels in the oocyte expression system. This current was activated by low external calcium, had a linear instantaneous current-voltage relationship, was inactivated at negative membrane potentials, and showed significant activation at potentials more positive than about -40 mV. Activation of this current was always associated with the activation of another linear non-selective cation conductance which depolarized the cells.

While hemichannel currents related to other connexins have been reported by a number of authors, to our knowledge, ours is the first study to report functional hemichannels in a mammalian tissue. Among the mammalian connexins, connexin46 appears to be unique in its ability to form hemichannels. It is selectively expressed in the lens fiber cells where it constitutes a mayor component of the total membrane protein. Lens fiber cells, therefore, could prove a valuable model system to study the physiological relevance of gap junction hemichannels in general.

## 1. Introduction

Among the mammalian connexins, Cx46 is remarkable for its ability to form functional hemichannels. To date, activation of hemichannel currents has mainly been observed in model systems such as oocytes or stably transfected cells which overexpress connexins. However, Cx46 is also one of the two lens specific gap junction proteins, which make up approximately 5% of the total lens fiber cell membrane protein. This situation closely resembles that in heterologous expression systems and, therefore, lens fiber cells seem to be an obvious target to look for hemichannel activation and its effects in the context of a normal tissue.

## 2. Lens Fiber Cell Preparation

Elongated fiber cells from the inner cortical portion of the lens are fragile, difficult to dissociate, and maintain in culture. We have recently succeeded in developing a preparation procedure that produces fiber cells suitable for electrophysiological recording [1] which has been modified for cortical fiber cells. In brief, 3-5 day old female Wistar rats were sacrificed by decapitation and the lenses dissected free from the extracted eyes and placed into a Petri dish containing EDTA-Ringer (100 mM NaCl, 4.7 mM KCl, 1 mM $MgCl_2$, 25 mM glucose, 60 mM mannitol, 10 mM EDTA, 10 mM HEPES pH 7.4, 320 mOsM) for at least 30 minutes at 37°C. Lenses were then placed under a dissecting microscope and the capsule removed by first rupturing it with a small tear in its posterior surface and then peeling it off using a pair of sharpened forceps. The remainder of the lens was dissected into a cortical portion and the lens core, a procedure which was simplified by the fact that the core region opacified when the temperature was not maintained at 37°C. Fiber cells were dissociated from the cortical portions by trypsinization for 40 min at 37°C. Trypsin was inactivated by addition of an equal volume of undiluted foetal calf serum and the cells were washed two to three times in EDTA-Ringer before plating into a recording chamber and use for electrophysiological experiments.

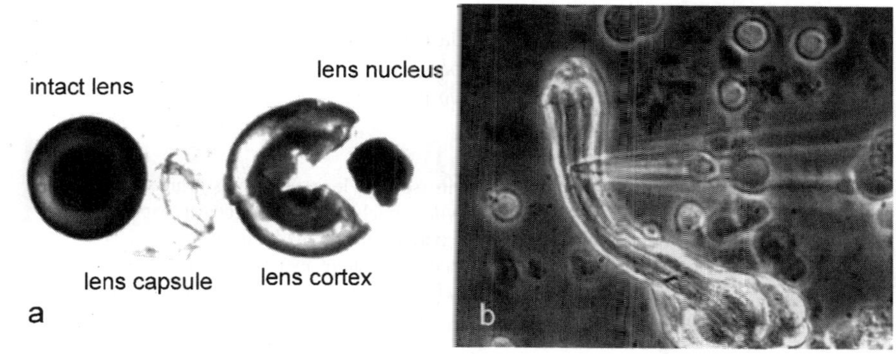


**Figure 1.** a) Preparation procedure for cortical lens fraction used for isolation of rat lens fiber cells. b) Whole-cell recording from cells of an isolated lens fiber bundle.

## 3. Electrophysiological Recording of Hemichannel-like Currents in Lens Fiber Cells

For current recording from isolated fiber cells standard whole-cell patch-clamp procedures were employed [2]. Patch pipettes were pulled to a final resistance of 2-5 MΩ and filled with CsCl pipette saline (130 mM CsCl, 2 mM $MgCl_2$, 10 mM HEPES, 10 mM EGTA/KOH pH 7.30). Whole cell currents from single cells were recorded using an Axopatch 1D patch clamp amplifier (Axon Instruments, Foster City, CA) in conjunction with the pClamp 5.5 software package (Axon Instruments) to generate voltage waveforms. For most of the recordings small fiber cell bundles were used rather than isolated fiber cells (Figure 1b), as they were usually more stable and easier to handle.

In $Ca^{2+}$ free conditions (EDTA-Ringer's) we have found a slowly activating current (h-

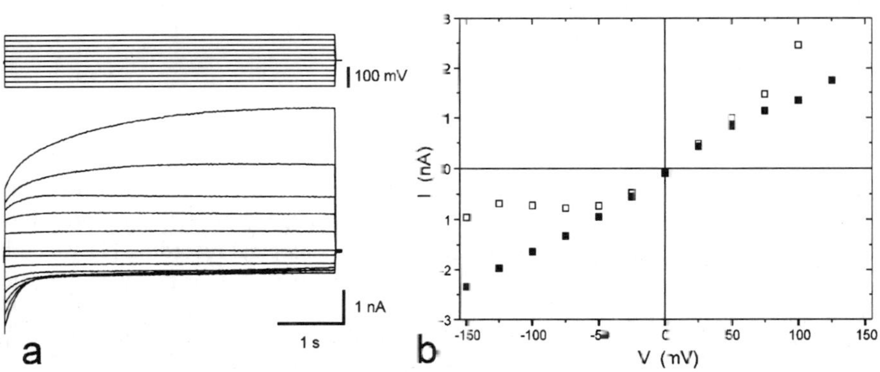


**Figure 2.** a) Hemichannel-like currents in elongated rat lens fiber cells. b) Current-voltage relation if the instantaneous(■) and steady-state current (□).

current) which closely resembles currents elicited by Cx46-hemichannels in the *Xenopus* oocyte expression system [3-5]. This h-current had a linear I/V relation for the instantaneous current with a reversal potential close to zero. It was inactivated at negative membrane potentials and still showed significant activation at potentials more positive than about -40 mV (Figure 2).

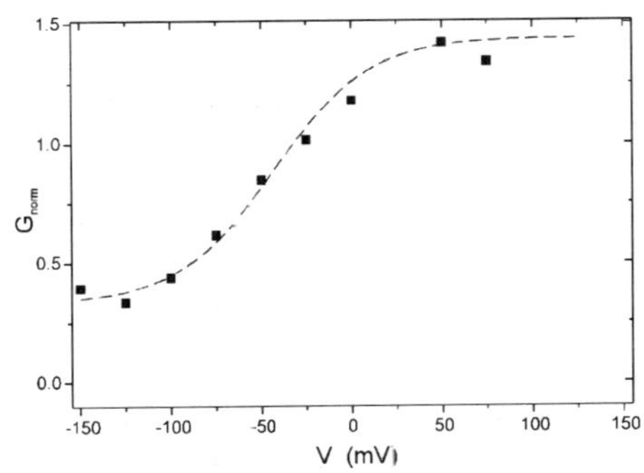


**Figure 3.** Steady-state activation curve for h-currents from a holding potential of 0 mV.

The steady-state activation of this current (Figure 3) could be described by a single Boltzmann function of the form

$$G_{norm} = G_{tot}\left(\frac{1-G_{\min}}{1+e^{(-A(V-V_0))}} + G_{\min}\right)$$

with the scale factor $G_{tot} = 1.4$, $G_{min} = 0.2$, $V_0 = -46$ mV, and $A = 0.036$ $mV^{-1}$. Similar results were obtained for measurements with slow linear voltage ramps (Figure 4) except that two superimposed Boltzmann functions were needed to fit the data due to a slight inactivation component at large positive voltages.

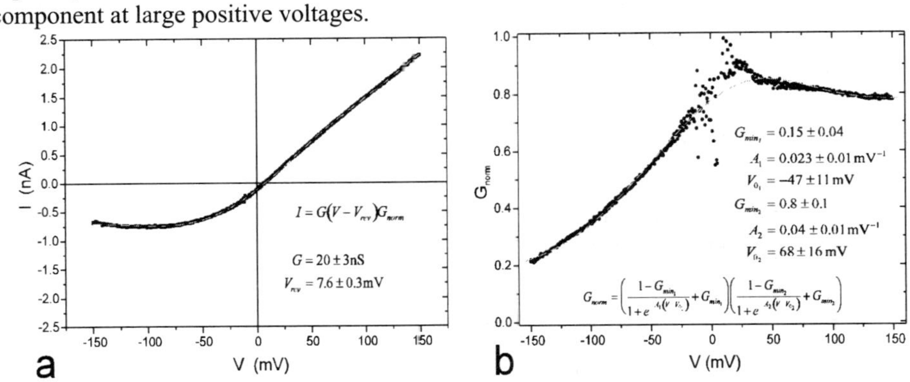


**Figure 4.** Steady state activation current measured in response to a slow linear voltage ramp (30 mV/s). a) I/V curve. b) Steady-state activation obtained by dividing by $G(V-V_{rev})$.

Similar to the hemichannel currents observed in oocytes, activation of the h-current was also dependent on the holding potential (Figure 5). Activation was significantly reduced when voltage pulses were applied from a holding potential of -70 mV as compared to -10 mV.

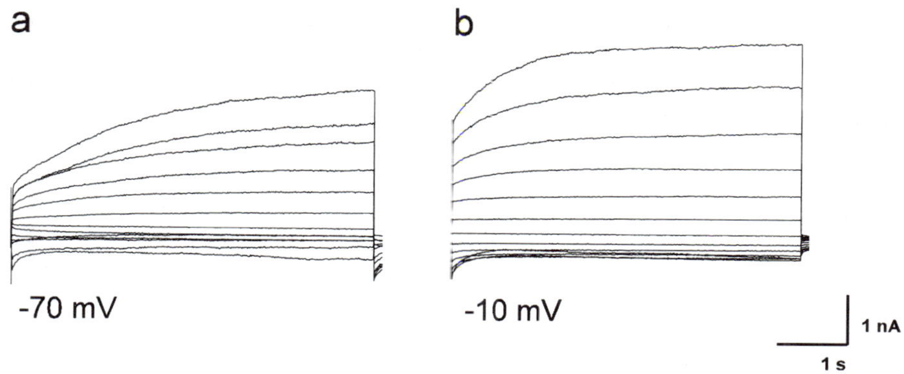


**Figure 5.** Effect of holding potential on the activation of h-currents.

Tail current analysis (Figure 6) revealed a linear current-voltage relation in the range of test voltages (-150 mV-150 mV) with a reversal potential close to zero but at slightly positive potentials. Reversal potentials obtained from tail currents and voltage ramps varied in the range of 4-7 mV. While in very rare cases we were able to see the onset of this current, in the majority of cells these currents were already activated and the membrane potential, therefore, depolarized.

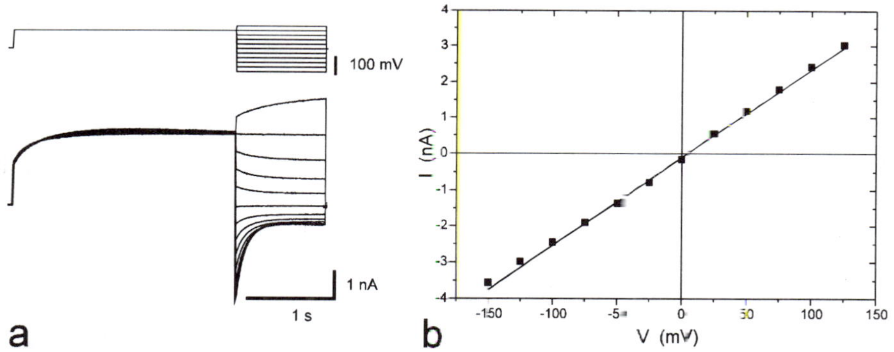


**Figure 6.** a) Tail currents in elongated rat lens fiber cells. b) Current voltage relation of tail currents.

These currents were always associated with a nonselective linear leak current which has been subtracted electronically. This leak component appeared to be a cation current. It was also blocked by external calcium albeit at lower concentrations than the hemichannel-like currents since this cation current was considerably reduced in Ringer bath solution containing nominal $Ca^{2+}$.

## 4. Discussion

We have identified a membrane current in the plasma membrane of elongated rat lens fiber cells which has similar properties to that of Cx46 expressed in *Xenopus* oocytes [3-5]. In lens fiber cells, these hemichannel-like currents are observed under conditions similar to those used for activation of hemichannel currents in the oocyte expression systems. Since in the lens Cx46 comprises a significant proportion of the total membrane protein, it seems likely that the observed current is the result of an activation of a population of unpaired hemichannels.

So far all evidence for gap junction hemichannels is purely by analogy of the slow activation kinetics and activation in low calcium conditions in oocytes. The final proof would require a pharmacological profile using specific activators and blockers for gap junction channels and hemichannels. Unfortunately, such specific reagents are currently not available. In addition, it should be noted that while the currents observed in lens fiber cells are *similar* to Cx46 hemichannels in oocytes, there are significant differences in some of their properties. However, lens fiber cells are known to express two connexins, Cx46 and Cx50, which form heteromeric hemichannels [6]. Any differences between the properties of hemichannel currents in fiber cells and oocytes may, therefore, be the result of heteromeric hemichannel formation in lens fiber cells.

The activation mechanism and permeability of Cx46 hemichannels are consistent with the depolarization of the membrane potential observed in whole lenses after removing external calcium. This depolarization is due to the activation of non-selective cation conductances and has also been implicated in the formation of cataract [7]. Since lens depolarization is a central event in cataract formation it is conceivable that this depolarization activates hemichannels even in the presence of external calcium. This hemichannel activation would lead to cell swelling and lysis which is characteristic of the changes that occur in osmotic cataract [8]. It is interesting to note, that in oocytes expressing Cx46 under similar conditions, the membrane potential is reduced and the cells become osmotically impaired and start to lyse [3]. Preliminary data also suggest an involvement or interference of h-currents in cell volume regulation.

These data so far are consistent with an involvement of gap junction hemichannels in pathophysiological processes (e.g. cataract). Whether fiber cells also maintain an independent pool of hemichannels for physiological processes remains to be seen.

## References


[1] P.J. Donaldson *et al.*, Changes in lens connexin expression lead to increased gap junctional voltage dependence and conductance. *Am. J. Physiol.* 269 (1995) C590-C600

[2] O.P. Hamill *et al.*, Improved patch-clamp techniques for high-resolution current recording from cells and cell-free membrane patches. *Pflügers Arch.* 391 (1981) 85-100

[3] D.L. Paul *et al.*, Connexin46, a novel lens gap junction protein, induces voltage-gated currents in nonjunctional plasma membrane of Xenopus oocytes. *J. Cell Biol.* 115 (1991) 1077-1089

[4] L. Ebihara E. Steiner, Properties of a nonjunctional current expressed from a rat connexin46 cDNA in Xenopus oocytes. *J. Gen. Physiol.* 102 (1993) 59-74

[5] E.B. Trexler *et al.*, Voltage gating and permeation in a gap junction hemichannel. *Proc. Natl. Acad. Sci. USA* 93 (1996) 5836-5841

[6] J.X. Jiang D.A. Goodenough, Heteromeric connexons in lens gap junction channels. *Proc. Natl. Acad. Sci. USA* 93 (1996) 1287-1291

[7] G. Duncan *et al.*, Calcium, cell signalling and cataract. *Prog. Retin. Eye Res.* 13 (1994) 623-652

[8] J. Bond *et al.*, Liquefaction of cortical tissue in diabetic and galactosemic rat lenses defined by confocal laser scanning microscopy. *Inv. Ophthal. Vis. Sci.* 37 (1996) 1557-1565

*Gap Junctions*
*R. Werner (Ed.)*
*IOS Press, 1998*



# Hypertension-Induced Modulation of Cx43 in Cardiovascular Muscle Cells

Jacques-Antoine HAEFLIGER, Gérard WAEBER, Pascal NICOD, Bernard WAEBER*, and Paolo MEDA**

*Department of Internal Medicine B, *Division of Hypertension, University Hospital, CHUV-1011 Lausanne, and **Department of Morphology, University of Geneva, CMU-1211 Genève4, Switzerland*

**Abstract**. The aim of this study was to investigate the effects of experimental hypertension on Cx43 expression in rat aorta and heart. To this end rats were made hypertensive by clipping one renal artery (two kidney, one-clip renal model) or by administration of deoxycorticosterone and salt (DOCA-salt model). After four weeks, all rats showed a similar increase in blood pressure, and in thickness of both aortic wall and heart. Northern blot analysis and immunolabeling for Cx43 showed that hypertensive rats expressed twice as much Cx43 than controls in aorta but not in heart. The data suggest that localized mechanical forces induced by hypertension regulate Cx43 expression in a tissue-specific manner. Since rats made similarly hypertensive by inhibiting nitric oxyde synthase with L-NAME, did not feature increased expression of Cx43 in aortic smooth muscle cells, this regulation appears related to the distensibility of conduit arteries, which differed in the 2K,1C and the L-NAME models.

*Key words*: Cell-Cell Communication, Connexins, Vasculature, Heart, Gap Junctions, Hypertension, Blood Pressure, Renin

## *1. Introduction*

The electrical and mechanical coupling of different types of muscle cells ensured by gap junctions [1,2] is critical in heart where proper propulsion of blood, obligatory depends on the coordinated contraction of atrial and ventricular cardiomyocytes [3].

Coordination of smooth muscle cells of vascular walls is also critical to the local modulation of vasomotor tone, thus contributing to the proper function of large vessels. The aorta, which is a sparsely innervated and electrically quiescent vascular tissue, is likely to be particulary dependent on gap junctional communication for coordinating the responses of smooth muscle cells to diverse neural and endothelial signals [4,5]. To assess whether conditions perturbing the function of the aortic wall are associated to alterations of connexins, gap junctions or coupling, we have studied the expression of Cx43, the physiologically predominant connexin of myocardial cells [3] and aortic smooth muscle cells [6,7] during chronic hypertension. To this end, we have investigated two rat models characterized by a similar degree of hypertension and of hypertrophy of both aorta and heart, but differing markedly by the mecanism causing these changes. In the two kidney, one-clip model (2K,1C), hypertension was produced by clipping one renal artery, leading to stimulation of renin secretion and to an angiotensin II-dependent elevation of blood pressure [8,9]. In the mineralocorticoid-salt induced model (DOCA-salt), hypertension results from sodium retention while renin secretion is suppressed [10,11].

## 2. Results

### 2.1. Effects of hypertension on the aorta in the 2K,1C and DOCA-salt models

Mean intra-arterial blood pressure of both 2K,1C and DOCA-salt animals was similarly elevated compared to that of normotensive controls (Table1).

As shown in Table 1 the thickness of the aortic wall (intima plus media layers) was significantly larger in hypertensive than in normotensive animals, resulting in a 30% increase of the vessel cross-sectional area, in spite of a constant lumen radius. These changes were due to an enlargment of smooth muscle cells, whose numerical density was slightly reduced in the media region of the two models studied. These cellular changes were paralleled by a significantly increase in the expression of α-skeletal actin.

Quantitative assessment of Cx43 gene expression by Northern blotting of total RNA, showed significantly higher values in the aorta of hypertensive than of normotensive rats (Figure 1). These changes were paralleled by a modest but sizeable increase in the amount of Cx43 as monitored by immunolabelling of aorta cryosections [12].

**Table 1. Morphometric Characteristics of Aorta**

| Groups | Mean Blood Pressure (mm Hg) | Lumen Radius (μm) | Media-Intima Thickness (μm) | Cross Sectional Area ($10^4$ μm$^2$) | Smooth Muscle Cells Numerical Density (μm$^{-2}$) | Smooth Muscle Cells Profile Area (μm$^2$) |
|---|---|---|---|---|---|---|
| 2K,1C | 203.4 ± 7.9*** | 632 ± 7 | 78.6 ± 2.8*** | 33.2 ± 1.2*** | 4.4±0.1 *** n=96 | 70.1±1.8*** m=399 |
| Sham | 128.4 ± 1.5 | 636 ± 10 | 55.6 ± 1.6 | 23.2 ± 0.7 | 4.9±0.1 n=88 | 61.4±1.3 m=320 |
| DOCA-salt | 192.3 ± 8.5*** | 626 ± 14 | 81.7 ± 1.8*** | 34.1 ± 0.5** | 3.8±0.1 *** n=24 | 68.5±1.8*** m=240 |
| Control | 134.3 ± 3.3 | 649 ± 21 | 56.3 ± 2.4 | 23.9 ± 1.2 | 4.3±0.1 n=24 | 60.6±1.6 m=240 |

2K,1C: two-kidney, one-clip hypertension model
DOCA-salt: deoxycorticosterone-salt hypertension model
Numerical Density = number of cells/1000 μm$^2$ ; (n = number of aortic fields measured)
Profile area = average size of smooth muscle cells; (m = number of smooth muscle cells measured)
** p<0.01 vs Control; *** p<0.001 vs Sham or Control

### 2.2 Effects of hypertension on the heart in the 2K,1C and DOCA-salt models

Hypertensive rats showed a 30% increase in heart index compared to normotensive counterparts [12]. In agreement with this change, Northern blot analysis showed a two fold increase in the expression of α-skeletal actin mRNA in the two groups of hypertensive rats, and histology revealed a thickening of the left ventricular wall [12].

Quantitative assessment by Northern blot analysis, showed that Cx43 expression was similar in the hypertrophied hearts of hypertensive rats and in those of normotensive controls (Figure 1). This finding correlated with the absence of significant differences in the amount of Cx43 which could be immunolabelled in heart muscle [12].

## *2.3 Effects of hypertension on the aorta in the L-NAME model*

Rats were made hypertensive, as in the 2K,1C and DOCA-salt models, by inhibiting nitric oxyde synthase with N-nitro-L-arginine-methyl ester (L-NAME) [13]. When compared to untreated controls, the aortas of these hypertensive animals did not feature increased levels of Cx43 mRNA (Haefliger et al., unpublished).

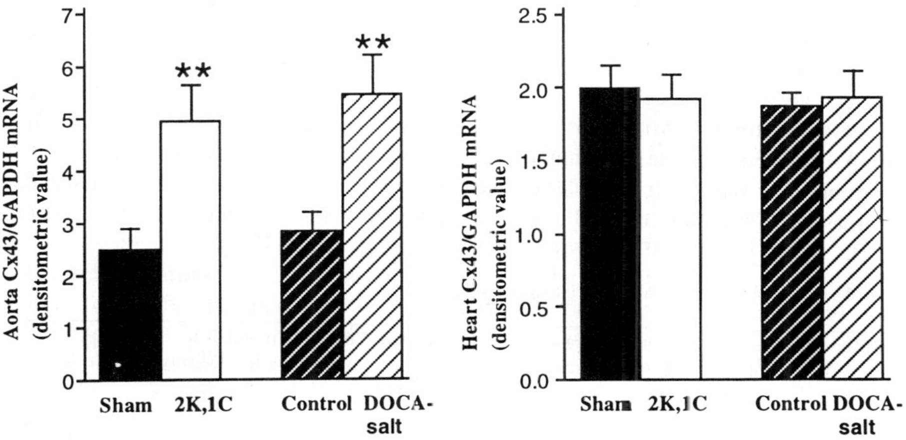


**Figure 1: Expression of Cx43 is increased in muscle cells of aorta but not of heart.**
*Left panel:* Analysis of aorta RNA revealed that the transcript for Cx43, which was mostly contributed to by smooth muscle cells, was increased about two fold in the hypertensive rats of the two models we investigated.
*Right panel:* A similar analysis of heart RNA revealed that the levels of the Cx43 transcript, which was mostly contributed to by cardiomyocytes, were not altered in the two types of hypertensive rats we investigated.
Values represent ratios of densitometric measurements of Cx43 and GAPDH mRNAs, and are expressed as mean ± SEM. Asterisks denote a difference significant at the $p < .01$ level.
Reproduced from *Circulation*, 1997, Vol.95, pp 1007-1014 copyright permission of the American Heart Association.

## *3. Discussion*

We have examined the effects of chronic experimental hypertension on the expression of Cx43, the major native connexin of the cardiovascular system [14]. We have found that hypertension was associated to increased levels of Cx43 in the wall of aorta, but not of heart, in two rat models characterized by comparable changes in blood pressure, due to different mechanisms.

In the 2K,1C model, the development of hypertension results from the constriction of one renal artery and the ensuing activation of the renin-angiotensin system [9], as reflected by enhanced renin mRNA levels in the hypoperfused kidney and by elevated plasma renin activity [15]. This change leads to the generation of the biologically active angiotensin II, a vasoconstrictor peptide which also plays a role in the development of vascular and cardiac hypertrophy [16,17]. Therefore, changes observed in 2K,1C rats could be due to an increase in blood pressure or in angiotensin II. To discriminate between these possibilities, we have studied the DOCA-salt model in which hypertension results from the administration of a salt-retaining mineralocorticoid in association with a high sodium intake [10]. After one month, the increase in blood pressure achieved in the two models was comparable, and the two groups of hypertensive animals also exhibited a similar degree of thickening of the aortic wall. All hypertensive rats also featured a comparable increase in the level of Cx43 expressed by smooth muscle cells of the aortic media, in spite of different circulating levels of angiotensin II in the 2K,1C and the DOCA-salt groups. Hence, the increase in Cx43 is likely to be associated to the elevated blood pressure.

The molecular mechanism leading to this change remains to be elucidated. The presence of multiple promoters in the 5' untranslated region of the Cx43 gene [18] raises the possibility that this increase is regulated in a tissue-specific way by distinct transcription factors [19,20]. Transcription of the Cx43 gene may be promoted by increased expression of c-fos [21], a transcription factor which accumulates in smooth muscle cells of rat aortas following exposure to angiotensin II [22]. The reasons why Cx43 increases between the smooth muscle cells of the aortic wall during hypertension also remains to be elucidated. Since gap junctions coordinate the mechanical contraction of smooth muscle cells [5], Cx43 may be implicated in modulating the vasomotor tone of aortic wall by allowing for the propagation of second messengers [6]. The finding that Cx43 is not upregulated between smooth muscle cells of rats treated with L-NAME, in spite of the fact that this drug increased mean blood pressure as in the 2K,1C and DOCA-salt models, further suggest that the Cx43-mediated cell-to-cell communication is somehow related to the distensibility characteristics of conduits arteries. In the 2K,1C model, carotid distensibility was increased, whereas in the L-NAME model, carotid distensibility was decreased [13,23].

Under the conditions we studied, all hypertensive animals featured cardiac hypertrophy [12] in the absence of differences in the levels of Cx43 between myocardial cells. This finding suggests that Cx43 is not involved in the myocardial adaptation that accompanies a hypertension-induced increase in heart load.

## Acknowledgements

Dr. Haefliger is supported by a career award from the Max Cloëtta Foundation. This work was supported by grants from the Swiss National Science Foundation (31-46770.96 to Dr. Haefliger, 32-29317.91 to Dr. Waeber, 32-43086.95 to Dr. Meda) and the European Union BMH4-CT96-1427 (to Dr. Meda).

## References

[1] D. C. Spray and J. M. Burt, Structure-activity relations of cardiac gap-junction channel. *Am J Physiol*, vol. 258, pp. C195-C205, 1990.

[2] G. J. Christ, Modulation of $a_1$-adrenergic contractility in isolated vascular tissue by heptanol: A functional demonstration of the potential importance of intercellular communication to vascular response generation. *Life Sci.* vol. 56, pp. 709-721, 1995.

[3] N. J. Severs, Pathophysiology of gap junctions in heart disease. *J. Cardiovasc. Electrophys.* vol. 5, pp. 462-475, 1994.

[4] S. S. Segal, Cell-to-Cell communication coordinates blood flow control *Hypertension*, vol. 23, pp. 1113-1120, 1994.

[5] G. J. Christ, D. C. Spray, M. El-Sabban, L. E. Moore, and P. R. Brink, Gap junction in vascular tissues. Evaluating the role of the intercellular communication in the modulation of vasomotor tone *Circ. Res.* vol. 79, pp. 631-646, 1996.

[6] G. J. Christ, P. R. Brink, W. Zhao, J. Moss, C. M. Gondré, C. Roy, and D. C. Spray, Gap junctions modulate tissue contractility and alpha adrenergic agonist efficacy in isolated rat aorta *J. Pharmacol. Exp. Ther*, vol. 266, pp. 1054-1065, 1993.

[7] R. Bruzzone, J.-A. Haefliger, R. L. Gimlich, and D. L. Paul, Connexin40, a component of gap junctions in vascular endothelium, is restricted in its ability to interact with other connexins. *Mol Biol Cell*, vol. 4, pp. 7-20, 1993.

[8] H. Goldblatt, J. Lynch, R. F. Hanzal, and W. W. Summerville, Studies on experimental hypertension: production of persistent elevation of systolic blood pression by means of renal ischemia. *J. Exp. Med.* vol. 59, pp. 347-379, 1934.

[9] F. H. H. Leenen, W. De Jong, and D. De Wied, Renal venous and peripheral plasma renin activity in renal hypertension. *Am J Physiol*, vol. 225, pp. 1513-1518, 1973.

[10] H. Gavras, H. R. Brunner, J. H. Larah, E. D. Vaughn, M. Koss, L. J. Cote, and I. Gavras, Malignant hypertension resulting from deoxycorticosterone acetate and salt excess. *Circ Res*, vol. 36, pp. 300-309, 1975.

[11] D. T. Liu, I. Birchall, T. Hewitson, P. Kincaid-Smith, and J. A. Whitworth, Effect of dietary calcium on the development of hypertension and hypertensive vascular lesions in DOCA-salt and two-kidney, one clip hypertensive rats. *Journal of Hypertension*, vol. 12 pp. 145-153, 1994.

[12] J.-A. Haefliger, E. Castillo, G. Waeber, G. E. Bergonzelli, J.-F. Aubert, E. Sutter, P. Nicod, B. Waeber, and P. Meda, Hypertension increases connexin43 in a tissue-specific manner *Circulation*, vol. 95, pp. 1007-1014, 1997.

[13] E. Delacrétaz, A. Zanchi, J. Nussberger, D Hayoz, J.-F. Aubert, H. R Brunner, and B. Waeber, Chronic nitric oxyde synthase inhibition and carotid artery distensibility in renal hypertensive rats *Hypertension*, vol. 26, pp. 332-336, 1995.

[14] E. C. Beyer, J. Kistler, D. L. Paul, and D. A. Goodenough, Antisera directed against connexin43 peptides react with a 43-KD protein localized to gap junctions in myocardium and other tissues *J. Cell. Biol.* vol. 108, pp. 595-605, 1989.

[15] J.-A. Haefliger, G. Bergonzelli, G. Waeber, J.-F. Aubert, J. Nussberger, H. Gavras, P. Nicod, and B. Waeber, Renin and angiotensin II receptor gene expression in kidneys of renal hypertensive rats *Hypertension*, vol. 26, pp. 733-737, 1995.

[16] B. I. Levy, J.-B. Michel, J.-L. Salzmann, M. Azizi, P. Poitevin, M. Safar, and J.-P. Camilleri, Effects of chronic inhibition of converting enzyme on mechanical and structural properties of arteries in rat renovascular hypertension. *circulation research*, vol. 63, pp. 227-239, 1988.

[17] R. Morishita, J. Higaki, M. Miyazaki, and T. Ogihara, Possible role of the vascular renin-angiotensin system in hypertension and vascular hypertrophy. *Hypertension*, vol. 19, pp. II-62-II-67, 1992.

[18] Z.-Q. Chen, D. L. Lefebvre, X.-H. Bai, A. Reaume, J. Rossant, and S. J. Lye, Identification of two regulatory elements within the promoter region of the mouse Cx-43 gene *J. Biol. Chem.* vol. 270, pp. 3863-3868, 1995.

[19] L. Chow and S. J. Lye, Expression of the gap junction protein Connexin-43 is increased of the human myometrium towards term and with the onset of labour. *Am. J. Obstet. Gynecol.* vol. 170, pp. 788-795, 1994.

[20] M. Piersanti and S. J. Lye, Increase in messenger ribonucleic acid encoding the myometrial gap junction protein, connexin-43, requires protein synthesis and is associated with increased expression of the activator protein-1, c-fos *Endocrinology*, vol. 136, pp. 3571-3578, 1995.

[21] D. L. Lefebvre, M. Piersanti, X.-H. Bai, Z.-Q. Chen, and S. J. Lye, Myometrial transcriptional regulation of the gap junction gene, connexin-43 *Reprod. Fertil. Devel.* vol. 7, pp. 603-611, 1996.

[22] A. J. Naftilan, R. E. Pratt, C. S. Eldridge, H. L. Lin, and V. J. Dzau, Angiotensin II induces c-fos expression in smooth muscle via transcriptional control *Hypertension*, vol. 13, pp. 706-711, 1989.

[23] E. Delacrétaz, D. Hayoz, M. C. Osterheld, C. Y. Genton, H. R. Brunner, and B. Waeber, Long-term nitric oxyde synthase inhibition and distensibility of carotid artery in intact rats *Hypertension*, vol. 23(part2), pp. 967-970, 1994.

*Gap Junctions*
*R. Werner (Ed.)*
*IOS Press, 1998*

# Signalling Between Immune Cells and Intestinal Epithelial cells *in Vitro*

Marwan E. El-Sabban[1], Carla A. Martin[2] and Fadia R. Homaidan[3]

[1]*Departments of Human Morphology and* [3]*Physiology, American University of Beirut, Lebanon; and* [2]*Division of Gastroenterology, Winthrop-University Hospital, Mineola, USA*

Mailing Address for Correspondence: Marwan E. El-Sabban (Ph.D.); Department of Human Morphology American University of Beirut, 850 Third Ave, 18th Floor, New York, NY 10022

**Abstract.** Activated macrophages found in the intestinal lesions of patients with inflammatory bowel disease secrete many regulators of intestinal cells function. However, little is known about direct macrophage-epithelial cell interactions. We investigated whether macrophages can adhere to intestinal epithelial cells in culture and form gap junction communication channels. Dye coupling between macrophages and epithelial cells was quantified by flow cytometry. Both homo- and hetero-cellular dye transfer were detected and were inhibited by heptanol. In response to mechanical stimulation, both epithelial cell homo-cellular, and macrophage-intestinal epithelial cells hetero-cellular propagagtion of calcium waves, typical of gap junctional communication was evident. These data indicate that epithelial cells and macrophages are coupled by gap junctions suggesting that gap junctional communication may provide additional means by which inflammatory cells might regulate epithelial cell function.

## 1. Introduction

Intestinal diseases change the normal pattern of fluid and electrolyte transport. These changes are the cause for diarrhea associated with many intestinal diseases especially inflammatory bowel disease (IBD). Diarrhea may be caused by a combination of a reduction in absorption or an increase in secretion. In IBD, the intestinal mucosa is infiltrated by inflammatory cells (1,2) which include activated macrophages (M$\phi$), naïve and memory T cells, and B cells (2,3). These cells are able to regulate intestinal epithelial cells (IEC) function by different mechanisms. The first is interaction through soluble mediators secreted by macrophage in response to inflammation. Many of these mediators are elevated in the inflamed intestinal tissues of patients of IBD and might contribute to the pathophysiology of IBD (4,5). These soluble mediators might directly or indirectly affect ion secretion in the inflamed tissue (6,7). The second mechanism by which M$\phi$ and IEC can interact is through macrophage binding of cellular adhesion molecules (8) or by receptor binding of antigens presented on the membranes of IEC (9,10). Direct consequences of adhesion include transepithelial migration of inflammatory cells, breaking of IEC tight junctions and/or disruption of ion transport (8). Recently, we have shown that a murine M$\phi$ line (P388D1) adhere to IEC lines and that this adhesion is partially dependent on $\beta_2$-integrin mediated binding (11). The last possible mechanism through which inflammatory cells can regulate IEC function is through cytoplasmic exchange of signals *via* gap junctions (12,13). Gap junctions are intercellular channels, formed by the coupling of two hemichannels, which allow for direct exchange of cytoplasmic substances between two adjacent cells (14-16). In this paper we present evidence of hetero-cellular gap junction formation between M$\phi$ and IEC by dye transfer and by propagation of calcium waves between the cells in response to mechanical stimulation.

## 2. Materials and methods

### *2.1. Cell isolation and cell culture*

The murine macrophage P388D1 and the rat small intestinal IEC-6 cell lines were obtained from ATCC and were maintained as recommended. The SV40 transformed murine intestinal epithelial cell line, Mode-K (17) was obtained from Dr. P.B. Ernst (University of Texas, Galveston) with the permission of Dr. D. Kaiserlian (Pasteur Institute, Lyon) and were maintained in DMEM (low glucose) containing 10%FBS, 10mM sodium pyruvate, and non-essential amino acids.

### *2.2. Adhesion and dye transfer*

Intestinal cells were cultured into 96 well plates until confluent. Mϕ were labeled in suspensions with calcein acetomethyl ester (calcein-AM)(Molecular Probes, Eugene, OR) at a concentration of 0.1μM calcein-AM for in DMEM 10% FBS for 15min at 37°C. Cells were washed, resuspended in medium and incubated for 45-60 min at 37°C to allow complete hydrolyses of the ester and for uncleaved ester molecules to leave the cytosol. Labeled Mϕ were washed and resuspended in MEM 2%FBS and added to IEC. The coculture was incubated at 37°C. At predetermined intervals nonadherent cells were removed by 3 washes with MEM 2%FBS, and adherent cells harvested using 0.25% trypsin containing 0.1mM EDTA. The cell population was resuspended in phosphate buffered saline containing 2% paraformaldehyde and analyzed on a FACScan (Becton Dickinson, San Jose, CA). Transfer of calcein between cell populations was measured as was detailed previously (11). Gated events (5000-10,000) were collected and the percent labeled-cells in the population (% Mϕ, adhesion) and mean fluorescence intensity (MFI, dye transfer) of both the labeled and the unlabeled populations were determined by quadrant analysis of dot plots.

### *2.3. Calcium waves*

Intestinal cells cultured in dishes fitted with coverslips were loaded with the $Ca^{+-}$ dual emission indicator, acetoxymethyl ester form of Indo-1 (Indo-1/AM; 1.5μM in DMSO) for 1hr at 37°C. The cells were then rinsed in MEM containing 2%FBS and 10mM HEPES and analyzed using Real Time Dual Emission Confocal Microscope (Nikon RCM-8000). All imaging experiments were performed in the presence of physiological calcium concentrations (1.2mM). Mechanical stimulus was applied to the apical membrane of a single cell by tapping it with a fine (0.5μm) glass microprobe. Pseudo-color scale shows high calcium concentration in red and low concentration in blue. Images were collected at a rate of 1/second.

### *2.4. Northern blot analysis*

Total RNA was extracted from confluent dishes of cultured cells by a rapid acid phenol-chloroform extraction method. RNA samples were electrophoresed, blotted and prehybridized according to established methods. Membranes were then hybridized with a labeled oligonucleotide probe complimentary to nucleotides sequences for Cx43 (G2, David Paul). The blot was developed by established chemeluminescence techniques and evaluated using a densitometer.

## 3. Results

### *3.1. Northern blot assay ofconnexin43 in IEC and in Mϕ.*

cDNA probe for Cx43 was used to test for its presence in IEC and Mϕ. Mode-K and Mϕ both possessed the message for Cx43 (Figure 1). Bovine aortic endothelial cells (BAEC) were used as a positive control. It has been established that epithelial cells lining the GI tract are coupled by gap junctions (18) but more recently, it has been shown that these gap junctions contain connexin 32 (Cx32) (19-21). Mϕ have been shown to express Cx43 mRNA and/or protein (22,23). The presence of Cx43 allows for homotypic communication. However, heteroypic communication should not be ruled-out.

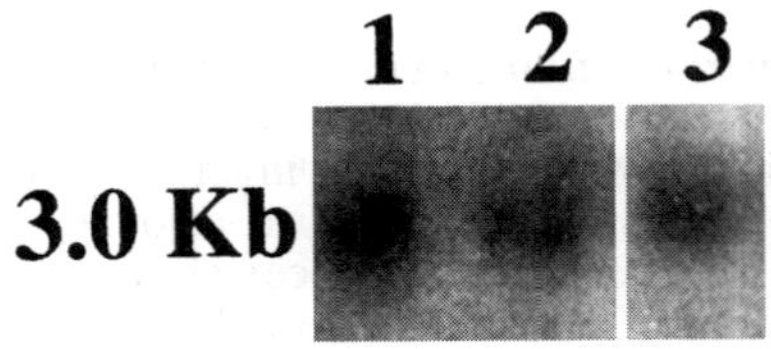


Figure 1. Northern blot analysis for Cx43 mRNA. Lane 1: Mode-K cells; lane 2: P388D1; lane 3: bovine aortic endothelial cells (BAEC) (positive control).

### 3.2. Adhesion and dye transfer between macrophages and IEC. Inhibition by Heptanol

Quantification of cytosolic dye transfer has become an established method for the detection of gap junctional communication (24,25). P388D1 was found to adhere specifically to Mode-K cell line (Figure 2a) as evident by the significant shift in Mode-K MFI. This suggested that the free acid form of calcein was transferred from P388D1 to IEC through cytoplasmic communication channels. Binding efficiency of P388D1 to Mode-K cells was dependent on the number of Mϕ in the coculture as well as the time of coculture (data not shown). Dye transfer between Mϕ and IEC was also shown to occurr in both directions, as is characteristic for gap junctional communication. Heptanol has been documented to inhibit gap junction formation (16). Mode-K were cocultured with calcein-labeled P388D1 in the presence of increasing concentrations of heptanol. As seen in Figure 2b, heptanol inhibited both adhesion and dye transfer in a dose dependent manner with 50% inhibition at a concentration between 1 and 2 mM. Although the two dose response curves for inhibition were very similar, it is evident that at low concentrations of heptanol, dye transfer inhibition proceeded that of adhesion, therefore suggesting that heptanol is indeed inhibiting gap junction formation. Similar results were obtained using α-glycerrhetinic acid which partially inhibited both adhesion and dye transfer when added to cocultures at concentrations of 100-500μM (data not shown).

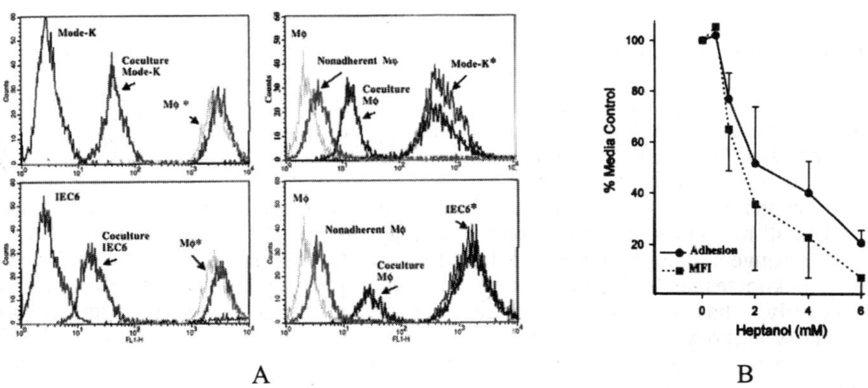


Figure 2. **a.** Labeled (*) or unlabeled P388D1 were cocultured with labeled or unlabeled Mode-K or IEC-6 cells and cocultures analyzed by flow cytometry. Controls included IEC alone, Mϕ alone and nonadherent Mϕ. **b.** Inhibition of dye transfer and adhesion by heptanol.

### 3.3. Homo-cellular and hetero-cellular calcium wave propagation

Mechanical stimulation of a single cell in the IEC monolayer induced a rise in intracellular calcium which then spread outward to adjoining cells in a wave like fashion. This propagation of signal from cell to cell has been shown to be transmitted, in many cell types, through gap junctions (26,27). In Mϕ-IEC cultures, $[Ca^{++}]_i$ increased in the stimulated cell and spread in the form of a wave to non stimulated adjacent cells (a shift in color from green/yellow to red, corresponds to an increase in relative fluorescence. This outwardly radiating $[Ca^{++}]_i$ wave involved progressively more distal cells to a radius of a few cells, depending on cell type and degree of confluence. In coculture experiments the focus was shifted between IEC and Mϕ since they were located in different focal planes. This highly ordered, sequential spread of signal occurred over a period of less than 60 sec, after which time, intracellular calcium reverted to resting levels.

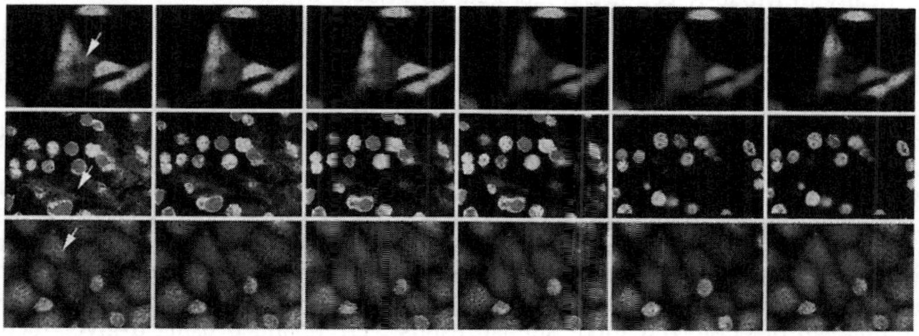


Figure 3. Propagation of calcium waves in response to mechanical stimulation. The relative increase in intracellular calcium is shown as a shift from green-yellow to red on a pseudo-color scale. Panels A: Mode-K; Panels B: Mode-K+P388D1; Panels C: IEC-6+P388D1. Arrow: site of stimulation by a micropipette.

## 4. Summary and conclusion

The formation of gap junctions requires close apposition of cell membranes and is highly dependent on the cell-cell binding mediated by adhesion molecules, cadherins or tight junction proteins (10,11,28,29). Gap junctions allow for bi-directional transfer of regulatory molecules between adjacent cells and are thought to play an important role in tissue organization and differentiation of cells (12,29). Cells coupled by gap junctions can exchange ions and small molecules up to approximately 1 Kd in size including cell metabolites and second messengers such as cyclic nucleotides, inositol triphosphate ($IP_3$) and calcium (14,30). Leukocytes, including Mϕ, form putative gap junctions with endothelial cells (22). More recently, it has been reported that gap junctions are formed between endothelium and inflammatory leukocytes as well as between endothelium and metastatic cancer cells (24). The role of gap junctions in inflammation and metastasis is currently unknown but it has been proposed that gap junctional communication may play a role in the transendothelial migration of invading cells (24,22).

In this study we presented evidence that Mϕ are able to form functional gap junctional channels with IEC that allow a physiological second messenger to be exchanged between th etwo cell types. This direct communication could play an important role in epithelial cell function in health and in disease.

## References

1. Podalsky, D.K. Inflammatory bowel disease. *New Eng. J. Med.* 325(1991) 928-937.

2. Braegger C.P. and MacDonald T.T. Immune mechanisms in chronic inflammatory bowel disease. *Ann. Allergy* 72 (1994) 135-141.

3. Burgio V.L., Stefano F., Boirivant M., Perrone A. and pallone F. Peripheral monocyte and naïve T-cell recruitment and activation in Crohn's disease. *Gastroenterology* 109 (1995) 1029-1038.

4. Fiocchi C. Cytokines and animal models: a combined path to inflammatory bowel disease pathogenesis. *Gastroenterology* 104 (1993) 1202-1205.

5. Sartor R.B. Cytokines in intestinal inflammation: pathophysiological and clinical considerations. *Gastroenterology* 106 (1994) 533-539.

6. Halm D.R. and Frizzell R.A. Intestinal chloride secretion. In 'Textbook of Secretory Diarrhea', eds E. Lebenthal and M. Duffey, New York, Raven, 1990, p.47-58.

7. Homaidan F.R., H. Desai, L. Zhao, G. Broutman and R. Burakoff. Regulation of electrolyte transport with IL-1b in rabbit distal colon. *Mediators of Inflammation* 4 (1995) 61-66.

8. Parkos C.A., C. Delp, M. Amin and J.L. Madara. Neutrophil migration across a cultured intestinal epithelium. Dependence on a CD11a/CD18-mediated event and enhanced efficiency in physiological direction. *J. Clin. Invest.* 88 (1995) 1605-1612.

9. Kaiserlan D. and K. Vidal. Antigen presentation by intestinal epithelial cells. *Immunol. Today* 14 (1993) 144.

10. Mayer L., A. Panja, Y. Li, E. Siden, A. Pizzimenti, F. Gerardi and N. Chandswang. Unique features of antigen presentation in the intestines. *Ann. N.Y. Acad. Sci.* 664 (1992) 39-46.

11. Martin C.A., M.E. El-Sabban, L. Zhao, R. Burakoff and F.R. Homaidan. Adhesion and cytosolic dye transfer between macrophages and intestinal epithelial cells. *Cell Adhesion and Communication.* 5 (1997) 1-7.

12. Spray D.C. and V.L. Bennett. Physiology and pharmacology of gap junctions. *Ann. Rev. Physiol.* 4 (1985) 281-303.

13. Beyer E.C., Paul D.L., and Goodenough D.A. Connexin 43: a protein from rat heart homologous to a gap junction protein from liver. *J. Membrane Biol.* 116 (1990) 187-194.

14. Bennett M.V.L., Barrio L.C., Bargiello T.A., Spray D.C. Hertzberg E. and Saez J.C. Gap junctions: new tools, new answers, new questions. *Neuron* 6 (1991) 305-320.

15. Bennett M.V.L. and Verselis V.K. Biophysics of gap junctions. *Semin. Cell Biol.* 3 (1992) 29-47.

16. Beyer E.C. Gap junctions. *Int. Rev. cytol.* 137C (1993) 1-37.

17. Vidal K., I. Grosjean, J.P. Revillard, C. Gespach and D. Kaiserlaian. Immortalization of mouse intestinal epithelial cells by SV40-large T gene. Phenotypic and immune characterization of the Mode-K cell line. *J. Immunol. Methods* 166 (1993) 63-73.

18. Bennett M.V.L. and Goodenough D.A. Gap junctions, electrical coupling and intercellular communication. *Neurosci. Res. Bull.* 16 (1978) 383.

19. Chanson M., White M.M. and Garber S.S. cAMP promotes gap junctional coupling in T84 cells. *Am. J. Physiol.* 271 (1996) C533-539.

20. Uchida Y., Matsuda K., Sasahara K., Kawabata H. and Nishioka M. Immunohistochemistry of gap junctions in normal and diseased gastric mucosa of humans. *Gastroenterology* 109(1995) 1492-1496.

21. Ueda F. Immunohistochemical alterations of gastric epithelial junctions induced by water-immersion stress in rats. *Ther. Res.*12 (1991) 135-139.

22. Jara P.I., M.P.Boric and J.C. Saez. Leukocytes express connexin43 after activation with lipopolysaccharide and appear to form gap junctions with endothelial cells after ischemic-reperfusion. *Proc. Natl. Acad. Sci. USA* 92 (1995) 7011-7015.

23. Beyer E.C. and Steinberg T.H. Evidence that the gap junction protein connexin-43 is the ATP-induced pore of mouse macrophages. *J. Biol. Chem.* 266 (1991) 7971-7974.

24. El-Sabban M.E. and Pauli B.U. Cytoplasmic dye transfer between metastatic tumor cells and vascular endothelium. *J. Cell Biol.* 115 (1991) 1375-1382.

25. Guinan, E.C., Smith, B.R., Davies, P.F. and Pober J.S. Cytoplasmic transfer between endothelium and lymphocytes: Quantitaion by flowcytometry. *Am. J. Pathol.* 132 (1988) 406-409.

26. Charles A.C., Naus C.C.G., Zhu D., Kidder G.M. and Dirksen E.R. Intercellular calcium signaling via gap junctions in glioma cells. *J. Cell Biol.* 118 (1992) 195-201.

27. Sanderson, M.J., Charles, A.C., and Dirksen, E.R. Mechanical stimulation and intercellular communication increases intercellular $Ca^{2+}$ in epithelial cells. *Cell Regulation* 1 (1990) 585-596.

28. Jongen W.M.F., J. Fitzgerald, M. Asamoto, C. Piccoli, T.J. Slaga, D. gros, M. Takeichi and H. Yamasaki. Regulation of connexin43-mediated gap junctional intercellular communication by $Ca^{2+}$ in mouse epidermal cells is controlled by E-Cadherin., *J. Cell Biol.* 114 (1991) 545-555.

29. Loewenstein W.R. and B. Rose. The cell-cell channel. *Seminars in Cell Biol.* 3 (1992) 59-79.

30. Wolburg H and Rohlman A. Structure-function relationship in gap junctions. *Int. Rev. Cytol.* 157 (1995) 315-373.

*Gap Junctions*
*R. Werner (Ed.)*
*IOS Press, 1998*



# Simultaneous Localization of Connexins 40, 37 and 43 in Pulmonary Artery Endothelial Gap Junctions

Yu-Shien KO, Hung-I YEH, Stephen ROTHERY,
Emmanuel DUPONT and Nicholas J. SEVERS
*National Heart and Lung Institute, Imperial College School of Medicine*
*Royal Brompton Hospital, Sydney Street, London, SW7 6NP, UK*

**Abstract.** Connexins 40, 37 and 43 have all been reported to be expressed in vascular endothelial cells, though information on the precise distribution of these connexins at the level of the individual gap junction is limited. In this study, we selected rat main pulmonary artery endothelium as a model system for multiple immunolocalization of these connexins by confocal microscopy and electron microscopy. Using affinity purified polyclonal anti-connexin 40 and 37 antibodies and a monoclonal anti-connexin43 antibody, pairs of the three connexins were localized to the same sites by confocal microscopy. By using three connexin specific antibodies raised in different species and gold markers of three sizes, it is now possible to undertake double and triple immunogold labeling at the electron-microscopical level, thereby demonstrating the presence of multiple connexin types in the same gap-junctional plaque.

## 1. Introduction

Gap junctions between vascular endothelial cells vary in abundance in different vessels, typically being more numerous and prominent on the arterial than the venous side of the vascular tree [1]. The potential roles for gap-junctional communication in endothelium include intercellular exchange of nutrients and signalling molecules, regulation of growth and differentiation (e.g., maintenance of the monolayer topology in the normal vessel and co-ordinating migration after injury), and mediating responses to exogenous stimuli [2-3]. Connexins 40, 37 and 43 have been reported to be expressed in vascular endothelial cells [3-9], but our knowledge of how these connexins are distributed at the level of the individual gap junction, and their relative quantities in different segments of the arterial system, remains limited. By applying affinity purified connexin specific antibodies and confocal microscopy, we found that whereas endothelial cells of rat aorta and coronary artery characteristically show extensive, prominent connexin40 and heterogeneous scattered connexin37, the aortic endothelium, unlike the coronary endothelium, also has abundant connexin43 (Table 1). The endothelium of the pulmonary artery reveals a similar expression pattern to that of the aorta, with all three connexin types being detected in relative abundance. As gap junctions composed of different connexins show distinctive properties in *in vitro* studies [10-15], it is important to establish whether these three connexins occur in separate gap junctions, or whether combinations of connexins co-exist in the same gap-junctional plaque. We therefore conducted multiple labeling experiments to localize connexins 40, 37 and 43 in the rat main pulmonary arterial endothelium by confocal microscopy and electron microscopy.

**Table 1. Summary of the Distribution of Connexins 40, 43 and 37 in Endothelium from Different Arteries as Observed by Immunoconfocal Microscopy***

| | Rat Pulmonary Artery | Rat Aorta | Rat Coronary Artery |
|---|---|---|---|
| **Connexin40** | + + + + + | + + + + + | + + + + + |
| **Connexin37** | + to + + + + + | + to + + + + + | + to + + + + + |
| **Connexin43** | + + + + + | + + + + + | + |

* Semiquantification by scoring + to + + + + + to indicate percentage of endothelial area that included distinct punctate connexin immunolabeling delineating the cell borders, as follows: +, <1%; + +, 1-25%; + + +, 26-50%; + + + +, 51-75%; + + + + +, 76-100% (Single labeling for each connexin type; modified from: [8]).

## 2. Anti-Connexin Antibodies and Confocal Microscopy

To undertake this study, new anti-connexin 40 and 37 antibodies were produced, as summarized in Table 2. Double labeling for confocal microscopy was carried out using pairs of these affinity purified primary antibodies, with appropriate CY3 and CY5- conjugated secondary antibodies. The results are illustrated in Figure 1.

Inspection of the results for double labeling for connexin37 with either connexin43 or connexin40 confirms that connexin37 has an overall lower abundance than either connexin40 or connexin43. The distribution of connexin37 spots in the endothelium in many instances overlaps with connexin43 or connexin40 spots. When double labeling for connexins 40 and 43 is carried out, the distribution of connexin43 in the endothelium coincides with that of connexin40. Only connexin43 is detected in medial smooth muscle.

The localization by immunoconfocal microscopy of fluorescent signals to spots in precisely the same position suggests that two or more connexins may be present in individual pulmonary artery endothelial gap-junctional plaques. However, because of the limited resolution of confocal microscopy compared with electron microscopy, such a superimposition could be attributable to signals from two closely associated plaques each composed of a different connexin.

**Table 2. Peptides Used for Generation of Anti-connexin 37 and 40 Antibodies***

| Connexin | Sequence (Carboxy Terminal) | Host | Antibody Designation |
|---|---|---|---|
| **Connexin37** | **RAT 266** Y L P M G E G P S S P P C P T Y **281** | **Rabbit** | **Y16Y (R4)** |
| **Connexin40** | **RAT 256** V Q G L T P P P D F N Q C L K **270** | **Guinea Pig** | **V15K (G139)** |

* Mouse monoclonal anti-connexin43 antibody was purchased from Chemicon.

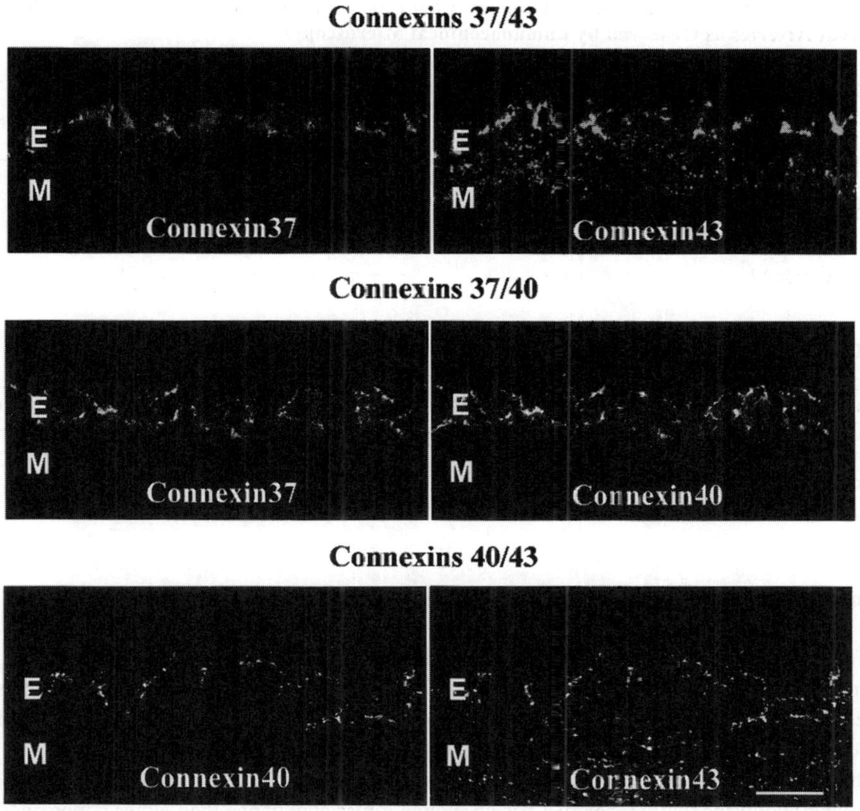


**Figure 1.** Double labeling for pairs of connexins in rat pulmonary artery by immunoconfocal microscopy. E, endothelium; M, media. Bar = 10 μm.

## 3. Immunogold Electron Microscopy

For detailed investigation of connexin make-up at the level of the individual gap-junctional plaque, we applied post-embedding immunogold labeling of rat main pulmonary artery specimens embedded at low temperature in Lowicryl K4M Sections were labeled with guinea pig anti-connexin40/10nm-gold, rabbit anti-connexin37/5nm-gold or mouse anti-connexin43/15nm-gold. Optimal conditions were anti-connexin40, 2 hours; anti-connexin37, overnight; and anti-connexin43, 4 hours. Negative controls consisted of (1) omission of the primary antibody and (2) using irrelevant secondary antibodies. All three secondary antibodies were confirmed to be specific to their individual primary antibody and without any cross-reactivity to other connexin types. Immunogold labeling for each connexin type demonstrated its presence specifically at morphologically identifiable gap junctions (Figure 2).

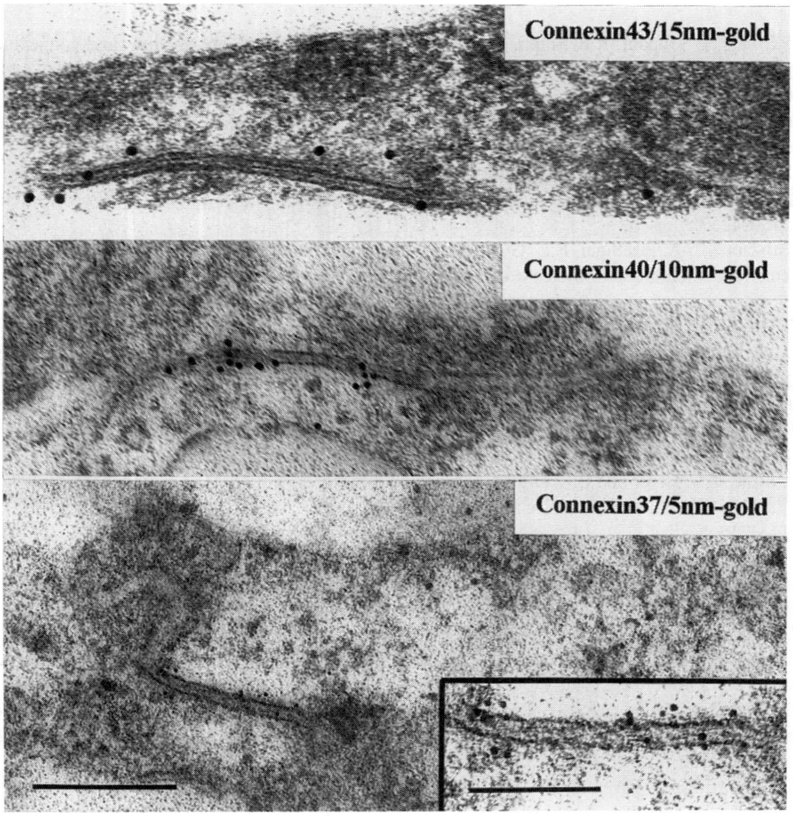


**Figure 2.** Single labeling immunogold electron microscopy for connexin43/15nm-gold, connexin40/10nm-gold and connexin37/5nm-gold in endothelial gap junctions of rat pulmonary artery. Gold markers are in each case specifically associated with morphologically recognizable gap-junctional plaques, and are absent from other portions of the cell. Main panels all at same magnification. Inset shows higher magnification view of connexin37/5nm-gold. Main panels, bar = 200 nm; inset , bar = 100 nm.

Two features were planned into this immunogold labeling work that facilitate the development of double and triple immunogold labeling. First, the use of antibodies produced in different species enables highly specific secondary detection of multiple connexins. Second, the use of gold markers of three distinct sizes that can be readily discriminated from one another (Figure 2) demonstrated the feasibility of simultaneously identifying up to three different connexins at the same time.

From this background we have successfully developed protocols for double and triple immunogold labeling of connexins 40, 37 and 43 in endothelial gap junctions of rat pulmonary artery. The results obtained, which will shortly be submitted for publication, suggest that the majority of pulmonary artery endothelial gap junctions contain all three connexin types. Given that functional studies indicate each of these connexin types confers distinct properties on gap-junctional channels, this complexity of connexin composition suggests a high potential for modulation of function within individual gap-junctional plaques.

## References


[1] N.J. Severs, E. Dupont, R.R. Kaprielian, H-I. Yeh and S. Rothery, Gap junctions and connexins in the cardiovascular system. In: M.H. Yacoub, A. Carpentier, J. Pepper and J-N Fabiani (eds.), *Annual of Cardiac Surgery 1996: 9th edition*. Current Science, London, 1996, pp. 31-44.

[2] D.M. Larson, Intercellular junctions and junctional transfer in the blood vessel wall. In: U. Ryan (ed.), *Endothelial Cells III*. CRC Press, Boca Raton, 1988, pp. 75-88.

[3] M.S. Pepper, R. Montesano, A. El Aoumari, D. Gros, L. Orci and P. Meda, Coupling and connexin 43 expression in microvascular and large vessel endothelial cells, *Am. J. Physiol (Cell Physiol.)* 262 (1992) C1246-C1257.

[4] R. Bruzzone, J.-A. Haefliger, R.L. Gimlich and D.L. Paul, Connexin40, a component of gap junctions in vascular endothelium, is restricted in its ability to interact with other connexins, *Mol. Biol. Cell* 4 (1993) 7-19.

[5] K.E. Reed, E.M. Westphale, D.M. Larson, H.-Z. Wang, R.D. Veenstra and E.C. Beyer, Molecular cloning and functional expression of human connexin37, an endothelial cell gap junction protein, *J. Clin. Invest.* 91 (1993) 997-1004.

[6] B. Bastide, L. Neyses, D. Ganten, M. Paul, K. Willecke and O. Traub, Gap junction protein connexin40 is preferentially expressed in vascular endothelium and conductive bundles of rat myocardium and is increased under hypertensive conditions, *Circ. Res.* 73 (1993) 1138-1149.

[7] T.L. Little, E.C. Beyer and B.R. Duling, Connexin 43 and connexin 40 gap junctional proteins are present in arteriolar smooth muscle and endothelium in vivo, *Am. J. Physiol Heart Circ. Physiol.* 268 (1995) H729-H739.

[8] H-I. Yeh, E. Dupont, S. Coppen, S. Rothery and N.J. Severs. Gap junction localization and connexin expression in cytochemically identified endothelial cells from arterial tissue, *J. Histochem. Cytochem.* 45 (1997) 539-550.

[9] H.V.M. Van Rijen, M.J.A. Van Kempen, L.J.S. Analbers, M.B. Rook, A.C.G. Van Ginneken, D. Gros and H.J. Jongsma, Gap junctions in human umbilical cord endothelial cells contain multiple connexins, *Am. J. Physiol. Cell Physiol.* 272 (1997) C117-C130.

[10] R.D. Veenstra, H.-Z. Wang, E.C. Beyer, S.V. Ramanan and P.R. Brink, Connexin37 forms high conductance gap junction channels with subconductance state activity and selective dye and ionic permeabilities, *Biophys. J.* 66 (1994) 1915-1928.

[11] F.F. Bukauskas, C. Elfgang, K. Willecke and R. Weingart. Biophysical properties of gap junction channels formed by mouse connexin40 in induced pairs of transfected human HeLa cells, *Biophys. J.* 68 (1995) 2289-2298.

[12] D.A. Beblo, H.Z. Wang, E.C. Beyer, E.M. Westphale and R.D. Veenstra, Unique conductance, gating, and selective permeability properties of gap junction channels formed by connexin40, *Circ. Res.* 77 (1995) 813-822.

[13] R.D. Veenstra, H.Z. Wang, D.A. Beblo, M.G. Chilton, A.L. Harris, E.C. Beyer and P.R. Brink, Selectivity of connexin-specific gap junctions does not correlate with channel conductance, *Circ. Res.* 77 (1995) 1156-1165.

[14] C. Elfgang, R. Eckert, H. Lichtenberg-Fraté, A. Butterweck, O. Traub, R.A. Klein, D.F. Hülser and K. Willecke, Specific permeability and selective formation of gap junction channels in connexin-transfected HeLa cells, *J. Cell Biol.* 129 (1995) 805-817.

[15] R.D. Veenstra, Size and selectivity of gap junction channels formed from different connexins, *J. Bioenerg. Biomembr.* 28 (1996) 327-337.

*Gap Junctions*
*R. Werner (Ed.)*
*IOS Press, 1998*

# Gap Junctional Communication Among Motor Neurons in Development and After Nerve Damage

Qiang CHANG[1], Alberto PEREDA[2], Martin J. PINTER[2] and Rita J. BALICE-GORDON[1]

[1]*Department of Neuroscience, University of Pennsylvania School of Medicine and*
[2]*Department of Neurobiology and Anatomy, Allegheny University of the Health Sciences, Philadelphia, PA, USA*

**Abstract** Gap junctional coupling among neurons during development appears to be wide-spread, although its functional significance is unknown. We are studying gap junctions among developing motor neurons as a model system for investigating the role of gap junctional communication in neuronal and synaptic differentiation. During development, dye injection of low molecular mass compounds into motor neurons in spinal cord slices shows that ventral horn motor neurons in the rodent lumbar spinal cord are dye coupled around the time of birth. Motor neurons have also been shown to be electrically coupled at this time. The percentage of dye- or electrically coupled motor neurons declines rapidly in the first postnatal weeks. RT-PCR analysis, *in situ* hybridization and immunohistochemistry show that at least three of the known rodent connexins, Cx37, Cx40, and Cx45, are expressed by motor neurons during this period. Thus it appears that there is gap junctionally mediated communication among developing motor neurons that disappears during early postnatal life. After nerve damage in adult animals, we have observed that adult motor neurons become re-coupled by gap junctions. Injection of low molecular mass compounds into axotomized motor neurons shows that axotomized but not control motor neurons are dye-coupled 3-4 weeks after nerve damage. Immunohistochemistry suggests that axotomized motor neurons re-express at least one connexin, Cx32. These preliminary observations suggest that transient gap junctional coupling among developing and axotomized motor neurons may mediate electrical or metabolic signaling that may be essential for the maintenance of function.

## Introduction

Gap junctional coupling is wide-spread throughout the developing nervous system, among neurons as well as among glia. Over the last several years, there have been demonstrations of electrical coupling, dye coupling, gap junction plaques and/or connexin expression in different types of neurons and glia in different parts of the developing nervous system [1, 2]. Roles for gap junctional communication in mediating pattern formation, cell migration, neuronal differentiation and circuit formation have been proposed [2-5]. Despite the apparent ubiquity of gap junctional communication during neural development, the functional significance of inter-neuronal or inter-glial communication mediated by gap junctions is largely unknown. We are studying gap junctions among developing motor neurons as a model system for investigating the role of gap junctional communication in neuronal and synaptic differentiation. Motor neurons, their axons and the synapses they make with skeletal muscle fibers, called neuromuscular junctions, are accessible to physiological and structural analyses at embryonic, neonatal and adult stages. In addition, pre- as well as postsynaptic elements are easily manipulated. Thus the neuromuscular axis is a useful model system for studying the mechanisms underlying neuronal differentiation, synapse formation and maintenance.

In development, neuromuscular junctions undergo a period of activity dependent synapse elimination. There is a transition from multiple to single innervation, a result of a competition among motor neuron inputs to the same muscle fiber which are initially structurally and functionally similar [6]. Neural activity has been shown to be a key determinant of the outcome of synaptic competition. Previous work from our lab and others has suggested that synchronous activity (or inactivity) among the synaptic sites within a neuromuscular junction results in all synapses being maintained, while asynchronous activity leads to the permanent loss of the least active synapses [7]. A similar activity dependent mechanism is believed to underlie synaptic rearrangements elsewhere in the developing nervous system [8]. The activity dependent elimination of excess synaptic connections can be recapitulated in adult animals after peripheral nerve injury. Nerve damage in adults induces axon degeneration; over time, motor axons regenerate and reinnervate denervated muscle fibers. During reinnervation, there is a transient period of multiple innervation of skeletal muscle fibers that is similar to that observed during development. This multiple innervation then disappears within a few weeks. Synapse elimination during adult reinnervation seems to occur via similar mechanisms to those occurring in development [9].

We are interested in understanding how the process of synaptic competition is modulated by activity in development and in adults after nerve injury at neuromuscular junctions. One hypothesis we are exploring is that synchronous motor neuron activity during early phases of synapse formation allows synapses formed from many different neurons to be maintained, possibly because the postsynaptic muscle fiber cannot discriminate among inputs that are structurally and functionally similar. Motor neuron activity may become progressively asynchronous over time, triggering synapse elimination (Figure 1).

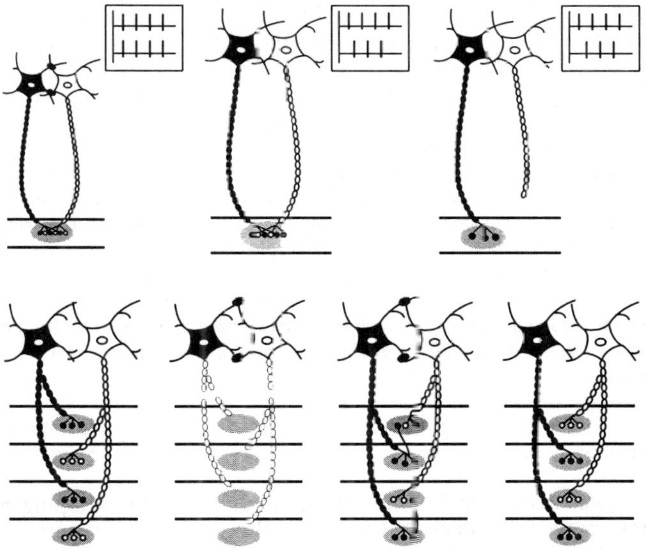

**Figure 1 A role for motor neuron gap junctional coupling in the maturation of neuromuscular synaptic connections? Upper:** In embryos and neonates (left panel), motor neuron gap junctions (small shaded ovals) may bias activity to be synchronous (boxed inset), allowing multiple innervation (by two or more motor axons) to be established at each neuromuscular junction (large shaded ovals). As gap junctions disappear (middle panel), activity may become more asynchronous, triggering the onset of synapse elimination (right panel). **Lower:** Adult neuromuscular junctions are singly innervated (left panel). After adult nerve damage, axons degenerate (panel 2); denervated fibers are then multiply reinnervated as axons re-grow (panel 3). Synapse elimination re-establishes the adult pattern of single innervation (right panel). Dye coupling and connexin expression among axotomized motor neurons has been observed, suggesting that gap junctional communication may be re-established after nerve injury.

While there are numerous cellular mechanisms that might bias activity to be relatively synchronous among neurons, electrical coupling is widespread in development and could play such a role. Electrophysiological studies have shown that, at birth, some motor neurons in the spinal cord are electrically coupled and that this coupling decreases in the first postnatal weeks [10, 11]. Functional gap junctions among motor neurons may introduce temporal correlation among the activity patterns of different motor neurons innervating the same junction, by rapidly and faithfully transmitting electrical activity from one cell to others coupled to it. This may be important for the transient maintenance of multiple innervation, while uncoupling may be related to the onset of synaptic competition. Thus gap junctional coupling might mediate electrical communication among neurons. On the other hand, gap junctions may facilitate biochemical communication essential for neuronal differentiation and survival. Because multiple innervation is maintained at times when motor neurons appear to be coupled by gap junctions, and is eliminated as coupling disappears, we are interested in understanding the functional role of gap junctional coupling among motor neurons during development and after nerve injury.

## Gap junctional coupling among motor neurons during embryonic and neonatal development

Intracellular recordings from motor neurons in neonatal spinal cord have detected short latency depolarizations evoked by antidromic stimulation of motor axons around the time of birth [10, 11]. These depolarizations are consistent with gap junction mediated coupling potentials, because they are resistant to fatigue during high frequency stimulation and persist after intracellularly evoked action potentials are used to collide out antidromically activated action potentials. In spinal cord slice preparations, and more recently in *in vitro* hemisected spinal cords, we have injected neonatal motor neurons with low molecular mass compounds, such as Neurobiotin, which are known to readily pass through gap junction pores [1]. From birth to postnatal day 4 (P4), clusters of several motor neurons are dye labeled following injection of a single motor neuron. By P7-P10, only one or two labeled neurons are detected, and after P10, only single labeled cells are observed. These preliminary data suggest that motor neurons are dye coupled at the times during which electrical coupling has been detected [10-13].

To determine the repertoire of connexins expressed by motor neurons during these times that may contribute to the formation of functional gap junctions, RT-PCR analysis was performed on mRNAs isolated from embryonic day 15 (E15), P1 and P14 rat spinal cord using primers homologous to unique regions within the cDNA of thirteen known rodent connexins. Cx26, Cx31, Cx31.1, Cx32, Cx37, Cx40, Cx43, and Cx45 were detected. To determine which connexins are expressed specifically in ventral horn motor neurons, E15 rat motor neurons were purified using a modified protocol of Henderson [14] and total RNA was isolated. Motor neuron preparations were assessed for homogeneity by immunostaining with antibodies against a motor neuron specific transcription factor, Islet-1 [15]. RT-PCR analysis showed that Cx37, Cx40 and Cx45 are expressed by motor neurons during embryonic development. PCR products were eluted from gels, cloned and sequenced to verify their identity.

*In situ* hybridization of rat brain and spinal cord sections using cRNA probes was used to further analyze the expression of the connexins found positive by RT-PCR. Cx37, 40 and 45 were found to be expressed by motor neurons as well as by other spinal cord cells in this analysis. Cx45 was observed to be expressed throughout the developing neural tube as early as E12. At later times, motor neurons and several other populations of cells in the spinal cord and dorsal root ganglia express Cx45. The number of Cx45 positive motor neurons decreases from E15 to P14 [16]. Some adult motor neurons continue to express Cx45 message.

Immunohistochemical staining with a polyclonal anti-Cx45 antibody (Chemicon) and motor neuron specific markers such as Islet-1 revealed a striking change in the cellular localization of Cx45 in developing motor neurons. From E15 until around the time of birth, confocal analysis of spinal cord sections showed that Cx45 is expressed by motor neurons and is localized to the cell membrane and, at late embryonic stages, also to the perinuclear region. After birth, Cx45 immunoreactivity disappears from the cell membrane and remains in perinuclear region. Some motor neurons lose Cx45 immunoreactivity after P14, while in other subpopulations of motor neurons in the ventral horn Cx45 can be detected in the perinuclear region through adulthood. Cx45 was not detected in the membrane of adult motor neurons. Cx45 immunoreactivity was also observed in the dorsal funiculi, near the floor plate, and in a subpopulation of dorsal root ganglia neurons. Taken together, these results suggest that Cx45, 37 and 40, and perhaps other connexins, are expressed in motor neurons *in vivo* and *in vitro* [16]. Moreover, these connexins appear to be developmentally regulated both in terms of their pattern of expression and their cellular localization.

## Recapitulation of gap junctional coupling among adult motor neurons after axotomy

The activity dependent elimination of synaptic connections observed during development is recapitulated in adult animals after peripheral nerve injury (Figure 1). Nerve damage in adults induces motor axon degeneration; over time, motor axons regenerate and reinnervate denervated muscle fibers. During the re-establishment of innervation, there is a transient period of multiple innervation of skeletal muscle fibers which disappears within a few weeks. Synapse elimination during adult reinnervation seems to occur via similar mechanisms as in development [9]. It is unknown why multiple innervation is recapitulated during reinnervation, but it is likely to be an important step in the re-establishment of functional synaptic connections. Thus we have begun to explore whether connexin expression and gap junctional communication are regulated among motor neurons after nerve damage and during reinnervation.

After nerve cut (axotomy) in adult animals, we have observed that motor neurons are transiently coupled by gap junctions during these times. Following nerve damage, intracellular dye fills and immunostaining suggest that gap junctional coupling among motor neurons is transiently re-expressed. Three weeks after damage to the medial gastrocnemius muscle nerve, clusters of 4-6 motor neurons are observed to be labeled following Neurobiotin injection of single identified medial gastrocnemius motor neurons. In contrast, intracellular dye fills in normal adults are observed to yield single labeled motor neurons. Motor neurons rapidly become immunopositive for connexins after nerve damage, with peak expression observed three to four weeks after nerve damage. Consistent with these observations, we have also found in a few experiment short latency depolarizations evoked by antidromic stimulation that are resistant to fatigue with high frequency stimulation and that persist after intracellularly evoked action potentials are used to collide out antidromically activated action potentials. These appear similar to the coupling potentials observed among motor neurons in the neonatal spinal cord. As may be the case during development, gap junctional coupling may lead to the establishment of multiple reinnervation because motor neuron activity may be biased toward relative synchrony. Thus motor neuronal gap junctional coupling may be important to ensure that denervated muscle fibers receive at least some innervation. As coupling disappears, progressively more asynchronous activity may drive synapse elimination during reinnervation in a similar fashion as it may in development.

## Summary

Taken together, our observations demonstrate that the extent of gap junctional coupling among motor neurons is dynamically regulated during development and reinnervation. While gap junctional coupling, as well as other aspects of immature spinal cord circuitry, may bias the electrical activity of motor neurons to be relatively synchronous, it is also possible that the importance of gap junctional coupling is that it allows biochemical signaling among coupled cells that is essential to their normal function. This form of intracellular communication may be important in the maturation of motor neurons, play a role in the formation of spinal neural circuits, and influence the formation and maintenance of motor neuron synapses in development and adult reinnervation. Such intercellular communication may allow the dissemination of retrograde, target-derived signals among the innervating population of neurons, exchange of calcium and/or second messengers, or perhaps mediate signaling during the period of naturally occurring cell death. Thus gap junctional coupling during motor neuron maturation may influence synapse formation and maintenance by mediating electrical signaling and/or by mediating biochemical signaling. We are presently exploring whether transient gap junctional coupling among developing and axotomized motor neurons may mediate electrical or metabolic signaling that are essential for neuronal function. Analysis of motor neuronal differentiation and synaptic connections in connexin knockout and transgenic mice may be helpful in this regard. Little information exists about the temporal aspects of motor neuron firing during perinatal life or adult reinnervation when synaptic connections are being established, how these change over time, and how these changes affect the emergence of mature patterns of innervation in muscle. These will be the focus of future work.

## Acknowledgements

We thank Drs. J. O'Malley and M. Gonzalez for helpful discussions and comments on the manuscript. This work was supported by grants from the NIH (NS34373) and the Paralyzed Veterans of America / Spinal Cord Research Foundation and a McKnight Neuroscience Scholar award to R.B.-G.

## References

[1] R. Dermietzel and D. C. Spray, Gap Junctions in the Brain: Where, What Type, How Many and Why? *Trends in Neuroscience* **16** (1993) 186-192.

[2] B. P. Fulton, Gap Junctions in the Developing Nervous System, *Perspectives in Developmental Neurobiology* **2** (1995) 327-334.

[3] E. Frank, New Life in an Old Structure: the Development of Synaptic Pathways in the Spinal Cord, *Current Opinion in Neurobiology* **3** (1993) 82-86.

[4] J. L. Ewart, M. F. Cohen, R. A. Meyer, G. Y. Huang, A. Wessels, R. G. Gourdie, A. J. Chin, S. M. Park, B. O. Lazatin, S. Villabon and C. W. Lo, Heart and Neural Tube Defects in Transgenic Mice Overexpressing the Cx43 Gap Junction Gene, *Development* **124** (1997) 1281-1292

[5] A. Peinado, R. Yuste and L. C. Katz, Gap Junctional Communication and the Development of Local Circuits in Neocortex, *Cerebral Cortex* **3** (1993) 488-498

[6] Q. Nguyen and J. W. Lichtman, Mechanism of Synapse Disassembly at the Developing Neuromuscular Junction, *Current Opinion in Neurobiology* **6** (1996) 104-112.

[7] R. J. Balice-Gordon and J. W. Lichtman, Long Term Synapse Loss Induced by Focal Blockade of Postsynaptic Receptors, *Nature* **372** (1994) 519-524.

[8] C. S. Goodman and C. J. Shatz, Developmental Mechanisms that Generate Precise Patterns of Neuronal Connectivity, *Neuron* **10** (1993) 77-98.

[9] M. M. Rich and J. W. Lichtman, *In Vivo* Visualization of Pre- and Postsynaptic Changes During Synapse Elimination in Reinnervated Mouse Muscle, *Journal of Neuroscience* **9** (1989) 1781-1805.

[10] B. P. Fulton, R. Miledi and T. Takahashi, Electrical Synapses Between Motoneurons in the Spinal Cord of the Newborn Rat, *Proceedings of the Royal Society, London, Series B* **208** (1980) 115-120.

[11] K. Walton and R. Navarrete, Postnatal Changes in Motoneurone Electronic Coupling Studied in the *In Vitro* rat Lumbar Spinal Cord, *Journal of Physiology (London)* **433** (1991) 283-305.

[12] R. J. Balice-Gordon, A. Pereda and M. J. Pinter, Functional Gap Junctions Couple Motor Neurons in Development and Reinnervation, *Society for Neuroscience Abstracts* **22** (1996) 1487.

[13] D. L. Becker and R. Navarrete, Dye Coupling Between Lumbar Motoneurones in the Embryonic and Neonatal Rat Spinal Cord: An *In Vitro* Study. *Journal of Physiology (London)* **423** (1990) 98P.

[14] W. Camu and C. E. Henderson, Rapid Purification of Embryonic Rat Motoneurons: An *In Vitro* Model for Studying MND/ALS Pathogenesis, *Journal of Neurological Sciences* **124** (1994) 73-74.

[15] T. Tsuchida, M. Ensini, S. B. Morton, M. Baldassare, T. Edlund, T. M. Jessell and S. L. Pfaff, Topographic Organization of Embryonic Motor Neurons Defined by Expression of LIM Homeobox Genes, *Cell* **79** (1994) 957-970.

[16] Q. Chang and R. J. Balice-Gordon, Developmental Regulation of Connexin Expression in Motor Neurons, *Society for Neuroscience Abstracts* **23** (1997) 1677.

*Gap Junctions*
*R. Werner (Ed.)*
*IOS Press, 1998*


# Gap Junctional Permeability Modulates Gene Transcription in Osteoblastic Cells

Fernando LECANDA, Thomas H. STEINBERG, and Roberto CIVITELLI
*Department of Medicine, Washington University School of Medicine, St. Louis, MO 63110*

**Abstract.** Osteoblasts are coupled through gap junctions formed primarily by connexin43 (Cx43). They also express connexin45 (Cx45) on the cell surface. Overexpression of Cx45 in a Cx43 background results in decreased dye transfer and transjunctional conductance. The changed gap junctional communication translates into lower steady state levels of mRNA for osteocalcin (OC) and bone sialoprotein (BSP), two proteins strictly linked to bone matrix deposition and mineralization. These effects are the consequence of transcriptional down-regulation of the relative promoters. Conversely, overexpression of Cx43 in cells which express primarily Cx45 results an increased molecular permeability and enhanced transcriptional activity of OC and BSP promoters. OC gene expression is also down-regulated in osteoblastic cells grown in sparse colonies, as compared to confluent cultures, despite equal Cx43 expression per cell. Thus, gap junctional communication regulates the transcriptional activity of osteoblast specific genes, presumably by modulating the signals permeating through the junctions. A functional communicating network is necessary for the full elaboration of the osteoblast phenotype.

## 1. Heterogeneity of connexins in osteoblasts

Deposition of new bone occurs throughout a lifetime, the result of a cooperative effort of osteoblasts within bone remodeling units. Osteoblasts form an epithelium-like layer on the bone surfaces, and are interconnected by abundant adherens and gap junctions [1, 2]. Human and rat osteoblasts express primarily Cx43, although they also express Cx45, albeit in lower abundance [3, 4]. Experiments with small-sized dyes and electrophysiologic measurements of gap junctional conductance have proven that cells that express prevalently Cx45, such as the osteogenic sarcoma cell lines, UMR 106-01 and SaOS-2 are electrically coupled although they diffuse dyes poorly [3]. When UMR 106-01 cells were transfected with Cx43 dye coupling increased, demonstrating that gap junction channels formed by Cx45 in osteoblasts have different molecular permeability than those formed by Cx43, and point to specific roles of different connexins in mediating intercellular communication between bone cells [3]

Further studies demonstrated that, when expressed in the same cells, Cx45 and Cx43 are able to interact in forming gap junctions, and that the gating properties of Cx45 prevail in the resulting channels. This type of interaction was directly assessed by overexpressing chick Cx45 in another osteosarcoma cell line, ROS 17/2.8, which express Cx43 only and are very highly coupled. Diffusion of Lucifer Yellow and transjunctional conductance were reduced by more than 50% in ROS/Cx45 transfectants, compared to the parent ROS 17/2.8 cells [5]. Therefore, expression of multiple connexins may enable cells to achieve different modalities of cell-cell communication that may not be possible by expression of a single connexin. This in turn may determine the type of molecules or ions that can diffuse through the intercellular channels, thus providing an additional mechanism for modulating signal exchange among cells in a network.

## 2. Gap Junctional communication modulates osteoblast gene expression

The studies discussed in the previous paragraph offer circumstantial evidence for the hypothesis that gap junctional communication may be involved in regulation of osteoblast function. To test the consequences of changes in gap junctional communication on osteoblast phenotypic expression, we initially analyzed the steady-state mRNA levels of several osteoblast phenotypic markers in stable transfectants ROS/Cx45, including osteocalcin (OC), and bone sialoprotein (BSP). Of the various proteins that define the osteoblast phenotype, only OC and BSP are highly specific for bone forming cells, and their expression is selectively detected at the onset of mineralization in bone [6, 7].

Steady-state mRNA levels for OC and to a lesser degree, BSP were lower in ROS/Cx45 compared to the parent cells and mock transfected cells, and they were associated with a lower degree of gap junctional communication. In addition, alkaline phosphatase activity, a less selective marker of mature osteoblasts, was greatly reduced in ROS/Cx45. Similar results were reproduced in 3 additional ROS/Cx45 clones, although some variability was observed.

These data suggest that changes in gap junction communication resulting from Cx43-Cx45 interactions affect the expression of several genes critical for osteoblastic phenotype. Interestingly, despite the different levels of bone matrix protein mRNA expression, ROS/Cx45 transfectants retained their responsiveness to $1{,}25(OH)_2D_3$, a known stimulator of OC secretion in the rat. The abundance of OC transcripts was considerably increased by 24 h incubation with $1{,}25(OH)_2D_3$ in each cell line, parents and transfectants, whereas BSP mRNA was undetectable in $1{,}25(OH)_2D_3$ treated cells. It is important to note that neither Cx45 mRNA expression, nor dye coupling were affected by $1{,}25(OH)_2D_3$. Thus, gap junctional communication regulates OC gene expression via a mechanism independent of $1{,}25(OH)_2D_3$.

To further prove that the relative abundance of Cx43 and Cx45 regulates osteoblast gene expression, we repeated the same experiments in the UMR 106-01 cells, which are poorly coupled and express predominantly Cx45 on the cell surface. The basal level of OC expression is also very low in these cells compared to the ROS 17/2.8 cells. As expected from the previous results, UMR/Cx43 transfectants exhibited an increased OC mRNA expression relative to parent cells, in association with an increase in dye coupling.

Although the data on the ROS/Cx45 and UMR/Cx43 transfectants strongly suggested that gap junctional communication alters gene expression in osteoblastic cells, clone selection represents a potential confounder that may influence steady-state mRNA levels in the transfectants. To avoid clone variability and to determine whether the changed steady-state mRNA levels were consequence of modulation of gene transcription, we transiently co-transfected the chick Cx45 expression construct and constructs containing the *cis*-regulatory regions of either the rat OC or BSP genes, upstream of the luciferase reporter gene (OCLUC and BSPLUC) into ROS 17/2.8 cells. In cells co-transfected with Cx45, OCLUC or BSPLUC activity was reduced by more than 70%, relative to cells co-transfected with the vector only. The same experiments were repeated in the phenotypically immature, immortalized MC3T3-E1 mouse osteoblastic cell line. Like ROS 17/2.8, the MC3T3-E1 cells are highly coupled and express abundant Cx43[8], as well as OC and BSP, and support activity of the OC promoter [9]. As in the ROS 17/2.8, OCLUC and ONLUC activity was >50% lower in MC3T3-E1 cells co-transfected with Cx45 relative to vector-transfected cells. On the contrary, transcriptional activity of an osteonectin promoter, which is less specific for the osteoblast, was unaffected by Cx45. Therefore, overexpression of chick Cx45, which alters gap junctional communication provided by Cx43, down-regulates the transcriptional activity of osteoblast-specific promoters.

Based on the above data, one would predict that overexpression of Cx43 into UMR 106-01 cells, which results in increased gap junctional communication, should up-regulate the activity of the OC promoter, mirroring the transcriptional down-regulation induced by Cx45 overexpression. Basal OCLUC and BSPLUC transcriptional activity were lower in UMR106-01 than in ROS 17/2.8, as expected from the different steady state OC and BSP mRNA levels in the two cell lines. When the two osteoblast-specific promoter-reporter constructs were co-transfected with Cx43, increased transcriptional activity was observed compared to vector transfected cells, whereas an SV40LUC construct was unaffected. Thus, the relative expression of Cx43 and Cx45 modulates the transcriptional activity of stage-specific promoters in osteoblasts.

## 3. Gap junctional communication is regulated by cell density

Lack of intercellular communication can be recapitulated by physically separating a cell population into isolated cells. Based on the above data, one would predict that in such condition, OC and BSP gene expression be very low, if not absent. Because extensive experimentation in isolated cells is impractical, we examined the phenotypic features of ROS 17/2.8 cells grown in sparse cultures constituted by very small colonies as compared to confluent cultures. Steady state levels of OC mRNA were very low in sparse cultures of ROS 17/2.8 cells, with individual colonies composed of <50 cells, whereas OC was abundantly expressed in confluent cultures. Interestingly, a lower relative expression of OC was observed even in the Cx45 transfectants grown in sparse cultures. Importantly, the expression and localization of Cx43 and Cx45 to appositional membranes were not affected by cell density, indicating that the number of functional gap junctions is presumably not decreased in small colonies. Although insufficient concentrations of a putative autocrine factor may account for the low level OC expression in sparse cultures, it is also quite possible that basal expression of OC is under the control of intracellular signal(s) that propagate across the cell network through gap junctions. Such a signalling mechanism may be sensitive to the number of interconnected cells, and/or changes in gap junctional permeability.

## 4. Conclusions

The nature of the mechanism that links gap junctional communication to gene expression remains elusive, but it certainly depends on the type of signals that permeate the junctional channel. Based on the pore size selectivity of Cx43 and Cx45 gap junctions, one could predict that intercellular diffusion of signalling molecules, such as cyclic nucleotides or inositol phosphates, may be impaired when Cx43 permeability is decreased by interaction with Cx45. In alternative, the affinity of transcriptions factors for specific DNA binding sites may be sensitive to spontaneous oscillations in intracellular ionic concentration, i.e. calcium, or in membrane polarity, whose propagation requires a certain type of gap junctional communication.

The observation that gap junctional communication serves a specific regulatory function on genes critical for the development and function of bone forming cells is consistent with observations in other cell systems pointing to a more general role of intercellular communication via gap junctions for the normal function of many highly differentiated tissues. For example, the steroidogenic response of adrenal cells to ACTH is dependent on cell density and gap junctional communication [10], and stimulation of smooth muscle contraction by $\alpha_1$-adrenergic agonists is blunted by pharmacologic inhibitors of cell coupling [11]. More to the point, insulin production in communication deficient insulinoma cells was restored by transfection of Cx43 [11, 12]. These observations demonstrate conclusively that gap junctions provide a fundamental regulatory mechanism controlling gene expression in tissues whose specialized function requires the synchronization of multicellular activity.

## References

[1] Doty SB: Morphological evidence of gap junctions between bone cells. *Calcif Tissue Int.* 1981;33:509-511

[2] Palumbo C, Palazzini S, Marotti G: Morphological study of intercellular junctions during osteocyte differentiation. *Bone* 1990;11:401-406

[3] Steinberg TH, Civitelli R, Geist ST, Robertson AJ, Hick E, Veenstra RD, et al: Connexin43 and connexin45 form gap junctions with different molecular permeabilities in osteoblastic cells. *EMBO J.* 1994;13:744-750

[4] Civitelli R, Beyer EC, Warlow PM, Robertson AJ, Geist ST, Steinberg TH: Connexin43 mediates direct intercellular communication in human osteoblastic cell networks *J.Clin.Invest.* 1993;91:1888-1896

[5] Koval M, Geist ST, Westphale EM, Kemendy AE, Civitelli R, Beyer EC, et al: Transfected connexin45 alters gap junction permeability in cells expressing endogenous connexin43. *J.Cell Biol.* 1995;130:987-995

[6] Mark MP, Butler WT, Prince CW, Finkelman RD, Ruch JV: Developmental expression of 44-kDa bone phosphoprotein (osteopontin) and bone gamma-carboxyglutamic acid (Gla)-containing protein (osteocalcin) in calcifying tissues of rats. *Differentiation* 1988;37:123-136

[7] Chen J, Shapiro HS, Sodek J: Developmental expression of bone sialoprotein mRNA in rat mineralized connective tissue. *J.Bone Miner.Res.* 1992;7:987-997

[8] Yamaguchi DT, Ma D, Lee A, Huang JT, Gruber HE: Isolation and characterization of gap junctions in the osteoblastic MC3T3-E1 cell line. *J.Bone Miner.Res.* 1994;9:791-803

[9] Boudreaux JM, Towler DA: Synergistic induction of osteocalcin gene expression. *J.Biol.Chem.* 1996;271:7508-7515

[10] Munari-Silem Y, Lebrethon MC, Morand I, Rousset B, Saez JM: Gap junction-mediated cell-to-cell communication in bovine and human adrenal cells. A process whereby cells increase their responsiveness to physiological corticotropin concentrations. *J.Clin.Invest.* 1995;95:1429-4139

[11] Christ GJ, Brink PR, Zhao W, Moss J, Gondré CM, Roy C, et al: Gap junctions modulate tissue contractility and *aplha*$_1$ adrenergic agonist efficacy in isolated rat aorta. *J.Pharmacol.Exp.Ther.* 1996;266:1054-1065

[12] Vozzi C, Ulrich S, Charollais A, Philippe J, Orci L, Meda P: Adequate connexin-mediated coupling is required for proper insulin production. *J.Cell Biol.* 1995;131:1561-1572

*Gap Junctions*
*R. Werner (Ed.)*
*IOS Press, 1998*

# Localization of Cx26, Cx32 and Cx43 in Myelinating Schwann Cells of Mouse Sciatic Nerve During Postnatal Development

Shumin Zhao and David C. Spray
*Department of Neuroscience, Albert Einstein College of Medicine, Bronx, New York 10461, USA*

**Abstract.** Expression and distribution of the gap junction proteins Cx26, Cx32 and Cx43 were examined in sciatic nerves of newborn, 1,2 and 4 wk old and adult mice using immunocytochemical techniques. In newborn mice, these connexins were expressed only weakly in the sciatic nerve: Cx32 was localized in the myelin sheaths, Cx43 was found in Schwann cell bodies and external membranes, and Cx26 was only in the external membrane. At 1 wk postnatally, Cx32 was primarily found in the myelin sheaths, Schmidt-Lanterman incisures and nodes of Ranvier; Cx43 remained localized to Schwann cell bodies and external membranes, but expression was stronger than in newborn animals; Cx26 expression remained low, with a pattern similar to Cx43. At 2 wk, Cx32 immunostaining was intense at the nodes of Ranvier, and also within the myelin sheaths and Schmidt-Lantermann incisures, whereas Cx43 and Cx26 were found on membranes throughout the Schwann cells, but not at the nodes of Ranvier. At 4 wk and in adult mice, patterns of Cx26, Cx32 and Cx43 expression were similar to those seen at 2 wk. Notably, Cx43 expression could be detected at the node of Ranvier in the adult, although its immunoreactivity was weak in comparison with that of Cx32. Cx26 was never found within the paranodal region. These results indicate that the gap junction proteins Cx26, Cx32 and Cx43 are expressed in the peripheral nervous system at early postnatal time points in mouse. The different localization and developmental profiles of these connexins in Schwann cells suggest that they provide distinct functions during the process of myelination.

## Introduction

Schwann cells myelinate large axons in the peripheral nervous system and, following nerve injury, proliferate to guide and repopulate regenerating axons. Previous studies have indicated that connexins 32, 43 and 46 are expressed in rodent Schwann cells. Connexin32 appears to be localized primarily at paranodal regions and Schmidt-Lantermann incisures, where the function appears to be the provision of a short circuit favoring the otherwise tortuous path for delivery of nutrients to adaxonal cytoplasmic processes of the Schwann cells [1,2,3,4,5]. A large number of primarily coding region mutations of Cx32 result in X-linked Charcot-Marie-Tooth disease, a progressive and slowly developing peripheral neuropathy. The presumed cause of this peripheral neuropathy is disruption of nutrient delivery to innermost cytoplasmic folds of the Schwann cell membrane[1]. Cx32 has also been reported to be present in compact myelin[3], where its function would also be expected to be the formation of homocellular junctional channels for ionic and molecular exchange across the myelin. Developmentally, Cx32 expression and confinement to Schmidt-Lantermann incisures and nodes of Ranvier correlates with attainment of myelinating phenotype by the Schwann cells[5]. Following crush injury, Cx32 expression radically declines distal to the injury[2,4].

Connexin43 expression in rat sciatic nerve has been reported to be restricted to perineurial cells, although endoneurial fibroblasts appear to express Cx43 following nerve injury[4]. However, in mouse sciatic nerve Schwann cells, Cx43 has been reported to be present[6], with Cx43 mRNA increasing rapidly after birth peaking at day 6, and declining thereafter.

Connexin46 expression in myelinating rat sciatic Schwann cells is detectable by Western and Northern blot analyses, with expression increasing dramatically in distal regions following nerve injury[4].

The studies described below were undertaken to systematically evaluate the connexin distribution in sciatic nerves of mice during the developmental process of myelination. Immunostianing of dissected nerves with antibodies specific for connexins 26, 32 and 43 indicate that each of these connexins is expressed in myelinating Schwann cells of mouse. The different temporal expression and spatial localization of these connexins observed suggests that each connexin may provide distinct functions during the myelination process.

## Material and Methods

Newborn and adult C57BL mice at 1,2, 4 postnatal week were used in this study. After painless death, sciatic nerves were dissected and cut into 0.1 to 0.2 cm pieces following removal of the adjacent connective tissue and epineurium. Using fine forceps, single nerve fibers were teased from the nerve pieces and mounted on glass coverslips which were pre-coated with poly-lysine(50$\mu$g/ml in $dH_2O$, Sigma). 4% formaldehyde was used to fix the nerve fibers for 10 min at room temperature. After three 10 min washes in PBS, the teased nerve fibers were blocked with 0.1% BSA and 0.5% Triton X-100 in PBS for 30 min at room temperature. Rabbit polyclonal primary antibodies monospecific for Cx26, Cx32 or Cx43 were then applied at 1:200 dilution. As controls, the primary antibody was omitted. After overnight incubation at 4° C, samples were washed in PBS (3x15 min with agitation), then incubated with 1:200 goat anti-rabbit IgG conjugated with FITC at 1:200 (Sigma) for 1 hr at room temperature in the dark. After 3 washes (10 min), the coverslips with the nerve fibers were mounted on glass slides and analyzed by fluorescence microscopy at 630x magnification.

## Results

In sciatic nerves of newborn mice, connexins 26, 32 and 43 were expressed only weakly: Cx26 was only in external membranes (upper left panel, Fig. 1), Cx32 was localized in the myelin sheath (leftmost panel, second row, Fig. 1), Cx43 was found in Schwann cell bodies and external membranes (leftmost panel, bottom row, and Fig. 1). At 1 week postnatally, Cx26 expression remained low and was confined to cell bodies and external membranes (second panel, top row,

**Fig. 1.** Developmental pattern of connexin expression in myelinating Schwann cells of mouse sciatic nerve. Broad arrowheads indicate Schwann cell bodies, narrow arrows point to nodes of Ranvier.

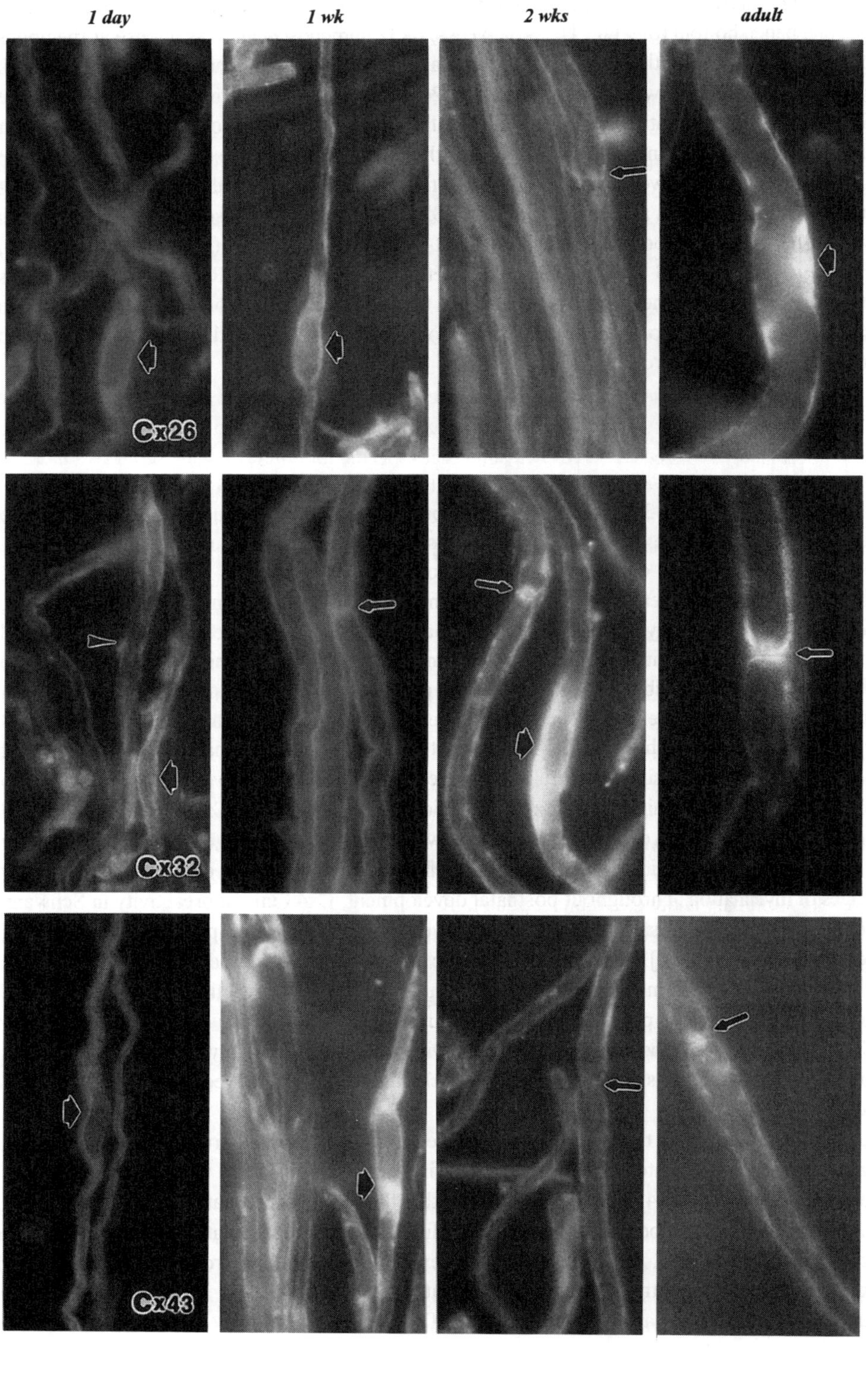

Fig. 1). Cx32 was primarily found in the myelin sheaths, Schmidt-Lantermann incisures and node of Ranvier (second panel, second row, Fig. 1) and Cx43 remained localized to Schwann cell bodies and external membrane, although the expression was stronger than in newborn animals (second panel, bottom row, Fig. 1). At 2 weeks, Cx32 immunostaining was intense at the nodes of Ranvier and also within the myelin sheaths and Schmidt-Lantermann incisures (third panel, second row, Fig. 1), whereas Cx43 and Cx26 were found on membranes throughout the Schwann cells, but not at the nodes of Ranvier (third panels, first and bottom rows, Fig. 1). At this stage, the expression of Cx43 was weaker than that at 1 week. At 4 weeks, all patterns of Cx26, Cx32 and Cx43 were similar to those seen at 2 weeks (data not illustrated). In the adult mouse, the localization and distribution of Cx32 were similar to that seen after 2 weeks, with strong staining in paranodal regions and Schmidt-Lantermann incisures (last panel, second row, Fig. 1). Notably, Cx43 expression could be detected at the nodes of Ranvier in sciatic nerves of adult mice (last panel, bottom row, Fig. 1), although its immunoreactivity was weak in comparison with that of Cx32. Cx26 immunostaining was weak and diffuse in adult sciatic nerves (last panel, top row, Fig. 1).

## Discussion

Connexins 26, 32 and 43 appear to be among the most abundantly expressed gap junction proteins in the developing nervous system[7,8]. The present results obtained during postnatal development of the peripheral nervous system reveal that these connexins have different expression patterns in sciatic nerve during the process of myelination. Until adulthood, Cx43 immunoreactivity was preferentially localized in Schwann cell bodies and external membranes, but no positive staining was seen in the myelin. This staining pattern is similar to that observed with antibodies to the Schwann cell cytoplasmic protein S-100 but is different from that with antibodies to myelin basic protein. Throughout postnatal development, Cx43 immunoreactivity in Schwann cells was found to be strongest at postnatal week 1, consistent with previous findings using Northern blot analyses[6]. In cultured Schwann cells, we have also observed cytoplasmic staining with Cx43 antibodies, and have recorded Cx43-like single channel currents from Schwann cellpairs (data not shown). These results suggest that Cx43 may play a role in the communication between Schwann cells or Schwann cells and other cells during early stages of myelination. Throughout postnatal development, Cx43 immunoreactivity in Schwann cells was found to be strongest at postnatal week 1, consistent with previous findings using Northern blot analyses[6]. In adult mouse sciatic nerve, Cx43 was found at nodes of Ranvier, a result that differs from previous studies in which Cx43 immunoreactivity in adult rat sciatic nerve was restricted to perineurium[4]. Although implications of this difference are unclear, Cx43 expression at paranodal regions could provide communication between cytoplasmic folds of single Schwann cells, between adjacent Schwann cells, or between Schwann cells and neighboring cell types.

Cx32 was found to be expressed only weakly in sciatic nerves from newborn mice, but immunoreactivity was intense at the end of the first postnatal week and thereafter. In contrast to the distribution of Cx43, Cx32 was found within the myelin sheath, at Schmidt-Lantermann incisures, and within nodes of Ranvier. This distribution coincides with that of myelin basic protein, but not of S100. Elaboration of Cx32 expression during the process of myelination is consistent with the distribution reported in rat sciatic nerve[2] and its expression pattern suggests that it is associated with the process of myelination.

Cx26 was found to show a developmental expression pattern similar to that of Cx32, although its expression was much weaker. Cx26 was also localized at myelin sheaths , Schmidt-Lantermann incisures, and on membranes throughout the Schwann cells. However, distinct from Cx32, Cx26 was not seen at the nodes of Ranvier at any time during development. Because of its lower expression, it appears that Cx26 is less important in the formation of gap junctions in Schwann cells and in the myelination process.

## Conclusions

The different patterns of temporal expression and localization of Cx26, Cx32 and Cx43 during myelination of mouse sciatic nerve fibers raises the possibility that the connexins may play distinct roles in the process. The weak expression of Cx26 and its association with myelin and Schmidt-Lantermann incisures but not paranodal regions implies that although it may be a myelin-associated protein, its participation is minor in the process of linking regions within the Schwann cell. By contrast, the pronounced expression of Cx32 in paranodal regions and in Schmidt-Lantermann incisures during myelination further emphasizes the likelihood that this connexin provides nutrient delivery to inner regions of the myelinating Schwann cell. The direct demonstration of such predicted deficits in Cx32 knockout mice is expected to be provided soon, and should serve to confirm the hypothesis that Cx32 dysfunction leads to X-linked Charcot-Marie-Tooth disease by this mechanism[1]. The expression of Cx43 in a pattern overlapping with S100 immunoreactivity suggests that it is not myelin associated, as does its maximal expression before myelination is complete. Whether this connexin may provide an avenue of intercellular communication with other Schwann cells or other cell types is an exciting possibility that awaits further experimentation.

## References

[1] J. Bergoffen, SS. Scherer, S. Wang, M. Oronzi-Scott, L. Bone, DL. Paul, K. Chen, MW. Lensch, P. Chance and K. Fischbeck, Connexin mutations in X-linked Charcot-Marie-Tooth disease. *Science* (1993) 262: 2039-2042.

[2] SS. Scherer, SM. Deschenes, YT. Xu, JB. Grinspan, KH. Fischbeck and DL. Paul, Connexin32 is a myelin-related protein in the PNS and CNS. *J Neuroscience* (1995) 15:8281-8294.

[3] T. Miyazaki, Y. Takeda, Y. Murakami, H. Kawano, T. Shimazu, S. Toya and K. Uyemura, Distribution of PAS II/PMP22 and connexin 32 proteins in peripheral nervous system. *Neurochem Int* (1995)27:377-383.

[4] KJ. Chandross, JA. Kessler, RI. Cohen, E. Simburger, DC. Spray, P. Bieri and R. Dermietzel, Altered connexin expression after peripheral nerve injury. *Mol Cell Neurosci* (1996) 7:501-518.

[5] M. Satake, T. Yoshimura, A. Ohnishi and T. Kobayashi, Connexin32 gene expression in rat sciatic nerves and cultured Schwann cells. *Dev Neurosci* (1997) 19:189-195.

[6] T. Yoshimura, M. Satake and T. Kobayashi T, Connexin43 is another gap junction protein in the peripheral nervous system. *J Neurochem* (1996) 67:1252-1258.

[7] R. Dermietzel, O. Traub, TK. Hwang, E. Beyer, MVL. Bennett, DC. Spray and K. Willecke, Differential expression of three gap junction proteins in developing and mature brain tissue. *Proc Natl Acad Sci USA* (1989) 86:10148-10152.

[8] R. Dermietzel, M. Farooq, JA. Kessler, H. Althaus, EL. Hertzberg and DC. Spray, Oligodendrocytes express gap junction proteins connexin32 and connexin45. *Glia* (1997) 20:101-114.

# Section 7:

# Regulation of Gap Junctional Communication

*Gap Junctions*
*R. Werner (Ed.)*
*IOS Press, 1998*



# Phosphorylation of chicken connexin56: possible implications

Viviana M. BERTHOUD*, Eileen M. WESTPHALE*, Wendy E. KURATA‡, Steven BASSNETT†, Alan F. LAU‡, Paul D. LAMPE|, Eric C. BEYER*

*Departments of *Pediatrics and †Ophthalmology, Washington University School of Medicine, St. Louis, MO 63110; ‡Cancer Research Center of Hawaii, University of Hawaii, Honolulu, HI 96813; |Fred Hutchinson Cancer Research Center, Seattle, WA 98109.*

**Abstract.** Phosphorylation has been implicated in connexin trafficking, assembly, insertion into the plasma membrane, degradation and retrieval from the plasma membrane and in gap junctional channel gating. Chicken connexin56 (Cx56) is a phosphoprotein that is abundantly expressed in the eye lens. In immunoblots of lens homogenates, Cx56 is detected at the 4$^{th}$ embryonic day as a single band that co-migrates with *in vitro* translated Cx56. By the 18$^{th}$ embryonic day, Cx56 is detected as multiple bands that collapse to a doublet after treatment with alkaline phosphatase. These results suggest that during development Cx56 is post-translationally modified by phosphate groups in a development-dependent process. Also, confocal immunofluorescence shows a developmental progression of immunoreactive Cx56 from large plaques to smaller dispersed punctate staining.

Lens cells in culture differentiate to form lentoids that express Cx56. In lentoids, modification of Cx56 resembles that observed in the intact lens. To identify possible sites of phosphorylation of Cx56 and their functional implications, lentoid-containing cultures were treated with 12-O-tetradecanoylphorbol-13-acetate (TPA), a direct activator of protein kinase C (PKC). Treatment with TPA decreased dye coupling between lentoid cells and increased the intensity of the Cx56 bands with an apparent molecular mass of 73-77 and 83 kDa. Incorporation of $^{32}P$ into immunoprecipitated Cx56 was observed in cultures labeled with $^{32}P$-orthophosphate. The two-dimensional pattern of Cx56 tryptic phosphopeptides differed between control and TPA-treated cultures. Treatment of lens cultures with TPA increased the abundance of two constitutively phosphorylated peptides that were mapped to the intracellular loop of Cx56 corresponding to the sequence RGSVK. We also identified a constitutively phosphorylated site in the carboxy terminal domain of Cx56 in the sequence RLSR, whose phosphorylation was unaffected by activation of PKC.

Pulse-chase experiments demonstrated the presence of two pools of Cx56 with half-lives of 2.4 h and 2 days. The Cx56 forms that appeared first were the ones with faster electrophoretic mobility; the forms with slower electrophoretic mobilities appeared later. These data suggest that phosphorylation may play a role in regulating Cx56 stability.

## 1. Lens Cx56 is a phosphoprotein

Many of the connexins studied are phosphoproteins [1-4]. This modification has been implicated in connexin trafficking, assembly, insertion into the plasma membrane, degradation and retrieval from the plasma membrane and in gap junctional channel gating [5-14]. Lens

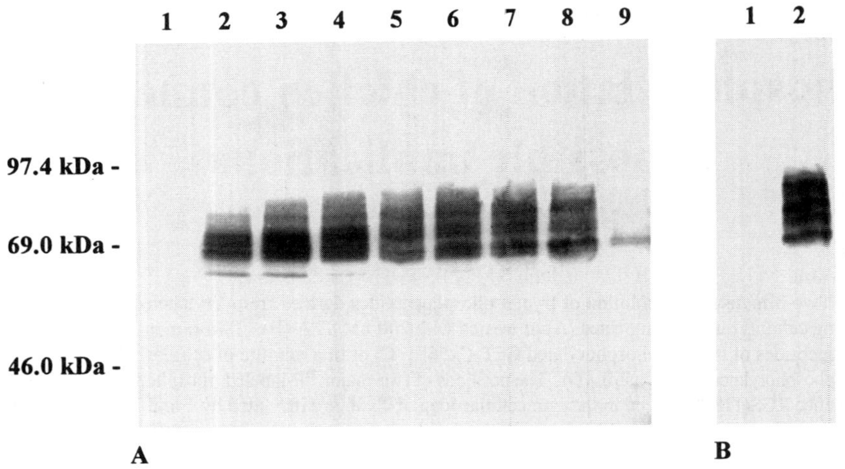


Fig. 1. Pattern of immunoreactive Cx56 bands in the lens during development. (A) Immunoblot of Cx56 in homogenates of chicken lens from 4- (lane 1), 6- (lane 2), 9- (lane 3), 12- (lane 4), 15- (lane 5), 16- (lane 6), and 20-day old (lane 7) embryos and from 1- (lane 8) or 112-day old (lane 9) chickens. (B) Immunoblot of Cx56 in homogenates from lenses of 4- (lane 1) and 18-day old (lane 2) embryos in which the detection reaction was allowed to proceed for longer to detect the Cx56 band present in the sample from embryonic day 4. The electrophoretic mobilities of the molecular weight standards are indicated on the left, and the mobility of the *in vitro* translated product is indicated by the short bar to the right of panel B. (Reproduced with permission from [17]).

connexin56 (Cx56) is a phosphoprotein. Its immunoblot pattern varies from a single band migrating with an apparent molecular mass of 66 kDa at embryonic day 4 to several immunoreactive bands at embryonic day 18 (Fig. 1). A higher degree of complexity in Cx56 forms becomes evident when the samples are resolved by two-dimensional electrophoresis [15, 16]. Treatment of 18th embryonic day lens homogenates with alkaline phosphatase reduces the immunoblot pattern of Cx56 to a doublet [17], suggesting that phosphorylation is responsible for most of Cx56 heterogeneity.

Chicken lens cells in culture differentiate into gap junction-containing structures called lentoids [18, 19] that express Cx56 [17]. In these cultures, Cx56 modification by phosphorylation is similar to that in the lens [16]. We have used this system to study the effects of activation of protein kinase C (PKC)-dependent pathways on Cx56 phosphorylation and intercellular communication.

## 2. Decrease in intercellular coupling between lentoid cells and phosphorylation of Cx56 after activation of PKC-dependent pathways

Treatment of lentoid-containing cultures with 12-O-tetradecanoylphorbol-13-acetate (TPA), a direct activator of PKC [20, 21] induces a decrease in dye coupling that correlates with an increase in the intensity of Cx56 bands migrating with apparent molecular mass of 73-77 and 83 kDa as detected by immunoblotting [22]. This effect is also associated with a modest increase in the amount of $^{32}P_i$ incorporated into Cx56 immunoprecipitated from lens cultures [16]. Analysis of two-dimensional maps from tryptic peptides of Cx56 reveals that TPA induces differential phosphorylation of the protein (Fig. 2).

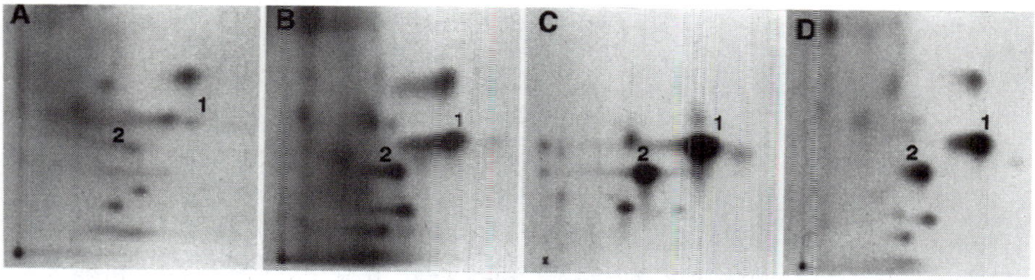


Fig. 2. Two-dimensional resolution of tryptic phosphopeptides derived from $^{32}$P-labeled Cx56 from lentoid-containing cultures that were untreated (A) or treated with 200 nM TPA (B). Two-dimensional maps of tryptic phosphopeptides of *in vitro* phosphorylated GST-Cx56IL (C) or of a mixture of cellular $^{32}$P-labeled Cx56 and *in vitro* phosphorylated GST-Cx56IL (D). The positions of two major $^{32}$P-labeled phosphopeptides derived from the sequence RGS(118)VK found in the intracellular loop of Cx56 are indicated by 1 and 2. (Reproduced from [16]).

## 3. TPA increases phosphorylation in the cytoplasmic loop of Cx56. Identification of the phosphorylation sites

To identify sites of phosphorylation of Cx56, two glutathione-S-transferase (GST) fusion proteins, one containing the carboxy terminus and another containing the cytoplasmic loop, were generated and phosphorylated *in vitro* by PKC. Analysis of their tryptic maps revealed incorporation of $^{32}P_i$ into one major peptide of the carboxy tail of Cx56 [16] and into two peptides of the cytoplasmic loop of Cx56 (Fig. 2, labeled 1 and 2). Analysis of a mixture of *in vitro* phosphorylated fusion protein and *in vivo* labeled Cx56 reveals that the major *in vitro* phosphorylated peptides co-migrate with constitutively phosphorylated peptides [16] and that TPA induces an increase in phosphorylation in the cytoplasmic loop of the protein (Fig. 2). The identified sequences of the phosphorylated peptides are: RLS(493)R in the carboxy terminus of Cx56 and RGS(118)VK in the cytoplasmic loop of Cx56 [16].

## 4. Two kinetic pools of Cx56 are detected in lens cultures

Previous reports on several connexins have shown the presence of a single kinetic pool with a rather rapid half-life of 1.5-4 h [2, 23-27]. When pulse-chase experiments are performed in lentoid-containing lens cultures, two kinetic pools of Cx56 can be detected [22]. One of them has a half-life of 2.4 h that is similar to the reported values for non-lens connexins [2, 23-27]. The other pool has a half-life of 2 days which is much longer than any other connexin half-life reported. Analysis of the time course of appearance of the different Cx56 forms reveals that the faster migrating forms appear first and that the slower migrating forms appear later. The slower migrating forms, which represent the main forms detected after long chase periods, might originate by post-translational modification of the faster migrating forms.

## 5. Appearance of Cx56 gap junctional plaques is dependent on lens fiber age

A change in appearance of Cx56 immunoreactivity is observed by confocal microscopy in lens sections. At embryonic day 15, young fiber cells in the superficial cortex show abundant discrete plaques on the broad and short faces of the fibers (Fig. 3A). In older fibers, located

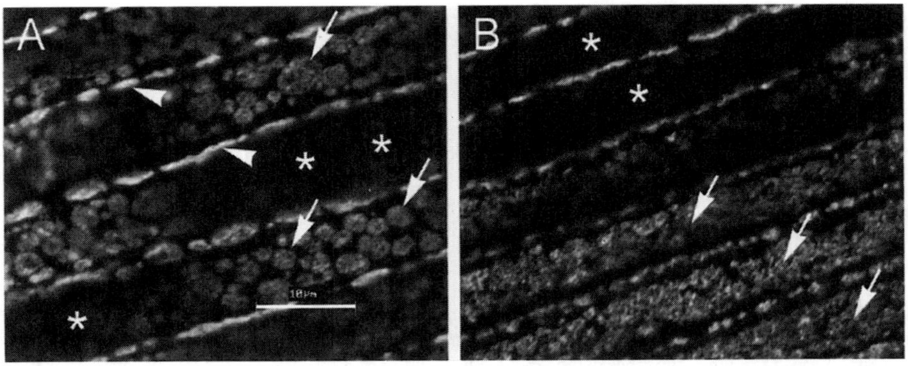


Fig. 3. Confocal images of Cx56 immunoreactivity. *En face* view of young lens fiber cells in the superficial cortex (A) or of older fiber cells (B) on a vibratome section of a lens from a 15 embryonic day old chicken. Broad and short faces of the fibers are indicated by arrows and arrowheads, respectively. In some cells (*), the optical section passes through the cytoplasm of the fibers rather than through the membrane.

toward the center of the section, Cx56 immunoreactivity has a granular appearance and discrete plaques are no longer evident (Fig. 3B).

## 6. Concluding remarks

Chicken Cx56 is first detected in the lens at embryonic day 4, when the first lens fibers and fiber-type gap junctions are detected [28]. Modification of Cx56 appears to be an early event, since samples prepared from lenses of the 4$^{th}$ embryonic day contain two isoforms detectable by two-dimensional electrophoresis. In older embryos, Cx56 is further modified as suggested by the numerous isoforms detected in immunoblots [16]. Most, if not all, of the heterogeneity observed by embryonic day 18 is due to phosphorylation [17], a modification that has been implicated in several steps of the life-cycle of a gap junctional protein [5-14].

Our studies suggest several different roles for phosphorylation of Cx56. (1) Involvement of phosphorylation in regulating gating of Cx56-containing gap junctional channels is suggested by the decrease in dye coupling induced by TPA treatment which activates PKCγ in lentoid cells. (2) The time course of appearance of the different Cx56 forms in pulse-chase experiments suggests that late phosphorylation events on Cx56 might be associated with stabilization of a pool of the protein. (3) Changes in phosphorylation of Cx56 may be associated with fiber cell differentiation, possibly including dispersal of Cx56-containing gap junctional plaques.

Unlike the data reported for other connexins, Cx56 is phosphorylated *in vivo* in both the intracellular loop and in the carboxy terminal region. The functional effects induced by treatment of lens cultures with TPA correlate with an increase in phosphorylation in Ser118 in the intracellular loop of Cx56. Although this regulation may be unique to Cx56, it is also possible that phosphorylation of the intracellular loop of connexins is a more generalized phenomenon, and that this part of the connexin molecule plays a role in gating of gap junctional channels.

## Acknowledgements

This research was supported in part by NIH grants CA52098 (to A.F.L.), EY09852 (to S.B.) GM55632 (to P.D.L.) and EY08368 (to E.C.B.).

## References

[1] J. C. Sáez *et al.*, cAMP increases junctional conductance and stimulates phosphorylation of the 27-kDa principal gap junction polypeptide, *Proc. Natl. Acad. Sci. USA* 83 (1986) 2473-2477.

[2] L.S. Musil *et al.*, Differential phosphorylation of the gap junction protein connexin43 in junctional communication-competent and -deficient cell lines, *J. Cell Biol.* 111 (1990) 2077-2088.

[3] A.F. Lau *et al.*, Evidence that heart connexin43 is a phosphoprotein, *J. Mol. Cell Cardiol.* 23 (1991) 659-663.

[4] J.G. Laing *et al.*, Characterization of the gap junction protein connexin45, *J. Membr. Biol.* 139 (1994) 31-40.

[5] K.L. Puranam *et al.*, Trapping an intermediate form of connexin43 in the Golgi, *Exp. Cell Res.* 206 (1993) 85-92.

[6] L.S. Musil and D.A. Goodenough, Biochemical analysis of connexin43 intracellular transport, phosphorylation, and assembly into gap junctional plaques, *J. Cell Biol.* 115 (1991) 1357-1374.

[7] P.D. Lampe, Analyzing phorbol ester effects on gap junctional communication: a dramatic inhibition of assembly, *J. Cell Biol.* 127 (1994) 1895-1905.

[8] K.I. Swenson *et al.*, Tyrosine phosphorylation of the gap junction protein connexin43 is required for the pp60v-src-induced inhibition of communication, *Cell Regul.* 1 (1990) 989-1002.

[9] A.F. Lau *et al.*, Epidermal growth factor disrupts gap-junctional communication and induces phosphorylation of connexin43 on serine, *Mol. Biol. Cell* 3 (1992) 865-874.

[10] A.P. Moreno *et al.*, Phosphorylation shifts unitary conductance and modifies voltage dependent kinetics of human connexin43 gap junction channels, *Biophys. J.* 62 (1992) 51-53.

[11] V.M. Berthoud *et al.*, Connexin43 in MDCK cells: regulation by a tumor-promoting phorbol ester and $Ca^{2+}$, *Eur. J. Cell Biol.* 57 (1992) 40-50.

[12] V.M. Berthoud *et al.*, On the mechanism of cell uncoupling induced by a tumor promoter phorbol ester in clone 9 cells, a rat liver epithelial cell line, *Eur. J. Cell Biol.* 62 (1993) 384-396.

[13] S.Y. Oh *et al.*, Phorbol ester induces phosphorylation and down-regulation of connexin 43 in WB cells, *Biochim. Biophys. Acta* 1094 (1991) 243-245.

[14] M. Elvira *et al.*, Phosphorylation of connexin-32 by protein kinase C prevents its proteolysis by mu-calpain and m-calpain, *J. Biol. Chem.* 268 (1993) 14294-14300.

[15] V.M. Berthoud *et al.*, Expression and regulation of Cx56 in chicken lens cultures, *Mol. Biol. Cell* 5 (1994) 1902a.

[16] V.M. Berthoud *et al.*, The gap-junction protein connexin 56 is phosphorylated in the intracellular loop and the carboxy-terminal region, *Eur. J. Biochem.* 244 (1997) 89-97.

[17] V.M. Berthoud *et al.*, Characterization of the gap junction protein connexin56 in the chicken lens by immunofluorescence and immunoblotting, *Invest. Ophthalmol. Vis. Sci* 35 (1994) 4109-4117.

[18] A.S. Menko *et al.*, Chicken embryo lens cultures mimic differentiation in the lens, *Dev. Biol.* 103 (1984) 129-141.

[19] A.S.Menko *et al.*, Junctions between lens cells in differentiating cultures: structure, formation, intercellular permeability, and junctional protein expression, *Dev. Biol.* 123 (1987) 307-320.

[20] M. Castagna *et al.*, Direct activation of calcium-activated, phospholipid-dependent protein kinase by tumor-promoting phorbol esters, *J. Biol. Chem.* 257 (1982) 7847-7851.

[21] J.E. Niedel *et al.*, Phorbol diester receptor copurifies with protein kinase C, *Proc. Natl. Acad. Sci.* 80 (1983) 36-40.

[22] V.M. Berthoud *et al.*, Regulation of lens gap junctions. In: R. Latorre and J.C. Sáez (eds.), From ion channels to cell-to-cell conversations. Plenum Press, New York, 1997, pp. 395-406.

[23] O. Traub *et al.*, Cyclic adenosine monophosphate stimulates biosynthesis and phosphorylation of the 26 kDa gap junction protein in cultured mouse hepatocytes, *Eur. J. Cell Biol.* 43 (1987) 48-54.

[24] O. Traub *et al.*, Comparative characterization of the 21-kD and 26-kD gap junction proteins in murine liver and cultured hepatocytes, *J. Cell Biol.* 108 (1989) 1039-1051.

[25] D.S. Crow *et al.*, Phosphorylation of connexin43 gap junction protein in uninfected and Rous sarcoma virus-transformed mammalian fibroblasts, *Mol. Cell Biol.* 10 (1990) 1754-1763.

[26] D.W. Laird *et al.*, Turnover and phosphorylation dynamics of connexin43 gap junction protein in cultured cardiac myocytes, *Biochem. J.* 273 (1991) 67-72.

[27] B.J. Darrow *et al.*, Expression of multiple connexins in cultured neonatal rat ventricular myocytes, *Circ. Res.* 76 (1995) 381-387.

[28] T.M. Miller and D.A. Goodenough, Evidence for two physiologically distinct gap junctions expressed by the chick lens epithelial cell, *J. Cell Biol.* 102 (1986) 194-199.

*Gap Junctions*
*R. Werner (Ed.)*
*IOS Press, 1998*

# Inhibition of Connexin43 junctional conductance by 2,3 Butanedione monoxime (BDM)

Gary S. Goldberg* and Alonso P. Moreno. *SUNY at Buffalo. Buffalo NY 14064 and Krannert Institute of Cardiology. Indianapolis IN 46202.

BDM has been shown to reduce coupling between isolated rat cardiocytes. We are currently presenting evidence that perfusion of BDM over cells in culture that express Cx43 also resulted in a dose-dependent reduction of the junctional conductance (gj) assayed by Lucifer Yellow transfer and double whole cell voltage clamp. Concentrations higher than 20 mM, quickly reduced gj to zero and single channel recordings unveiled channel blocking activity. Lower concentrations induced a slow irreversible reduction of GJ until cells became totally uncoupled after 20-30 minutes. Voltage gating properties and unitary conductance were not significantly affected. We conclude that BDM uncouple cells through a reversible fast blocking of the gap junction channels without evidence of Cx43 phosphorylation, and in a long term (possibly through dephosphorylation), BDM uncouples without altering transjunctional voltage gating.

## 1. Introduction

2,3 butanedione monoxime (BDM) is considered a non enzymatic dephosphorylating agent. It has been widely used to eliminate motion artifacts during studies on the propagation of action potential by means of optical mapping of isolated Langendorff-perfused or superfused preparations of the heart [1][2]. It has been shown that this drug can block, and also can modify the activity of various types of channels including those forming gap junctions between isolated rat cardiocytes [3]. We decided first to study the characteristics of the proposed blockage of Cx43 channels, using cell lines that endogenously express Cx43. Then, we tried to correlate this apparent blockage with changes in Cx43 phosphorylation by western blot assays.

## 2. Methods

Novikoff Hepatoma cells and Normal Rat Kidney (NRK) cells were used in this study. They were grown 37 °C in supplemented D-MEM media and seeded on 12 mm circular glass coverslips. Dye transfer assay involved injecting cells with 5 % Lucifer Yellow in 150 mM LiCl through a glass micropipette of 30-50 MΩ.

Junctional conductance ($g_j$) between cell pairs was calculated using the dual whole cell voltage clamp technique. The micropipettes were filled with patch solution with ATP and high CsCl (130 mM CsCl). During recording, cells were kept at room temperature and in cesium-containing bathing solution. Voltage sensitivity of junctional channels was calculated by recording the transjunctional current ($I_j$) in one of the cells (held at constant voltage) while a protocol of voltage was applied to the contiguous cell. Time constants for voltage inactivation currents were obtain using single exponential pClamp algorithms. Single channel currents were measured after gj was reduced using either Halothane (2 mM) or 2,3Butanedione monoxime (BDM).

To determine protein phosphorylation levels, lysates of cells from confluent mono-layers of 60 mm dishes were sonicated in SDS/PAGE sample buffer, boiled, clarified by centrifugation, assayed for protein concentration, and analyzed by Western blotting with an antibody directed against amino acids 302-319 of Connexin43.

## 3.Results

### *3.1 Uncoupling by BDM assayed by Lucifer Yellow injections*

As shown in Figure 1, dye transfer was abolished using high concentrations of BDM. Superfusion of this compound at concentrations of 5 and 10 mM produced a rapid uncoupling in less than 3 min, with partial recovery in the next 5 minutes. At concentrations of 15 mM or higher, an immediate effect was observed without complete recovery.

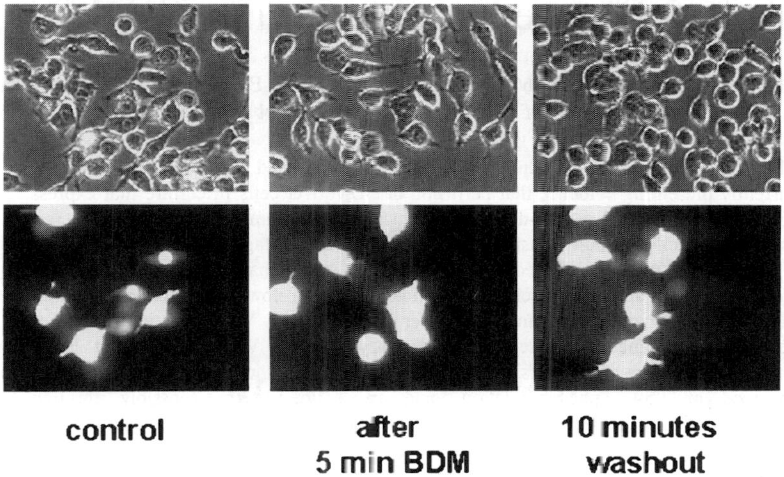


Figure 1 Phase contrast and fluorescent images showing that BDM reduces Lucifer Yellow transfer between Novikoff cells. In control conditions (left), these cells are profusely coupled (80%). After five minutes of 30 mM BDM, dye transfer between cells is completely abolished (MIDDLE PANEL). After washout, dye coupling is partially recovered (right).

### *3.2 Uncoupling by BDM assayed by double whole cell voltage clamp*

After dual whole cell voltage clamp was achieved, cell pairs were challenged with saline solution containing various concentrations of BDM. The time course of uncoupling by superfusion of a pair of cells with 30 mM BDM is shown in Figure 2. During the first Phase, junctional current became drastically reduced. The behavior of this junctional current was observed at higher gain, using a protocol of ramps as shown in Figure 3.

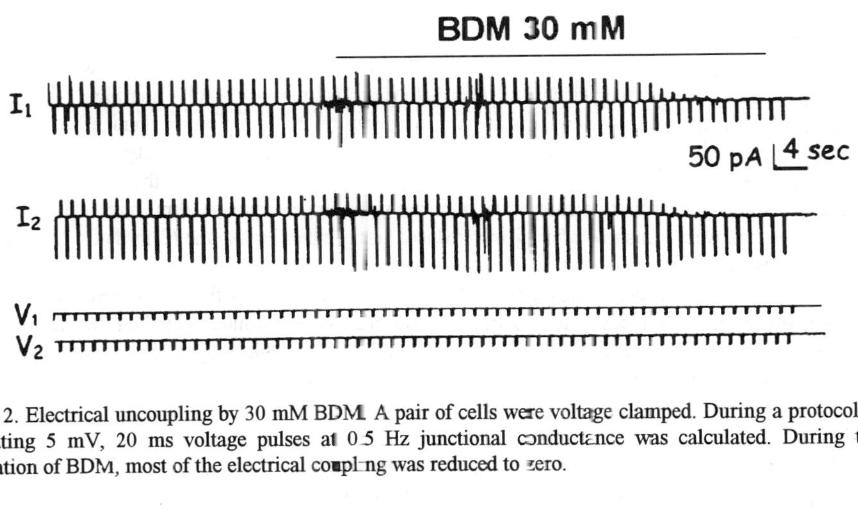


Figure 2. Electrical uncoupling by 30 mM BDM. A pair of cells were voltage clamped. During a protocol of alternating 5 mV, 20 ms voltage pulses at 0.5 Hz junctional conductance was calculated. During the application of BDM, most of the electrical coupling was reduced to zero.

***3.3 Blocking by BDM at the single channel level.*** Single channel gap junctional events were recorded before and after total uncoupling occurred during applications of high concentrations of BDM. Figure 3 (top) presents the (Ij) recorded during a -100/100 mV voltage ramp. The events recorded have a short open time duration, particularly in the regions of higher voltage sensitivity (60-100 mV). In order to compare the effects of BDM with other uncouplers, we also treated sister pairs of cells with 2 mM Halothane. As shown in Figure 3 (lower panels), when junctional conductance reduction is achieved, single channel recordings expose long open times. These recordings were also obtained during washout after total uncoupling. Noteworthy is the fact that these treatments did not affect the unitary conductance of Cx43 channels. This can be observed by calculating the unitary conductance of the current traces of Figure 3.

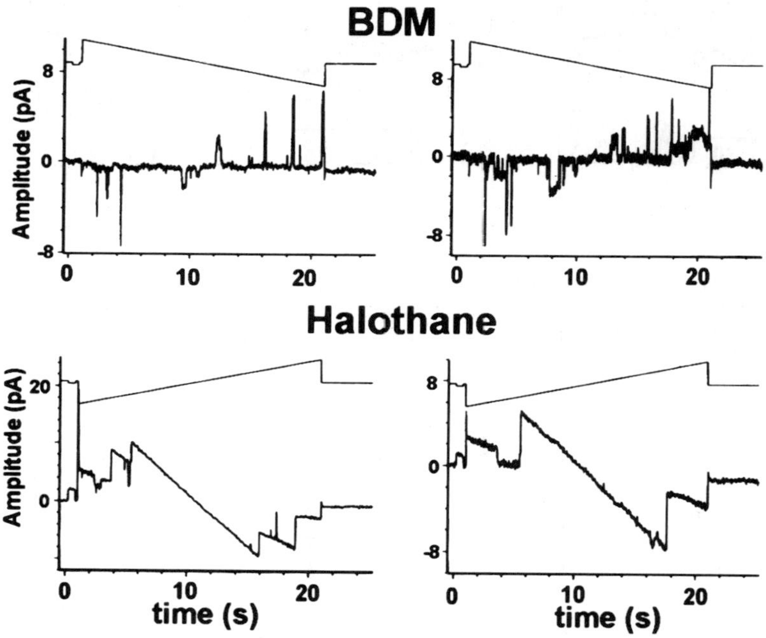


Figure 3. Single channel activity of Cx43 gap junction channels. On the top panels, a ramp protocol allows us to observe fast transitions between open and closed states In this conditions, the channels spend very small time in the open state. In both recordings, cells were uncoupled and slowly washed out for recovery. Note that the open times are still voltage dependent and the non-voltage dependent sub-state is only present after longer washout stages. In the lower panels, uncoupling was induced by 2mM Halothane. It is clear that the open times are much longer if compared with those observed after BDM.

### *3.4 Low doses of BDM do not affect inactivation kinetics*

To test the ability of this compound to affect other electrophysiological parameters, we decided to test the inactivation time constant during the process of uncoupling and re-coupling when 15 mM of BDM was added to the extracellular media. As it is shown in Figure 4, the voltage pulses applied (±30, ±50 and ±70 mV) inactivated the gap junction channels in a voltage dependent manner. To compare it with control, we measured the steady state current and normalized with the initial current, just before the pulse was imposed. We also fit the inactivation of Ij recorded at various voltages to a single exponential. The average of four different experiments is shown in Figure 5. Here the steady state conductance was calculated. The plot on the left shows that there is not a significant difference on the Gj/voltage curve with control experiments done in similar cells.

Furthermore, according to the panel on the right, there is also no significant difference in the inactivation time constants.

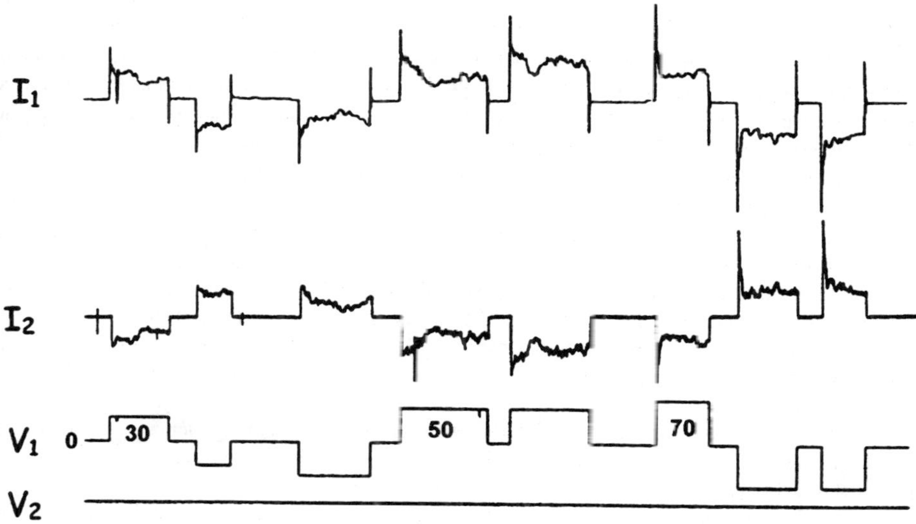


Figure 4. Cx43 maintains its voltage dependent characteristics during the application and recovery from 15mM of BDM. In this experiment, the voltage protocol applied to one of the cells is shown at the bottom. Note that the steady state current is not stable, because BDM was constantly perfused during the experiment.

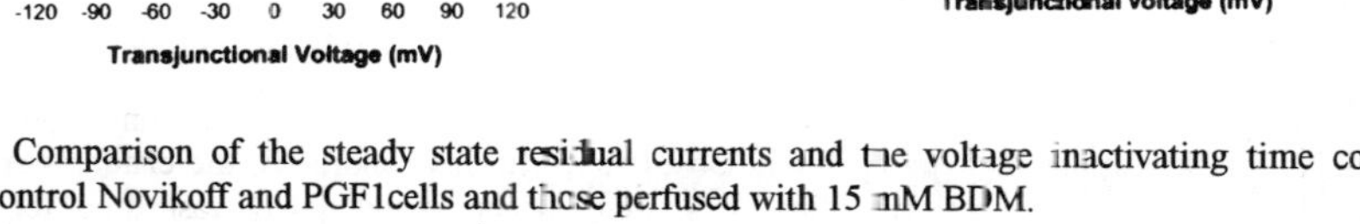


Figure 5. Comparison of the steady state residual currents and the voltage inactivating time constants between control Novikoff and PGF1cells and those perfused with 15 mM BDM.

### *3.5 Changes in phosphorylation detected by Western blots analysis*

Dephosphorylating properties of BDM on connexin43 were examined during the first 20 minutes (when the fast uncoupling effect of BDM is observed) by evaluation of changes in the mobility of Cx43 produced by phosphorylation. As presented in Figure 6, no quantifiable modification on the relative amount for the different Cx43 isoforms was apparent.

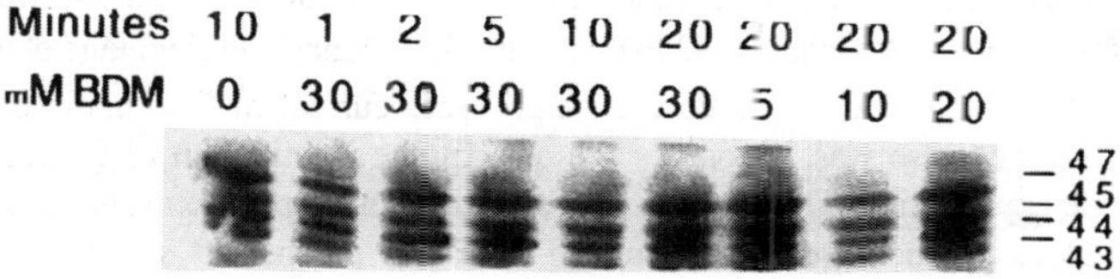


Figure 6. Western blots showing no apparent effect of BDM on Cx43 phosphorylation. Experiments were performed with sister dishes treated with concentrations and times as indicated. There were no significant

changes in the relative amount of the different subspecies of connexin43 for each lane. Differences observed in overall signals may be due to lane loading.

## 4. Discussion

BDM is a drug involved in multiple cell processes, although the mechanism of action is not well understood. It is clear that at high concentrations, this drug induces irreversible effects on the junctional communication through gap junction. In intact cells, this can be observed after testing fluorescent dye permeation through the junctions (Fig 1). A more rigorous assay proved a substantial decrease in gj, and not a change in the permeability to Lucifer Yellow (Figure 2).

We found that during the application of high concentrations of this drug, there is a fast and partially reversible reduction of junctional communication. According to Figure 3, the channels stay in the open state for extremely short periods of time. This type of behavior suggests that the active agent could be blocking the channel in its open state. This blocking effect has been suggested for the delayed rectifyer channel [4], the glycine-gated $Cl^-$ channel [5] and Ca++ channels [6].

For gap junctions, it has been proposed that BDM reduces gj without the requirement of phosphorylation for channel closing or reopening and without the intervention of cytosolic Ca++ released from intracellular stores [3]. Altogether, the evidence obtained for gap junctions and other channels points toward a direct action of BDM on the channel, by blocking their pores. Our results add single channel evidence for this mechanism, specially when the drastic reduction on the channel open time by BDM is compared with Halothane applications.

Cx43 protein has been shown to be dephosphorylated under prolonged applications (>1 hr) of non-blocking concentrations of BDM in dog heart wedges (See abstract by Beermann et al.), suggesting that long term uncoupling is due to a dephosphorylating action. Nevertheless, we could not detect in our cultured cells, any change in the mobility of Cx43 isoforms, even during the application of 30 mM for 20 minutes. In a long term, it is possible that the irreversible effect is due to dephosphorylation that uncouples the cells without altering transjunctional voltage gating. This was previously shown in SKHep1 cells using dephosphorylating enzymes [7,8]. Other drugs, like glycyrrhetinic acid derivatives are known to uncouple by the inducing changes in the structure of the junctions, though further studies are needed to elucidate if there act through a blocking mechanism [9]. So far, BDM may be the first substance reported to reduce junctional coupling by means of a direct channel blocking. Future studies of this drug and homologous substances could lead to the characterization of connexin specific blockers that may help in the study of the channel structure and its permeability properties.

## References

1. Sellin LC. McArdle JJ. Multiple effects of 2,3-butanedione monoxime. Pharmacology & Toxicology. 74(6):305-13, 1994

2. Rosenbaum, D. S.; Kaplan, et al. Repolarization inhomogeneities in ventricular myocardium change dynamically with abrupt cycle length shortening. Circulation, 84; 1331, 1991

3. Verrecchia F. Herve JC. Reversible blockade of gap junctional communication by 2,3-butanedione monoxime in rat cardiac myocytes. American Journal of Physiology. 272:C875-85, 1997

4. Lopatin AN. Nichols CG. Block of delayed rectifier (DRK1) K+ channels by internal 2,3-butanedione monoxime in Xenopus oocytes. [Journal Article] Receptors & Channels. 1(4):279-86, 1993.

5. Ye JH. McArdle JJ. 2,3-Butanedione monoxime modifies the glycine-gated chloride current of acutely isolated murine hypothalamic neurons. Brain Res. 735(1):20-9, 1996.

6. Zhu Y. Ikeda SR. 2,3-butanedione monoxime blockade of Ca2+ currents in adult rat sympathetic neurons does not involve 'chemical phosphatase' activity. Neuroscience Letters. 155(1):24-8, 1993

7. Moreno AP. Fishman GI. Spray DC. Phosphorylation shifts unitary conductance and modifies voltage dependent kinetics of human connexin43 gap junction channels. Biophysical Journal. 62(1):51-3, 1992

8. Moreno AP. Saez JC. Fishman GI. Spray DC. Human connexin43 gap junction channels. Regulation of unitary conductances by phosphorylation. Circ. Res. 74(6):1050-7, 1994

9. Goldberg GS. Moreno AP. et al. Evidence that disruption of connexon particle arrangements in gap junction plaques is associated with inhibition of gap junctional communication by a glycyrrhetinic acid derivative. Exp. Cell Res. 222(1):48-53, 1996.

Supported by NIH grants HL50485 and P50HL52323-1

*Gap Junctions*
*R. Werner (Ed.)*
*IOS Press, 1998*



# Enhanced Gap Junction Assembly Following the Elevation of cAMP Requires Full Length Cx43

TenBroek, E.[1], P. Lampe[2], S. Taffet[3], J. Reynhout[4], K. Martyn[5], W. Kurata[5], A. Lau[5], and R. Johnson[1]

*[1]Genetics and Cell Biology, University of MN; [2]Fred Hutchinson Cancer Research Institute; [3]Microbiology and Immunology, SUNY Health Science Center, Syracuse, [4]Bethel College, St. Paul, MN, [5]Molecular Carcinogenesis, Cancer Research Institute of Hawaii*

**Abstract.** In several cell systems, the formation of new gap junctions, or *de novo* gap junction assembly (GJA), is enhanced by agents which elevate intracellular cAMP. We have obtained evidence that such cAMP-enhanced GJA is the result of increased trafficking of connexin hemichannels (connexons) to the plasma membrane. To further examine the basis for enhanced GJA, immortalized fibroblast cell lines derived from embryonic Cx43-knockout mice were transfected with full length rat Cx43 (KO fibroblasts + $^{Full}$Cx43) and GJA between these cells was compared with that between immortalized fibroblasts derived from wild type embryonic mice (WT fibroblasts). GJA between both Cx43-expressing WT and KO fibroblasts + $^{Full}$Cx43 was increased 2-6 fold by 8Br-cAMP, and this increase was inhibited by the cAMP-dependent protein kinase (PKA) inhibitor H89. Both basal and enhanced GJA were significantly depressed by the protein synthesis inhibitor cycloheximide in spite of the continued presence of significant levels of Cx43, indicating that an accessory protein may be required for GJA. In addition, enhanced, but not basal, GJA was blocked by the microtubule inhibitor nocodazole, suggesting that two separate pathways are utilized for enhanced and basal GJA. A change in the number or distribution of Cx43 phosphoresidues on the C-terminus of Cx43 could play a role in enhanced GJA. Thus, Cx43-knockout fibroblasts were transfected with rat Cx43 truncated at amino acid 257 (KO fibroblasts + $^{M257}$Cx43) and GJA examined. GJA between the KO fibroblasts + $^{M257}$Cx43 was not enhanced by 8Br-cAMP. To confirm that the C-terminus was required for enhanced GJA, neuro2A (N2A) cells were obtained which had been transfected with either Cx43 or M257 (Cx43- or M257-N2A cells). In contrast to a 3-4 fold enhancement of GJ communication between Cx43-N2A cells, GJ communication between M257-N2As was not enhanced by 8Br-cAMP treatment. Although these results indicate that the C-terminus is necessary for enhanced GJA, it is unlikely that PKA-mediated phosphorylation of Cx43 occurs, since treatment of the WT fibroblasts with either forskolin or 8Br-cAMP failed to increase the incorporation of [$^{32}PO_4$] into Cx43. A Cx43 C-terminal glutathione sulfotransferase (GST) fusion protein was also not phosphorylated *in vitro* by purified PKA. Thus, PKA-mediated phosphorylation of another protein which interacts with specific residues of the Cx43 C-terminus may be required for the positive regulation of gap junction assembly.

## 1. Introduction

Increases in gap junction communication [1, 2], the size and number of gap junctions [3, 4, 5], and the assembly of new gap junctions (Paulson, manuscript in preparation) have all been shown to be enhanced by agents which elevate intracellular cAMP. Evidence indicates that cAMP-enhanced gap junction assembly (GJA) is the result of increased trafficking of connexons to the plasma membrane [6, 7]. Although elevation of intracellular cAMP results in the activation of cAMP-dependent protein kinase (PKA), to date there have been no reports demonstrating a direct phosphorylation of Cx43 by this enzyme. The C-terminus of Cx43 possesses the majority of the predicted Cx43 phosphorylation sites, certain of which appear to be required for changes in gating following the activation of different kinases [8, 9, 10]. It is not known whether PKA-mediated phosphorylation of Cx43 occurs during enhanced GJA. A C-terminal truncation mutant of Cx43 (M257) lacking most of these phosphorylation sites, assembles into functional gap junctions in the paired Xenopus oocyte system [11]. Although gap junctions formed by full length Cx43 in this system are sensitive to pH, those formed by the truncation mutant are not [11]. As an initial step in determining whether phosphorylation of the C-terminus is required for enhancement of GJA, two null cell types, Neuro2A (N2A) and Cx43-knockout fibroblasts, were transfected with the M257 truncation mutant and the regulation of GJA by cAMP was examined.

## 2. Results and Discussion

Several fibroblast cell lines have been cloned from the tissue of wild type (WT) and Cx43-knockout fetal mice and characterized with regard to gap junction protein expression and function [12]. Neither Lucifer yellow nor 5,6-carboxyfluorescein dye coupling is detected between the knockout fibroblasts, although the three WT fibroblast cell lines transfer both dyes (not shown). As the WT fibroblasts are well coupled and express significant levels of Cx43 ([12] and not shown), the Cx43-knockout (KO) fibroblasts could provide an optimal cellular milieu for the expression of mutant Cx43 and the study of regulation of GJA. Prior to examining mutant Cx43 constructs in this system, GJA was characterized between knockout fibroblasts transfected with full length Cx43 (KO fibroblasts + $^{Full}$Cx43). Stably transfected clones which were well coupled and expressed significant levels of Cx43 were isolated using standard techniques [12]. Of two transfection trials, 72% (36/50 clones) of the stably transfected KO fibroblast clones + $^{Full}$Cx43 were found to transfer Lucifer yellow dye (Fig 1). Well coupled KO fibroblast clones + $^{Full}$Cx43 (*) expressed significant levels of Cx43 with mobilities on Western immunoblots resembling those of Cx43 expressed by WT fibroblasts (Fig 2). The uppermost two bands detected with this C-terminal antibody to Cx43 in Figure 2 are unrelated to Cx43 (data not shown).

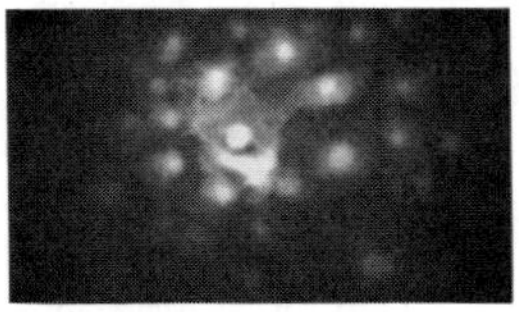

**Figure 1.**

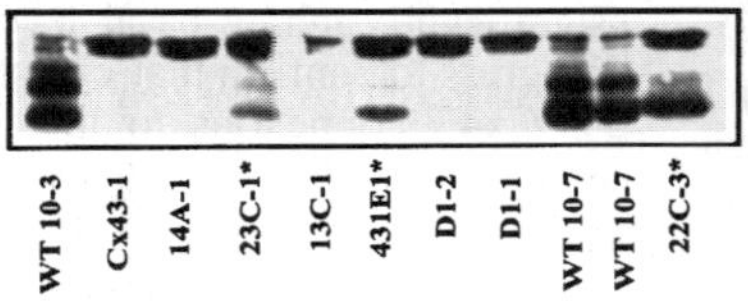


**Figure 2.**

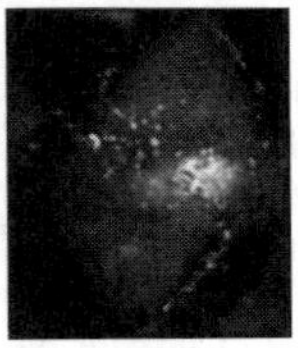

**Figure 3.**

Antibodies specific for Cx43 significantly labeled interfaces between the well coupled KO fibroblasts + $^{Full}$Cx43 (Fig 3), but not between the untransfected cells. Only intracellular labeling was observed in Cx43 transfectants which expressed Cx43 but did not transfer Lucifer yellow dye (not shown). Labeling of the WT fibroblasts revealed larger Cx43 puncta at cell-to-cell interfaces and little intracellular labeling (not shown).

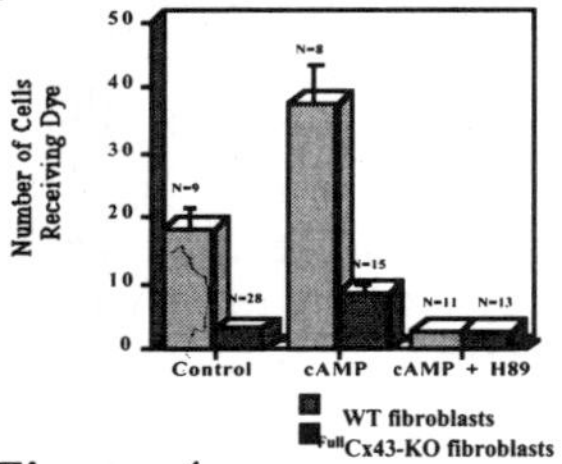


**Figure 4.**

To characterize *de novo* GJA, WT or KO fibroblasts + $^{Full}$Cx43 were trypsinized and allowed to recover as a single cell suspension for 3 hours. Gap junction remnants are internalized during this recovery period, as indicated by immunofluorescence antibody labeling for Cx43 (not shown). Cells were then reaggregated as a confluent monlayer for 1 hour and injected with Lucifer yellow dye (Fig 4). To determine whether GJA could be enhanced by elevation of cAMP between either the WT or KO fibroblasts + $^{Full}$Cx43, cells were treated with 300μM of 8Br-cAMP, a cell permeant cAMP analogue, for the entire 4 hour period of the GJA assay. In this set of experiments, such a treatment resulted in a two to threefold increase in dye transfer between both cell types, and the increase was blocked when 30 μM H89, a cAMP-dependent protein kinse (PKA) inhibitor, was included with the 8Br-cAMP (Fig 4). H89 inhibited basal levels of GJA between the WT fibroblasts but not the KO fibroblasts + $^{Full}$Cx43, indicating that the WT fibroblasts may have a higher level of basal PKA activity. Cx43 levels, as detected by Western immunoblotting, were slightly elevated (only 10-20%) in 1 out of 4 experiments following a 4 hour treatment with 8Br-cAMP. Although increased synthesis of Cx43 following extended exposure to cAMP-elevating agents may explain cAMP-mediated enhancement of gap junction communication in some systems [4], the occasional increase in Cx43 we observed does not explain the consistent 2-6 fold enhancement of GJA. (In Figure 6, for example, the 6-fold increase in GJA was not accompanied by an increase in Cx43 protein.)

In Novikoff hepatoma cells, a 12 hour treatment with 100 μg/ml cycloheximide, an inhibitor of protein synthesis, did not have a significant effect on basal GJA, in spite of inhibiting protein synthesis by 99% [13]. When present during dissociation (30 min), recovery (90 min) and reaggregation (1 hour) of Novikoff cells, cycloheximide only slightly inhibited basal GJA, but completely inhibited enhanced GJA [14]. In contrast, if cycloheximide (100 μg/ml) is added during the 1 hour assembly of KO fibroblasts + $^{Full}$Cx43, *both* basal and enhanced GJA are inhibited by 50% (not shown). If cycloheximide is introduced at the beginning of the recovery period, control and enhanced levels of GJA are completely inhibited (Fig 5). By light microscopy, the treated fibroblasts are 100% viable and appear to make excellent contact with one another and with the substratum. In both fibroblast experiments significant levels of Cx43 were detected in whole cell homogenates (Western, Fig 5), making it likely that the synthesis of another protein is required for both control and enhanced GJA.

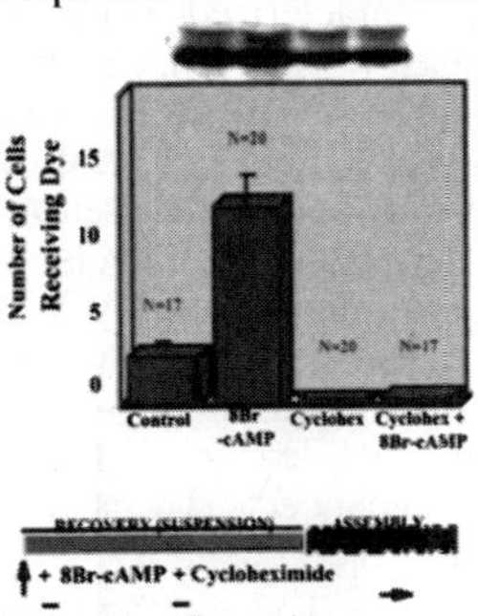


**Figure 5.**

The discrepancy between the results obtained in this study and those obtained in the Novikoff studies are likely due to procedural differences. In the Novikoff experiments, cycloheximide was added prior to the EDTA-mediated dissociation of the cells. Inhibition of protein synthesis may have blocked the subsequent internalization of Cx43 (which occurs during recovery), allowing the untrypsinized Cx43 connexons to participate in GJA upon reaggregation of the cells. Thus, in the Novikoff system cycloheximide did not significantly inhibit basal levels of GJA. Experiments are in progress to confirm that cycloheximide inhibits the trafficking of Cx43 connexons to the plasma membrane, rather than assembly of the gap junction once the connexon is in the membrane. Microtubules are required for the transport of certain membrane proteins to the cell surface [15] and could be involved in enhanced GJA, if the trafficking of Cx43 to the plasma membrane is also increased during this process. Indeed, nocodazole (13.5 μg/ml), added during reaggregation of the KO fibroblasts + $^{Full}$Cx43, significantly blocked enhanced, but not basal, levels of GJA (Fig 6). The same degree of inhibition was observed when the agent was added at the beginning of recovery, rather than at the beginning of assembly (not shown). Microtubules also have been found to be required for enhanced but not basal levels of GJA in the Novikoff system (Paulson, manuscript in preparation). These results are in agreement with early studies in which basal levels of GJA were not disrupted by microtubule dissociating agents [16] and indicate that a second trafficking pathway may be utilized by Cx43 on its way to the plasma membrane during enhanced GJA.

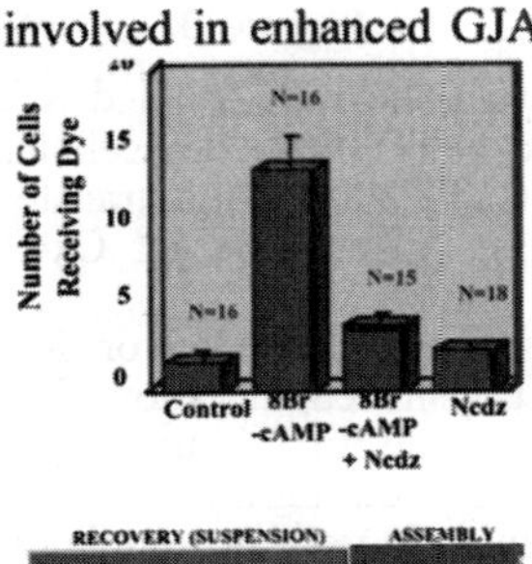


**Figure 6.**

Phosphorylation of one or more amino acids on the C-terminus of Cx43 could play a role in enhanced GJA. To determine whether C-terminal phosphorylation of Cx43 occurs during enhanced *de novo* GJA, Cx43 truncated at amino acid 257 (M257) was transfected into the knockout fibroblasts. Of two transfection trials, only 31% (29/93) of the M257-transfected knockout fibroblast (KO fibroblasts + $^{M257}$Cx43) clones were found to transfer Lucifer yellow dye (Fig 7 and * clones in Fig 8).

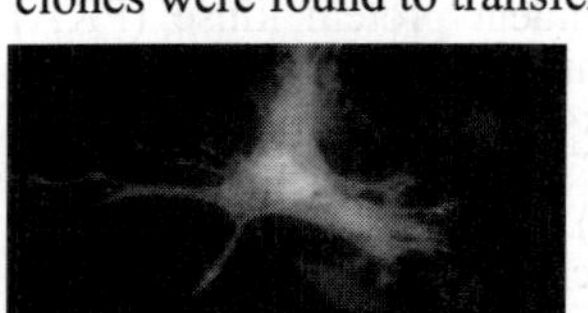

**Figure 7.**

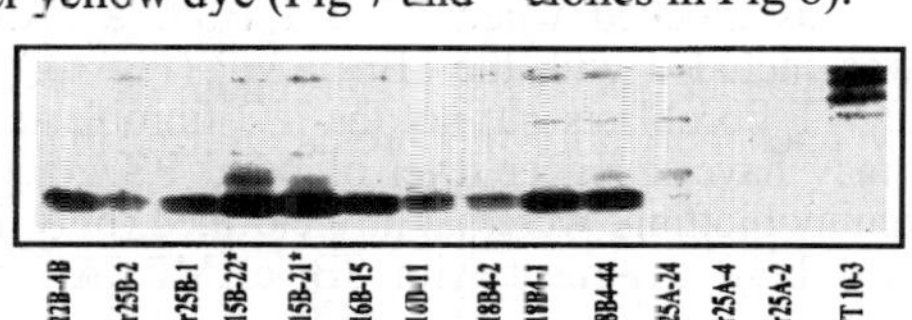


**Figure 8.**

A number of stable transfectants expressed significant levels of truncated protein and yet were not coupled with respect to dye transfer (Fig 8, non * clones). Those clones which communicated, rarely transfered dye to more than three cells (within three minutes of injection), and with successive passages the dye transfer was found to diminish. As in the previous experiments, the KO fibroblasts + $^{M257}$Cx43 were recovered and reaggregated in the presence or absence of 300 μM 8Br-cAMP for studies of enhanced GJA. 8Br-cAMP was found to have no effect on the low level of dye transfer between these cells (not shown).

Because the M257 transfectants were not well coupled, we also obtained and tested N2A cells transfected with full length and truncated rat Cx43. Prior to transfection, N2A cells have a very low level of coupling which can only be measured electrophysiologically and is reportedly the result of Cx45 expression [17].

Confluent monolayers of Cx43- and M257-N2A transfectants were equally well coupled, as measured by microinjection of Lucifer yellow dye (not shown). Treatment of established monolayers with 8Br-cAMP for 4.5 hours increased coupling between the Cx43-N2As 3-4-fold, but coupling between the M257-N2As remained at control levels (not shown). Similarly, when cells were dissociated, allowed to recover in suspension with or without 8Br-cAMP for 1.5 hours, and reaggregated for 2.5 hours also with or without 8Br-cAMP, GJA between the full-length transfectants was increased 2-fold by 8Br-cAMP, but that between truncation transfectants remained unchanged (Fig 9). In spite of communicating equally well in established monolayers, the M257-N2A cells assembled gap junctions more slowly than did the Cx43-N2A cells. As with the WT and KO fibroblasts + $^{Full}$Cx43, enhanced GJA between Cx43-N2As was inhibited by the PKA inhibitor H89 (Fig. 9).

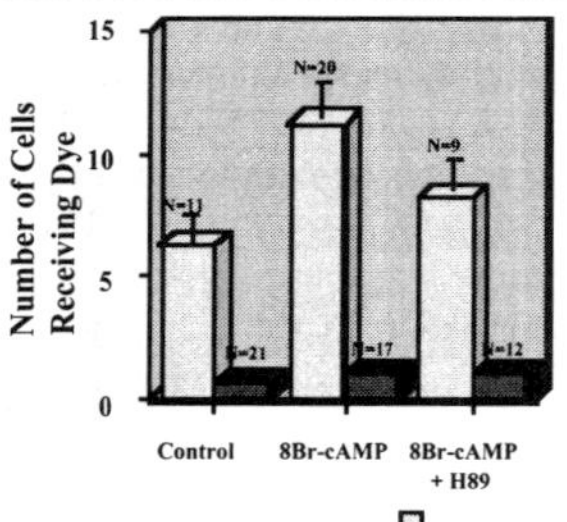


**Figure 9.**

These studies indicate that the C-terminus of Cx43 is required for enhanced but not basal levels of GJA. Because PKA was also required for enhanced GJA, it was possible that PKA was acting through the phosphorylation of the C-terminal domain of Cx43. Thus, a Cx43 C-terminal glutathione sulfotransferase (GST) fusion protein [10] was used to test for direct phosphorylation by PKA. The fusion protein was phosphorylated *in vitro* by purified protein kinase C (PKC) but not by the catalytic subunit of PKA (Fig 10). The C-terminus of Cx56 was phosphorylated under the same conditions (Fig 10 and [18]), indicating that the PKA subunit was active.

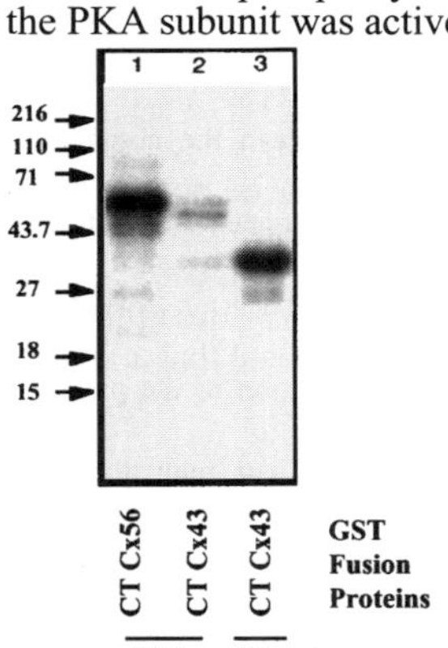


**Figure 10.**

Similarly, Cx43 was not phosphorylated in [$^{32}PO_4$]-labeled monolayers of WT fibroblasts following a four hour treatment with forskolin (5.0 μM), an activator of adenylate cyclase, or of 8Br-cAMP (300 μM). Treatment of parallel cultures with the same agents resulted in significant elevation of dye coupling between the cells (not shown). Thus, PKA activation in the WT fibroblasts does not appear to result in an increased phosphorylation of Cx43. We have yet to determine whether a particular pool of Cx43 is transiently phosphorylated by PKA, or whether the distribution of phosphoresidues on Cx43 is altered by the activation of PKA. The studies described here are in agreement with an earlier report documenting the inability of PKA to phosphorylate synthetic peptides of Cx43 *in vitro* [19].

Thus, although activation of PKA is required for enhanced GJA, it is likely that PKA is phosphorylating another protein rather than Cx43. This protein could subsequently interact with the C-terminus of Cx43 and facilitate enhanced GJA. Alternatively, phosphorylation of a protein utilized in the trafficking pathway could affect the transit of Cx43 to the plasma membrane. This later possibility is supported by the fact that microtubules are required for enhanced GJA, and also that certain microtubule-associated proteins involved in trafficking are phosphorylated by PKA [20, 21].

## References

[1] Banoub R, Fernstrom M, Malkinson A, Ruch R, Enhancement of gap junctional intercellular communication by dibutyryl cyclic AMP in lung epithelial cells, *Anticancer Res. 16* (1996) 3715-3720.
[2] Darrow B, Fast V, Kleber A, Beyer E, Saffitz J, Functional and structural assessment of intercellular communication: Increased conduction velocity and enhanced connexin expression in dibutyryl cAMP-treated cultured cardiac myocytes., *Circ. Res. 79* (1996) 174-183.
[3] Flagg-Newton J, Dahl G, Loewenstein W, Cell junctions and cyclic AMP: I. Upregulation of junctional membrane permeability and junctional membrane particles by administration of cyclic nucleotide or phosphodiesterase inhibitor, *J. Membr. Biol. 63* (1981) 105-121.
[4] Mehta P, Yamamoto M, Rose B, Transcription of the gene for the gap junctional protein connexin43 and expression of functional cell-to-cell channels are regulated by cAMP, *Mol. Biol. Cell 3* (1992) 839-50.
[5] Murray S, Taylor F, Dibutyryl cyclic AMP modulation of gap junctions in SW-13 human adrenal cortical tumor cells, *Am. J. Anat. 181* (1988) 141-148.
[6] Atkinson M, Lampe P, Lin H, Kollander R, Li X-R, Kiang D, Cyclic AMP induces a persistent increase in the junctional permeability of mouse mammary tumor cells by modifying the cellular distribution of connexin 43, *J. Cell Sci. 108* (1995) 3079-3090.
[7] Burghardt R, Barhoumi R, Sewall T, Bowen J, Cyclic AMP induces rapid increases in gap junction permeability and changes in the cellular distribution of connexin 43, *J. Membr. Biol. 148* (1995) 243-253.
[8] Berthoud V, Rook M, Traub O, Hertzberg E, Saez J, On the mechanisms of cell uncoupling induced by a tumor promoter phorbol ester in clone 9 cells, a rat liver epithelial cell line, *Eur. J. Cell Biol. 62* (1993) 384-96.
[9] Swenson K, Piwnica-Worms H, McNamee H, Paul D, Tyrosine phosphorylation of the gap junction protein connexin43 is required for the pp60 v-src-induced inhibition of communication, *Cell Regulation 1* (1990) 989-1002.
[10] Warn-Cramer B, Lampe P, Kurata W, Kanemitsu M, Loo LWM, Eckhart W, Lau A, Characterization of the mitogen-activated protein kinse phosphorylation sites on the connexin-43 gap junction protein, *J. Biol. Chem. 271* (1996) 3779-3786.
[11] Dunham B, Liu S, Taffet S, Trabka-Janik E, Delmar M, Petryshyn R, Zheng S, Perzova R, Vallano ML, Immunolocalization and expression of functional and nonfunctional cell-to-cell channels from wild-type and mutant rat heart connexin43 cDNA, *Cir. Res 70* (1992) 1233-43.
[12] Martyn K, Kurata W, Warn-Cramer B, Burt J, TenBroek E, Lau A, Immortalized connexin43 knockout cell lines display a subset of biological properties associated with the transformed phenotype, *Cell Growth Develop.* (1997) In press.
[13] Epstein M, Sheridan J, Johnson R, Formation of low-resistance junctions *in vitro* in the absence of protein synthesis and ATP production, *Exp. Cell Res. 104* (1977) 25-30.
[14] Meyer R, Malewicz B, Baumann W, Johnson R, Increased gap junction assembly between cultured cells upon cholesterol supplementation, *J. Cell Sci. 96* (1990) 231-238.
[15] Tousson A, Fuller C, Benos D, Apical recruitment of CFTR in T-84 cells is dependent on cAMP and microtubules but not $Ca^{2+}$ or microfilaments, *J. Cell Sci. 109* (1996) 1325-34.
[16] Preus D, Li X-R, Sheridan J, Balson J, Wagner M, Johnson R, Gap junction formation in the presence of colchicine and cytochalsin B, *J. Cell Biol. 101* (1985) 179a.
[17] Veenstra R, Wang H, Westphale E, Beyer E, Multiple connexins confer distinct regulatory and conductance properties of gap junctions in developing heart, *Cir. Res 71* (1992) 1277-83.
[18] Berthoud V, Beyer E, Kurata W, Lau A, Lampe P, The gap junction protein connexin 56 is phosphorylated in the intracellular loop and the carboxy-terminal region. *Eur. J. Biochem. 244* (1997) 89-97.
[19] Saez J, Nairn A, Czernik A, Spray D, Hertzberg E, Rat connexin43: Regulation by phosphorylation in heart. In: J. Hall (ed), Progress in Cell Research. vol 3, Elsevier, 1993.
[20] Perez J, Tinelli D, Cagnoli C, Pecin P, Brunello N, Racagni G, Evidence for the existence of a cAMP-dependent protein kinase phosphorylation system associated with specific phosphoproteins in stable microtubules from rat cerebral cortex, *Brain Res. 602* (1993) 77-83.
[21] Sato-Harada R, Okabe S, Umeyama T, Kanai Y, Hirokawa N, Microtubule-associated proteins regulate microtubule function as the track for intracellular membrane organelle transports, *Cell Structure and Function 21* (1996) 283-95.

*Gap Junctions*
*R. Werner (Ed.)*
*IOS Press, 1998*

# 17β-estradiol-induced blockade of cell-to-cell communication in neonatal rat cardiomyocytes is not mediated by activation of tyrosine kinase, protein kinases A or C

Franck Verrecchia, Brenda R. Kwak* and Jean-Claude Hervé

*Laboratoire de Physiologie Cellulaire, UMR CNRS n° 6558, Poitiers, France; *Medical Physiology and Sports Medicine, Utrecht University, The Netherlands*

**Abstract** Short term exposures to 17β-estradiol propionate (1 to 25 μM) completely abolish cell-to-cell communication between cardiac myocytes, but the mechanism(s) of this action remain(s) to be elucidated. As several 17β-estradiol effects have been found affected by protein kinases inhibitors, we have examined the influence of this compound, in combination with several protein kinase activators and inhibitors, on junctional permeability and on macroscopic and single channel conductances. Protein kinases A or C as well as protein tyrosine kinase do not seem involved in this 17β-estradiol effect. This compound, known able to reduce membrane fluidity, might, like several other lipophilic junctional uncouplers (*e.g.* heptanol, gossypol, tamoxifen), interfere with the plasma membrane structure, leading to perturbations of junctional channel conformation, with accompanying effects on their functions.

## 1. Introduction

We have shown previously that some steroids can abolish cell-to-cell communication in different types of primary cells and, among them, 17β-estradiol appeared as the most potent uncoupler [13]. This compound, in the concentration range 1 to 25 μM (with an $IC_{50}$ of approximately 2 μM) reversibly impaired cell-to-cell communication, but the mechanism of this acute effect, which takes place without concomitant changes in intracellular $[H^+]_i$ or $[Ca^{2+}]_i$, remains to be elucidated. As several 17β-estradiol effects were seen, in different target tissues, seriously hindered by protein kinase inhibitors, particularly cAMP-dependent protein kinase (PKA), protein kinase C (PKC) and protein tyrosine kinase (PTK) inhibitors, the possible involvement of these enzymes in 17β-estradiol-induced uncoupling has been examined, since a large body of evidence exists that cell-to-cell communication is modulated in response to the activation and/or inhibition of protein kinases [reviewed in 21]. In the mammalian heart ventricle, the most abundant junctional channel-forming protein is connexin43 (Cx43), although Cx40 and Cx45 have been reported to be expressed as well [see 10]. Cx43 is a phosphoprotein, and one dephosphorylated (41 kDa) and two phosphorylated (43 and 47 kDA, respectively) forms have been identified [for review, see 21]. Single channel experiments revealed that Cx43 channels exhibit two or three distinct conductance states of about 20, 40-45 and 70 pS [see 16]. Changes in the Cx43 phosphorylation state, with subsequent effects on intercellular coupling, have been reported to occur in the action of some hormones (*e.g.* luteinizing hormone action on rat ovarian follicle; [9]). In order to gain insight in the mechanism of 17β-estradiol-induced uncoupling, we have examined the effects of this hormone in combination with several protein kinase activators and inhibitors on permeability and on macroscopic and single channel conductances of cardiac junctional channels.

## 2. Materials and Methods

Cells were enzymatically dispersed from neonatal rat ventricles and cultured for 1 to 3 days [3]. Macroscopic junctional conductance ($g_j$) and unitary conductance ($\gamma_j$) were investigated using the dual whole-cell voltage-clamp technique, as previously described [3, 23]. Low resistance patch-pipettes (2-5 MΩ) were backfilled with solutions containing respectively (in mM) : KCl 140, MgATP 5, HEPES 10, EGTA 5, glucose 10 (pH 7.2) for gj determination, and CsCl 135, $CaCl_2$ 0.5, $MgCl_2$ 1, $Na_2ATP$ 5, HEPES 10, EGTA 10 (pH 7.1) for the detection of single channel events. The gap junction permeability for the fluorescent dye 6-carboxyfluorescein (6-CF) was determined by analysing the Fluorescence Recovery After Photobleaching (gap FRAP) as previously described [19]. In selected cells the fluorescence was photobleached and, when recovery took place, it followed (at least in its initial part) an exponential time course, allowing to determine the rate constant ***k*** (the inverse value of the time constant) of the fluorescence recovery

## 3. Results

### 3.1. Susceptibility of junctional communication to 17β-estradiol or its ester

Under control conditions, $g_j$ was 35.5 ± 1.8 nS (mean ± s.e.m., *n* = 215). When a continuous flow of 17β-estradiol propionate (EP, 25 µM) containing solution was applied by means of a capillary (∅ 30 µm) positioned in the bath near the investigated cell pair, a progressive decrease of gj was observed, leading within 5 to 12 minutes to complete closure of the junctional channels, as exemplified in Fig 1A. In the bath medium, unesterified 17β-estradiol did not affect gj, even at concentrations up to 100 µM. However, when added to the pipette filling solution (Fig. 1B), 250 nM unesterified estradiol reduced intercellular coupling (- 49.3 ± 5.5 % after 25 min, n = 12), and a 500 nM concentration was sufficient to lead to complete intercellular uncoupling. The uncoupling was progressive in both cases, in contrast with the sharp closure observed in the presence of heptanol, complete within <0.5 s [2], butanedione monoxime, achieved in <10 sec [24] or fatty acids such as arachidonic or oleic acids, total within 1 to 2 min [4, 8]. This relatively slow time course is however comparable with junctional uncoupling induced by some other lipophilic compounds, *e.g.* oleoyl acetyl glycerol [3], gossypol [12] or testosterone propionate [19].

Under control conditions, photobleach reduced the light emission of the selected cells to approximately 30 to 50% of its initial level; then a rise in fluorescent emission took place, and ***k*** was estimated to 0.24 ± 0.02 $min^{-1}$ (n = 470). In other words, the fluorescence

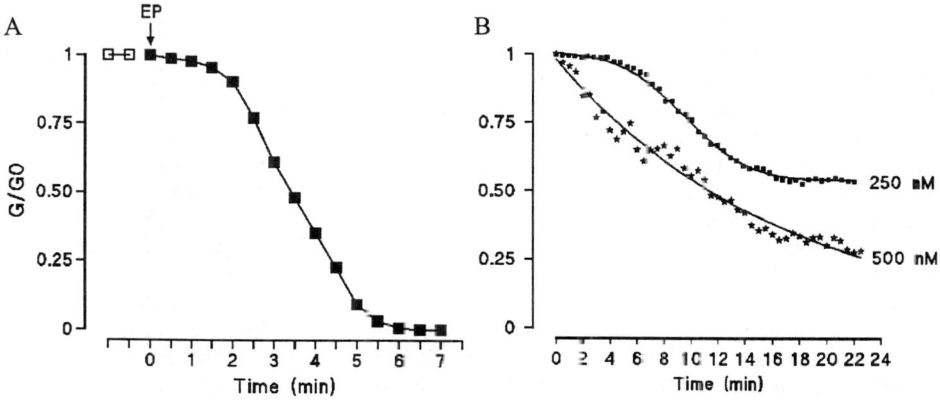


Fig. 1. Time courses of the inhibitory effects of 17β-estradiol or its ester on cell-to-cell communication between ventricular myocytes. **(A)** The cells were bathed in Tyrode's solution, then a stream of EP-containing (25 µM) solution was applied in the bath (arrow) in the close vicinity of the cell pair. **(B)** Unesterified estradiol (250 or 500 nM) was added to the pipette solution, and gj were measured as soon as the double whole cell configuration was obtained. At the highest concentrations, 500 nM, gj decayed with a monoexponential time course (time constant of approximately 16 min). At 250 nM, a partial junctional conductance persisted for the duration of the experiment. Results were obtained from respectively 12 (A and B, 250 nM) and 14 (B, 500 nM) myocytes pairs

recovery followed a monoexponential time course with a mean time constant of 4 min. This recovery did not occur in presence of EP (25 μM; $k = 0.007 \pm 0.001$ min$^{-1}$; $n = 50$) but was progressively restored after EP removal ($k = 0.10 \pm 0.01$ and $0.19 \pm 0.02$ min$^{-1}$ after respectively 15 and 30 min, $n = 16$), reflecting the reopening of the intercellular channels (and the absence of short-term toxicity of EP at this concentration). This steroid effect, occurring within a few minutes only at concentrations far above the known physiological concentrations of the sex hormone, almost certainly represents a nongenomic action. This probability was ascertained by examining EP effects after incubation of the cells for 24 h in the presence of tamoxifen (1 μM), an anti-estrogen that prevents the genomic effects of the hormone by binding to the estrogen nuclear receptor. Even after 24h, tamoxifen had no influence on ***k***, and it also failed to prevent the uncoupling EP action (Fig. 2).

### 3.2. Involvement of protein kinases in 17β-estradiol-induced junctional uncoupling

Several biological actions of 17β-estradiol have been found, in different target tissues, seriously hindered by protein kinase inhibitors, particularly those of PKA, PKC and PTK. The activity of many channels can be modulated by cycles of phosphorylation-dephosphorylation of the pore-forming proteins. Connexin phosphorylation could lead to conformational changes, involving subsequent modifications of the channel conductance state. The presence of phosphotyrosine in Cx43 was correlated with the loss of GJIC in fibroblasts [6]. As shown in Table 1, the lack of effect of a PTK inhibitor, genistein, together with the persistence of the uncoupling effect in presence of genistein allowed to rule out the possibility of PTK involvement in the estradiol-induced uncoupling. ***k*** was not modified during short-term exposures of the cells to a PKC inhibitor, calphostin C, which could have promoted a protein dephosphorylation. Calphostin C also failed to prevent the EP-induced junctional uncoupling. An agent that increases intracellular cAMP concentration by inhibiting phosphodiesterase, IBMX and a membrane-permeant cAMP derivative, 8Br-cAMP, had no effect on ***k*** and, when co-added with 17β-estradiol, they both failed to preserve gap junctional communication.

The possibility of involvement of serine or threonine kinase(s) in estradiol action was also examined by γj measurements. Indeed, phosphorylating or dephosphorylating treatments of neonatal rat cardiomyocytes have been observed to alter the γj frequency

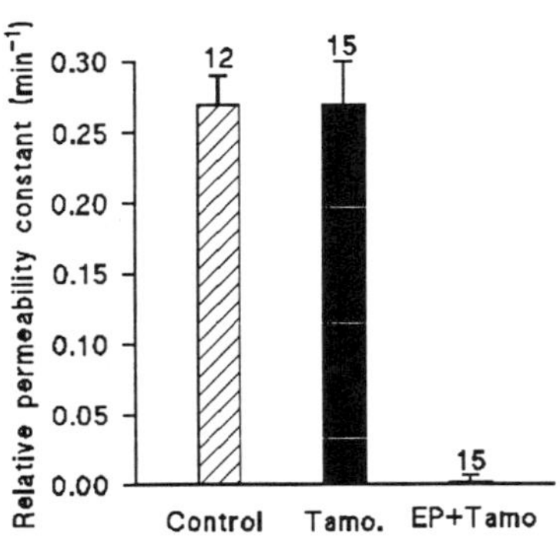


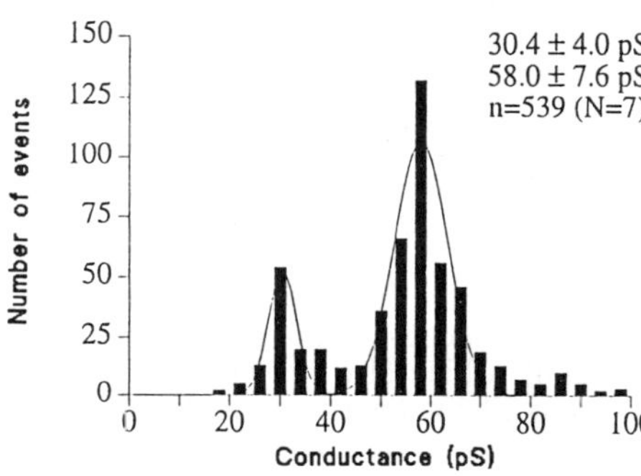


**Left** : Fig. 2. Normalised relative permeability constants (***k***) of dye diffusion, measured in control conditions and in cells of the same batch cultured in presence of tamoxifen (1 μM) for 24 h, then exposed to EP (25 μM for 15 min) (means ± s.e.m. with (*n*) numbers on top).

**Right** : Fig. 3. Effects of 17β-estradiol propionate on single gap junctional channel conductance. Cell pairs were treated for 10 min with 25 μM EP after which they were further uncoupled by additional treatment with 1 mM heptanol. A frequency histogram of conductance changes was determined from the amplitudes of single channel events in junctional current in response to a transjunctional voltage of 100 mV. Mean and SD of the single-channel conductances were obtained by fitting the data to a double/triple gaussian distribution using a least-squares method and are indicated in the figure. N = number of experiments; n = number of events.

Table 1. *k* values (means ± s.e.m.) in control conditions, in presence of IBMX (100 μM) or 8Br-cAMP (1 mM), or of inhibitors of PKC (calphostin C, 0.2 μM), or PTK (genistein, 370 μM), and in the same cells after EP exposure.

| | Control | Test | Test + EP | *n* |
|---|---|---|---|---|
| IBMX | 0.27 ± 0.01 | 0.27 ± 0.01 | 0.002 ± 0.001 | 11 |
| 8 Br-cAMP | 0.25 ± 0.01 | 0.24 ± 0.01 | 0.002 ± 0.001 | 9 |
| Calphostin C | 0.27 ± 0.02 | 0.26 ± 0.03 | 0.002 ± 0.001 | 13 |
| Genistein | 0.23 ± 0.01 | 0.23 ± 0.01 | 0.002 ± 0.001 | 17 |

distribution [see 15]. Under control conditions, neonatal rat ventricular cell pairs exhibited channels of two distinct sizes that could be separated in frequency distribution of channel transitions as two conductance states of around 30 and 60 pS, with the more numerous events being of the larger size [data not shown, see *e.g.* 16]. Addition of 17β-estradiol (25μM for 15 min) did not affect this frequency distribution (Fig. 3).

## 4. Discussion

17β-estradiol was seen, under physiological conditions, to up regulate the cell-to-cell communication by a genomic action, particularly in the myometrium of many mammals [for review, see 11]. At higher concentrations, it is a potent inhibitor of GJIC; in the present study, acute introductions of unesterified 17β-estradiol into the cytosol impaired junctional coupling at moderate concentrations : a partial uncoupling was already obtained at 250 nM, a concentration not far above the range observed during pregnancy, where the serum level concentration of 17β-estradiol can reach up to 105 nM [22]. However, doses can be considerably higher in steroid drug abuses [18] and the concentration of this lipophilic compound could be higher in membrane than in serum. Steroid hormone action has usually been described as the modulation of nuclear transcription, thus triggering genomic events that are responsible for physiological effects. However, increasing evidence shows that sex hormones have also rapid, non-genomic actions involving phospholipase C, phosphoinositide turnover, intracellular pH and calcium, PKA, PKC, PTK, etc... [for recent review, see 25]. Moreover, 17β-estradiol abolished cell-to-cell communication even after the cells had been exposed to a prolonged pre-treatment with an antiestrogen, tamoxifen, which binds to the specific nuclear receptor.

As the activation of several protein kinases was observed in some 17β-estradiol effects, the possibility of their involvement in EP-induced junctional uncoupling was examined. A PTK inhibitor, genistein, neither modify the junctional permeability nor prevented the estradiol effect. An enhancement of the cellular cAMP level protected the cells against phorbol ester-induced inhibition of GJIC [20]. In the present study, the degree of junctional coupling was not modified by 8Br-cAMP or IBMX, which also failed to prevent the junctional uncoupling promoted by 17β-estradiol. Similarly, a PKC inhibitor, calphostin C, did not modify the junctional permeability, nor prevented the estradiol-induced uncoupling. The γj frequency distribution was not significantly modified after treatment with 17β-estradiol, which also makes the involvement of a serine/threonine kinase in this steroid action unlikely. Interruptions of GJIC without concomitant changes in the phosphorylation state of Cx43 have already been reported, for example after treatments with several other classes of lipophilic compounds, *e.g.* heptanol [17] or fatty acids [14]. Several of these uncouplers are able to decrease membrane fluidity, for example tamoxifen [5], heptanol [1] or gossypol [7]. 17β-estradiol was also seen, in the uncoupling concentration range, to reduce membrane fluidity [5]. The incorporation of lipophiles of appropriate shape and size into the lipidic domains might alter the microenvironment of various membrane proteins, with consequences on their tertiary structure and on their functions.

## References


[1] E.M.L. Bastiaanse *et al.*, Heptanol-induced decrease in gap junctional conductance is mediated by a decrease in the fluidity of membranous cholesterol-rich domains. *J Membrane Biol* **136** (1993) 135-145.

[2] B. Bastide *et al.*, Rapid onset and calcium independence of the gap junction uncoupling induced by heptanol in cultured heart cells. *Pflügers Arch* **429** (1995) 386-393.

[3] B. Bastide *et al.*, The uncoupling effect of diacylglycerol on gap junctional communication of mammalian heart cells is independent of Protein Kinase C. *Exp Cell Res* **214** (1994) 519-527.

[4] J.M. Burt *et al.*, Uncoupling of cardiac cells by fatty acids: structure-activity relationships. *Am J Physiol* **260** (1991) C439-C448.

[5] R. Clarke *et al.*, Pharmacological and supra pharmacological concentrations of both 17β estradiol and tamoxifen reduce the membrane fluidity of MCF-7 and MDA-MB-436 human breast cancer cells. *Biochem Soc Trans* **15** (1987) 243-244.

[6] D.S. Crow *et al.*, Phosphorylation of Connexin43 gap junction protein in uninfected and Rous sarcoma virus-transformed mammalian fibroblasts. *Mol Cell Biol* **10** (1990) 1754-1763.

[7] A. Cuéllar and J. Ramírez, Further studies on the mechanism of action of gossypol on mitochondrial membrane. *Int J Biochem* **25** (1993) 1149-1155.

[8] C. Giaume *et al.*, Arachidonic acid closes gap junction channels in the rat lacrymal glands. *Pflügers Arch* **413** (1989) 273-279.

[9] I. Granot and N. Dekel, Phosphorylation and expression of connexin43 ovarian gap junction protein are regulated by luteinizing hormone. *J. Biol. Chem.* **48** (1994) 30502-30509.

[10] D. Gros and H.J. Jongsma, Connexins in mammalian heart functions. Bioessays **18** (1996) 719-730.

[11] E.M. Hendrix *et al.*, Steroid hormone regulation of rat myometrial gap junction formation : effects on Cx43 levels and trafficking. *Biol Reproduct* **52** (1995) 547-560.

[12] J.C. Hervé *et al.*, Contraceptive gossypol blocks cell-to-cell communication in human and rat cells. *Eur J Pharmacol* **313** (1996) 243-255.

[13] J.C. Hervé *et al.*, Influence of the molecular structure of steroid compounds on their ability to interrupt gap junctional communication. *J Membrane Biol* **149** (1996) 179-187.

[14] C.S.T. Hii *et al.*, Inhibition of gap junctional communication by polyunsaturated fatty acids in WB cells : evidence that connexin 43 is not hyperphosphorylated. *Carcinogenesis* **16** (1995) 1505-1511.

[15] B.R. Kwak and H.J. Jongsma, Regulation of cardiac gap junction channel permeability and conductance by several phosphorylating conditions. *Mol. Cell. Biochem.* **157** (1996) 93-99.

[16] B.R. Kwak *et al.*, Effects of cGMP-dependent phosphorylation on rat and human connexin43 gap junction channels. *Pflügers Arch.* **430** (1995) 770-778.

[17] L.S. Musil *et al.*, Differential phosphorylation of the gap junction protein connexin43 in junctional communication-competent and -deficient cell-lines. *J Cell Biol* **111** (1990) 2077-2088.

[18] W.A. Narducci *et al.*, Anabolic steroids : a review of the clinical toxicology and diagnostic screening. *Clin Toxicol* **28** (1990) 287-310.

[19] F. Pluciennik *et al.*, Reversible interruption of gap junctional communication by testosterone propionate in cultured Sertoli cells and cardiac myocytes. *J Membrane Biol.* **149** (1996) 169-177.

[20] L.E. Roseng *et al.*, Effect of cAMP elevating compounds on inhibition of gap junctional communication and induction of morphological transformation in Syrian hamster embryo cells. *Carcinogenesis* **13** (1992) 1803-1809.

[21] J.C. Sáez *et al.*, Gap junctions. Multiplicity of controls in differentiated and undifferentiated cells and possible functional implications. In S. Shenolikar and A.C. Nairn (eds.) Advances in Second Messenger and Phosphoprotein Research, vol. 27, Raven Press, NY, 1993, pp. 193-198.

[22] T. Tabei *et al.*, Serum levels of inhibitin in maternal and umbilical blood during pregnancy. *Am J Obstet Gynecol* **164** (1991) 896-900.

[23] B.R. Takens-Kwak and H.J. Jongsma, Cardiac gap junctions : three distinct single channel conductances and their modulation by phosphorylating treatments. *Pflügers Arch.* **422** (1992) 198/200.

[24] F. Verrecchia and J.C. Hervé, Reversible blockade of gap junctional communication by 2,3-butanedione monoxime in rat cardiac myocytes. *Am. J. Physiol.* **272** (1997) C875/C885.

[25] M. Wehling, Specific, non-genomic actions of steroid hormones. *Ann. Rev. Physiol* **59** (1997) 365/393.

*Gap Junctions*
*R. Werner (Ed.)*
*IOS Press, 1998*



# Inhibition of Gap Junctional Communication by Platelet-derived Growth Factor

Mohammad Z. HOSSAIN, Peng AO and Alton L. BOYNTON

*Molecular Medicine, Northwest Hospital, Seattle, WA 98125, USA*

**Abstract.** Gap junctional communication (GJC) is inhibited by various mitogens including platelet-derived growth factor (PDGF). To understand the mechanism of PDGF-induced GJC blockade, we transfected a rat liver epithelial cell line T51B with either a wild type human PDGFrβ ($Kin^+$) or a mutant PDGFrβ lacking receptor tyrosine kinase activity ($Kin^-$). In $Kin^+$ cells, within 15-20 min. of addition PDGF caused a complete but transient interruption of cell communication and phosphorylation of connexin43 (Cx43), the only known gap junction protein expressed in these cells. These effects were not exhibited in either parental T51B cells or in clones expressing tyrosine kinase-dead mutant of PDGFrβ ($Kin^-$), indicating a requirement of the receptor tyrosine kinase activity. While the inhibition of GJC appears to be caused by phosphorylation of Cx43, the re-establishment of GJC following PDGF treatment was dependent on protein synthesis. This report describes a cell system suitable for the characterization of the PDGF signaling pathway responsible for the inhibition of GJC.

## 1. Introduction

Gap junctions are narrow, hydrophilic channels connecting adjacent cells [1, 2] composed of a closely related group of membrane proteins termed connexins (Cx) and serve as conduits for the intercellular transfer of small molecules and ions. Some of these molecules are believed to act as regulatory signals controlling contraction of smooth muscle [3], embryonic development [4] and most notably, control of cell proliferation [5, 6, 7]. The growth regulatory function of GJC suggests its crucial involvement in neoplasia as evidenced by numerous studies showing disruption of GJC in tumor cells and by mitogens [8-13]. Conversely, establishment of GJC correlated with tumor suppression [14, 15] and reversal of tumor phenotype [7, 16-19].

Growth factors are an important class of mitogens that exert inhibitory effects on GJC [12, 13, 20]. The mode of action of growth factors includes the activation of their respective receptor tyrosine kinases followed by initiation of complex cascades of signal transduction pathways involving binding and subsequent activation of protein kinases and phosphatases [21-23]. The disruption of GJC by EGF was transient and associated with phosphorylation of Cx43 [12]. This phosphorylation was later shown to be mediated by mitogen-activated protein kinase (MAPK) [24, 25].

Although signal transduction pathways have been outlined for several actions of PDGF [22, 26], so far no detailed study has been conducted to delineate platelet-derived growth factor (PDGF)-induced interruption of GJC [13, 27]. To accomplish our goal, wild-type and mutant human PDGFrβ were retrovirally introduced into a T51B rat liver epithelial cell line which is devoid of any endogenous PDGFr expression [28]. This report demonstrates that a rapid and transient disruption of GJC concomitant with the phosphorylation and subsequent degradation of Cx43 can be induced by PDGF which requires PDGFr tyrosine kinase activity. Moreover, the restoration of GJC is dependent on the synthesis of protein(s) presumably crucial for the re-establishment of gap junctions.

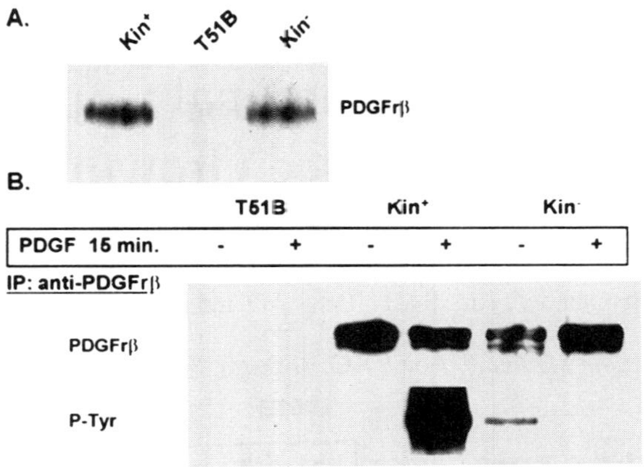


Figure 1. Expression of functional human PDGFrβ in Kin+ cells.
(A) Western blot of PDGFrβ in total cellular lysates from confluent cells. (B) Confluent cells treated with buffer (-) or PDGF (10 ng/ml) (+) for 15 min. were lysed and PDGFrβ was immunoprecipitated using monoclonal anti-PDGFrβ antibody. The immunoprecipitates were separated by SDS-PAGE followed by Western blotting and probed with either polyclonal anti-PDGFrβ or anti-phospho-tyrosine (p-Tyr) antibodies.

## 2. Results

T51B cells were retrovirally infected with either wild-type (Kin+) or mutated receptor tyrosine kinase-inactive (Kin-) PDGFrβ constructs [29] (kindly provided by Dr. J. Cooper, FHRC, Seattle, WA). Expression of a ~190 kDa PDGFrβ band in infected but not in uninfected T51B cells was confirmed by Western blotting utilizing anti-PDGFrβ antibody (Fig. 1A). To verify whether this expressed receptor was functionally active, cells were treated with PDGF and several manifestations of receptor activation were investigated. Increased levels of phosphotyrosine in the immunoprecipitated PDGFr indicating activation of its tyrosine kinase was observed only in PDGF-treated Kin+ cells, but not in kinase-inactive mutant Kin- cells (Fig. 1B). As expected [26, 30], recruitment of several signaling molecules by the activated PDGFr was only seen in PDGF-treated Kin+ cells [28].

To examine the effect of PDGF on GJC, Kin+ cells were treated with PDGF for various lengths of time and GJC was assayed by dye microinjection [14, 28]. Within 5 minutes of treatment, the disruption of GJC was evident (Fig. 2A), which reached its maximal levels after 15-20 minutes. Similar to EGF [12], this inhibition was also transient in nature. Within 30 minutes of treatment, the number of communicating cells began to rise which continued until 90 minutes post-treatment when the communicating network returned to its normal level. Immunostaining of PDGF-treated Kin+ cells with anti-Cx43 antibody did not detect any gross alterations [28] indicating that the mechanism of PDGF-induced GJC blockade did not include a structural destabilization of gap junctional plaques or a cellular re-distribution of Cx43.

Since studies with tumor promoters and growth factors strongly suggest a regulatory role of Cx43 phosphorylation/dephosphorylation in gap junctional channel opening and closure [12, 24, 31], we then investigated whether a similar phenomenon occurs in PDGF-mediated disruption of GJC. Western blots of untreated Kin+ cells showed that two phosphorylated forms of Cx43, noted as Cx43-P1 and Cx43-P2 [32] constituted the majority of the gap junction protein (Fig. 2B). A faint low mol. wt. band (Cx43-NP) representing the non-phosphorylated form of Cx43 was also seen. Following PDGF treatment, Cx43-NP disappeared completely along with a partial decrease in the Cx43-P1 and Cx43-P2 forms. Additionally, a hyper-phosphorylated form of Cx43 (Cx43-P3) gradually appeared which peaked at 15 min. (when GJC was maximally inhibited) and disappeared after 45-60 min. post-treatment. At later time points, Cx43 profile continued to return to its pre-stimulation state. This close association between phosphorylation of Cx43 and disruption of GJC was also seen in PDGF dose-response studies [28]. Neither parental

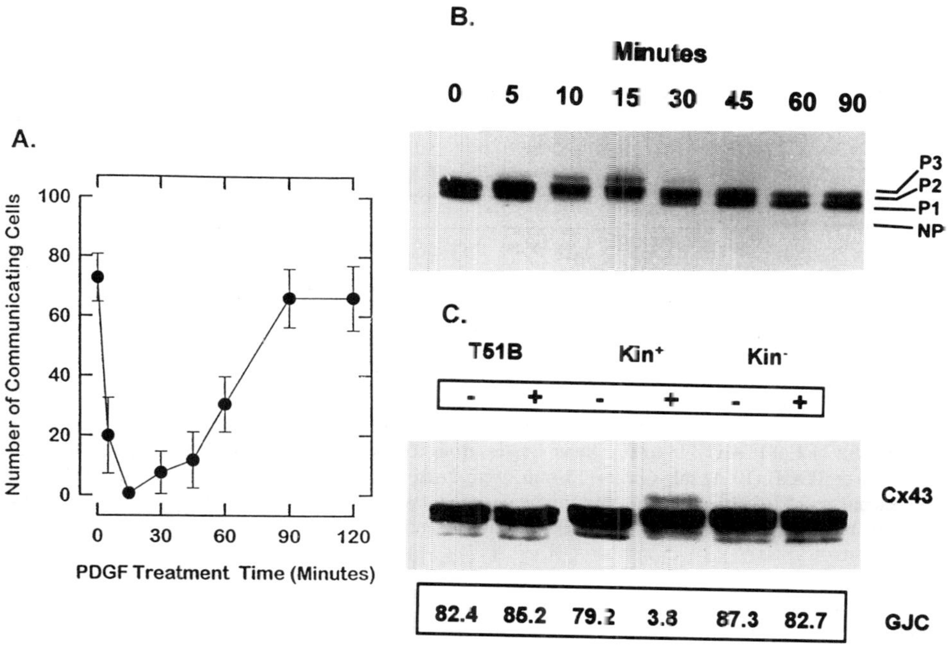


Figure 2. Effect of PDGF on GJC and Cx43 phosphorylation.
A. Confluent Kin$^+$ cells were treated with PDGF (10 ng/ml) and GJC was assayed. Data points represent the mean $\pm$ SD of cells to which lucifer yellow was transferred from the injected cell within 10 min. of injection. B. Western blot analysis of Cx43 in confluent Kin$^+$ cells following PDGF treatment. C. Western blot analysis of Cx43 after a 15 min. treatment with buffer (-) or PDGF (+). The numbers in the GJC box represents the mean number of communicating cells in the indicated treatment group.

T51B nor kinase-inactive mutant Kin$^-$ cells showed any effects of PDGF on either GJC or Cx43 phosphorylation (Fig. 2C). Taken together, these results strongly suggest that the phosphorylation of Cx43 may be causally related to the inhibition of GJC by PDGF which requires activation of PDGFr tyrosine kinase.

The recovery of GJC and restoration of the Cx43 profile after EGF treatment was mediated via an okadaic acid (OA)-sensitive phosphatase [12], however, in Kin$^+$ cells, OA-pretreatment did not interfere with the recovery of GJC or of the Cx43 profile after PDGF addition [28]. Interestingly, pretreatment with the protein synthesis inhibitor cycloheximide (CHX) completely abolished the re-establishment of junctional communication after 90 minutes of PDGF treatment (Fig. 3). Although the molecular identity of the required protein(s) is unknown, this result clearly shows that protein synthesis is essential for the formation of gap junctions and recovery of GJC after PDGF treatment.

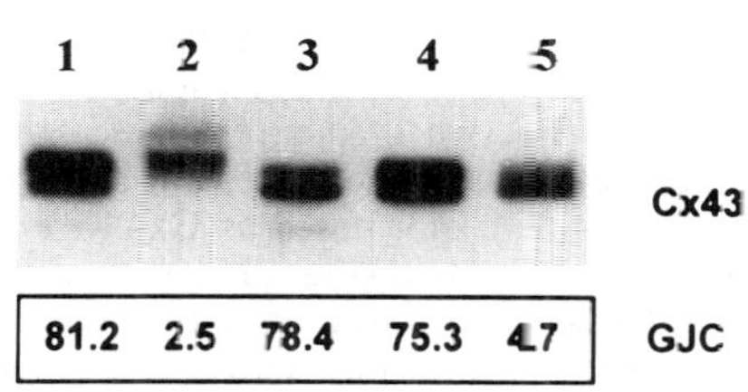


Figure 3. Re-establishment of GJC requires protein synthesis.
Confluent Kin$^+$ cultures were pre-treated with buffer or cycloheximide (CHX) (10 μg/ml) for 1 h and then treated with PDGF for the indicated lengths of time. Cx43 profile and GJC were examined as described in Fig. 2. Lanes: 1, control; 2, PDGF-15$'$; 3, PDGF- 90$'$; 4, CHX-60$'$ and 5, CHX-60$'$ + PDGF-90$'$.

## 3. Discussion

Interruption of GJC by growth factors has been previously described [12, 13, 26] although the mechanism remains unknown. In this report we describe a suitable cell system where PDGF-mediated interruption of GJC and phosphorylation of Cx43 can be demonstrated. We also show that the signaling pathway requires the activation of PDGFr tyrosine kinase activity suggesting involvement of additional factors. Although interruption of GJC appears to be linked with Cx43 phosphorylation, the recovery, in contrast, requires synthesis of new proteins.

Phosphorylation of Cx43 has been implicated for the acute inhibition of GJC by various factors [2]. The mechanism by which Cx43 phosphorylation causes interruption of GJC is currently unknown although a phosphorylation-dependent alteration of junctional gating has been reported [34]. While some oncogenes induced tyrosine phosphorylation of Cx43 [10], Cx43 phosphorylation on ser/thr residues were responsible for the disrupted GJC by growth factors [12, 13, 28]. Our study [28] supports that this transient PDGF effect was possibly mediated by a ser/thr kinase [13]. Further examination of the PDGF signaling pathway revealed that activation of two important protein kinases namely, protein kinase C (PKC) and MAPK, are required in PDGF-induced Cx43 phosphorylation (unpublished data). Activation of MAPK but not PKC was shown to be involved the interruption of GJC by EGF [24]. At present, we are attempting to characterize the kinase(s) directly responsible for Cx43 phosphorylation.

In addition to phosphorylation, Cx43 also undergoes lysosomal degradation following PDGF treatment [28]. Although not responsible for GJC disruption, an increased Cx43 degradation by PDGF treatment strongly indicates that PDGF-induced phosphorylation of Cx43 possibly triggers the activation of a proteolytic pathway. While re-establishment of GJC in EGF-treated cells correlated with the dephosphorylation of Cx43 [12], a requirement for protein synthesis in the present experimental system indicated that the restoration of GJC following PDGF treatment was mediated by newly constructed gap junctions. Pretreatment with cycloheximide and Brefeldin A [28] suggests that synthesis of protein(s) aiding the establishment of gap junctions are also necessary for the recovery of cell communication. At present, our knowledge about gap junction-associated proteins is rudimentary; only cell adhesion molecules have been so far reported to be involved in facilitating gap junctional assembly [33]. Further studies will be required for the characterization of such factor(s).

In summary, our results demonstrate that PDGF-mediated biological responses including interruption of GJC and phosphorylation of Cx43 can be demonstrated in a normally PDGF-nonresponsive cell line after insertion of wild-type human PDGFrβ. The next step of our investigation will involve generation of a series of clones derived from T51B cells expressing a variety of mutant PDGFrβ to determine the signaling pathway resulting in disrupted GJC and Cx43 phosphorylation and to examine whether interruption of cell-cell coupling is necessary for the growth factor-induced cell proliferation.

## Acknowledgment

This work was supported by NIH/NCI grant # 57064. The figures were reproduced from [28] with permission of Wiley-Liss, a division of John Wiley & Sons, Inc.

## References

[1] N.M. Kumar and N.B. Gilula, The gap junction communication channel. *Cell* 84 (1996) 381-388.

[2] R. Bruzzone *et al.*, Connection with connexins: the molecular basis of direct intercellular signalling. Eur. J. Biochem. 238 (1996) 1-27.

[3] R.E. Garfield et al., Gap junction formation in myometrium: control by estrogens, progesterone and prostaglandins. Amer. J. Physiol. 238 (1980) C81-C89.

[4] S. Caveney, The role of gap junctions in development. Annu. Rev. Physiol. 47 (1985) 319-335.

[5] W.R. Loewenstein and B. Rose, The cell-cell channel in the control of growth. Semin. Cell Biol. 3 (1992) 59-79.

[6] H. Yamasaki and C.C.G. Naus, Role of connexin genes in growth control. Carcinogenesis 17 (1996) 1199-1213.
[7] P.P. Mehta *et al.*, Growth inhibition of transformed cells correlates with their junctional communication with normal cells. Cell 44 (1986) 187-196.
[8] H. Yamasaki, Gap junctional intercellular communication and carcinogenesis. Carcinogenesis 11 (1990) 1051-1058.
[9] S.W. Lee *et al.*, Transcriptional downregulation of gap junction proteins blocks junstional communication in human mammary tumor lines. J. Cell Biol. 118 (1992) 1213-1221.
[10] D.S. Crow *et al.*, Phosphorylation of Cx43 gap junction protein in unfected and Rous sarcoma virus-transformed mammlian fibroblasts. Mol. Cell. Biol. 10 (1990) 1754-1763.
[11] L.P. Yotti *et al.*, Elimination of metabolic cooperation in chinese hamster cells by a tumor promoter. Science 206 (1979) 1089-1091.
[12] A.F. Lau *et al.*, Epidermal growth factor disrupts gap junctional communication and induces phosphorylation of connexin43 on Serine. Mol. Biol. Cell 3 (1992) 865-874.
[13] D.B. Pelletier and A.L. Boynton, Dissociation of PDGF receptor tyrosine kinase activity from PDGF-mediated inhibition of gap junctional communication. J. Cell Physiol. 158 (1994) 427-434.
[14] M.Z. Hossain *et al.*, Enhancement of gap junctional communication by retinoids correlates with their ability to inhibit neoplastic transformation, Carcinogenesis 10 (1989) 1743-1748.
[15] L-X. Zhang *et al.*, Carotenoids enhance gap junctional communication and inhibit lipid peroxidation in C3H/10T1/2 cells: Relationship to their cancer chemopreventive action Carcinogenesis 12 (1991) 2109-2114.
[16] B. Eghbali *et al.*, Involvement of gap junctions in tumorigenesis: Transfection of hepatoma cells with connexin32 retards growth in vivo. Proc. Natl. Acad. Sci. USA 88 (1991) 10701-10705.
[17] D. Zhu *et al.*, Transfection of C6 glioma cells with connexin43 cDNA: Analysis of expression, intercellular coupling and cell proliferation. Proc. natl. Acad. Sci. USA 88 (1991) 1883-1887.
[18] S.C. Chen *et al.*, Connexin43 reverses the phenotype of transformed cells and alters their expression of cyclin/cyclin-dependent kinases. Cell Growth Differentiation 6 (1995) 681-690.
[19] M. Mesnil *et al.*, Negative growth control of HeLa cells by connexin genes: connexin species specificity. Cancer Res. 55 (1995) 629-639.
[20] C.D. Moorby *et al.*, HGF/SF inhibits junctional communication. Exp. Cell Res. 219 (1995) 657-663.
[21] C.H. Heldin, Dimerization of cell surface receptors in signal transduction. Cell 80 (1995) 213-223.
[22] L. Claesson-Welsh, Platelet-derived growth factor receptor signals. J. Biol. Chem. 269 (1995) 32023-32026.
[23] P. van der Geer and T. Hunter, Receptor protein-tyrosine kinases and their signal transduction pathways. Annu. Rev. Cell Biol. 10 (1994) 251-337.
[24] M.Y. Kanemitsu and A.F. Lau, Epidermal growth factor stimulates the disruption of gap junctional communication and connexin43 phosphorylation independent of 12-O-tetradecanoylphorbol 13-acetate-sensitive protein kinase C: The possible involvement of mitogen-activated protein kinase. Mol. Biol. Cell 4 (1993) 837-848.
[25] B.J. Warn-Cramer *et al.*, Characterization of the mitogen-activated protein kinase phosphorylation sites on the connexin-43 gap junction protein. J. Biol. Chem 271 (1996) 3779-3786.
[26] A. Kazlauskas *et al.*, Functions of the major tyrosine phosphorylation site of the PDGF receptor β subunit. Cell Regul. 2 (1991) 413-425.
[27] P.E. Maldonado *et al.*, Growth factors modulate junctional cell-to-cell communication. J. Membr. Biol. 106 (1988) 203-210.
[28] M.Z. Hossain *et al.*, Rapid disruption of gap junctional communication and phosphorylation of connexin43 by platelet-derived growth factor in T51B rat liver epithelial cells expressing PDGF receptor. J Cell Physiol (1997) in press.
[29] A. Kazlauskas and J. Cooper, Autophosphorylation of the PDGF receptor in the kinase insert region regulates interaction with cell proteins. Cell 58 (1989) 1121-1133.
[30] M. Valius and A. Kazlauskas, A. Phospholipase C-γ1 and phosphatidyinositol 3 kinase are the downstream mediators of the PDGF receptor's mitogenic signal. Cell 73 (1993) 321-334.
[31] V.M. Berthoud *et al.*, On the mechanisms of cell uncoupling induced by a tumor promoter phorbol ester in clone 9 cells, a rat liver epithelial cell line. Eur. J. Cell Biol. 61 (1993) 384-396.
[32] L.S. Musil *et al.*, Differential phosphorylation of the gap junction protein connexin-43 in communication-competent and -deficient cell lines. J. Cell Biol. 111 (1990) 2077-2088.
[33] B.R. Kwak *et al.*, Differential regulation of distinct types of gap junctional channels by similar phopshorylating conditions. Mol. Biol. Cell 6 (1995): 1707-1719.

*Gap Junctions*
*R. Werner (Ed.)*
*IOS Press, 1998*

# Characterization of a Novel SH3-containing Protein that May Interact with Connexin43

Chengshi JIN and Alan F. LAU. *Molecular Carcinogenesis, Cancer Research Center, and Department of Genetics and Molecular Biology, School of Medicine, University of Hawaii at Manoa, Honolulu, Hawaii 96813*

We searched for putative Cx43 interacting proteins using a two-hybrid screen of an embryonic mouse cDNA library. A cDNA fragment (CID62) encoding a SH3 domain was identified in this screening. We isolated a 2.7-Kb long full length cDNA (CIP62) using CID62 as a probe. The CIP62 cDNA was predicted to encode a 760 aa protein without significant homology to proteins of known function. To further characterize the protein, flag-tagged CIP62 was expressed in *Spodoptera frugiperda* (Sf9) cells. We detected an approximately 83-kDa protein by Western blot analysis using anti-Flag antibody. We observed that this protein was predominantly localized to the plasma membrane of Sf9 cells at the immunofluorescence microscopic level.

## Introduction

Specific protein-protein interactions form the basis of many essential biological processes. For example, Ras-Raf interactions are critical for the activation of the Raf kinase during signal transduction [1]. Whether the metabolism and function of Cx43 are regulated through similar mechanisms is not fully understood. Previous work from our laboratory have shown that both epidermal growth factor and v-Src markedly disrupt gap junctional communication which was correlated with increased phosphorylation of the carboxyl terminus of Cx43 on serine and tyrosine residues, respectively [2,3,4]. These data indicated that a dynamic association of protein kinases with Cx43 may be involved in the alteration of channel permeability in response to these signaling events. In addition, a proline-rich region and phosphotyrosine residues present in the C-terminal region of Cx43 provide structural bases for its interaction with SH3 and SH2 domains of v-Src [5].

Traditionally, Cx43 interacting proteins could be detected through biochemical approaches such as coimmunoprecipitation experiments in which antibody to a known protein is used to co-precipitate associated proteins. However, it is difficult to identify unknown interacting proteins by such a biochemical approach. A genetic method for detecting protein interactions *in vivo* called the yeast two-hybrid system circumvents this problem. The two-hybrid system originally described by Fields and Song [6] is based upon the reconstitution of the activity of a transcriptional activator *in vivo* from separated DNA-binding and transcriptional activation domains provided these domains are in close physical proximity. This system can be used to investigate whether two known proteins interact, if the genes encoding both proteins are available or to identify previously unknown proteins that interact with a protein of interest using a specially constructed cDNA library. Most importantly, compared to *in vitro* methods, two-hybrid assays detect *in vivo* interactions, where the proteins involved are more likely to be in their native conformation. Therefore, we undertook the two-hybrid screen to identify proteins potentially associated with Cx43.

## Results

**Expression of the bait hybrid protein LexA-Cx43CT in yeast.** As the transmembrane domains of Cx43 may interfere with targeting the hybrid proteins to the nucleus where gene transcription can occur, full length Cx43 is probably not a good candidate for "bait". Instead, the cytoplasmic tail of Cx43 (Cx43CT, aa 221-382) was chosen as target in the two-hybrid screen because of the following reasons: 1) Cx43CT is thought to be responsible for the distinct biophysical and regulatory properties exhibited by gap junction channels composed of Cx43 [7]. 2) Many of the phosphorylation sites of regulatory nature are also located in this region [8,9] which indicated that protein kinases may modulate the channel activity by interacting with this region. Therefore, we constructed a bait plasmid encoding a LexA-Cx43CT fusion protein.

Expression of the LexA-Cx43CT fusion protein in yeast transformed with the bait plasmid was confirmed by Western blot analysis using anti-Cx43 antibody (Figure 1). A protein of about 40 kDa, corresponding to the predicted molecular weight of the LexA-Cx43CT reacted with the antibody. In contrast, no signal was detected from the yeast lysate expressing LexA alone. This suggested that LexA-Cx43CT was expressed in yeast.

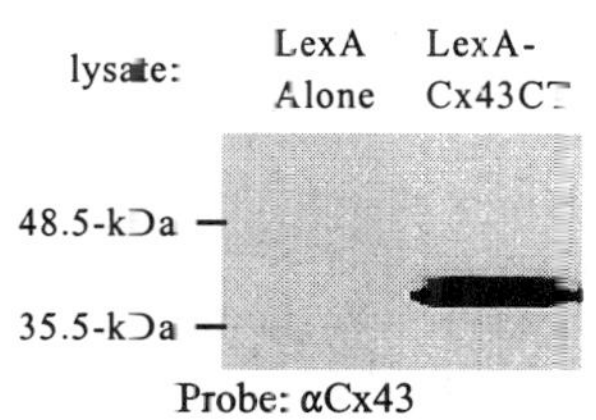


Figure 1. Expression of LexA-Cx43CT in yeast. *S. cerevisiae* strain, L40 expressing either LexA alone (first lane) or expressing LexA-Cx43CT were lysed and analyzed by immunoblotting with anti-Cx43 antibody. Approximate molecular weights are shown at the left.

**Screen for Cx43-interacting proteins.** The yeast reporter strain L40 expressing LexA-Cx43CT was transformed with a mouse 9-10 day embryo cDNA library [10]. Approximately 12 million yeast transformants were screened by this approach (Figure 2). 132 colonies grew on histidine minus plates and 42 out of these expressed detectable level of β-galactosidase activity. The individual library plasmids were isolated and further tested for their interaction with LexA-Cx43CT by directed two-hybrid assays. Fifteen clones activated the *His3* and *LacZ* reporter genes when coexpressed with LexA-Cx43CT but not with a control plasmid encoding LexA-lamin. These clones are referred to as CID representing Cx43-interacting domain.

12 million yeast transformants screened

↓

132 His+

↓

42 His+/LacZ+

↓

15 Cx43CT-dependent His+/LacZ+

Figure 2. Summary of library screen. As described in the text, of the 12 million yeast transformants screened, 132 colonies grew in the absence of histidine and 15 out of these also expressed detectable level of β-galactosidase activity in Cx43CT-dependent manner.

Sequencing and analysis of one of these positive clones, CID62, showed that it contained a SH3 domain found in numerous proteins. Because the SH3 domain may mediate its binding to a putative proline rich region of Cx43, CID62 was selected for further investigation. To obtain the full length cDNA encoding CID62, the CID62 cDNA fragment was used as a probe to screen an oligo dT primed cDNA library derived from a day 16 mouse embryo. A 2.7-kb full length cDNA was isolated and sequenced in both directions using an automated DNA sequencer. The complete nucleotide sequence contained an open reading frame of 740 amino acids with a molecular mass of 82-kDa. This cDNA was designated as Cx43-interacting protein 62 (CIP62). Sequence analysis of the derived peptide sequence revealed that a 48 amino acid region ( amino acids 497-545) of CIP62 was 41.6% identical and 64.5% similar to the SH3 domain (amino acids 163-210) of human Grb2 protein (Figure 3). The rest of the sequence did not show any significant similarity to proteins of known function.

```
        497                                                    545
CIP62   ALLDFERHDDDELGFRKNDI IT I I S QKDEHCWVGE LNGLRGWFPAKFV
        **: **:  :   *:** * **:  *:*  ::       * :  *  *   :*   *  **   :*
hGrb2   ALFDFDPQE DGELGFRRGDFIHVMDNSDPNWWKGACHGQTGMFPRNYV
        163                                                    210
```

Figure 3. Comparison of the amino acid sequence of a region (amino acids 497-545) of CIP62 with the SH3 domain (amino acids 163-210) of human Grb2. Amino acid identity and similarity are indicated by asterisks (*) and colon (:), respectively.

**Northern Blot analysis of CIP62 mRNA.** To detect mRNA encoding CIP62 protein in mammalian Rat1 fibroblasts, Northern blotting was performed using CID62 cDNA fragment as the probe. As shown in Figure 4, a RNA of approximately 2.7-kb was detected.

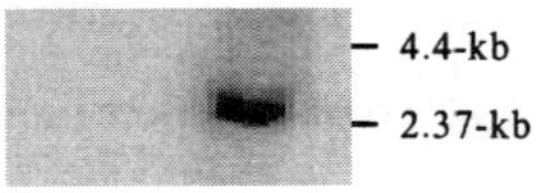


Figure 4. Northern blot of total RNA isolated from Rat1 fibroblasts. Total RNA was probed at high stringency with CID62 cDNA. The positions of RNA markers are shown at the right.

**Expression of Flag-CIP62 in Sf9 cells.** Since CIP62 is a potentially novel protein, antibody against this protein is unavailable. In order to establish both *in vitro* and *in vivo* binding assays, as well as to characterize the biochemical properties of this protein, we constructed a recombinant baculovirus expressing CIP62 tagged at its amino-terminus with the Flag epitope for infection of Sf9 cells. Infected Sf9 cells were harvested, lysed and the resulting cellular extracts were separated by SDS-PAGE and probed with anti-Flag antibody. As illustrated in Figure 5, a protein of approximately 83-kDa was detected in the Sf9 cell lysate infected with Flag-CIP62 baculovirus. The expression of this protein was not observed in uninfected Sf9 cells.

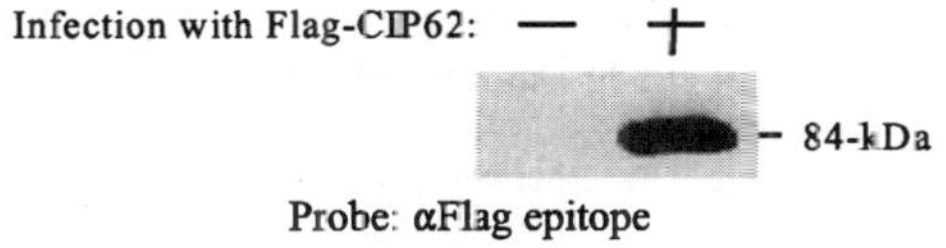


Figure 5. Expression of Flag-CIP62 in Sf9 cells. Sf9 cells were uninfected (lane 1) or infected with recombinant baculovirus encoding Flag-CIP62 (lane 2). After infection, cells were lysed, proteins were resolved by SDS-PAGE, and submitted to Western blot analysis using Flag epitope antibody.

**Subcellular localization of Flag-CIP62 in Sf9 cells.** To determine the subcellular localization of Flag-CIP62, immunofluorescence microscopy studies were conducted on Flag-CIP62 baculovirus-infected Sf9 cells Sf9 cells were grown for 48 hrs after infection, permeabilized with Triton X-100, and incubated with anti-Flag monoclonal antibody. Immune complexes were detected with fluorescein-conjugated goat anti-mouse secondary antibody. A strong signal was observed in the cell periphery with only a faint signal in the cytoplasm (Figure 6D) in the Flag-CIP62 baculovirus infected Sf9 cells suggesting that Flag-CIP62 was predominantly localized to the plasma membrane of infected Sf9 cells. The absence of signal in the uninfected cells also confirmed the specific reaction of the antibody (Figure 6B).

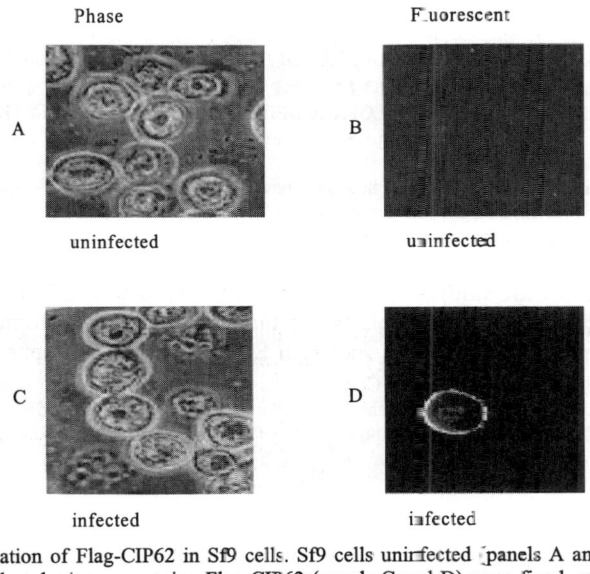


Figure 6. Localization of Flag-CIP62 in Sf9 cells. Sf9 cells uninfected (panels A and B) and infected with recombinant baculovirus expressing Flag-CIP62 (panels C and D) were fixed, permeabilized, then examined by indirect immunofluorescent staining with anti-Flag antibody and fluorescein isothiocyanate-conjugated goat anti-mouse antibody. Phase (panels A and C) and fluorescent (panels B and D) images were photographed with a Zeiss Axioplane Universal microscope equipped with epifluorescence.

## Discussion

We have isolated a CIP62 cDNA encoding an 82-kDa protein during a yeast two-hybrid screen designed to search for proteins capable of binding to the carboxyl-terminal tail of Cx43. Sequence analysis revealed that the amino acid sequence of CIP62 did not show any significant similarity to proteins from database except a 48 aa region (aa 497-545) similar to SH3 domains of numerous proteins (41.6% identity and 64.5% similarity to the SH3 domain of human Grb2 protein ).

The interaction of CID62 with Cx43CT observed in two-hybrid assays, the presence of a proline-rich region in Cx43CT and a SH3 domain in CIP62, as well as the membrane localization of both proteins suggested a possible interaction between full length Cx43 and CIP62. However, the physical association of these two proteins *in vivo* and its significance remain to be investigated.

## Acknowledgments

We thank Dr. Stanley M. Hollenberg for generously providing the two-hybrid plasmids and the cDNA library, and Dr. Megan Brown for providing the oligo-dT primed cDNA library and technical advice. We also thank Dr. Robert Eisenman and Jonathan A. Cooper for their insightful discussions. This research was supported in part by Grant CA52098 from the National Cancer Institute, NIH (to A.F.L.) and Predoctoral Fellowship from American Heart Association, Hawaii Affiliate to (C.J.).

## References

[1] A.B. Vojtek *et al.*, Mammalian Ras interacts directly with the serine/threonine kinase Raf, *Cell* 74 (1993) 205-214.

[2] A.F. Lau *et al.*, Epidermal growth factor disrupts gap-junctional communication and induces phosphorylation of connexin43 on serine, *Mol. Biol. Cell* 3 (1992) 865-874.

[3] D.S. Crow *et al.*, Phosphorylation of Cx43 gap junction protein in uninfected and Rous sarcoma virus-transformed mammalian fibroblasts, *Mol. Cell. Biol.* 10 (1990) 1754-1763.

[4] D.S Crow *et al.*, Phosphorylation of connexin43 in cells containing mutant src oncogenes, *Oncogene* 7 (1992) 999-1003.

[5] M.Y. Kanemitsu *et al.*, Tyrosine phosphorylation of connexin43 by vSrc is mediated by SH2 and SH3 domain interactions, *J. Biol. Chem.* (1997) In Press.

[6] Fields, S. and O. Song, A novel genetic system to detect protein-protein interactions, *Nature* 340 (1989) 245-247.

[7] E.C. Beyer *et al.*, Connexin family of gap junction proteins, *J. Membr. Biol.* 116 (1990) 187-194.

[8] A.J. Godwin *et al.*, In situ regulation of cell-cell communication by cAMP-dependent protein kinase and protein kinase C, *Mol. Cell. Biochem.* 127/128 (1993) 293-307.

[9] Kennelly, P.J. and E.G. Krebs, Consensus sequences as substrate specificity determinants for protein kinases and protein phosphatases, *J. Biol. Chem.* 266 (1991) 15555-15558.

[10] S.M. Hollenberg *et al.*, Identification of a new family of tissue-specific basic helix-loop-helix proteins with a two-hybrid system, *Mol. Cell. Biol.* 15 (1995) 3813-3822.

*Gap Junctions*
*R. Werner (Ed.)*
*IOS Press, 1998*



# Biologically Active Compounds Capable of Stimulating Gap Junctional Communication

Helmut Sies, Swantje Nicolai, Anke Stock, Annette Clairmont and Wilhelm Stahl
*Institut für Physiologische Chemie I, Heinrich-Heine-Universität Düsseldorf, P.O.Box. 10 10 07, D-40001 Düsseldorf, Germany*

**Abstract.** Compounds with tumor-promoting or tumor-inhibiting activity such as phorbol esters, retinoic acid or ß-carotene, have been reported to modulate gap junctional communication (GJC). Chemicals exhibiting physiological or pathological properties were also tested in the dye-transfer assay to assess for their effects on GJC in human fibroblasts. Vitamin D induces GJC at concentrations between $10^{-6}$ and $10^{-8}$ M. The extent of induction is similar to that observed with retinoic acid. The physiologically active metabolite of vitamin D, 1,25-dihydroxy vitamin D, stimulates GJC at concentrations of $10^{-8}$ to $10^{-11}$ M. In fibroblasts devoid of a functional vitamin D receptor (VDR) there is no effect on GJC. Parallel to the increase in GJC there is a VDR-dependent increase in connexin43 protein and connexin43 mRNA levels. Apart from ß-carotene a variety of related carotenoids induce GJC at $10^{-6}$ M, with six-membered ring compounds being more active than five-membered ring analogs. 4-Oxo-retinoic acid, a decomposition product of canthaxanthin in cell culture is active at even $10^{-8}$ M. The data suggest that oxidation products of carotenoids are the ultimately active compounds responsible for the effects of carotenoids on GJC. The human teratogen thalidomide induces GJC in fibroblasts after activation by liver microsomes in concentrations as low as $10^{-7}$ M. Treatment of cells with the thalidomide analog EM-12 stimulated GJC without prior activation. No effects on GJC were detected for phthalimide and glutamate, the components of thalidomide. However, 2-phthalimido glutaric acid (PGA), a hydrolysis product of thalidomide, raised GJC without activation at concentrations between $10^{-10}$ M and $10^{-5}$ M. It might be speculated that disturbances in GJC are in part related to the biological properties of thalidomide.

## 1. Vitamin D

The physiologically active metabolite of vitamin D, 1α,25-dihydroxy-vitamin $D_3$, is known for its role in calcium homeostasis and as a regulator of cell growth and differentiation [1]. It binds to the vitamin D receptor, a nuclear acting transcription factor. The activated receptor binds as a homodimer or a heterodimer with the RXR receptor to VDRE, for regulation of transcription [2].

1α,25-dihydroxyvitamin $D_3$ induces gap junctional intercellular communication in human skin fibroblasts 161 BR at a concentration of $10^{-7}$ M [3]. In human skin fibroblasts, FIB5, devoid of a functional nuclear vitamin D receptor (VDR), there is no effect on GJC. Parallel to the increase in cell-cell communication, we observed a VDR-dependent increase in connexin43 protein and connexin43 mRNA levels. These results suggest that 1α,25-dihydroxyvitamin $D_3$ affects GJC at the level of transcription or of mRNA via the nuclear VDR.

Vitamin D3 (cholecalciferol) induces GJC in 10T ½ cells at concentrations between 0.01 and 1.0 μM, as assayed by the dye transfer method [4]. The extent of induction is similar to that obtained with canthaxanthin or retinoic acid at 10 and 1 μM, respectively. Vitamin D2 also induces GJC. At elevated concentrations of vitamin D3 (5 μM) there is a suppression of GJC, reversible upon exposure to all-trans retinoic acid (1 μM) after removal of vitamin D3 from the medium. Conversely, communication between cells prestimulated with all-trans retinoic acid (1 μM) rapidly decreases when the retinoid is replaced by Vitamin D3 (5μM). The

inductory effects of vitamin D at low concentrations are rather slow as compared to the rapid inhibition of GJC at high vitamin D levels. This implies two different vitamin D dependent regulatory mechanisms of GJC, one being activatory, possibly by regulation of connexin gene expression. The other mechanism is inhibitory, likely operating through changes in cellular calcium concentration.
The data demonstrate a role of vitamin D in the regulation of GJC. This novel property of Vitamin D may contribute to the antiproliferative effects of vitamin D exhibited in some types of cancer.

Table 1: Effects of all-trans retinoic acid and 1α,25-dihydroxy vitamin $D_3$ on GJC in different cell lines

| Compound | VDR-negative cells GJC (% of control | VDR-positive cells GJC (% of control |
|---|---|---|
| Control | 100 ± 55 | 100 ± 33 |
| all-trans retinoic acid ($10^{-6}$ M) | 195 ± 49 | 216 ± 47 |
| 1α,25-dihydroxy vitamin $D_3$ ($10^{-7}$ M) | 107 ± 38 | 161 ± 44 |

GJC was measured on day 2 after incubation by the dye transfer assay. Control was 0.5% THF. See Ref. [3].

## 2. Carotenoids

Induction of GJC has been discussed as a possible biochemical mechanism underlying the cancer preventive properties of carotenoids [5]. ß-Carotene, echinenone, canthaxanthin, cryptoxanthin, and 4-hydroxy-ß-carotene efficiently induce GJC in murine fibroblasts by up to 5-fold in comparison to solvent control [6]. The analog of ß-carotene, retro-dehydro-ß-carotene is as effective as the parent compound. The position and nature of the substituent at the six-membered ring appears to have little influence on GJC. Echinenone and 4-OH-ß-carotene exhibit about similar activity. Cryptoxanthin with an OH-group at the 3-position is somewhat less active than the corresponding 4-OH-ß-carotene.
Carotenoids containing a five-membered ring system are less active. No stimulatory effects were found for capsorubin or violerythrin. A direct comparison of five- and six-membred ring carotenoids can be taken from canthaxanthin and dinor-canthaxanthin. The six-membered ring carotenoid is about twice as active as its five-membered ring analog. No influence on GJC was observed for the open-chain polyene compounds C-20, C-30, and C-40 dialdehyde containing 7, 11, and 15 carbon-carbon double bonds, respectively.

Table 2: GJC in C3H/10T1/2 mouse fibroblasts

| Compound | GJC (% of control) | Compound | GJC (% of control) |
|---|---|---|---|
| Control | 100 ± 15 | | |
| ß-Carotene | 530 ± 21 | Capsorubin | 107 ± 11 |
| ß-Carotene | 509 ± 27 | Dinor-Canthaxanthin | 183 ± 23 |
| Canthaxanthin | 428 ± 24 | Violerythrin | 148 ± 21 |
| 4-OH-ß-Carotene | 388 ± 30 | C-20-Dialdehyde | 120 ± 17 |
| Cryptoxanthin | 307 ± 24 | C-30-Dialdehyde | 113 ± 14 |
| Retrodehydro-ß-C | 522 ± 32 | C-40-Dialdehyde | 125 ± 16 |

GJC was measured on day 5 after incubation with the respective carotenoid (10 μM); capsorubin was used in 1 μM concentration in the dye transfer assay. Control was 0.5% THF. See Ref. [6].

These effects could be related to the intact carotenoids or to their decomposition products. Decomposition fractions of canthaxanthin were isolated by preparative HPLC and shown to be active in the cell communication assay [7].

Both isomers of 4-oxo-retinoic acid (all-trans and 13-cis) enhanced GJC in C3H/10T 1/2 cells accompanied by an increase of the expression of connexin 43 mRNA. Decomposition fractions of canthaxanthin were isolated by preparative HPLC and shown to be active in the cell communication assay. Two of the decomposition products were identified as all-trans and 13-cis 4-oxo-retinoic acid. Therefore, it is concluded that the biological activity of canthaxanthin is, at least in part, due to the formation of active decomposition products such as 4-oxo-retinoic acid. To obtain a significant induction of cell-cell-communication in 10T1/2 cells only 0.1% of the 10 μM incubation mixture with canthaxanthin needs to be converted to 4-oxo-retinoic acid, which is active at levels as low as 10 nM.

It is interesting to note that the non-provitamin A carotenoid canthaxanthin decomposes to yield the non-vitamin A retinoid 4-oxo-retinoic acid which exhibits regulatory activity. Thus, other carotenoids might serve as precursors for retinoid derivatives with interesting biochemical properties.

## 3. Thalidomide and related compounds

Gap junctional intercellular communication (GJC) can be involved in embryogenesis by coupling compartments of the same developmental potential [8]. We found that thalidomide induces GJC in human fibroblasts after activation by liver microsomes in concentrations as low as $10^{-7}$ M [9]. Treatment of cells with the thalidomide analog EM-12 increased GJC without prior activation. No alteration of GJC was detected with phthalimide and glutamate, the components of thalidomide. However, 2-phthalimido glutaric acid (PGA), a hydrolysis product of thalidomide, stimulated GJC without activation at concentrations between $10^{-10}$ M and $10^{-5}$ M.

Table 3: GJC induced by thalidomide and related compounds in human skin cell lines

| | Human skin | | Human fetal skin | |
|---|---|---|---|---|
| Compound | day 1 | day 3 | day 1 | day 3 |
| Control | 100±15 | 100±10 | 100±19 | 100±20 |
| Retinoic acid | 136±11 | 136±11 | 137±13 | 190±16 |
| Thalidomide | 108±17 | 100±15 | 94±15 | 98±13 |
| Thalidomide (activated) | 172±12 | 147± 8 | 151±19 | 165±13 |
| EM-12 | 141± 8 | 134±11 | 137±11 | 167±13 |
| EM-12 (activated) | 147±11 | 142±13 | 143±19 | 167±13 |
| Supidimide | 100±10 | 112±12 | 103±17 | 102±18 |
| Supidimide (activated) | 113±11 | 141±16 | 157±16 | 167±12 |
| 2-Phthalimido glutaric acid | 153±18 | 194±12 | 137±16 | 190±17 |

Confluent cells were incubated with the indicated concentrations of the compounds. On days 1 and 3, junctional communication was measured by the dye transfer assay using the fluorescent dye Lucifer yellow. See Ref. [9].

These data [9] are the first reporting on the stimulation of intercellular communication via gap junctions by thalidomide. The stimulation of GJC is not mediated by the parent compound itself, and at present the nature of the ultimately active compound is yet unknown.
Both thalidomide and EM-12 are teratogenic in primates, causing typical limb malformations [10]. GJC plays a role in embryonal development, and the expression of gap junction proteins (connexins) varies during fetal development. Other teratogenic compounds such as retinoic acid and phorbol esters also affect GJC (see above). In most cases, retinoids induce while phorbol esters inhibit GJC. It is not known whether these properties are responsible for the teratogenic potential of these compounds. However, it might be speculated that disturbances in GJC, either by induction or by inhibition, could influence embryonal development.

**Acknowledgement**

Our studies have been supported by the Bundesministerium für Bildung, Wissenschaft, Forschung und Technologie (Bonn) and the Deutschen Forschungsgemeinschaft (Bonn; Project Si255/10-1).

**References**

[1] A.W. Norman, Vitamin D, In: E. E. Ziegler and L. J. Filer (eds.), Present Knowledge in Nutrition. ILSI Press: Washington, DC., 1996. pp. 120-129.
[2] C. Carlberg, Mechanisms of Nuclear Signalling by Vitamin D3. Interplay with Retinoid and Thyroid Hormone Signalling. *Eur. J. Biochem.* 231 (1995) 517-527.
[3] A. Clairmont *et al*, Induction of Gap Junctional Intercellular Communication by Vitamin D in Human Skin Fibroblasts is Dependent on the Nuclear Vitamin D Receptor. *Carcinogenesis* 17 (1996) 1389-1391.
[4] W. Stahl *et al*, Vitamin D Influences Gap Junctional Communication in C3H/10T1/2 Murine Fibroblast Cells. *FEBS Lett.* 352 (1994) 1-3
[5] L.-X. Zhang *et al*, Carotenoids Enhance Gap Junctional Communication and Inhibit Lipid Peroxidation in C3H/10T1/2 Cells: Relationship to their Cancer Chemopreventive Action. *Carcinogenesis* 12 (1991) 2109-2114
[6] W. Stahl *et al*, Biological Activities of Natural and Synthetic Carotenoids: Induction of Gap Junctional Communication and Singlet Oxygen Quenching. *Carcinogenesis* 18 (1997) 89-92
[7] M. Hanusch *et al*, Induction of Gap Junctional Communication by 4-Oxoretinoic acid Generated from its Precursor Canthaxanthin. *Arch. Biochem. Biophys.* 317 (1995) 423-428
[8] E. Dahl *et al*, Expression of Gap Junction Genes, Connexin40 and Connexin43, During Fetal Mouse Development. *Anat. Embryol.* 191 (1995) 267-278
[9] S. Nicolai *et al*, Stimulation of Gap Junctional Intercellular Communication By Thalidomide and Thalidomide Analogs in Human Skin Fibroblasts. *Biochem. Pharmocol.* 53 (1997) 1553-1557
[10] W. Lenz, A Short History of Thalidomide Embryopathy. *Teratology,* 38 (1988) 203-215

*Gap Junctions*
*R. Werner (Ed.)*
*IOS Press, 1998*



# A Role for Mitogen Activated Protein Kinase (MAPK) During RAS-Induced Down Regulation of Gap Junctional Intercellular Communication

Melinda R. Wilson, Ifeanyi Umeh, Kyung-Sun Kang, and James E. Trosko
*Michigan State University, Department of Pediatrics and Human Development, East Lansing, MI 48824, USA*

**ABSTRACT**

Over-expression of the RAS oncogene in rat liver epithelial cells leads to a severe reduction in gap-junctional intercellular communication. In addition, the connexin 43 protein is hypo-phosphorylated as compared to normal control cells. Communication can be partially restored by blocking farnesylation of the RAS protein. Since RAS is known to activate downstream elements such as RAF, which activates MEK which activates MAPK, we wanted to investigate if activation of kinases downstream of RAS was involved in the altered gap-junctional intercellular communication displayed by RAS over-expressing cells. In addition, other investigators have shown that MAPK can phosphorylate connexin 43 *in vitro*, and connexin 43 contains potential MAPK phosphorylation sites. Over-expression of the RAS downstream effector, RAF, in rat liver epithelial cells did not greatly effect homologous gap-junctional intercellular communication. Therefore we chose to investigate the role of MAPK in RAS over-expressing cells. Treatment of cells with the MEK inhibitor, PD98059, led to a dose and time dependent augmentation of gap-junctional intercellular communication in RAS over-expressing cells. In addition, there was a partial restoration of the phosphorylated CX43 species in cells treated with the MEK inhibitor. The MEK inhibitor had little effect on gap-junctional intercellular communication of normal control cells. These results suggest that activation of MAPK may be partially responsible for the reduction of gap-junctional intercellular communication in RAS over-expressing cells. It is possible that RAS activation of MAPK ultimately leads to activation of phosphatases and/or inactivation of kinases which directly act on the connexin 43 protein.

## *1. Introduction*

The etiology of cancer is a complex multi-step mechanism that has been defined to occur in three major stages: initiation, promotion and progression [1]. During initiation, single cells have acquired mutations that potentiate their ability to become cancer cells. Theoretically, initiated cells may be kept in control by normal neighboring cells through exposure to small regulatory molecules which pass through gap junctions [2]. Upon exposure to tumor promoters, initiated cells are promoted to a higher propensity towards cancer. These initiated cells clonally expand and become available to further genetic changes. In 1979, Yotti Trosko and Chang [3] and separately Murray and Fitzgerald [4] found that tumor promoters such as the phorbol ester TPA blocked gap junction intercellular communication (GJIC). Subsequently, it has been suggested that down-regulation of gap junction function facilitates escape of initiated cells from normal regulatory growth signals. Since 1979, there is ample evidence that tumor promotion is accompanied by alteration of gap junction function [5]. During progression, further genetic or epigenetic damage results in complete loss of regulatory controls and promoted cells acquire the ability to progress to metastatic disease [1].

Oncogene activation can occur during any of the three stages of carcinogenesis. Most oncogenes encode proteins that participate in various cascades of events by which growth factors stimulate normal cell division. One of the most widely studied is the RAS family of oncogenes [6]. RAS is a 21 kd guanosine triphosphate binding protein with GTPase activity. Point mutations in the RAS family of genes are detected in human carcinomas, leukemias, and melanomas. RAS is an integral component of a membrane signal transduction system that modulates phosphatidyl inositol-4,5 diphosphate. In-creased levels of diacylglycerol (DAG), which is one of the $PIP_2$ by-products is known to activate phosphokinase C (PKC), a step considered by many to be critical in cell transformation [7]. RAS activates the Mitogen Activated Protein Kinase (MAPK) pathway [8]. Activation of MAPK can lead to cell division or cell proliferation, and was suggested to participate in oncogene transformation [9]. Extracellular signals such as growth factors, cytokines, hormones, and tumor promoters will activate transmembrane receptors to initiate signal transduction pathways (Figure 1). Activation of RAS leads to an accumulation of DAG, which in turn activates PKC [10]. RAS will recruit RAF-1 to the membrane, and PKC may activate RAF-1 by phosphorylation [11]. RAF-1 then activates MAPK/Erk kinase (MEK). MEK, a dual specific kinase, can catalyze serine/threonine and tyrosine phosphorylation which activates MAPK (Extracellular receptor kinase 1, 2; Erk 1, Erk 2)[12]. The relationship between the MAPK pathway and GJIC has been recently described [13]. In rat liver epithelial (WB-F344) cells, lysophosphatitic acid induces a concomitant reduction of GJIC and activation of MAPK [14]. Also, potential MAPK phosphorylation sites have been mapped to the carboxyl tail of CX43 [15].

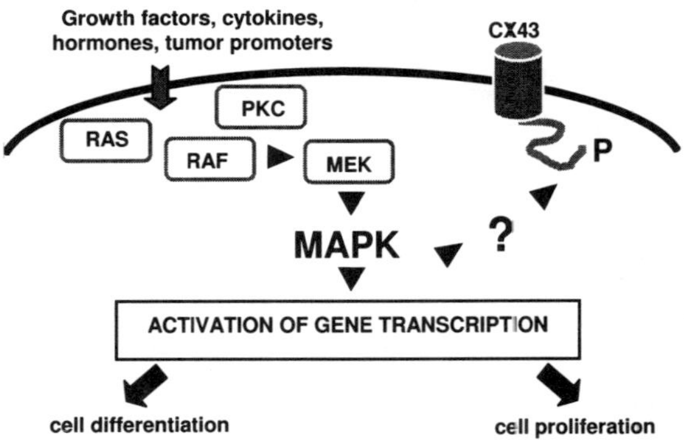


**Figure 1**. Relevant features of the MAPK pathway from cell surface to nucleus and beyond.

## 2. *Hypothesis*

Reduction of gap junction function has been suggested to occur at the promotion/ progression stage of carcinogenesis [2], implying activation of MAPK would led to reduced gap junction function. Since tumor promoters such as TPA activate MAPK and block GJIC, we hypothesized that MAPK down-regulates CX43 function during the tumor promotion stage of carcinogenesis. Thus, our goal was to understand the relationship between gap junctions and MAPK during carcinogenesis. The MAPK pathway was suggested to be essential for the growth and maintenance of the RAS transformed phenotype [16]. Therefore we used an extensively characterized rat liver epithelial (WB-F344) cell line,

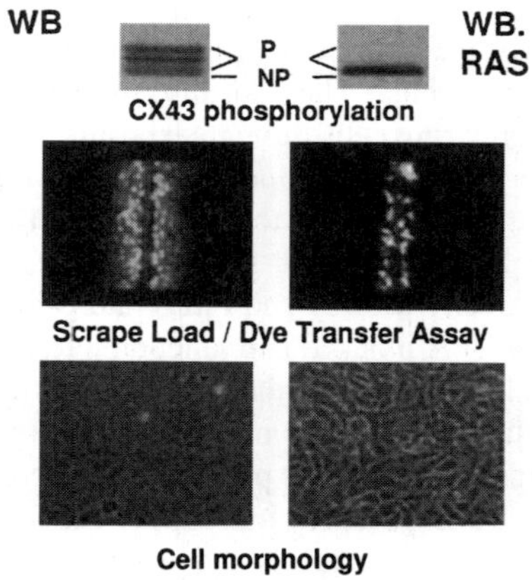


**Figure 2**. Comparison of CX43 protein phosphorylation, gap junction function, and morphology between normal WB and WB.RAS2a.

WB.RAS2a, that over-expressed the RAS oncogene Figure 2). In WB.RAS2a cells, GJIC, assessed by their ability to transfer Lucifer Yellow dye, is reduced to approximately 20% that of normal WB cells and are tumorgenic [17]. In addition, connexin 43 (CX43) is hypo-phosphorylated and protein levels are not significantly altered [18]. Blockage of GJIC can be reversed by inhibition of RAS farnesylation [19].

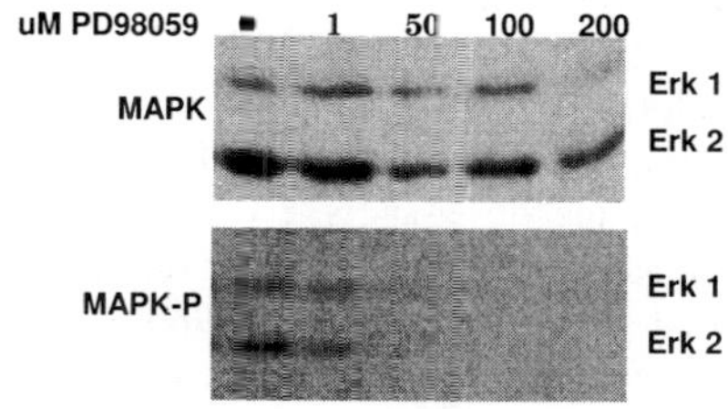


**Figure 3.** Immunoblot analysis of MAPK activity in WB.RAS2a cells treated with

### 3. *Results and Discussion*

If MAPK is involved in the down-regulation of GJIC in RAS-transformed cells, then inhibition of MAPK activity should lead to an up-regulation of GJIC. Therefore, a specific inhibitor for MEK 1 [16] was used to inhibit MAPK activity in WB.RAS2a cells. The MEK 1 inhibitor (PD98059; 2-(2-amino-3-methoxyphenyl)-4H-1-Benzopyran-4-one) exerts its effect by binding to the inactive form of MEK 1 and subsequently prevents phosphorylation of ERK 1 and ERK 2 by MEK 1 [16]. WB.RAS2a cells were treated with PD98059 for varying times and concentrations. Changes in MAPK activity, cell morphology, GJIC, CX43 phosphorylation, and CX43 subcellular localization were determined. MAPK activity was assayed on western blots using antibodies specific for the activated (phosphorylated) form of the MAPK isoforms ERK 1 and ERK 2. Population GJIC was assayed using the scrape load/dye transfer assay [20].

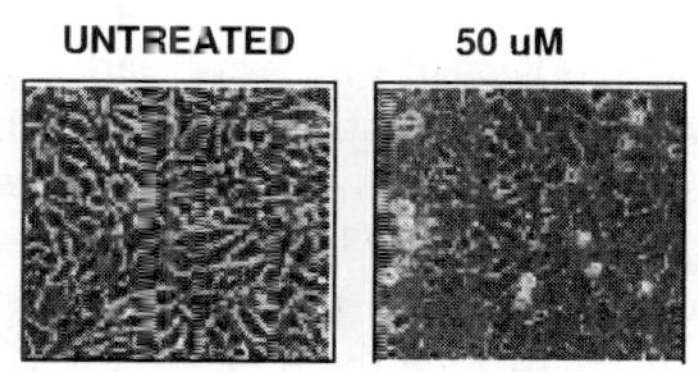


**Figure 4.** Morphology of WB.RAS2a cells treated with PD98059.

In WB.RAS2a cells that were treated for 24 hours with 1 μM, PD98059, there was a two-fold decrease in MAPK activity (Figure 3). At 50 μM PD98059, MAPK activity was reduced to undetectable levels and there was no detectable change in gap junction function (data not shown). However, the cells began to flatten, resembling normal WB cells. Some apoptotic bodies were evident (Figure 4). At concentrations of 1-10 μM PD98059, there was some augmentation of GJIC. The highest level of GJIC occurred seven hours after treatment with 1-10 μM PD98059 (Figure 5). This was coincident with enhanced levels of the phosphorylated CX43 species (Figure 6). The augmentation of gap junction function and CX43 phosphorylation was transient and both were reduced by 24 hours.

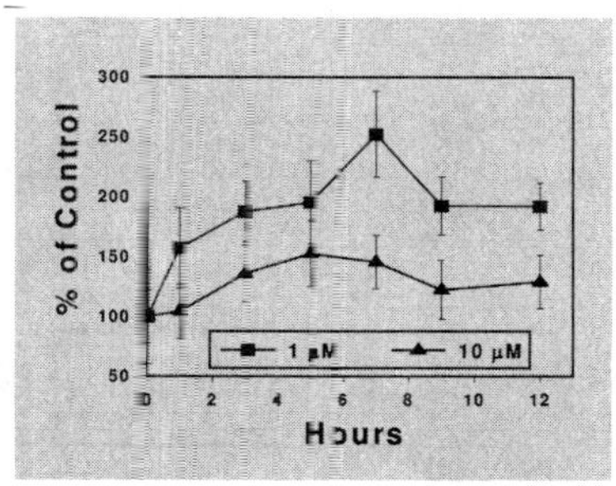


**Figure 5** Change in GJIC of WB.RAS2a cells treated with PD98059.

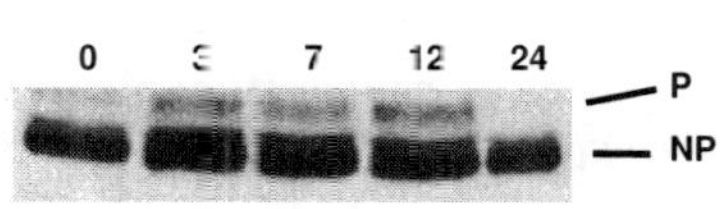


**Figure 6.** Alteration in gap junction phosphorylation in Wb.RAS2a treated with PD98059. Hours after treatment are indicated above each lane.

To determining if the MEK inhibitor would alter the subcellular localization of the CX43 protein in WB.RAS2a cells, normal WB and WB.RAS2a cells were treated with 1 and 10 μM of PD98059 for 5-7 hours and immunostained using antibodies specific for CX43. In normal WB cells, PD98059 had little effect on cytoplasmic membrane staining. However, there seemed to be a slight increase in peri-nuclear staining (data not shown). In RAS transformed WB cells, CX43 staining was very disorganized, with few punctate spots on the cytoplasmic membrane. In cells treated with 1 and 10 μM PD98059, there seemed to be a subtle increase in the organized punctate spots on the cytoplasmic membrane.

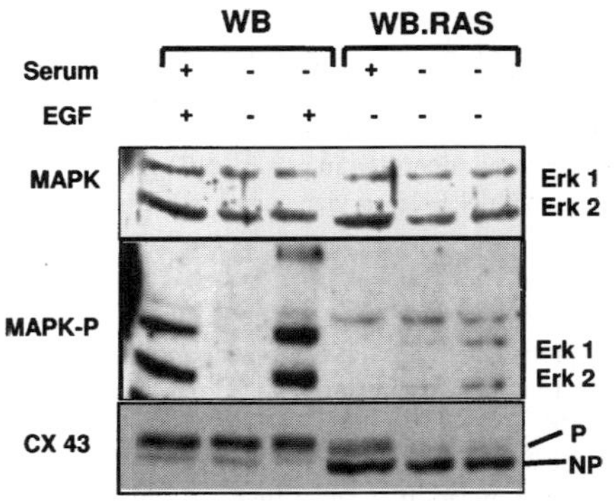


**Figure 7.** EGF responsiveness of the MAPK pathway in WB and WB.RAS2a cells. Cells were cultured with (+) or without (-) serum for 24 hours, 5 minute EGF treatment.

The data presented above indicate that inhibition of MAPK activity resulted in some transient restoration of GJIC. Thus, it was possible that the MAPK pathway was partially involved in regulating gap junction channel activity. However, we expected that PD98059 would exert a much greater effect. A possible explanation is that the MAPK pathway was regulated differently in the WB.RAS2a cells as compared to normal WB cells. To test this hypothesis, we compared the ability of epidermal growth factor (EGF) to activate MAPK in normal WB and WB.RAS2a cells as a way to assess the responsiveness of the MAPK pathway (Figure 7). This experiment indicated: 1) The MAPK pathway was down-regulated in WB.RAS2a cells, 2) A slight response to EGF stimulation in WB.RAS2a cells was evident, 3) Since experiments were performed within a 5 minute EGF treatment, any change in CX43 phosphorylation was slight.

## 4. A Model to Test

What is the mechanistic role of MAPK in WB.RAS2a cells? We propose a testable model for the role of MAPK in down regulation of gap junction function in RAS transformed WB cells (Figure 8). This model is based upon the observation that over-expression of RAS **(1)** initially leads to the activation of MAPK **(2)**. Eventually the MAPK pathway is down-regulated by activation of phosphatases and/or inactivation of kinases **(3)**. In a cell over-expressing RAS, there may be a continual "tug of war" between activation and down-regulation of MAPK. The activated phosphatases, such as PP2A, may act directly to dephosphorylate CX43 **(4)**, resulting in a reduction of gap-junction function **(5)**. This model is based upon results presented here combined with previous observations that a) MAPK is inactivated by protein phosphatase PP2A and CD45 [11], and b) the phosphatase inhibitor, okadaic acid blocks the de-phosphorylation of CX43 after EGF-induced hyper-phosphorylation [13].

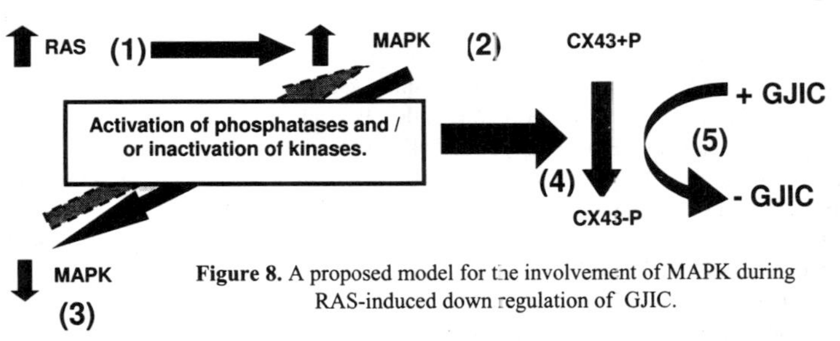


**Figure 8.** A proposed model for the involvement of MAPK during RAS-induced down regulation of GJIC.

Experiments are in progress to test this model and elucidate the mechanisms involved in RAS down regulation of GJIC by MAPK in WB.RAS2a cells.

***Acknowledgements***

We gratefully acknowledge D. Dudley, H. Camp, and A. Saltiel, of Parke Davis, Inc. for helpful discussions and providing the MEK inhibitor, PD98059. We also thank K. Guan of the Univ. of Michigan for supplying MAPK antibodies, R. Ruch and K. Cesen-Cummings for helpful discussion. This work was supported in part by NCI grant #CA21104, Air Force grant #95NL271, awarded to J.E. Trosko, and a NIEHS service award to M.R. Wilson.

***References***

[1] Pitot HC *et al.*, (1981) The natural history of carcinogenesis: Implications of experimental carcinogenesis in the genesis of human cancer. J. Supramol. Struct. Cellul. Biochem. 17:133-146.

[2] Trosko JE *et al.*, (1983) A mechanism of tumor promotion: Potential role of intercellular communication. Cancer Invest. 1:5 1-526.

[3] Yotti LP *et al.*, (1979) Elimination of metabolic cooperation in Chinese hamster cells by a tumor promoter. Science 206:1089-1091.

[4] Murray AW and Fitzgerald DJ (1979) Tumor promoters inhibit metabolic cooperation in coculture of epidermal and 3T3 cells. Biochem. Biophys. Res. Commun. 91:395-401.

[5] Trosko JE *et al.*, (1993) Oncogenes tumor suppressor genes and intercellular communication in the oncogeny as partially blocked ontogeny hypothesis. pp 181-192. In OH Iverson (ed.), New Frontiers in Cancer Causation. Taylor & Francis, Ltd, London.

[6] Denhardt DT, (1996) Signal-transducing protein phosphorylation cascades mediated by Ras/Rho proteins in the mammalian cell: the potential for multiplex signalling. Biochem J. 318:729-747.

[7] Fleischman LF *et al.*, (1986) Ras-transformed cells: Altered levels of phosphatidyl inositol 4,5-bisphosphate and catabolites. Science 231:407-410.

[8] Marshall CJ, (1995) Specificity of receptor tyrosine kinase signaling: Transient versus sustained extracellular signal-regulated kinase activation. Cell 80(2):179-185.

[9] Mansour SJ *et al.*, (1994) Transformation of mammalian cells by constitutively active MAP kinase kinase. Science 265:966-970.

[10] Nishizuka Y, (1986) Studies and perspectives of protein kinase C. Science 233:305-312.

[11] Seger R and Krebs E, (1995) The MAPK kinase signaling cascade. FASEB J. 9:726-735.

[12] Guan K-L, (1994) The mitogen activated protein kinase signal transduction pathway: From the cell surface to the nucleus. Cellular Signaling 6:581-589.

[13] Lau AF *et al.*, (1996) Regulation of connexin43 function by activated tyrosine protein kinases. J. Bioenergetics and Biomembranes 28(4)359-368.

[14] Hill CS *et al.*, (1994) Lysophosphatidic acid inhibits gap-junctional communication and stimulates phosphorylation of connexin-43 in WB cells: Possible involvement of the mitogen-activated protein kinase cascade. Biochem. J. 303(2):475-479.

[15] Warn-Cramer BJ *et al.*, (1996) Characterization of the mitogen-activated protein kinase phosphorylation sites on the connexin-43 gap junction protein. J.B.C. 271:3779-3786.

[16] Dudley DT *et al.*, (1995) A synthetic inhibitor of the mitogen-activated protein kinase cascade. Proc. Natl. Acad. Sci. USA 92:7686-7689.

[17] de Feijter AW *et al.*, (1990) Infection of rat liver epithelial cells with v-Ha-ras: Correlation between oncogene expression, gap junctional communication, and tumorigenicity. Molec. Carcinogenesis 3:54-67.

[18] de Feijter AW *et al.*, (1996) Localization and function of the connexin 43 gap-junction protein in normal and various oncogene-expressing rat liver epithelial cells. Molec. Carcinogenesis 16:203-212.

[19] Ruch RJ *et al.*, (1993) Reversal of ras-induced inhibition of gap-junctional intercellular communication, transformation and tumorigenesis by lovastatin. Molec. Carcinogenesis 7:50-59.

[20] El-Fouly MH *et al.*, (1987) Scrape-loading and dye transfer: A rapid and simple technique to study gap junctional intercellular communication. Exp. Cell Res. 168:428-430.

*Gap Junctions*
*R. Werner (Ed.)*
*IOS Press, 1998*

# Arachidonic Acid-Induced Cell Uncoupling in Rat Astrocytes is Dependent on Cyclo- and Lypo-Oxygenase Pathway and is Blocked by Melatonin

Agustín D.MARTÍNEZ and Juan C. SÁEZ

*Departamento de Ciencias Fisiológicas, Facultad de Ciencias Biológicas, Pontificia Universidad Católica de Chile. Santiago. Chile*

**Abstract** . During high neuronal activity, astrocyte networks play an important role in the spatial buffering of extracellular potassium. Gap junctions (GJs) between astrocytes contribute to this function by allowing diffusion of ions from the cytoplasm of one cell to that of neighboring cells. Little is known about mechanisms that may control the degree of astrocyte coupling. Arachidonic acid (AA) is a signaling molecule that might modulate glial GJs under physiological and/or pathological conditions, thus its effect and mechanism of action on astrocyte GJs was studied. After 60 min treatment with 5 or 30 μM AA dye coupling (Lucifer yellow) was reduced by 70 to 80 %. The effect was spontaneously reversed 3 h after treatment. No significant changes in levels and state of phosphorylation of Cx43, the main astrocyte GJs protein, were detected. Cell uncoupling was associated with and increase and a decrease in Cx43 labelling in cytoplasmic compartments and membrane appositions, respectively. Indomethacin (50 μM), an inhibitor of the cyclooxygenase, prevented completely the AA-induced cell uncoupling. Nordihydroguaiaretic acid (5 μM), a lipoxygenase inhibitor, increased the recovery rate of the AA-induced effect. A partial inhibition of the AA-induced cellular distribution of Cx43 was observed in cells pretreated with indomethacin. Moreover, melatonin completely prevented the AA-induced effects. Since melatonin also blocks cyclooxygenase, these results suggest that a product of the AA metabolism, and not AA itself, mediates the AA-induced uncoupling between astrocytes.

## 1. Introduction

Astrocytes communicated by gap junctions (GJs), formed mainly by Cx43 [5,9], are believed to form networks which favor the spatial buffering of extracellular potassium. Moreover, GJs permit the propagation of calcium waves through astrocyte networks providing an extraneuronal signaling pathway [3].

During physiological and pathological conditions, arachidonic acid (AA) may be an important signaling molecule that modulates various astrocyte functions via alteration of gap junctional communication. AA is known to inhibit GJs in hepatocytes [18], lacrimal glands [10], neonatal cardiac myocytes [15] and retinal horizontal cells [16]. Recently, its has been found that anandamide, an AA-derivative, blocks astrocyte GJs, and the propagation of calcium wave [23]. But the AA effect on astrocyte GJs and its mechanism of action remain unknown.

AA is a fatty acid that diffuses readily across cell membranes. It is metabolized to protaglandins and thromboxanes and to leukotrienes and hydroxyeicosatetraenoic acid, by cyclooxygenase- and lipoxygenase-dependent pathways, respectively [11,20]. Moreover, the AA-metabolization generates oxygen free radicals [20].

AA or its metabolites could affect various cellular functions, such as, membrane currents ($Ca^{2+}$ and $K^{+}$) [12,19] and some signal transduction pathways ($Ca^{2+}$, nitric oxide, cAMP and $IP_3$) [4, 11,24].

The aim of the present work was to study the effects and mechanism of action of AA on astrocyte GJs.

## 2. Methods

Astrocytes were prepared from neocortex of 19 day old rat embryos. More than 95% of the cells showed GFAP positive immunoreactivity.

In confluent cell cultures the functional state of GJs was tested by the transfer to neighboring cells of Lucifer yellow microinjected into one cell. The number of coupled cells was scored by counting neighboring cells that accumulated the dye.

The levels and state of phosphorylation of Cx43 were evaluated by Western blot analysis using the alkaline phosphatase developing method and $^{32}$Pi-incorporation, as previously described [1].

The cellular distribution of Cx43 was studied by indirect immunofluorescence.

## 3. Results

After 60 min treatment with 5 or 30 μM AA cell coupling was reduced by 80 % and 70 %, respectively. The effect was spontaneously reversed 3 h after the AA-addition (Fig.1A).

To study the role of nitric oxide (NO) production and oxidative stress on the AA-induced astrocyte uncoupling, cells were pretreated for 30 min with 10 mM L-w-nitroarginine or 50 mM dimethylthiourea (DMTU), a NO synthase blocker and a free radical scavenger, respectively. It was found that these compounds did not affect the dye coupling inhibition induced by 5 μM AA .

The role of oxidative stress was also studied by pretreating the cells with melatonin, a potent free radical scavenger [17]. Previously, it has been shown that physiological doses of melatonin induced a late increase in the gap junctional comunication of mouse fibroblasts that migth be dependent on DNA synthesis [22]. Now we found that after 60 min treatment with 500 μM melatonin cell coupling was increased by 60% with respect to the control condition (Fig. 1B). This effect was spontaneously reversed 3 h after treatment. To study whether the melatonin-induced increase in dye coupling depended on the activation of melatonin membrane receptor, astrocyte cultures were incubated with 500 μM luzindole, a competitive melatonin receptor antagonist [6]. Luzindole was found to inhibit the increase in dye coupling induced by melatonin (Fig. 1B). On the other hand, 500 μM melatonin completely prevented the AA-induced astrocyte uncoupling but this effect was not prevented by luzindole (Fig. 1C).

To study if the AA-induced astrocyte uncoupling depended on the AA metabolization, cultures were incubated with indomethacin or nordihydroguaiaretic acid, cyclo and lipoxigenase blockers, respectively. It was found that, indomethacin (50 μM) prevented completely the AA-induced cell uncoupling (Fig. 1D). Moreover, nordihydroguaiaretic acid (5 μM), reduced the duration of the AA-induced effect predominantly by increasing the recovery rate (Fig. 1D).

After treatment with 5 μM AA no significant changes in levels of the unphosphorylated (NP) and phosphorylated (P1, P2 and P3) forms of Cx43 were detected by Western blots analysis (Fig. 2A). In addition, after 30 min treatment with 5 or 30 μM AA no significant changes in $^{32}$Pi-incorporation into Cx43 were detected (Fig. 2B).

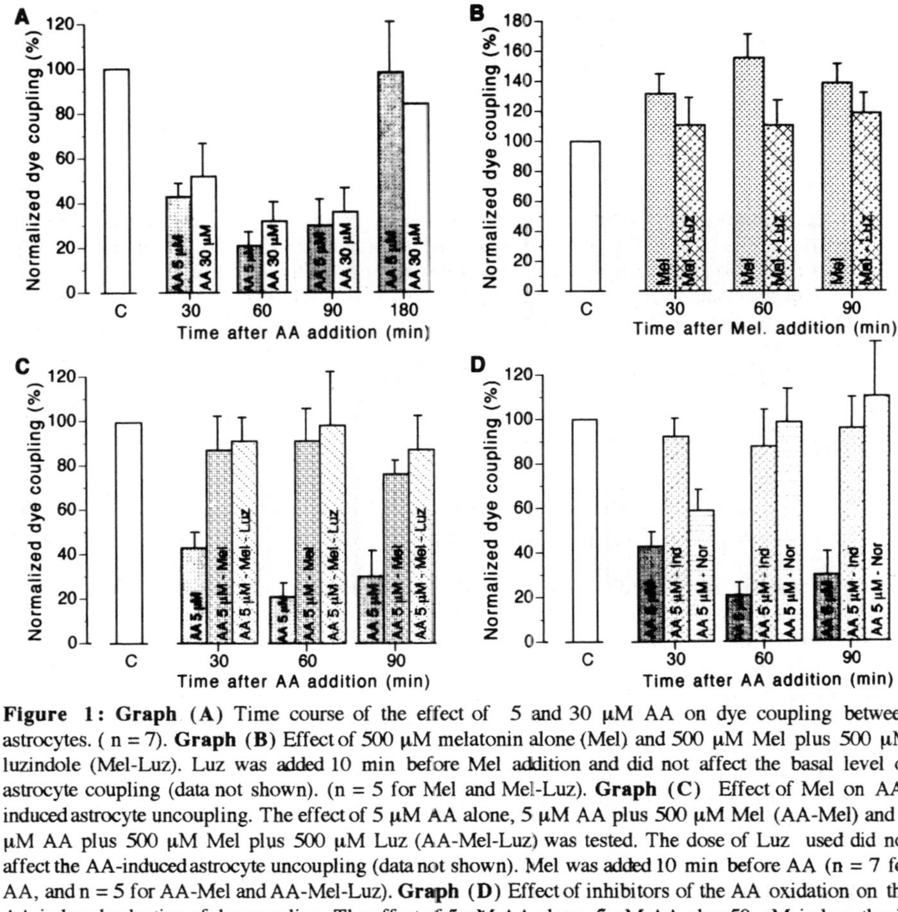


**Figure 1: Graph (A)** Time course of the effect of 5 and 30 µM AA on dye coupling between astrocytes. ( n = 7). **Graph (B)** Effect of 500 µM melatonin alone (Mel) and 500 µM Mel plus 500 µM luzindole (Mel-Luz). Luz was added 10 min before Mel addition and did not affect the basal level of astrocyte coupling (data not shown). (n = 5 for Mel and Mel-Luz). **Graph (C)** Effect of Mel on AA-induced astrocyte uncoupling. The effect of 5 µM AA alone, 5 µM AA plus 500 µM Mel (AA-Mel) and 5 µM AA plus 500 µM Mel plus 500 µM Luz (AA-Mel-Luz) was tested. The dose of Luz used did not affect the AA-induced astrocyte uncoupling (data not shown). Mel was added 10 min before AA (n = 7 for AA, and n = 5 for AA-Mel and AA-Mel-Luz). **Graph (D)** Effect of inhibitors of the AA oxidation on the AA-induced reduction of dye coupling. The effect of 5 µM AA alone, 5 µM AA plus 50 µM indomethacin (AA-Ind) and 5 µM AA plus 5 µM nordihydroguaiaretic acid (AA-Nor) was tested. Nor and Ind at the used concentration did not affect the basal level of coupling within 3 h after addition (data not shown). Each inhibitors was addedto the astrocyte culture 30 min before the addition of AA. (n = 7 for AA, and n = 3 for Nor or Ind plus AA). Dye coupling is expressed as normalized value with respect to the control condition (C). Each value represents X $\pm$ S.E.

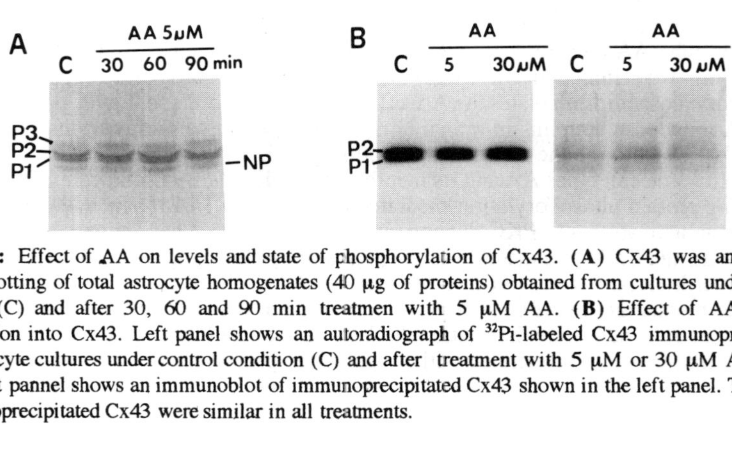


**Figure 2:** Effect of AA on levels and state of phosphorylation of Cx43. (**A**) Cx43 was analyzed by immunoblotting of total astrocyte homogenates (40 µg of proteins) obtained from cultures under control condition (C) and after 30, 60 and 90 min treatmen with 5 µM AA. (**B**) Effect of AA on $^{32}$Pi incorporation into Cx43. Left panel shows an autoradiograph of $^{32}$Pi-labeled Cx43 immunoprecipitated from astrocyte cultures under control condition (C) and after treatment with 5 µM or 30 µM AA for 30 min. Right pannel shows an immunoblot of immunoprecipitated Cx43 shown in the left panel. The levels of immunoprecipitated Cx43 were similar in all treatments.

After 30 min treatment with 5 μM AA the immunofluorescence revealed abundant Cx43-immunoreactivity in vesicle-like structures and a reduce Cx43 labeling at membrane appositions (Fig. 3B). A partial inhibition of the AA-induced cellular redistribution of Cx43 was observed in cells pretreated with 50 μM indomethacin (Fig. 3C).

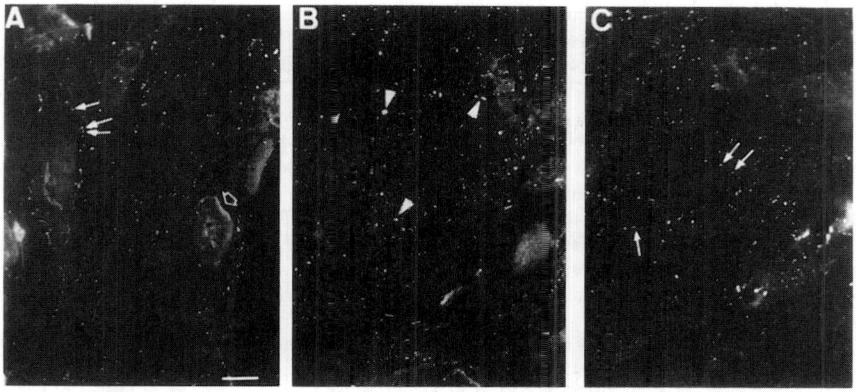


**Figure 3:** AA induces cellular redistribution of Cx43 which is prevented by indomethacin. Cultured astrocytes were fixed and permeabilized with ethanol, incubated with anti-Cx43 antibody and binding sites were detected with goat anti-rabbit FITC-conjugated antibody. (**A**) Cx43 immunoreactivity under control condition, (**B**) after 30 min treatment with 5 μM AA and, (**C**) after 30 min treatment with 5 μM AA in astrocyte cultures pretreated for 30 min with 50 μM indomethacin. In control condition, Cx43 was detected as puncta at cell appositions (arrows) and as faint and diffuse perinuclear staining (open arrow). After treatment with AA the diffuse perinuclear labeling remained unaffected, but labeling at membrane appositions was reduced. Moreover, scatter vesicle-like structures were detected (arrow heads). (Bar = 10 μm).

## 4. Discussion

Although NO synthesis has been associated to astrocyte uncoupling [2] and AA metabolites could activate NO synthesis [24], our results indicate that the AA-induced astrocyte uncoupling is not associated to NO production since it was not prevented by L-w-nitroarginine.

The protective effect of melatonin migth be explained by its free radical scavenger effect [17]. Nonetheless, the lack of protection seen after treatment with DMTU, ruled out this possibility. Alternatively, melatonin could act as cyclooxygenase blocker [8], preventing the AA metabolization. Consistently, it was found that AA-induced astrocyte uncoupling was inhibited by cyclo- and lipoxygenase blockers, suggesting that an unknown eicosanoid mediates the AA-effect. This is in agreement with previous reports. The AA-induced uncoupling in retinal horizontal cells and cardiomyocytes depends on its metabolization through the lipoxygenase pathway [15,16].

In some cell types AA and its metabolites could affect various cellular functions by activating protein phosphorylation mediated by PKC [13,14,21]. In addition, in various cell types activation of PKC-dependent pathways has been shown to induce cell uncoupling and Cx43 phosphorylation [1]; in astrocyte, activation of PKC also block GJs [7]. Neverthelless, our results indicate that the AA-induced astrocyte uncoupling is not mediated by changes in the state of phosphorylation of Cx43.

The AA-induced redistribution of Cx43 to intracellular compartments migth reflect a reduce insertion into or increase withdrawal of the protein from the plasma membrane.

## References

[1].- V.M. Berthoud et al, On the mechanism of cell uncoupling induced by a tumor promoter phorbol ester in clone 9 cells, a rat liver epithelial cell line, Eur. J. Cell Biol. (1993) 62: 384-396.

[2].- J.P. Bolaños and J.M. Medina Induction of nitric oxide synthase inhibits gap junction permeability in cultured rat astrocytes. J. Neurochem. (1996) 66: 2091-2099.

[3].- A.H. Cornell-Bell et al, Glutamate induces calcium waves in cultured astrocytes: long-range glial signalling, Science (1990) 247: 470-473.

[4].- D.S. Damron et al, Arachidonic acid and endothelin potentiate Ca2+ transients in rat cardiac myocytes via inhibition of distinct K+ channels, J. Biol. Chem. (1993) 268: 27335-273344.

[5].- R. Dermietzel et al, Gap junctions between cultured astrocytes: immunocytochemical, molecular, and electrophysiological analysis, J. Neurosci. (1991) 11: 1421.

[6].- M.L. Dubocovich, Melatonin receptors: are there multiple subtypes?, TIPS (1995) 16: 50-56.

[7].- M.O.K. Enkvist and K.D. McCarthy., Activation of protein kinase C blocks astroglial gap junction communication and inhibits the spread of calcium waves, J. Neurochem.(1992) 59: 519-526.

[8].- A.M. Franchi et al, Melatonin, 5-methoxytryptamine and some of their analogs as cyclooxigenase inhibitors in rat medial basal hypothalamus, Brain Res. (1987) 405: 384-388.

[9].- C. Giaume et al, Gap junctions in cultured astrocytes: single-channel currents and characterization of channel-forming protein. Neuron (1991) 6: 133-143.

[10].- C. Giaume et al, Arachidonic acid closes gap junction channels in rat lacrimal glands, Pflüegers Arch. (1989) 413: 273-279.

[11].- E.J. Goetzl et al, Specificity of expression and effects of eicosanoid mediators in normal physiology and human disease, FASEB J. (1995) 9: 1051-1058.

[12].- Y. Koyama et al, Evidence that prostaglandins activate calcium channels to enhance basal and stimulation-evoked catecholamine release from bovine adrenal chromaffin cells in culture, Biochem. Pharmacol. (1988) 37: 1725-1730.

[13].- D.S. Lester et al, Arachidonic acid and diacylglycerol act synergistically to activate protein kinase C in vitro and in vivo, Biochem. Biophys. Res. Comm. (1991) 179: 1522-1558.

[14].- Y. Luo and M. L. Vallano, Arachidonic acid, but not sodium nitroprusside, stimulates presynaptic protein kinase C and phosphorylation of GAP-43 in rat hippocampal slices and synaptosomes, J. Neurochem. (1995) 64: 1808-1818.

[15].- K.D. Massey et al, Arachidonic acid and lipoxigenase metabolites uncouple neonatal rat cardiac myocyte pairs, Am. J. Physiol. (1992) 263: C494-C501.

[16].- E. Miyachi et al, Arachidonic acid blocks gap junctions between retinal horizontal cells, NeuroReport (1994) 5: 485-488.

[17].- R.J. Reiter et al, Melatonin as free radical scavenger: implications for aging and age-related processes, Ann. NY Acad. Sci. (1994) 719: 1-12.

[18].- J.C. Sáez et al, Oxidant stress blocks hepatocyte gap junctions independently of changes in intracellular pH, Ca2+ or MP27 phosphorylation state, Biophys J. (1987) 51: 39a.

[19].- P. Schweitzer et al, Arachidonic acid metabolites as mediators of somatostatin-induced increase of neuronal M-current, Nature (1990) 346: 464-467.

[20].- N.A. Simonian and J.T. Coyle, Oxidative stress in neurodegenerative diseases, Annu. Rev. Pharmacol. Toxicol. (1996) 36: 83-106.

[21].- D.G. Tang et al, The lipoxygenase metabolite, 12(S)-HETE, induces a protein kinase C-dependent cytoskeletal rearrangement and retraction of microvascular endothelial cells, Exp. Cell Res. (1993) 207: 361-375.

[22].- A. Ubeda et al., A 50 Hz magnetic field blocks melatonin-induced enhancement of junctional transfer in normal C3H/10T1/2 cells, Carcinogenesis (1995) 16: 2945-2949.

[23].- L. Venance et al., Inhibition by anandamide of gap junctions and intercellular calcium signalling in striatal astrocytes, Nature (1995) 376: 590-594.

[24].- J.B. Warren et al, PGD2 is an intermediate in agonist-stimulated nitric oxide release in rabbit skin microcirculation, Am. J. Physiol. (1994) 266: H1846-H1853.

This study was supported by FONDECYT grants 2960001 (to A.D.M.) and 1960559 (to J.C.S.).

*Gap Junctions*
*R. Werner (Ed.)*
*IOS Press, 1998*



# Mitotic Cells Contain a Distinct Connexin43 Phosphoisoform

Paul D. Lampe[§], Wendy E. Kurata*, Bonnie J. Warn-Cramer*, and Alan F. Lau*

*Fred Hutchinson Cancer Research Center[§], Seattle, WA 98109*
*Cancer Research Center*, University of Hawaii at Manoa, Honolulu, HI 96813*

**Abstract** The gap junction protein connexin43 (Cx43) is a phosphoprotein that typically migrates as three bands (nonphosphorylated or NP, P1 and P2) during polyacrylamide gel electrophoresis. Mitotic cells, prepared by shakeoff of either nocodazole-treated or untreated cultures, contain Cx43 that has a distinctly reduced electrophoretic mobility (P3) that migrated slower than P2. Mitotic FT210 cells, which contain a temperature sensitive mutation in the $p34^{cdc2}$ kinase, showed abundant levels of the P3 Cx43 when maintained at the permissive temperature where $p34^{cdc2}$ is active. In contrast, nocodozole-treated FT210 cells grown at the nonpermissive temperature did not contain P3 Cx43. These results indicated that generation of the P3 Cx43 was dependent upon active $p34^{cdc2}$/cyclin B kinase. Mitotic cells exhibited a marked redistribution from cell-cell plasma membrane interfaces to multiple, distinctly-stained cytoplasmic structures. This event may be part of the dramatic structural changes observed in mitotic cells undergoing cell rounding and cytokinesis.

## Introduction

All of the connexin proteins, except Cx26, have been demonstrated to be phosphoproteins [1]. Cx43 appears to be phosphorylated at multiple serine residues in vivo [2-6]. Upon phosphorylation, the mobility of Cx43 in sodium dodecylsulfate (SDS)-polyacrylamide gels decreases and immunoprecipitation of Cx43 usually yields multiple bands that probably represent different phosphorylation states since treatment with alkaline phosphatase collapses the broad band into a sharp, faster migrating band [4, 7]. Some of these phosphorylation events appear to occur prior to the arrival of Cx43 at the plasma membrane [8, 9] and some may occur after its localization at the cell surface [10]. The protein kinases that are responsible for the phosphorylation of Cx43 in unstimulated cells are currently unidentified.

Among other processes, phosphorylation of gap junction proteins could regulate channel open time (gating), channel assembly, or connexin protein turnover. Many laboratories have shown that tumor promoting phorbol esters specifically activate protein kinase C (PKC) [11], increase Cx43 phosphorylation, and decrease gap junctional communication in a wide variety of cell types [12]. 12-*O*-tetradeconylphorbol-13-acetate (TPA) has been shown to affect parameters related to gap junction channel gating [13] and has been shown to block completely assembly of newly forming gap junction structures [14]. Both TPA-induced alteration of channel gating and assembly of gap junctions have been associated with increased phosphorylation on Cx43. Transfection of mammalian cells with a temperature-sensitive $pp60^{v\text{-}src}$ increased phosphorylation of Cx43 on tyrosine, decreased gap junctional communication, and induced the transformed cell phenotype when the cells were shifted from the non-permissive to the permissive temperature [15-18]. Cx43 appears to be a direct target of the v-Src tyrosine kinase which phosphorylates Y265 in Cx43 [19] and possibly additional sites [20]. Stimulation of epithelial cells by epidermal growth factor (EGF), which activates the EGF receptor tyrosine kinase, induced a transient disruption of gap junctional communication that correlated with a marked increase in serine phosphorylation of Cx43 followed by its subsequent dephosphorylation [21]. The EGF-induced phosphorylation of Cx43 appeared to occur at three sites (S255, S279, and S282) which may result from the activation of the mitogen-activated protein (MAP) kinase [22, 23].

Cyclin-dependent kinases play key roles at multiple "checkpoints" during progression of the cell cycle [24]. The entry of cells into mitosis is regulated by the $p34^{cdc2}$/cyclin B kinase complex. Multiple $p34^{cdc2}$/cyclin B kinase substrates have been identified including nuclear lamins, histone H1, cdc25, protein phosphatase, caldesmon, pp60$^{src}$, and c-Abl [25]. During mitosis, a number of dramatic structural changes occur in the cell including the breakdown of the nuclear envelope, chromatin condensation, formation of the mitotic spindle, cell rounding and cytokinesis [24]. Phosphorylation of the nuclear lamin proteins by activated p34$^{cdc2}$/cyclin B kinase results in the dissolution of the nuclear matrix and the eventual breakdown of the nuclear envelope. In addition, intercellular communication levels have been reported to decrease markedly [26].

**Mitotic cells contain a unique phosphoisoform of Cx43**

Nocodazole-treated Rat1 cell cultures contained numerous rounded-up mitotic cells which were greater than 90% viable by trypan blue exclusion. These rounded-up mitotic cells were easily collected by shaking the dish and triturating the growth medium over the culture five to ten times.

The SDS-PAGE mobility of Cx43 from untreated, nocodazole-treated, and the "shake-off" Rat1 cells was compared by immunoblotting (Figure 1A). As has been shown by many laboratories, untreated cells (CON) exhibited a nonphosphorylated (NP) band at approximately 40 kDa and multiple phosphorylated bands at 42 and 42.5 kDa (often labeled as P1 and P2). Untreated Rat1 cells (CON) exhibited primarily the NP and P1 forms of Cx43. When these cells were treated with nocodazole (NOC) a slower migrating Cx43 band appeared (labeled P3) which was the predominate Cx43 form detected in nocodazole-treated Rat1 cells. Remarkably, mitotic Rat1 cells prepared by the shake-off procedure (SO) contained primarily the P3 Cx43 form. Phosphorylation of Cx43 was the likely cause of the retardation in SDS-PAGE mobility because treatment of the shake-off cells with alkaline phosphatase (SO/AP) increased the Cx43 mobility to the NP form.

Phosphorylation of Cx43 to the P3 form was not a non-specific effect of the nocodazole treatment because the P3 Cx43 was also present in mitotic cells shaken off from untreated cultures (UT/SO, Figure 1A, additional bands are present due to the low numbers of mitotic cells in untreated cultures). Furthermore, generation of the P3 Cx43 was not merely the consequence of cell rounding because treatment of Rat1 cells with EDTA which caused cell rounding and release from the culture dish did not produce the P3 Cx43 (EDTA, Figure 1A). Therefore, phosphorylation of Cx43 to produce the P3 form appeared to be a unique and specific event in mitotic cells.

To directly demonstrate the phosphorylated state of the P3 Cx43, Rat1 c-src cells were metabolically-labeled with [$^{32}$P]-orthophosphate. Immunoprecipitation analysis showed that only the phosphorylated P3 Cx43 was present in the mitotic shakeoff cells and phosphoamino acid analysis demonstrated that the P3 Cx43 from mitotic cells and the P2 Cx43 from non-mitotic cells contained only phosphoserine (data not shown).

**Appearance of mitotic P3 Cx43 is dependent upon p34$^{cdc2}$ kinase activity**

Because p34$^{cdc2}$/cyclin B kinase plays a key role in the G2/M transition of the cell cycle, we investigated its potential role in the phosphorylation of Cx43 in mitotic cells. FT210 cells have a temperature-sensitive mutant p34$^{cdc2}$ kinase that is active at the permissive temperature of 32°C allowing cells to divide normally [27]. However, at the nonpermissive temperature of 39°C the p34$^{cdc2}$ kinase is inactive and the FT210 cells arrest late in G2. The parental FM3A cells which contain wild type p34$^{cdc2}$ kinase grew well at either temperature. Immunoblotting demonstrated that both cell lines contained Cx43. As with NRK, T51B and Rat1 cells, nocodazole treatment of the parental FM3A cells, grown at either 32°C or 39°C induced the accumulation of P3 Cx43 compared to the untreated control cells (data not shown). FT210 cells treated with nocodazole at 32°C (32°-N) where p34$^{cdc2}$ kinase is active also contained P3 Cx43 (Figure 1B). However, FT210 cells treated with nocodazole at 39°C (39°-N) where p34$^{cdc2}$ is inactive did not contain P3 Cx43 (Figure 1B). Treatment of FT210 cells with nocodazole for 12 hr at 32°C was sufficient to induce the appearance of the P3 Cx43 isoform (32° N-12h). However, if FT210 cells were first treated with nocodazole at 32°C for 12 hr then shifted to 39°C for 4 hr (32°/39°-N), nocodazole-induced phosphorylation of Cx43 was lost. These results are consistent with the report that p34$^{cdc2}$ kinase activity and the level of hyperphosphorylation of histone H1, a known p34$^{cdc2}$/cyclin B substrate, were dramatically decreased in FT210 cells within 4 hr of incubation at 39°C [28].

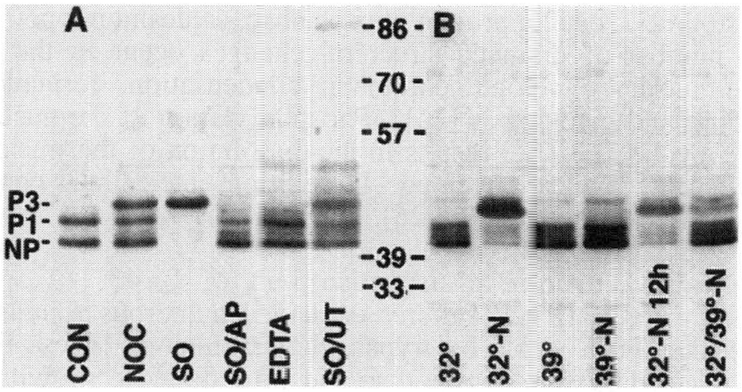


Figure 1. **A.** Mitotic cells contain a distinct phosphorylated form of Cx43. The immunoblot pattern of untreated control (CON) cells, 0.1 μg/ml nocodazole-treated (NOC) cells, cells shaken off after nocodazole treatment (SO), shakeoff cells treated with alkaline phosphatase (SO/AP), 5 mM EDTA-treated/released (EDTA) cells, and cells shaken off from untreated cultures (UT/SO). **B.** Phosphorylation of Cx43 in a p34$^{cdc2}$/cyclin B kinase temperature-sensitive cell line. Cx43 was detected in an immunoblot of the p34$^{cdc2}$/cyclin B kinase temperature-sensitive FT210 cell line incubated for 18 hr at 32°C (32°) and 39°C (39°) in the presence (N) or absence of 0.1 μg/ml nocodazole. In one case, nocodazole treatment of FT210 cells was terminated after only 12 hour (32°-N 12h), and in another, the nocodazole-treated cells were shifted from 32°C after 12 hr to 39°C for 4 hr (32°/39°-N).

## Cx43 is localized intracellularly in mitotic cells

The subcellular distribution of Cx43 in mitotic cells was determined by immunofluorescence microscopy of nocodazole treated Rat1 cells. Although it is not possible to focus simultaneously on the nonmitotic (flat) and mitotic (rounded) cells with conventional fluorescence microscopy at this magnification, the nonmitotic cells exhibited detectable Cx43 staining at cell-cell interfaces and less non-interface staining than nonmitotic nocodazole treated cells (Figure 2A). In marked contrast, when the focus was adjusted to view the rounded up mitotic cells, punctate Cx43 staining was primarily at non-interface intracellular locations (Figure 2B).

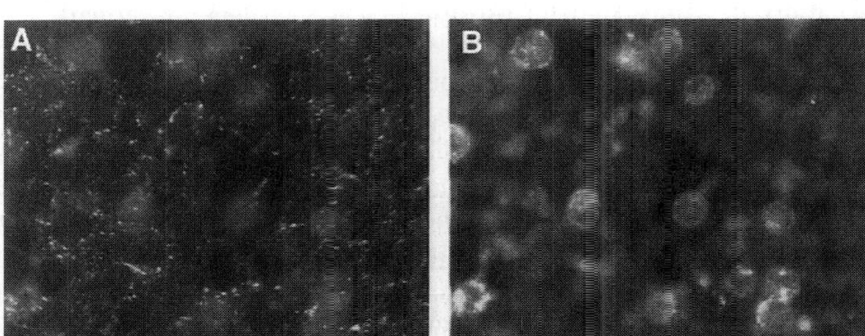


Figure 2. Intracellular distribution of Cx43 in Rat1 cells. Extensive gap junction immunostaining at cell-cell interfaces and little cytoplasmic staining was observed in the nocodazole-treated cells, nonmitotic cells (panel A). Viewing the rounded-up mitotic cells by focusing upward in the same field of view showed only cytoplasmic staining in the mitotic cells (panel B).

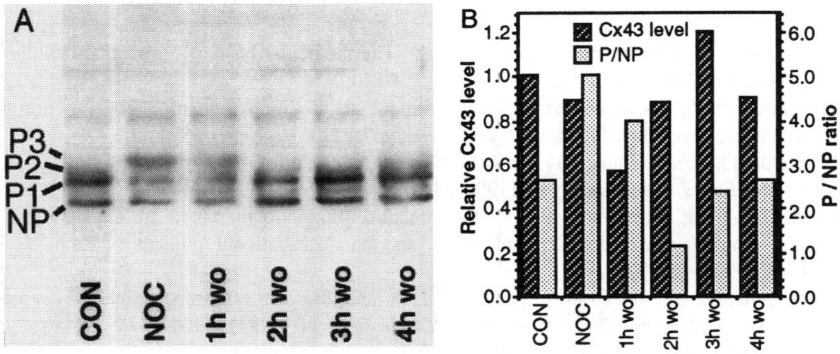


Figure 3. A. Immunoblot analysis of the Cx43 phosphoisoform content of Cx43 from control (CON), nocodazole-treated cultures (NOC), and nocodazole-treated cultures that remained in fresh drug-free media for 1-4 hr (1h wo-4h wo, equal cell numbers were loaded). B. Densitometric quantitation of the results in A.

## The fate of Cx43 in mitotic cells

To investigate the possible fate of Cx43, Rat1 cells were treated with nocodazole for 12 hr and then the medium was replaced with fresh medium lacking nocodazole (washout). As was previously shown in Figure 1A, the P3 form of Cx43 becomes the predominate species upon nocodazole treatment (Figure 3A, NOC). The level of P3 was reduced significantly after 1 hr of washout (1h wo) and was undetectable at 2-4 hr of washout. Densitometry of the different phosphospecies was determined to compare their abundance after nocodazole treatment and washout. Figure 3B shows that the level of total Cx43 protein relative to the control level was reduced slightly after nocodazole treatment and to less than 60% of the control level after 1 hr of washout. After 2-4 hr of washout the level of Cx43 returned to control values. These results indicated that the level of total cellular Cx43 may be reduced during mitosis and is consistent with the observation that the level of gap junctional communication drops dramatically during mitosis [26]. In addition, the ratio of the nonphosphorylated (NP) to phosphorylated (P) isoforms of Cx43 was maximal after 2 hr of washout. The most likely explanation for these results is that primarily the phosphorylated species were lost by dephosphorylation and/or degradation after 1 h of washout.

## Conclusions

In this study, we have shown that mitotic cells contain Cx43 which is uniquely and specifically phosphorylated in a $p34^{cdc2}$/cyclin B kinase-dependent manner that correlates with the loss of gap junctions from cell-cell interfaces. The SDS-PAGE mobility of Cx43 was reduced in a phosphorylation-dependent manner when mitotic cells were prepared by different procedures. P3 Cx43 was not evident in cells released from the culture dish by EDTA treatment. The FT210 cell line which contains a temperature-sensitive $p34^{cdc2}$/cyclin B kinase showed nocodazole-dependent phosphorylation of Cx43 to the P3 isoform only at the permissive temperature where the kinase is active. Immunofluorescence studies showed that the changes in phosphorylation were correlated with the redistribution of Cx43 from cell-cell interfaces to cytoplasmic locations in mitotic cells. Release of cells from a nocodazole block led to a dramatic reduction in the P3 form and a significant reduction in the total level of Cx43.

The $p34^{cdc2}$/cyclin B kinase is the main regulatory element that controls the transition of cells from $G_2$ to M phase of the cell cycle and, therefore, it was a logical candidate as one of the protein kinases that induced the phosphorylation of Cx43 during mitosis. Our evidence is consistent with the idea that the $p34^{cdc2}$/cyclin B kinase is not only necessary for the appearance of the P3 form of Cx43, but that it may phosphorylate Cx43 directly in mitotic cells.

## Acknowledgments

This research was supported in part by grants GM55632 (P.D.L.) and CA52098 (A.F.L.) from the National Institutes of Health and grant HIFW-09-95 from the American Heart Association, Hawaii Affiliate (B.J.W.C.).

## References


1. R. Bruzzone, *et al.*, Connections with connexins: The molecular basis of direct intercellular signaling, *Eur. J. Biochem.* **238** (1996), 1-27.
2. D.W. Laird, *et al.*, Turnover and phosphorylation dynamics of connexin43 gap junction protein in cultured cardiac myocytes, *Biochem. J.* **273** (1991), 67-72.
3. V.M. Berthoud, *et al.*, Connexin43 in MDCK cells: Regulation by a tumor-promoting phorbol ester and calcium, *Eur. J. Cell Biology* **57** (1992), 40-50.
4. L.S. Musil, *et al.*, Expression of the gap junction protein connexin43 in embryonic chick lens: Molecular cloning, ultrastructural localization, and post-translational phosphorylation, *J. Membrane Biol.* **116** (1990), 163-175.
5. J.L. Brissette, *et al.*, The tumor promoter TPA and the *ras* oncogene modulate expression and phosphorylation of gap junction proteins, *Mol. Cell. Biol.* **11** (1991), 5364-5371.
6. R. Kadle, *et al.*, Tissue-specific distribution of differentially phosphorylated forms of Cx43, *Mol. Cell. Biol.* **11** (1991), 363-369.
7. D.S. Crow, *et al.*, Phosphorylation of connexin43 gap junction protein in uninfected and Rous sarcoma virus-transformed mammalian fibroblasts, *Mol. Cell. Biol.* **10** (1990), 1754-1763.
8. L.S. Musil, *et al.*, Differential phosphorylation of gap junction protein connexin43 in junctional communication-competent and –deficient cell lines, *J. Cell Biol.* **111** (1990), 2077-2088.
9. D.W. Laird, *et al.*, Identification of intermediate forms of connexin43 in rat cardiac myocytes, *Prog. Cell Res.* **3** (1993), 263-268.
10. L.S. Musil, *et al.*, Biochemical analysis of connexin43 intracellular transport, phosphorylation and assembly into gap junctional plaques, *J. Cell Biol.* **115** (1991), 1357-1374.
11. Y. Nishizuka, Studies and perspectives of protein kinase C, *Science* **233** (1986), 305-312.
12. A.W. Murray, *et al.*, Regulation of gap junctional communication by protein kinases, in *Cell Interactions and Gap Junctions*, N. Sperelakis and W.C. Cole, Editor. 1989, CRC Press, Inc.: Boca Raton. p. 97-106.
13. A.P. Moreno, *et al.*, Human connexin43 gap junction channels. Regulation of unitary conductances by phosphorylation, *Circ. Res.* **74** (1994), 1050-1057.
14. P.D. Lampe, Analyzing phorbol ester effects on gap junction communication: A dramatic inhibition of assembly, *J. Cell Biol.* **127** (1994), 1895-1905.
15. M.M. Atkinson, *et al.*, Rapid and reversible reduction of junctional permeability in cells infected with a temperature-sensitive mutant of avian sarcoma virus., *J. Cell Biol.* **91** (1981), 573-578.
16. R. Azarnia, *et al.*, Intercellular communication and control of cell growth. X. Alteration of junctional permeability by the *src* gene, *J. Membr. Biol.* **82** (1984), 191-205.
17. A.J. Filson, *et al.*, Tyrosine phosphorylation of a gap junction protein correlates with inhibition of cell-to-cell communication, *Cell Growth Diff.* 1 (1990), 661-668.
18. D.S. Crow, *et al.*, Phosphorylation of connexin43 in cells containing mutant *src* oncogenes, *Oncogene* **7** (1992), 999-1003.
19. K.I. Swenson, *et al.*, Tyrosine phosphorylation of the gap junction protein connexin43 is required for pp60$^{src}$-induced inhibition of communication, *Cell Regul.* **1** (1990), 989-1002.
20. L.W.M. Loo, *et al.*, pp60$^{src}$-mediated phosphorylation of connexin 43, a gap junction protein, *J. Biol. Chem.* **270** (1995), 12751-12761.
21. A.F. Lau, *et al.*, Epidermal growth factor disrupts gap-junctional communication and induces phosphorylation of connexin43 on serine, *Mol. Biol. Cell* **3** (1992), 865-874.
22. M.Y. Kanemitsu, *et al.*, Epidermal growth factor stimulates the disruption of gap junctional communication and connexin43 phosphorylation independent of TPA-sensitive protein kinase C: The possible involvement of mitogen-activated protein kinase, *Mol. Biol. Cell* **4** (1993), 837-848.
23. B.J. Warn-Cramer, *et al.*, Characterization of the MAP kinase phosphorylation sites on the connexin43 gap junction protein, *J. Biol. Chem.* **271** (1996), 3779-3786.
24. X. Grana, *et al.*, Cell cycle control in mammalian cells: role of cyclins, cyclin dependent kinases, growth suppressor genes and cyclin-dependent kinase inhibitors, *Oncogene* **11** (1995), 211-219.
25. J. Pines, Cyclins and cyclin-dependent kinases: a biochemical view, *Biochem. J.* **308** (1995), 697-711.
26. L.S. Stein, *et al.*, Reduced cell-cell communication between mitotic and nonmitotic cells, *Exper. Cell Res.* **198** (1992), 1-7.
27. J.P.H. Th'ng, *et al.*, The FT210 cell line is a mouse G2 phase mutant with a temperature-sensitive *cdc2* gene product, *Cell* **63** (1990), 313-324.
28. L.L. Hall, *et al.*, A brief staurosporine treatment of mitotic cells triggers premature exit from mitosis and polyploid cell formation, *Cancer Res.* **56** (1996), 3551-3559.

*Gap Junctions*
*R. Werner (Ed.)*
*IOS Press, 1998*

# Modulation of human umbilical vein endothelial gap junctions by phosphorylating treatments

Harold V.M. van Rijen, Sjoerd Postma & Habo J. Jongsma
*Department of Medical Physiology and Sports Medicine, Faculty of Medicine, Utrecht University, Universiteitsweg 100, 3584 CG Utrecht, The Netherlands*

**Abstract.** We investigated the effect of activation of protein kinase A (PKA), G (PKG), and C (PKC) on the electrophysiological characteristics and dye-permeability of gap junction channels in human umbilical vein endothelial cells (HUVEC). We have previously determined by immunohistochemistry that gap junctions in HUVEC are composed of three different gap junction proteins (connexins, Cx), i.e., Cx37, Cx40, and Cx43. Using heptanol to resolve single gap junction channel events, conductances ranging from 10 to 100 pS were measured under control conditions. Activation of PKC, PKG, or PKA increased the frequency of events around 20 pS, most likely representing the smallest state of Cx43. Interestingly, PKA and PKG also induced a large increase in events of 60-160 pS, probably due to upregulation of Cx40 and/or Cx37 channels. Lucifer Yellow dye coupling assays showed that PKC activation decreased dye transfer by 20% relative to control. PKA activation decreased dye transfer by 8%.These findings suggest a differential modulation of Cx43 and Cx40 by second messenger metabolites.

## 1. Introduction

The entire vascular system is lined by endothelial cells. They form the selective barrier between the vessel lumen and the rest of the body, and have to perform multiple tasks, many of which are regulated via second messenger systems, which activate protein kinases (for a review of vascular physiology see, e.g., [1]). In non-excitable tissue, like vascular endothelium, gap junction channels allow metabolic cooperation between cells [2] and are involved in propagation of vasomotor signals along the vessel wall [3,4]. In a previous immunocytochemical study, we have shown that gap junctions between human umbilical vein endothelial cells (HUVEC) contain Cx43, Cx40 and some Cx37 [5]. Cx43, Cx40, and Cx37 all contain consensus sequence motifs for protein kinase C (PKC), cAMP dependent protein kinase (PKA), and cGMP-dependent protein kinase (PKG) [6-9] . Electrical conductance and permeability of gap junctions formed of Cx43 have been shown to be sensitive to phosphorylation or dephosphorylation (see [10]). Cx40 was shown to be a phosphoprotein [11], and can be phosphorylated by cAMP [12]. No data on phosphorylation of Cx37 are available as yet. Using dual voltage clamp and dye-coupling techniques, we have investigated whether electrical properties and permeability of HUVEC gap junctions, composed of three Cx's, are modulated by phosphorylation.

## 2. Materials & Methods

*Isolation of endothelial cells:* HUVEC were isolated and cultured as described previously [5] Passages 2-4 of HUVEC cultures from multiple seperate isolations were used for both dye coupling and electrophysiological experiments and were equally distributed in the different treatment groups.
*Phosphorylating treatments:* In order to activate PKC, PKA, or PKG, the cells were incubated for 10 min with different drugs (all purchased from Sigma) before the actual experiments were started. To activate protein kinase C, 100 nM 12-*O*-tetradecanoylphorbol-13-acetate (TPA) was added to the bathing medium (stock concentration 10 μM in

dimethylsulfoxide, DMSO). As an control, cells were treated with 100 nM of the inactive phorbol ester 4α-phorbol 12,13-didecanoate (α-PDD; stock concentration 10 μM in dimethylsulfoxide, DMSO). Protein kinase A and G were activated by addition of 1mM 8-bromo-adenosine 3'-5'-cyclic monophosphate (8-Br-cAMP) or 1mM 8-bromo-guanosine 3'-5'-cyclic monophosphate (8-Br-cGMP) to the bathing medium, respectively.

*Dye coupling experiments:* Dye coupling experiments were performed as described previously [5].After insertion of the electrode in a confluent HUVEC monolayer the dye was allowed to diffuse out of the pipette into the impaled cell and adjacent cells for 2 min. The electrode was then retracted and the number of stained cells (including the impaled one) was counted.

*Electrophysiological measurements:* Single gap junctional channel conductances between HUVEC were measured as described previously [5]. Transjunctional voltage was always set to 50 mV. Composition of the internal pipette solution (in mM): 130 K-gluconate, 10 KCl, 10 $MgCl_2$, 0.6 $CaCl_2$, 10 EGTA, 5 $Na_2ATP$, 10 HEPES, pH 7.2 (KOH).

## 3. Results

*Modulation of single channel distribution:* Fig. 1 shows the collective amplitude histograms of several experiments for each treatment. Under control conditions, the single channel conductances ranged from 10 to 100 pS (Fig. 1A). Only very few events between 60 an 100 pS were observed under control conditions. Under α-PDD conditions, no differences were observed as compared to control conditions (Fig. 1B). Under TPA conditions however, the amplitude histogram changed to a two peak histogram (Fig. 1C). Gaussian curve fitting resulted in two mean conductances of 18.1 and 42.5 pS (Fig. 1F). Under TPA conditions, a marked relative increase in the number of '20 pS' events is found, almost equal in frequency to the 40 pS events.

Exposure to cAMP had a large effect on the single channel distribution (Fig. 1D). Besides a relative increase in events between 20 and 30 pS, we also found a large increase in events between 60 and 160 pS, especially around 65 and 80 pS. In this range hardly any events were observed under control conditions. Incubation with cGMP had an effect which is similar to that observed under cAMP conditions (Fig. 1E): A relative increase of events between 20 and 30 pS, and of events between 60 and 150 pS, that were scarcely observed under control conditions (Fig. 1J). Again, as under cAMP conditions, a new population of channel conductances around 60-65 pS is apparent.

*Dye-coupling:* We have used a dye-injection protocol that provides for an internal control by performing 10 injections of 2 min before, and 10 injections of 2 min after a 10 min incubation with one of the control or phosphorylating agents. The 'dye-ratio' was calculated by dividing the number of stained cells after incubation, by the number of stained cells before incubation. For each treatment 10 of such experiments were performed. The results are shown in Fig. 2. Ideally, the mean dye-ratio under control condition would equal 1 (dashed line in Fig. 2). Under control conditions, however, there is some decrease in dye coupling in time, resulting in a mean dye-ratio of 0.95±0.09 (mean±S.D.). The dye-ratio under α-PDD and cGMP conditions were very close to, and did not significantly differ from the dye-ratio under control conditions. An ANOVA-test showed that there were highly significant differences between the treatments (P<0.0001): Exposure to TPA led to a decrease of the dye-ratio with 20% as compared to control (Student's *t*-test, P<0.01). Incubation with cAMP decreased the dye-ratio by 8% (Student's *t*-test, P<0.01) as compared to control.

## 4. Discussion

*Modulation of single channel conductance distribution:* In HUVEC, where Cx37, Cx40, and Cx43 are coexpressed, it is hard to assign, without additional data, a conductance value to a channel composed of a specific connexin. When HUVEC are cultured in normal medium, from which FCS is omitted, they dedifferentiate within two days, and at this stage, only Cx43 can be detected immunohistochemically (data not shown). The results of single channel experiments on such cells are shown in Fig. 3A.

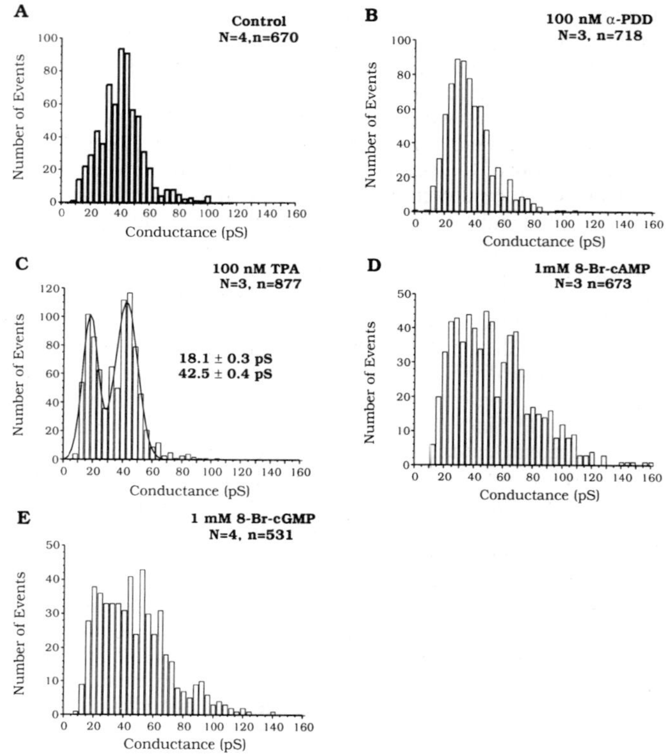


*Figure 1:* Single gap junction channel conductance distribution in HUVEC under various phosphorylating conditions. The transjunctional voltage was 50 mV in all experiments. Solid curve in panel C is the two gaussian fit giving mean single channel conductances of 18.1±0.3 and 42.5±0.4 pS (±S.E.) under TPA conditions. A: Control. B: 100 nM α-PDD. C: 100 nM TPA. D: 1 mM 8-Br-cAMP. E: 1mM 8-Br-cGMP.

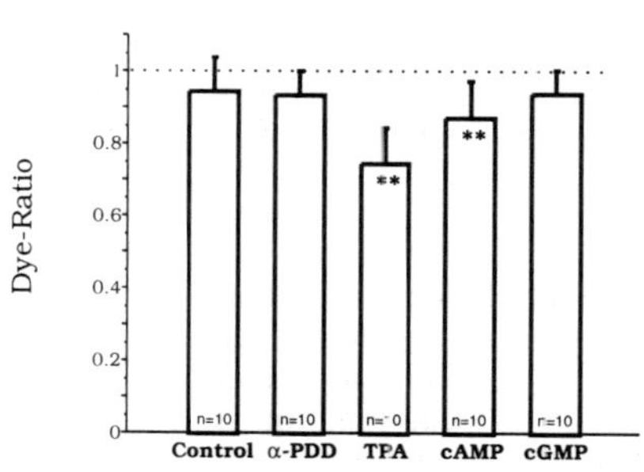


Figure 2: Permeability of HUVEC gap junctions for Lucifer Yellow CH under various conditions Error bars indicate S.D. Asteriks: significantly different from control (P<0.01).

Only two conductances of about 16 and 37 pS are now present (Fig. 3A), which compare well to those reported for Cx43 in the literature [13-16]. Single channel measurements on SKHep1 cells, transfected with human Cx40, under identical experimental conditions as used on HUVEC show two conductances of 82 and 117 pS (Fig. 3B). These values also fit very well to published values for Cx40 gap junction channel conductance, when the effect of the pipette solution is taken into account [12,17-19]. The single channel conductance of 30 pS in Fig. 3B is of the endogeneous Cx45 gap junction channel of SKHep1 cells [20-22].

Most channel conductances measured in HUVEC under control conditions are in the range of Cx43 with few in the range of Cx40, whilst no evident channel conductances of Cx37, which were reported to be 63 and 297 pS [23], are measured. When exposed to TPA, an increase in frequency of an 18 pS conductance, most likely being the smallest state of Cx43, was observed as has been described earlier for Cx43 in neonatal rat heart [24], and SKHep1 cells transfected with rat Cx43 [22] or human Cx43 [25].

Incubation with cAMP resulted in an increase of 20-30 pS events, revealed a channel population between 60 and 65 pS, which was not present under control conditions and increased the frequency of single channel conductances of 80 pS and larger. It has been shown that exposure to cAMP increased the phosphorylation of Cx40, but not of Cx43 [26]. Exposure to cAMP did not affect the single channel conductance distribution of Cx43 gap junctions [22,27]. From Fig. 3B it can be deduced that the upregulation of the conductances of 80 pS and larger most likely reflect channels formed of Cx40. The population of conductances between 60 and 65 pS might reflect channels formed of Cx37 [23].

Incubation of HUVEC with cGMP changes the single channel distribution similar to the change by cAMP: An increase of events between 20 and 30 pS and between 60 and 140 pS. It was previously reported that the single channel distribution of human Cx43 is not affected by exposure to cGMP [28], although multiple consensus sites for protein kinase G are present [6,7]. When cells expressing rat Cx43 were exposed to cGMP, the relative frequency of '20 pS' events was found to be increased [22,28]. As after treatment with cAMP, the increase in frequency of events around 65, and between 80 and 120 pS by cGMP treatment can be attributed to modulation of Cx37 and Cx40, respectively.

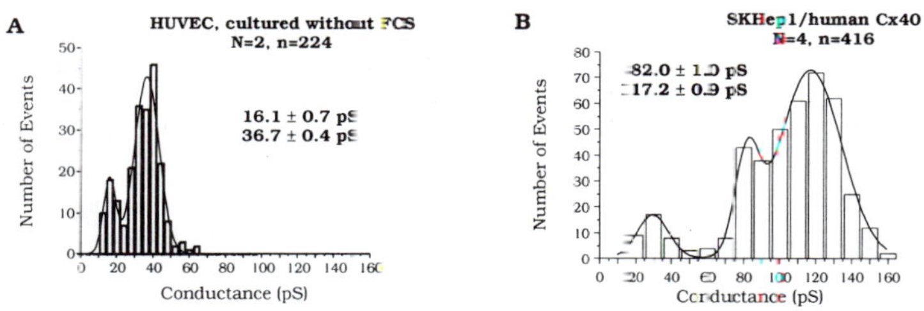


*Figure 3:* A: Amplitude histogram from single channel events between HUVEC, cultured for 2 days in normal culturing medium, from which FCS was omitted. B: Amplitude histogram of experiments on SKHep1 cells transfected with human Cx40.

*Modulation of dye-permeability:* In HUVEC, treatment with TPA or cAMP significantly decreased the dye permeability. TPA is known to strongly decrease dye permeability of Cx43 gap junction channels [22,26,27]. The inhibitory effect of TPA is not so large in HUVEC however. The permeability of rat Cx43 gap junctions is not altered by exposure to cAMP [22,27]. In HUVEC however, a small decrease (8%) was found in dye permeability. This suggests modulation of Cx40 of Cx37, thereby reducing the dye-permeability. No effect on dye permeability was found after treatment with cGMP, although it was reported that rat Cx43 gap jucntion channels become less permeant to the dyes 6-carboxyfluorescein [27] and LYCH [22] after exposure to cGMP. The absence of an effect of cGMP might be caused by the increased gating frequency of Cx40 and Cx37 channels, by which the inhibitory effect on Cx43 by cGMP is counteracted, giving no change in dye permeability as a net result.

## References


[1] A.C. Shore, Vascular biology and physiology. In: J. E. TookeandG. D. Lowe (ed.), A textbook of vascular medicine. Arnold, London, 1996, pp. 7-42.

[2] W.R. Loewenstein, Junctional intercellular communication: The cell-to-cell membrane channel. *Physiol Rev* **61** (1981) 829-913.

[3] S.S. Segal and B.R. Duling, Flow control among microvessels coordinated by intercellular conduction, *Science* **234** (1986) 868-870.

[4] S.S. Segal and B.R. Duling, Conduction of vasomotor responses in arterioles: a role for cell-to-cell coupling, *Am J Physiol* **256** (1989) H838-H845.

[5] H.V.M. Van Rijen *et al.*, Gap junctions in human umbilical cord endothelial cells contain multiple connexins, *Am J Physiol* **272** (1997) C117-C130.

[6] B.E. Kemp and R.B. Pearson, Protein kinase recognition sequence motifs, *Trends Biochem Sci* **15** (1990) 342-346.

[7] G.I. Fishman *et al.*, Molecular characterization and functional expression of the human cardiac gap junction channel, *J Cell Biol* **111** (1990) 589-598.

[8] P.J. Kennely and E.G. Krebs, Consensus sequences as substrate specificity determinants for protein kinases and protein phosphatases, *J Biol Chem* **266** (1991) 15555-15558.

[9] K.E. Reed *et al.*, Molecular cloning and functional expression of human connexin37, an endothelial cell gap junction protein, *J Clin Invest* **91** (1993) 997-1004.

[10] D. Gros and H.J. Jongsma, Connexins in mammalian heart function, *BioEssays* **18** (1996) 719-730.

[11] B. Delorme *et al.*, Developmental regulation of Connexin40 gene expression in mouse heart correlates with the differentiation of the conduction system, *Develop Dynam* **204** (1995) 358-371.

[12] O. Traub *et al.*, Immunochemical and electrophysiological characterization of murine connexin40 and -43 in mouse tissues and transfected human cells, *Eur J Cell Biol* **64** (1994) 101-112.

[13] M.B. Rook *et al.*, Properties of single gap junctional channels between isolated neonatal rat heart cells, *Am J Physiol* **255** (1988) H770-H782.

[14] M.B. Rook *et al.*, Differences in gap junction channels between cardiac myocytes, fibroblasts, and heterologous pairs, *Am J Physiol* **263** (1992) C959-C977.

[15] B.R. Takens-Kwak and H.J. Jongsma, Cardiac gap junctions: Three distinct single channel conductances and their modulation by phosphorylating treatments, *Pflügers Arch* **422** (1992) 198-200.

[16] A.P. Moreno *et al.*, Human connexin43 gap junction channels, *Circ Res* **74** (1994) 1050-1057.

[17] D.A. Beblo *et al.*, Unique conductance, gating, and selective permeabilities properties of gap junction channels formed by Connexin40, *Circ Res* **77** (1995) 813-822.

[18] F.F. Bukauskas *et al.*, Biophysical properties of gap junction channels formed by mouse Connexin40 in induced pairs of transfected human HeLa cells, *Biophys J* **68** (1995) 2289-2298.

[19] P. Hellmann *et al.*, Properties of Connexin40 gap junction channels endogeneously expressed and exogeneously overexpressed in human choriocarcinoma cell lines, *Pflügers Arch* **432** (1996) 501-509.

[20] J.G. Laing *et al.*, Characterization of the gap junction protein, Connexin45, *J Membrane Biol* **139** (1994) 31-40.

[21] A.P. Moreno *et al.*, Properties of gap junction channels formed of connexin45 endogenously expressed in human hepatoma (SKHep1) cells, *Am J Physiol* **268** (1995) C356-C365.

[22] B.R. Kwak *et al.*, Differential regulation of distinct types of gap junction channels by similar phosphorylating conditions, *Mol Biol Cell* **6** (1995) 1707-1719.

[23] R.D. Veenstra *et al.*, Connexin37 forms high conductance gap junction channels with subconductance state activity and selective dye and ionic permeabilities, *Biophys J* **66** (1994) 1915-1928.

[24] B.R. Kwak *et al.*, TPA increases conductance but decreases permeability in neonatal rat cardiomyocyte gap junction channels, *Exp Cell Res* **220** (1995) 456-463.

[25] A.P. Moreno *et al.*, Phosphorylation shifts unitary conductance and modifies voltage dependent kinetics of human connexin43 gap junction channels, *Biophys J* **62** (1992) 51-53.

[26] V.M. Berthoud *et al.*, Connexin43 in MDCK cells: regulation by a tumor-promoting phorbol ester and $Ca^{2+}$, *Eur J Cell Biol* **57** (1992) 40-50.

[27] B.R. Kwak and H.J. Jongsma, Regulation of cardiac gap junction channel permeability and conductance by several phosphorylating conditions, *Mol Cell Biochem* **157** (1996) 93-99.

[28] B.R. Kwak *et al.*, Effects of cGMP-dependent phosphorylation on rat and human Connexin43 gap junction channels, *Pflügers Arch* **430** (1995) 770-778.

# Section 8:

# Intercellular Communication by Calcium Signalling

*Gap Junctions*
*R. Werner (Ed.)*
*IOS Press, 1998*



# Intercellular $Ca^{2+}$ Signaling Between Glial and Endothelial Cells

**Michael J. Sanderson, Koen Paemeleire*, Andreja Strahonja and Luc Leybaert***
*Department of Physiology, University of Massachusetts Medical Center, Worcester, MA, USA and *Laboratory of Physiology, University of Gent, Gent, Belgium*

**Abstract.** To explore the hypothesis that intercellular $Ca^{2+}$ waves are a mechanism of heterotypic cell communication and convey information across biological interfaces, such as the blood-brain barrier, intercellular $Ca^{2+}$ signaling was investigated in co-cultures of glial and endothelial cells. Mechanically-initiated, intercellular $Ca^{2+}$ waves were found to propagate from glial to endothelial cells or vice versa. To determine if glial-endothelial $Ca^{2+}$ waves propagate by the intercellular diffusion of inositol trisphosphate ($IP_3$) through gap junctions, the multicellular $Ca^{2+}$ responses to increases in intracellular concentrations of $IP_3$ and $Ca^{2+}$, generated by the photolysis of caged compounds, were examined. Caged-$IP_3$ was microinjected into a single cell but was free to diffuse to adjacent cells. Photolytic release of $IP_3$ in the injected cell or in a distal, non-injected cell, initiated an intercellular $Ca^{2+}$ wave. By contrast, photolysis of caged-$Ca^{2+}$ induced an increase in $[Ca^{2+}]_i$ in the illuminated cell but did not initiate an intercellular $Ca^{2+}$ wave. These results indicate that 1) intercellular $Ca^{2+}$ waves propagate over short distances by the intercellular diffusion of $IP_3$ via gap junctions, 2) $Ca^{2+}$ itself does not serve as an intercellular messenger in these cells and 3) $Ca^{2+}$ waves may contribute to the regulation of blood-brain barrier function.

## 1. Introduction: Intercellular $Ca^{2+}$ Waves

Intercellular $Ca^{2+}$ waves occur in a diversity of tissues and consist of increases in intracellular $Ca^{2+}$ concentration ($[Ca^{2+}]_i$) that propagate between adjacent cells[1, 2]. $Ca^{2+}$ waves also propagate between different cell types, e.g. between glia and neurons [3-5] or between airway epithelial and smooth muscle cells [6]. Thus, intercellular $Ca^{2+}$ waves appear to be an important mechanism of heterotypic cell communication and may carry information across biological interfaces that are composed of different cell types such as the blood-brain barrier. To explore this hypothesis, intercellular $Ca^{2+}$ signaling was investigated in co-cultures of glial and endothelial cells. In co-culture, endothelial cells often were surrounded by glial cells and together formed an easily recognized cell interface or boundary. Mechanical stimulation of a single glial cell near the cell boundary induced an intercellular $Ca^{2+}$ wave that propagated from the glial cells, across the boundary, and through neighboring endothelial cells (Fig. 1). A prolonged delay in wave propagation was observed at the cell boundary. Mechanical stimulation of an endothelial cell induced the reverse response.

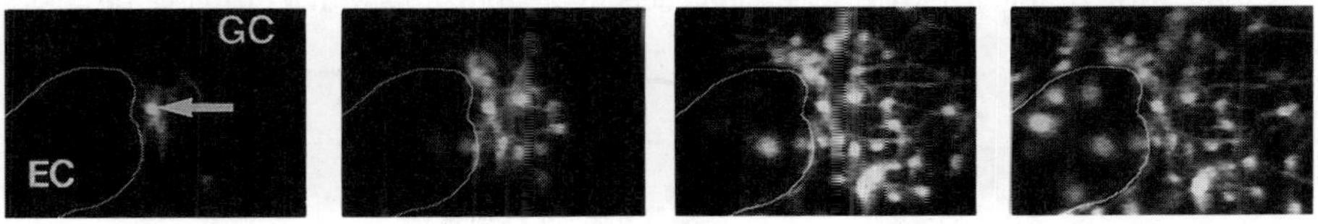


**0.6 sec** **2.4 sec** **6 sec** **12 sec**

**Figure 1:** An intercellular $Ca^{2+}$ wave evoked by mechanical stimulation (arrow) of a single glial cell (GC) propagates to adjacent glial cells, crosses the cellular interface (white line) and propagates through neighboring endothelial cells (EC). Images presented represent the changes in $[Ca^{2+}]_i$ as measured with fluo-3.

### *a) The passive diffusion hypothesis of intercellular $Ca^{2+}$ wave propagation*

The mechanism of mechanically-induced intercellular $Ca^{2+}$ waves has been investigated extensively in epithelial and glial cells [7, 8] and these studies have led to the "passive diffusion" hypothesis [2]. In brief, this hypothesis states that $Ca^{2+}$ waves propagate from cell to cell by the diffusion of inositol trisphosphate ($IP_3$) through gap junctions. As $IP_3$ enters each cell it releases $Ca^{2+}$ from intracellular stores in the endoplasmic reticulum via the $IP_3$ receptor. Because this process is amplified by $Ca^{2+}$-induced $Ca^{2+}$ release (CICR), an $IP_3$ concentration gradient can stimulate a near maximal $Ca^{2+}$ response in each cell as long as the $[IP_3]_i$ exceeds the activation threshold of the $IP_3$ receptor (Fig. 2).

The reliance of wave propagation on a diffusive rather than a regenerative process was predicted by the limited propagation distances (5-15 cells in any direction) of mechanically-stimulated $Ca^{2+}$ waves. A role for gap junctions was implicated by the circuitous path and propagation delays at cell borders of the $Ca^{2+}$ wave [7, 9] and was confirmed by the inability of C6 glioma cells that lack connexin 43 gap junctions to propagate $Ca^{2+}$ waves. C6 glioma cells transfected to express connexin 43 displayed dye-coupling and propagated $Ca^{2+}$ waves [10]. Gap junction blockers inhibited wave propagation [6, 7, 11] while agonists that influenced gap junction conductivity altered the extent of wave propagation [12].

The idea that $IP_3$ was the diffusing messenger was indicated by the facts that $Ca^{2+}$ waves a) are propagated by the release of intracellular $Ca^{2+}$ stores, b) are initiated by iontophoresis of $IP_3$ into a cell [7, 8] and c) are inhibited by heparin, an antagonist of the $IP_3$ receptor [13]. The diffusion hypothesis requires that $IP_3$ is produced only in the stimulated cell and the inhibition of $Ca^{2+}$ waves by U73122, a phospholipase C (PLC) antagonist [14], suggests that mechanical stimulation may activate PLC [7, 15]. Although $Ca^{2+}$ is a potential intercellular messenger and can pass though gap junctions [16], $Ca^{2+}$ does not appear to play a major role as an intercellular messenger. $Ca^{2+}$ waves can be initiated without an increase in $[Ca^{2+}]_i$ in the stimulated cell [7] and can propagate similar distances when the elevation of $[Ca^{2+}]_i$ is reduced by dantrolene, an inhibitor of CICR [17]. Furthermore, $Ca^{2+}$ waves themselves can initiate asynchronous $Ca^{2+}$ oscillations but the $Ca^{2+}$ elevation associated with the oscillations does not initiate $Ca^{2+}$ waves even though gap junction communication is present [8, 13, 18]. In comparison to $IP_3$, $Ca^{2+}$ is a poor candidate for a long range intercellular messenger because its diffusion rate is greatly impaired by cytoplasmic $Ca^{2+}$ buffering proteins [19].

The diffusion hypothesis also explains variations in experimentally observed $Ca^{2+}$ waves. Because the characteristics of $Ca^{2+}$ waves are based on the $Ca^{2+}$ signaling machinery of the each cell, variations in PLC activity, $Ca^{2+}$ stores, $IP_3$ receptor sensitivity and distribution, and the extent of gap junction coupling will all influence $Ca^{2+}$ wave initiation and propagation. However, in certain preparations of glial cells, brain slices and liver lobules, $Ca^{2+}$ waves can travel over hundreds [20, 21] or thousands of cells [22, 23]. The propagation of long-distance $Ca^{2+}$ waves cannot be explained by diffusion alone and probably requires some form of regenerative process that acts in combination with a diffusive process.

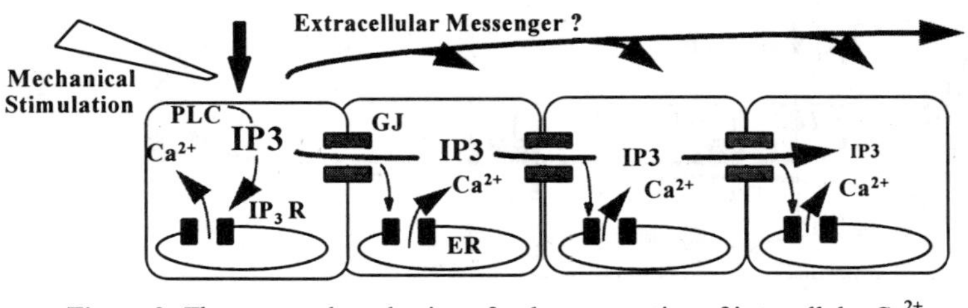


**Figure 2:** The proposed mechanisms for the propagation of intercellular $Ca^{2+}$ waves.

### *b) $Ca^{2+}$ wave propagation by the diffusion of an extracellular messenger*

In certain cases, intercellular $Ca^{2+}$ waves associated with mast, glial, liver and insulinoma cells [3, 21, 24-26] involve the release and diffusion of an extracellular messenger (Fig. 2). The messenger (e.g. ATP in mast cells) stimulates adjacent cells via membrane receptors to release internal $Ca^{2+}$ stores and to propagate $Ca^{2+}$ waves without cell contact. Although an extracellular fluid flow biases the $Ca^{2+}$ wave propagation in one direction in a variety of cell types, a limited amount of upstream propagation remains indicating that intercellular diffusion may also occur simultaneously [11]. In mast cells, the ATP is released by each cell but it is not clear if a regenerative process also occurs in other cell types or if a messenger is released just from the initiating cell. An involvement of extracellular ATP in $Ca^{2+}$ waves of epithelial cells was ruled out because $Ca^{2+}$ wave propagation was not influenced by an external fluid flow or by suramin, an antagonist of purinergic receptors [27]. A combination of intercellular and extracellular messenger diffusion may help explain why some waves propagate very large distances.

## 2. Results: The intercellular messenger

To test the hypothesis that $IP_3$ acts as an intercellular messenger between glial and endothelial cells, the effect of increasing $[IP_3]_i$ in a single cell by the photolysis of caged-$IP_3$ was examined. Based on the premise that caged-$IP_3$ (MW 635 Daltons) diffuses through gap junctions (permeability <1000 Daltons) in a manner similar to $IP_3$ (MW 437 Daltons) [28], caged-$IP_3$ should diffuse through a cell culture from a single injected cell (Fig. 3). Caged-$IP_3$ can be indirectly detected when exposed to UV light; caged-$IP_3$ liberates active $IP_3$ which in turn releases intracellular $Ca^{2+}$ and this can be detected with the $Ca^{2+}$-specific fluorescent dye fluo-3.

### *a) Photolysis of Caged-$IP_3$*

Following the pressure microinjection of a glial cell with caged-$IP_3$, a focused beam of UV light was used to illuminate briefly (*flash*) the injected cell. A one to two second UV flash increased $[Ca^{2+}]_i$ in the flashed cell and initiated an intercellular $Ca^{2+}$ wave that propagated to adjacent glial cells up to 80 μm away. The propagation distance of the $Ca^{2+}$ wave was dependent on the duration of the UV flash. This indicates that only a limited amount of $IP_3$ is released in the exposed cell and that a $Ca^{2+}$-dependent regeneration of $IP_3$ does not occur. Similar $Ca^{2+}$ waves were induced by photolysis in endothelial cells but the distance travelled was generally smaller, perhaps due to differences in gap junctional coupling or sensitivity to $IP_3$.

Approximately 10 minutes, but up to two hours, after microinjection of caged-$IP_3$, a UV flash focused on a non-injected glial cell up to 100 μm away from the injected cell again evoked an increase in $[Ca^{2+}]_i$ and initiated an intercellular $Ca^{2+}$ wave (Fig. 4). These results indicate that caged-$IP_3$ had reached the distal cells by diffusion through the inter-

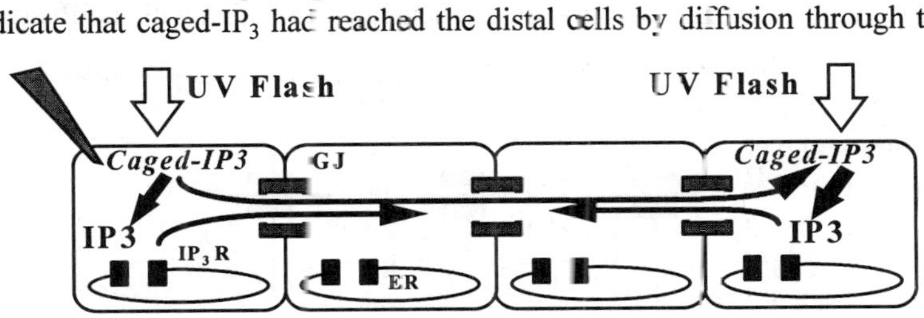


**Figure 3:** The experimental approach for using caged compounds to study wave propagation.

**Figure 4:** An intercellular calcium wave initiated by the photolytic release of $IP_3$ (long arrow) from caged-$IP_3$ in a distal, non-injected cell. Position of injected cell is indicated by the short arrow.

vening cells. The charged phosphate groups of caged-$IP_3$ and the retention of caged-$IP_3$ in the cells after two hours makes it unlikely that caged-$IP_3$ reached the distal cells by leakage across cell membranes. Furthermore, the ability to initiate multiple waves from the same cell by successive flashes indicates that the cell is replenished with caged-$IP_3$ from surrounding cells. A UV flash applied to cells that did not contain caged-$IP_3$ did not initiate any changes in $[Ca^{2+}]_i$.

$Ca^{2+}$ waves, initiated by the photolysis of caged-$IP_3$, also propagated from endothelial cells to glial cells. In this case, caged-$IP_3$ was injected into a glial cell and allowed to diffuse to a distal endothelial cell. Interestingly, a $Ca^{2+}$ wave initiated by photolysis in a glial cell did not propagate into adjacent endothelial cells. This uni-directional communication of a $Ca^{2+}$ wave can be explained by either retification of gap junction permeability for $IP_3$, a faster rate of $IP_3$ metabolism or a different sensitivity of the $Ca^{2+}$ signaling structures of the cell.

### *b) Photolysis of Caged-$Ca^{2+}$*

Because the photolysis of caged-$IP_3$ increases the $[Ca^{2+}]_i$, the possibility that $Ca^{2+}$ is initiating the $Ca^{2+}$ waves remains. To test this hypothesis, experiments were performed with caged-$Ca^{2+}$, a photosensitive $Ca^{2+}$ buffer (NP-EGTA) that releases $Ca^{2+}$ upon exposure to UV light [29]. All cells were loaded with an ester form of caged-$Ca^{2+}$ and when a UV flash was applied to a single cell, an increase in $[Ca^{2+}]_i$ was observed only in the flashed cell. By contrast, mechanical stimulation of the same cell evoked an intercellular $Ca^{2+}$ wave indicating that the cells were capable of cell communication and that the caged-$Ca^{2+}$ present in adjacent cells did not prevent $Ca^{2+}$ wave propagation. The increases in $[Ca^{2+}]_i$ in the flashed cell were comparable, as judged by relative changes in dye fluorescence, with those induced by the photolytic release of $IP_3$. The increase in $Ca^{2+}$ in only the flashed cell also confirmed, because all cells contained caged-$Ca^{2+}$, that the UV light flash was restricted to one cell.

## 3. Conclusions

We conclude 1) that intercellular diffusion of $IP_3$ through gap junctions can mediate the limited propagation of intercellular $Ca^{2+}$ waves, 2) that $Ca^{2+}$ does not act as an intercellular messenger in these $Ca^{2+}$ waves and 3) that gap junction-mediated intercellular $Ca^{2+}$ signaling occurs between endothelial and glial cells. We speculate that intercellular $Ca^{2+}$ waves may serve as a regulatory signal in the blood-brain barrier.

**This work was supported by NIH grant HL 49288 to MJS and the Belgium NFWO**

## References


[1] M.J. Sanderson *et al.*, Mechanisms and function of intercellular calcium signaling, *Mol Cell Endocrinol,* 98 (1994) 173-187.

[2] M.J. Sanderson, Intercellular waves of communication, *News Physiol Sci,* 11 (1996) 262-269.

[3] A.C. Charles, Glia-neuron intercellular calcium signaling, *Dev Neurosci,* 16 (1994) 196-206.

[4] M. Nedergaard, Direct signaling from astrocytes to neurons in cultures of mammalian brain cells, *Science,* 263 (1994) 1768-1771.

[5] E.A. Newman and K.R. Zahs, Calcium waves in retinal glial cells, *Science,* 275 (1997) 844-847.

[6] L. Moore and M.J. Sanderson, Intercellular $Ca^{2+}$ signaling between airway epithelial and smooth muscle cells, *J Cell Sci,* Submitted (1997).

[7] M.J. Sanderson *et al.*, Mechanical stimulation and intercellular communication increases intracellular $Ca^{2+}$ in epithelial cells, *Cell Regul,* 1 (1990) 585-596.

[8] A.C. Charles *et al.*, Intercellular signaling in glial cells: calcium waves and oscillations in response to mechanical stimulation and glutamate, *Neuron,* 6 (1991) 983-992

[9] M.J. Sanderson *et al.*, Intercellular communication between ciliated cells in culture, *Am J Physiol,* 254 (1988) C63-74.

[10] A.C. Charles *et al.*, Intercellular calcium signaling via gap junctions in glioma cells, *J Cell Biol,* 118 (1992) 195-201.

[11] L. Venance *et al.*, Mechanism involved in initiation and propagation of receptor-induced intercellular calcium signaling in cultured rat astrocytes, *J Neurosci,* 17 (1997) 1981-1992.

[12] M.O. Enkvist and K.D. McCarthy, Astroglial gap junction communication is increased by treatment with either glutamate or high $K^+$ concentration, *J Neurochem,* 62 (1994) 489-495.

[13] S. Boitano *et al.*, Intercellular propagation of calcium waves mediated by inositol trisphosphate, *Science,* 258 (1992) 292-295.

[14] M. Hansen *et al.*, A role for phospholipase C activity but not ryanodine receptors in the initiation and propagation of intercellular calcium waves, *J Cell Sci,* 108 (1995) 2583-2590.

[15] J.A. Felix *et al.*, Stretch increases inositol 1,4,5-trisphosphate concentration in airway epithelial cells, *Am J Respir Cell Mol Biol,* 14 (1996) 296-301.

[16] J.C. Saez *et al.*, Hepatocyte gap junctions are permeable to the second messenger, inositol 1,4,5-trisphosphate, and to calcium ions, *Proc Natl Acad Sci U S A,* 86 (1989) 2708-2712.

[17] A.C. Charles *et al.*, Mechanisms of intercellular calcium signaling in glial cells studied with dantrolene and thapsigargin, *Glia,* 7 (1993) 134-145.

[18] A. Strahonja and M.J. Sanderson, Intracellular $Ca^{2+}$ oscillations induced in glial by intercellular $Ca^{2+}$ waves, Submitted (1997).

[19] N.L. Allbritton and T. Meyer, Localized calcium spikes and propagating calcium waves, *Cell Calcium,* 14 (1993) 691-697.

[20] A.H. Cornell-Bell and S.M. Finkbeiner, $Ca^{2+}$ waves in astrocytes, *Cell Calcium,* 12 (1991) 185-204.

[21] T.D. Hassinger *et al.*, An extracellular signaling component in propagation of astrocytic calcium waves, *Proc Natl Acad Sci U S A,* 93 (1996) 13268-13273

[22] M.H. Nathanson *et al.*, $Ca^{2+}$ waves are organized among hepatocytes in the intact organ, *Am J Physiol,* 269 (1995) G167-171.

[23] L.D. Robb-Gaspers and A.P. Thomas, Coordination of $Ca^{2+}$ signaling by intercellular propagation of $Ca^{2+}$ waves in the intact liver, *J Biol Chem,* 270 (1995) 8102-8107.

[24] Y. Osipchuk and M. Cahalan, Cell-to-cell spread of calcium signals mediated by ATP receptors in mast cells, *Nature,* 359 (1992) 241-244.

[25] D. Cao *et al.*, Mechanisms for the coordination of intercellular calcium signaling in insulin-secreting cells, *J Cell Sci,* 110 (1997) 497-504.

[26] M.K. Frame and A.W. de Feijter, Propagation of mechanically induced intercellular calcium waves via gap junctions and ATP receptors in rat liver epithelial cells, *Exp Cell Res,* 230 (1997) 197-207.

[27] M. Hansen *et al.*, Intercellular calcium signaling induced by extracellular adenosine 5'- triphosphate and mechanical stimulation in airway epithelial cells, *J Cell Sci,* 106 (1993) 995-1004.

[28] T.D. Carter *et al.*, Porcine aortic endothelial gap junctions: identification and permeation by caged InsP3, *J Cell Sci,* 109 (1996) 1765-1773.

[29] G.C. Ellis-Davies and J.H. Kaplan, Nitrophenyl-EGTA, a photolabile chelator that selectively binds $Ca^{2+}$ with high affinity and releases it rapidly upon photolysis, *Proc Natl Acad Sci U S A,* 91 (1994) 187-191.

*Gap Junctions*
*R. Werner (Ed.)*
*IOS Press, 1998*

# Increased Intracellular $Ca^{++}$ Enhances Gap Junctional Conductance

Alberto PEREDA[1], Theodore BELL[1], Thomas SODERLING[2], Andrew J. CZERNIK[3], Angus C. NAIRN[3], and Donald S. FABER[1]
[1]*Allegheny University of the Health Sciences, Philadelphia, PA*; [2]*The Vollum Institute, Portland, OR*; [3]*Rockefeller University, New York, NY*

**Abstract.** Modifiability is an essential feature of chemical synapses. In contrast, synaptic transmission mediated by gap junctions is considered to be relatively inflexible. Mixed, electrical and chemical, synapses between goldfish auditory afferents and the Mauthner cell allow a detailed analysis of the properties of gap junctional transmission in the central nervous system. Patterned electrical activity of these afferents produces a synaptically-induced NMDA receptor-dependent enhancement of gap junctional conductance that is triggered by a postsynaptic increase in intracellular [$Ca^{++}$]. Thus, electrical synapses are modifiable in a manner analogous to the plasticity of their chemical counterparts and, in contrast to the generally accepted dogma, an increase in intracellular [$Ca^{++}$] can lead to an enhancement of gap junctional conductance, presumably by acting through an intermediate effector.

## 1. Introduction

The term *electrical synapse* is used when gap junctions occur between two neurons [1]. Synapses are more often chemically-mediated, and it is now well established that these sites of functional interaction are very plastic and modifiable. Therefore, the influence of a given neuron on a second one may vary, either due to the action of modulatory transmitters or as a result of the cell's own electrical activity. On the other hand electrical synapses classically have been perceived as less important, mostly restricted to fast reflex propagation in phylogenetically lower organisms, and less plastic or modifiable. This misconception probably reflects the lack of detailed information regarding the extent and properties of electrical synapses in the central nervous system, due largely to their poor accessibility to experimental manipulation.

Because of the higher incidence and widespread distribution of electrical synapses, studies of electrical transmission in the central nervous system have focused on lower vertebrates [1]. In some synaptic terminals, both electrical (gap junctions) and chemical synaptic transmission can coexist. This is the case for an identifiable class of eighth nerve afferents (Fig. 1A) that terminate on the lateral dendrite of the goldfish Mauthner (M-) cells as individual "large myelinated club endings" [2]. M-cells are a pair of large reticulospinal neurons that are located in the medulla of teleosts and play a central role in the organization of the fish escape response [3]. Stimulation of the posterior branch of the eighth nerve evokes a mixed excitatory synaptic response in the M-cell (Fig 1 B), a fast electrical response being followed by a chemical excitatory synaptic potential (EPSP) mediated by the amino acid glutamate [4, 5]. Due to the short-membrane time constant of the M-cell (about 400 µs), both components can be easily distinguished and reliably measured, providing a rare opportunity to explore modifications of junctional conductance under different physiological conditions *in situ* [3]. As a result of their bidirectionality, these electrical synapses provide i) fast anterograde transfer of sensory information from the periphery matching the requirement that the startle reflex mediated by the M-cell be rapid, and ii) a substrate for retrograde intercellular electrical communication, a phenomenon that produces backfiring and synchronization of presynaptic terminals and may lead to a specific enhancement of the synaptic potential evoked by these fibers [6].

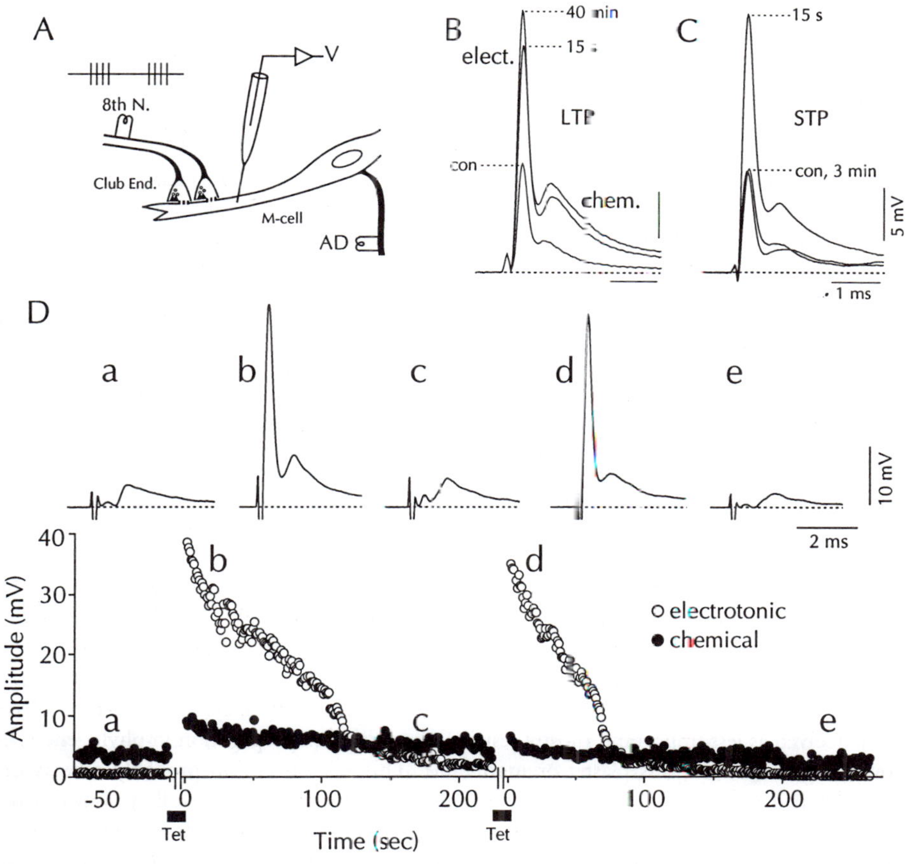


Figure 1. Discontinuous tetanic stimulation can lead to long- or short lasting potentiations of the eighth nerve evoked synaptic response. A, Experimental arrangement. Responses elicited by stimulating the posterior eighth nerve (8th n.; Club End.) or the spinal cord (AD) were recorded intracellularly (V) at the distal part of the M-cell lateral dendrite (M-cell). Eighth nerve stimulation evokes a mixed synaptic response consisting of a fast electrotonic potential (B, elect.) followed by a chemical EPSP (B, chem.). B–C, Results from two experiments showing that discontinuous tetanic stimulation of this input produces an early (B, con vs. 15 s; C, con vs. 15 s) potentiation of both components of this response, which is either long- (B, 40 min, LTP) or short-lasting (C, 3′, STP). Superimposed traces in B and C represent averages of 15 or more consecutive responses. D, Tetanic stimulation (Tet) transiently restores electrotonic coupling in spontaneously decoupled afferents. The tetanic paradigm repeatedly restored coupling transiently and evoked short-term potentiation of chemical transmission. Below, Plots of the amplitudes of the electrotonic (o) and chemical (•) components of the synaptic response versus time, encompassing effects of two conditioning periods. Above, representative averaged records ($n \geq 12$) obtained at the corresponding times labeled below.

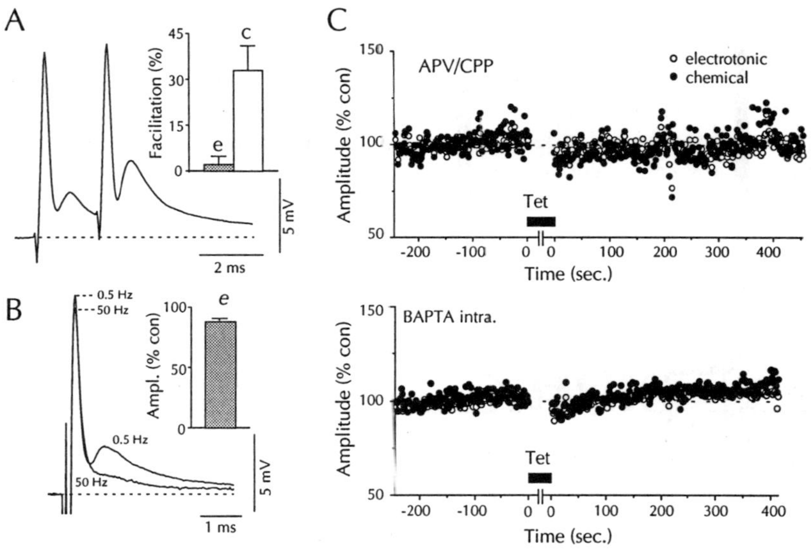


Figure 2. Role of postsynaptic $Ca^{++}$ in potentiated electrotonic tranmission. A-B, Comparison of effects of different stimulating paradigms known to increase presynaptic $[Ca^{++}]$ on the amplitudes of the electrically and the chemically mediated synaptic potentials. A, Paired-pulse stimulation of the eighth nerve facilitates chemical transmission but minimally affects electrotonic coupling (average of 25 responses). Inset, percent changes in amplitude of the electrotonic coupling potential (e) and EPSP (c) relative to the first responses (n=16, ± SEM). B, Effect of continuous tetanic stimulation on the synaptic response. Average of 12 consecutive traces obtained, at 0.5 Hz and 50 Hz. During high frequency stimulation the chemical component fatigued, while the electrotonic potential remained almost the same (Inset, coupling potential amplitude during high frequency stimulation, n=7, ± SEM). C, The increase in junctional conductance depends on an NMDA receptor-mediated post-synaptic increase in intracellular $Ca^{++}$. Above and below, Absence of potentiation when tetanic stimulation was applied in the presence of NMDA receptor antagonists in the superfusate (APV/CPP, 50 μM each), or after the $Ca^{++}$ chelator BAPTA had been injected intradendritically through the recording electrode (5 mM), respectively.

## 2. Electrical synapses are also modifiable

As with other vertebrate and invertebrate electrical synapses, junctional conductance at these eighth nerve afferents is modulated by dopamine [7]. Dopamine simultaneously enhances both components of the synaptic response by a mechanism that follows a postsynaptic cAMP-dependent phosphorylation route [8]. Unlike any other system however, these electrical synapses also exhibit an activity-dependent plasticity, similar to that of their chemical counterparts [9]. Discontinuous tetanic (high frequency) stimulation of the posterior eighth nerve produces homosynaptic potentiations (Fig. 1B) of both the electrical and glutamatergic components [9]. In most cases these modifications are long lasting (long-term potentiation or LTP) and persist throughout an experiment (up to 2 hours).

Interestingly, in some cases these changes can be transient [10], lasting for a period of only 3 to 10 minutes (short-term potentiation or STP; Fig. 1C). This rapid and reversible potentiation emphasizes that junctional conductance in these terminals is tightly regulated, probably via post-translational modifications of gap junction channels. STP is better illustrated in Fig. 1D, from an experiment in which the amplitude of the electrotonic component was spontaneously reduced to nearly zero during control, thereby reversing the typical 3 to 1 ratio between the sizes of the coupling potential and the chemical EPSP (compare Fig. 1B, C with Fig. 1D, control). Tetanic stimulation then produced, along with a potentiation of the chemical EPSP, a dramatic restoration of electrotonic coupling that decayed rapidly over 1 to 2 min-

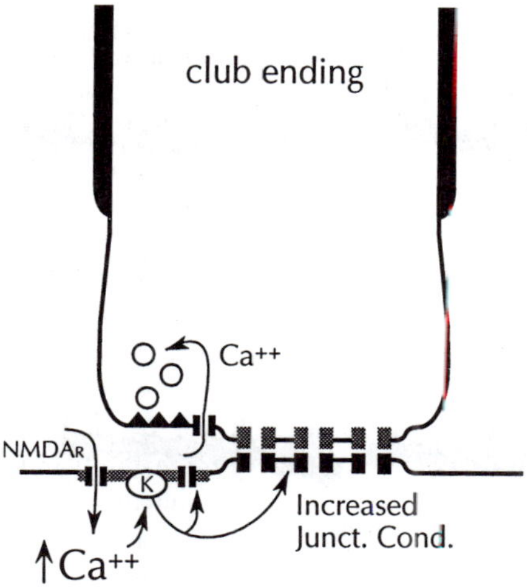


Figure 3. Scheme proposed for the activity-dependent regulation of the strength of the 8th nerve input. An elevated-presynaptic [$Ca^{++}$]increases transmitter release without affecting junctional conductance. Activation of NMDA receptors by discontinuous tetanic stimulation, instead promotes an elevation in postsynaptic [$Ca^{++}$], which in turns activates an intracellular calcium sensor (that we speculate should be located in the proximity of the NMDA receptor channel) leading to an enhancements of both components of the synaptic response. Preliminary results indicate this sensor is $Ca^{++}$/calmodulin-dependent protein kinase II (K).

utes, and these enhancements could be repeatedly induced.

Thus, not only chemical synapses but also electrical ones are subject to short-term changes in efficacy [10, 11]. Unlike chemical synapses, however, these changes are not due to an increased presynaptic [$Ca^{++}$]. A paired pulse stimulating paradigm that increases presynaptic [$Ca^{++}$] [11], "facilitates" chemical transmission but has no effect on the electrotonic component (Fig. 2A). Moreover, presynaptic [$Ca^{++}$] does not induce changes in junctional conductance (Fig.2B) even during a continuous train of presynaptic action potentials evoked by high frequency stimulation of the posterior eighth nerve [10], a paradigm that produces higher concentrations of residual internal $Ca^{++}$ [11]. These findings argue against a primary role for presynaptic $Ca^{++}$ in the STP of coupling.

The induction of both short and long-term changes in junctional conductance depends on activation of a glutamate receptor subtype, the NMDA receptor. The discontinuous, burst-like, tetanizing paradigm used to induce these activity-dependent potentiations optimizes the activation of NMDA receptors by presynaptically released glutamate [5, 10]. Accordingly, both forms of potentiations can be prevented by superfusing with a saline solution containing the NMDA receptor antagonists [9, 11] APV and CPP (Fig. 2C, above). These results suggested that plasticity of coupling requires a postsynaptic increase in [$Ca^{++}$], since NMDA receptor activation causes an influx of this ion. As illustrated in Fig. 2C (below), induction of the activity-dependent changes was prevented by prior intradendritic injections of the $Ca^{++}$ chelator BAPTA [9, 11]. Thus, we postulated that gap junctional conductance in the eighth nerve/M-cell synapses can be potentiated as a consequence of postsynaptic, but not presynaptic, elevations in [$Ca^{++}$]. Furthermore, given the well known inhibitory action of $Ca^{++}$ on junctional conductance [12], this ion may act as a messenger molecule to activate an intracellular modulatory element or pathway.

Preliminary data from experiments in which $Ca^{++}$ was injected pre- and postsynaptically show that this indeed could be the case. For this purpose, $CaCl_2$ (2-4 mM) was added to the electrode recording solution. Presynaptic injections, performed while recording either simultaneously from a single eighth nerve afferent and the M-cell dendrite or from afferent terminals in the medulla, did not induce any changes in the amplitudes of ortho- or antidromic coupling. In contrast, intradendritic $Ca^{++}$ injections triggered long-term enhancements of both components of the mixed synaptic response, and these potentiations occluded those induced by eighth nerve tetani. Based on several preliminary experimental observations, the $Ca^{++}$-dependent pathway most likely implicated in this potentiation involves activation of $Ca^{++}$/calmodulin-dependent protein kinase II (CamKII): i) induction of LTP is prevented by prior intradendritic injections of KN-93, an inhibitor of $Ca^{++}$-calmodulin-dependent protein kinases, ii) immunohistochemical studies with the CamKII-specific antibodies G-301 and RU-16 revealed the presence of this enzyme in the M-cell lateral dendrite, and iii) intradendritic injections of a constitutively active form of the $\alpha$-subunit of CamKII caused a long-term enhancement of both components of the synaptic response.

## 3. Discussion

Because of its well established direct effect on gap junctional conductance, an elevation of [$Ca^{++}$] is generally expected to reduce intercellular coupling. In the case of the M-cell system, however, an increase in postsynaptic [$Ca^{++}$] can cause an enhancement of junctional conductance. Specifically, $Ca^{++}$ acts as a messenger molecule, activating an intracellular pathway which preliminary data suggest involves CamKII (Fig. 3). Strikingly, this effect is asymmetric, due to either different properties of gap junction channels on each side of the junctional plaque (heterologous junctions) or the existence of different regulatory mechanisms in the M-cell and the afferent fibers. These results suggest that each side of the gap junction plaque may be independently regulated. Finally, $Ca^{++}$-dependent NMDA receptor-mediated modulation of gap junctions could be widespread, since both channel types are co-expressed during development and in several mammalian adult central nervous system structures.

**Acknowledgments.** Supported by NIH grants DC03186 to A.P. and NS15335 to D.S.F.

## References

[1] Bennett, M.V.L. Electrical transmission: a functional analysis and comparison with chemical transmission, in Cellular Biology of Neurons, vol. I, sect I, Handbook of Physiology. The Nervous System (Kandel E.R ed.).Williams and Wilkins, Baltimore, 1977, pp 357-416.

[2] Tuttle, R., Masuko, S. and Nakajima Y. Freeze fracture study of the large myelinated club ending synapse on the goldfish Mauthner cell: special reference to the quantitative analysis of gap junctions. *J. Comp. Neurol.* 246 (1986) 202-211.

[3] Faber, D.S. and Korn, H. Electrophysiology of the Mauthner cell: basic properties, synaptic mecha nisms, and associated networks, in Neurobiology of the Mauthner cell, (Faber, D.S., Korn, H., eds.) Raven, New York, 1978, pp 47-132

[4] Lin, J.W. and Faber, D.S. Synaptic transmission mediated by single club endings on the goldfish Mauthner cell. I. Characteristics of electrotonic and chemical postsynaptic potentials. *J. Neurosci.* 8 (1988)1302-1312.

[5] Wolszon, L., Pereda A., and Faber, D. S. A fast synaptic potential mediated by NMDA and non-NMDA receptors. *J. of Neurophysiol.* (1997), in press.

[6] Pereda, A., Bell, T., and Faber, D.S. Retrograde synaptic communication via gap junctions coupling auditory afferents to the Mauthner cell. *J. Neurosc.* 15 (1995) 5943-5955.

[7] Pereda, A., Triller, A., Korn, H., and Faber, D S. Dopamine enhances both electrotonic coupling and chemical excitatory postsynaptic potentials at mixed synapses. *Proc. Natl. Acad. Sci. USA* 89(1992) 12088-12092.

[8] Pereda, A., Nairn, A., Wolzson, L., and Faber, D.S. Postsynaptic modulation of synaptic efficacy at mixed synapses on the Mauthner cell. *J. Neurosc.* 14(1994)3704-3712.

[9] Yang, X.D., Korn, H., and Faber, D.S. Long-term potentiation of electrotonic coupling at mixed synapses. *Nature* 348(1990) 542-545.

[10] Pereda, A., and Faber, D.S. Activity-dependent short-term enhancement of intercellular coupling. *J. Neurosc.* 16(1996) 983-992.

[11] Zucker, R.S. Short-term synaptic plasticity. *Ann. Rev. Neurosci.* 12 (1989) 13-31.

[12] Baux, G., Simmoneau, M., Tauc, L. and Segundo, J.P. Uncoupling of electrotonic synapses by $Ca^{++}$. *Proc. Natl. Acad. Sci. USA* 75 (1978) 4577-4581.

*Gap Junctions*
*R. Werner (Ed.)*
*IOS Press, 1998*



# Multiple mechanisms for intercellular calcium waves

Thomas H. Steinberg, Roberto Civitelli, Eric C. Beyer, Niklas R. Jørgensen, Dongrong Cao, Steven T. Geist, and George Lin
*Washington University School of Medicine, St. Louis, MO, and Osteoporosis Research Center, Copenhagen Municipal Hospital, Copenhagen, Denmark*

**Abstract.** Many cells coordinate calcium signaling by propagating intercellular calcium waves. We have studied mechanically-induced intercellular calcium waves in osteoblastic cells, insulinoma cells, and tracheal epithelial cells, and have detected three distinct mechanisms for the propagation of these waves. The most widespread mechanism for the propagation of intercellular calcium waves in these cells is stimulation of $P_{2U}$ purinergic receptors by extracellular ATP, which appears to be secreted by the stimulated cells. Thus in the UMR rat osteoblastic cell line, the RIN rat insulinoma cell line, and hamster tracheal epithelial cells, intercellular calcium waves do not require gap junctional communication and are inhibited by the $P_{2U}$-blocker suramin or desensitization of P2U receptors by prior addition of ATP. We have also identified two gap junction-dependent mechanisms for intercellular calcium waves, neither of which require IP3-mediated release of intracellular calcium stores. In RIN cells transfected with connexin43, gap junction-mediated calcium waves are blocked by inhibiting L-type calcium channels or by preventing membrane depolarization and appear to be mediated by gap junctional electrical coupling. In contrast, ROS cells propagate gap junction-mediated calcium waves that require influx of calcium across the plasma membrane, but do not require membrane depolarization. Gap junction-mediated and ATP-mediated mechanisms may coordinate calcium waves in diverse settings.

## 1. Introduction: Intercellular calcium signaling

Transient elevations in the cytosolic calcium concentration trigger many cellular responses, and can propagate from cell to cell as intercellular calcium waves. These intercellular calcium waves have been identified in respiratory tract ciliated cells [1,2], neurons [3], glial cells and cell lines [4-9], smooth muscle cells [10], osteoblastic cells [11], chondrocytes [12], mast cells [13], PC12 cells [14], lens [15], and hepatocytes [16]. Two different mechanisms for the propagation of calcium responses among cells have been previously identified, both of which involve the release of calcium from intracellular stores via inositol trisphosphate ($IP_3$). In one model, best characterized in rabbit tracheal epithelia, mechanical stimulation of a single cell results in both cytosolic calcium transients and in the generation of $IP_3$[1,17,18]. This $IP_3$ traverses gap junctions and results in release of calcium stores in neighboring cells. In the second model, stimulation of a single cell results in release of ATP and activation of ATP receptors on neighboring cells via an autocrine mechanism [13]. In the experiments described below, we sought to identify the mechanisms responsible for mechanically-induced intercellular calcium waves in several different cell lines and transfectants.

## 2. Mechanisms of calcium waves in osteoblastic cells

*1. ROS and UMR rat osteoblastic cell lines propagate calcium waves with different kinetics*

We have studied gap junctional communication in bone cells and in particular have focused on two rat osteoblastic cell lines, ROS 17/2.8 (ROS) and UMR 106-01 (UMR). ROS and UMR cells express different gap junction proteins and differ in their ability to pass microinjected dye: ROS cells express Cx43 on the cell surface and are well dye-coupled, while UMR cells express predominantly Cx45 on the cell surface and are poorly dye coupled [19]. We examined the ability of ROS and UMR cells to propagate intercellular calcium waves induced by mechanical stimulation of a single cell [20]. ROS and UMR cell calcium waves were quite different, but not in ways we had anticipated: the well dye-coupled ROS cells propagated calcium waves that were much slower and extended to many fewer cells than did the poorly dye-coupled UMR cells. In ROS cells calcium waves spread to 8-15 cells, and there was frequently a time lag of 15-30 sec between the calcium transient in adjacent cells. In contrast, UMR cell calcium waves spread within a few seconds to 30-50 cells.

*2. Intercellular calcium waves in ROS cells require gap junctional communication and influx of extracellular calcium*

The dramatic differences between the kinetics of ROS and UMR calcium waves suggested that these phenomena were mediated by distinct mechanisms. To understand these mechanisms, we first asked whether calcium wave propagation required intracellular or extracellular calcium in ROS and UMR cells. After intracellular calcium stores were depleted using thapsigargin, ROS cells were still able to propagate mechanically-induced intercellular calcium waves, but UMR cells failed to propagate calcium signals even though the stimulated cell underwent a rise in calcium. Conversely, removal of extracellular calcium immediately before mechanical stimulation prevented calcium waves in ROS cells, but not in UMR cells.

We next asked whether gap junctional communication was required for ROS and UMR cell calcium waves. Intercellular calcium waves in ROS cells, but not in UMR cells, were totally inhibited by 2 mM heptanol, a gap junction inhibitor. We demonstrated more conclusively that UMR cell calcium waves did not require gap junctional communication by showing that these waves could spread between islands of cells that were not in physical contact with each other. In ROS cells, we were unable to demonstrate spread of calcium waves unless the cells were in physical contact.

These results demonstrated that calcium waves in ROS cells occur by a novel mechanism. They do require gap junctional communication as do intercellular calcium waves in rabbit tracheal epithelia, but unlike previously defined mechanisms for the propagation of intercellular calcium, they do not require release of intracellular calcium stores and therefore are unlikely to involve diffusion of $IP_3$ through gap junctional pores.

*3. Intercellular calcium waves in UMR cells require activation of P2U purinergic receptors*

Because UMR cell calcium waves did not require gap junctional communication, but did require release of intracellular calcium stores, we examined whether these calcium waves occurred via activation of P2 receptors by extracellular ATP. We first showed that UMR cells, but not ROS cells expressed ATP-induced calcium transients and mRNA for the P2U purinergic receptor. We next demonstrated that activation of these P2 receptors were involved in UMR cell calcium wave propagation. Calcium waves in UMR cells, but not in ROS cells, were inhibited by concentrations of suramin that blocked ATP-induced calcium transients in those cells. In addition, desensitization of P2 receptors blocked mechanically-induced calcium waves in UMR cells. In these experiments, 10 μM ATP was added to monolayers of UMR cells, causing a calcium transient in all cells. After the cytosolic calcium concentration returned to the resting level, a single cell was mechanically stimulated. No calcium wave was propagated to other cells, even though a calcium rise was seen in the stimulated cell. To determine whether the failure of calcium wave propagation was indeed caused by P2 receptor desensitization and not be depletion of calcium stores, we subsequently added thrombin, which elicited a calcium rise dependent on intracellular calcium store release. Thus, these experiments demonstrated that intercellular calcium waves in UMR cells required activation of P2 purinergic receptors.

*4. Calcium waves in ROS/P2U transfectants*

We next asked whether expression of P2U receptors in ROS cells was sufficient to allow these cells to propagate calcium waves with the kinetic and pharmacologic properties seen in UMR cells. Upon mechanical stimulation of a single cell, ROS/P2U transfectants propagated calcium waves that were nearly as fast as those seen in UMR cells, and were blocked by P2 desensitization.

## 3. ATP-dependent and gap junction-dependent calcium waves in the RIN rat insulinoma cell line

*1. RIN rat insulinoma cells propagate ATP-mediated calcium waves*

Intercellular communication mediated by gap junctions appears to be required for normal insulin secretion by pancreatic beta cells [21]. Cytosolic calcium transients are known to be important effectors of insulin secretion. We therefore studied calcium waves in the gap junction-deficient RIN insulinoma cell line and in RIN cells transfected with Cx43, which is normally expressed in beta cells [22]. Although RIN cells were not coupled as assessed by dye transfer and electrophysiologic studies, these cells were able to propagate intercellular calcium waves. RIN cell calcium waves required activation of P2 receptors, which we demonstrated in experiments similar to those described in UMR cells above. Thus RIN cell calcium waves did not occur after depletion of intracellular calcium stores, after addition of suramin, or after desensitization with ATP. Furthermore, calcium waves in these cells were able to cross a disruption in the monolayer that did not allow cell contact.

*2. Gap junction-dependent calcium waves in RIN/Cx43 cells require activation of L-type calcium channels*

In addition to ATP-mediated calcium waves, RIN/Cx43 transfectants propagated calcium waves that occurred by a gap junction-dependent mechanism. After ATP desensitization, RIN/Cx43 cells propagated calcium waves whereas RIN cells did not. These gap junction-dependent calcium waves were blocked by heptanol, and did not require intact intracellular calcium stores, but did require extracellular calcium. These findings indicated that gap junctional communication mediated intercellular calcium signaling in RIN/Cx43 cells by permitting the intercellular transfer of a signal that allowed activation of plasma membrane calcium channels. Inhibition of L-type calcium channels with 10 μM nifedipine prevented gap junction-dependent calicum waves in RIN/Cx43 cells; this experiment suggested that the signal that spreads from cell to cell via gap junctions is a signal that propagates changes in the plasma membrane potential. To test this hypothesis, we prevented membrane depolarization with diazoxide, which prevents closure of $K_{ATP}$ channels. The addition of 150 μM diazoxide after intracellular calcium depletion blocked mechanically-induced calcium waves just as nifedipine had.

These experiment suggest the following mechanism for the propagation of intercellular calcium waves. Mechanical stimulation of a single cell leads to both plasma membrane depolarization and a calcium transient in the stimulated cell. The change in membrane potential is transmitted to other cells via gap junctional communication, presumably by ionic coupling. In adjacent cells, this leads to activation of voltage-gated calcium channels, which produces the calcium transient in the coupled cell.

## 4. ATP-mediated intercellular calcium waves in hamster trachea epithelial cells

We also studied mechanically-induced calcium waves in a primary cell model, hamster tracheal epithelial (HTE) cells. Mechanical stimulation of a single cell in a monolayer of tracheal epithelial cells resulted in propagation of calcium waves with kinetics similar to those of RIN and UMR cells. These calcium waves were inhibited by desensitization of P2U receptors with ATP, but not by inhibition of gap junctions with heptanol. Calcium waves were seen to spread to cells that were not directly in physical contact with the stimulated cell. Thus in the HTE cells, mechanically-induced intercellular calcium waves appeared to be propagated by an ATP-dependent mechanism.

## 5. Mechanisms of intercellular calcium waves

These studies demonstrated three different mechanisms for the propagation of mechanically induced calcium waves. In UMR cells, RIN cells and HTE cells, calcium signals were coordinated among cells by activation of P2 purinergic receptors, presumably by released ATP, in a mechanism that did not require gap junctional communication or even cell contact. Of the three mechanisms we observed, this mechanism was the only one that required the release of intracellular calcium stores. The other two mechanisms we observed, those seen in ROS cells and in RIN/Cx43 transfectants, both required gap junctional communication and influx of extracellular calcium. Although we have not defined the calcium influx channel in ROS cells, it is clear that this pathway differs from that of RIN/Cx43 cells. In the RIN/Cx43 cells calcium influx is mediated by L-type voltage gated calcium channels, but ROS cells lack a voltage-gated calcium influx pathway. Thus while the signal traversing gap junctions in RIN/Cx43 cells is likely to be ionic traffic that mediates membrane depolarization, in ROS cells this is unlikely to be the case.

In previous studies with rabbit tracheal epithelia and other cells, gap junction mediated calcium waves have been defined that are thought to involve transjunctional passage of inositol trisphosphate and release of intracellular calcium stores. We have not detected this pathway in the cell models we have studied above: in none of these cells was there a requirement both for gap junctional communication and for release of intracellular calcium stores. Explanations for our inability to demonstrate this mechanism might include 1) differences in the gap junction proteins expressed in the cells we have studied; 2) differences in cell models; 3) differences in the techniques used to induce the calcium waves. Whatever the reason for inability to detect gap junction-mediated intercellular calcium waves that require intercellular transfer of $IP_3$, our studies demonstrate that several different mechanisms exist by which cells can coordinate their calcium signaling.

## References


[1] S. Boitano, E. R. Dirksen, and M. J. Sanderson, Intercellular propagation of calcium waves mediated by inositol trisphosphate, *Science* **258** (1992) 292-295.

[2] M. Hansen, S. Boitano, E. R. Dirksen and M. J. Sanderson, Intercellular calcium signaling induced by extracellular adenosine 5'-triphosphate and mechanical stimulation in airway epithelial cells, *Journal of Cell Science* **106** (1993) 995-1004.

[3] A. C. Charles, S. K. Kodali, and R. F. Tyndale, Intercellular calcium waves in neurons, *Molecular and Cellular Neuroscience* **7** (1996) 337-353.

[4] M. O. K. Enkvist and K. D. McCarthy, Activation of protein kinase C blocks astroglial gap junction communication and inhibits the spread of calcium waves, *J.Neurochem.* **59** (1992) 519-526.

[5] A. C. Charles, J. E. Merrill, E. R. Dirksen, and M. J. Sanderson, Inercellular signaling in glial cells: calcium waves and oscillations in response to mechanical stimulation and glutamate, *Neuron* **6** (1991) 983-992.

[6] A. C. Charles, C. C. G. Naus, D. Zhu, G. M. Kidder, E. R. Dirksen, and M. J. Sanderson, Intercellular calcium signalling via gap junctions in glioma cells, *Journal of Cell Biology* **118** (1992) 195-201.

[7] T. D. Hassinger, P. B. Guthrie, P. B. Atkinson, M. V. L. Bennett, and S. B. Kater, An extracellular signaling component in propagation of astrocytic calcium waves, *Proc.Natl.Acad.Sci.USA* **93** (1996) 13268-13273.

[8] L. Venance, N. Stella, J. Glowinski, and C. Giaume, Mechanism involved in inititiation and propagation of receptor-induced intercellular calcium signaling in cultured rat astrocytes, *J.Neurosci.* **17** (1997) 1981-1992.

[9] L. Venance, D. Piomelli, J. Glowinski, and C. Giaume, Inhibition by anandamide of gap junctions and intercellular calcium signalling in striatal astrocytes, *Nature* **376** (1995) 590-594.

[10] S. H. Young, H. S. Ennes, and E. A. Mayer, Propagation of calcium waves betweeen colonic smooth muscle cells in culture, *Cell Calcium* **20** (1996) 257-271.

[11] S.-L. Xia and J. Ferrier, Propagation of a calcium pulse between osteoblastic cells, *Biochemical and Biophysical Research Communications* **186** (1992) 1212-1219.

[12] P. D'Andrea and F. Vittur, Gap junction mediate intercellular calcium signalling in cultured articular chondrocytes, *Cell Calcium* **20** (1996) 389-397.

[13] Y. Osipchuk and M. Cahalan, Cell-to-cell spread of calcium signals mediated by ATP receptors in mast cells, *Nature* **359** (1992) 241-244.

[14] B. D. Lynn, C. A. Marotta, and J. I. Nagy, Propagation of intercellular calcium waves in PC12 cells overexpressing a carboxy-terminal fragment of amyloid precursor protein, *Neuroscience Letters* **199** (1995) 21-24.

[15] G. C. Churchill, M. M. Atkinson, and C. F. Louis, Mechanical stimulation initiates cell-to-cell calcium signaling in ovine lens epithelial cells, *Journal of Cell Science* **109** (1996) 355-365.

[16] S. F. Schlosser, A. D. Burgstahler, and M. H. Nathanson, Isolated rat hepatocytes can signal to other hepatocytes and bile duct cells by release of nucleotides, *Proc Natl.Acad.Sci.USA* **93** (1996) 9948-9953.

[17] M. J. Sanderson, A. C. Charles, and E. R. Dirksen, Mechanical stimulation and intercellular communication increases intracellular $Ca^{2+}$ in epithelial cells, *Cell Regulation* **1** (1992) 585-596.

[18] M. Hansen, S. Boitano, E. R. Dirksen, and M. J. Sanderson. A role for phospholipase C activity but not ryanodine receptors in the initiation and propagation of intercellular calcium waves, *Journal of Cell Science* **108** (1995) 2583-2590.

[19] T. H. Steinberg, R. Civitelli, S. T. Geist, A. J. Robertson, R. D. Veenstra, H.-Z. Wang, P. M. Warlow, E. Hick, J. G. Laing, and E. C. Beyer, Connexin43 and connexin45 form gap junctions with different molecular permeabilities in osteoblastic cells, *EMBO J.* **13** (1994) 744-750.

[20] N. R. Jørgensen, S. T. Geist, R. Civitelli, and T. H. Steinberg, ATP-dependent and gap junction-dependent intercellular calcium signaling in osteoblastic cells, *Journal of Cell Biology* **in press** (1997).

[21] P. Meda, Gap junction involvement in secretion: the pancreas experience, *Clinical & Experimental Pharmacology & Physiology* **23** (1996) 1053-1057.

[22] D. Cao, G. Lin, E. M. Westphale, E. C. Beyer, and T. H. Steinberg, Mechanisms for the coordination of intercellular calcium signaling in insulin-secreting cells, *Journal of Cell Science* **110** (1997) 497-504.

*Gap Junctions*
*R. Werner (Ed.)*
*IOS Press, 1998*

# Connexin43 is a Target of the MAP Kinase Signalling Pathway: Functional Consequences of MAP Kinase Activation

Bonnie J. WARN-CRAMER and Alan F. LAU. *Molecular Carcinogenesis, Cancer Research Center, University of Hawaii at Manoa, Honolulu, HI 96813.*

Earlier studies demonstrated that epidermal growth factor (EGF) induced a rapid fall in gap junctional communication (GJC) and an increase in serine phosphorylation on connexin43 (Cx43) in treated T51B cells. Mitogen-activated protein (MAP) kinase phosphorylated Cx43 *in vitro* on tryptic peptides that comigrated with phosphotryptic peptides obtained from EGF-treated cells, suggesting a role for MAP kinase in the regulation of Cx43 function. We have identified the *in vitro* sites of MAP kinase-mediated phosphorylation on Cx43 as the consensus MAP kinase phosphorylation sites, Ser255, Ser279, and Ser282. In this study, we have directly examined the participation of MAP kinase *in vivo* in the phosphorylation of Cx43 and the regulation of GJC in EGF-treated cells. The results of these studies confirm that the activation of MAP kinase through a signal transduction pathway initiated from the activated EGF receptor results in serine phosphorylation on Cx43 and the disruption of GJC. Phosphorylation at one or more of the consensus MAP kinase phosphorylation sites (Ser255, Ser279, and/or Ser282) is sufficient to disrupt GJC.

## Introduction

Previous studies from this laboratory demonstrated that EGF induced a rapid and transient increase in serine phosphorylation on Cx43 in T51B rat liver epithelial cells and a downregulation of GJC [1,2]. The EGF-induced phosphorylation on Cx43 and decreased GJC were not dependent on the activation of phorbol ester-sensitive isoforms of protein kinase C [2]. However, purified MAP kinase was demonstrated to phosphorylate a recombinant Cx43 baculovirus preparation *in vitro* on phosphotryptic peptides that comigrated with phosphotryptic peptides of Cx43 obtained from EGF-treated cells [2]. These studies supported a potential role for MAP kinase in mediating the EGF-induced phosphorylation on Cx43. We identified the *in vitro* sites of MAP kinase phosphorylation on Cx43 as Ser255, Ser279, and Ser282 [3], the consensus MAP kinase recognition sites. Since the fall in GJC in EGF-treated cells followed a time course coincident with the increase in serine phosphorylation on Cx43, the data suggested that phosphorylation at one or more of these specific serine sites could be responsible for the downregulation of GJC.

To directly examine the role of MAP kinase in regulating Cx43 function, we expressed wild type (wt) Cx43 using a retroviral expression system in communication-deficient cells, Cx43 knockout (K/O) cells [4] (derived from Cx43 K/O mouse embryos) or human HeLa cells. We used these Cx43 expressing cells to determine the effects of activating the MAP kinase pathway with different agents (EGF, lysophosphatidic acid (LPA) [5] and pervanadate [6]) on Cx43 phosphorylation and on GJC. We also utilized a specific MEK inhibitor, PD98059 [7], to prevent the activation of MAP kinase in stimulated cells.

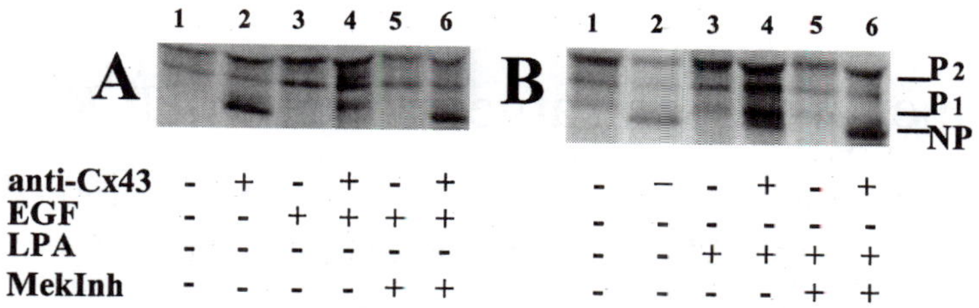


Figure 1. Autoradiograph of a SDS-PAGE gel showing EGF (panel A) and LPA (panel B) induced increased phosphorylation on Cx43 expressed in the K/O cells. Cells were metabolically labeled with [$^{35}$S]-methionine and treated with EGF or LPA in the presence or absence of the MEK inhibitor (MekInh) as described in the text. Cx43 was immunoprecipitated with a Cx43 antiserum or with non-immune rabbit serum (control).

## Results

Confluent monolayers were metabolically labeled with [$^{35}$S]-methionine in methionine-free medium for 2-3 hr to label all isoforms of the Cx43 protein. Cx43 was immunoprecipitated from the clarified cell lysates with a Cx43 antiserum (to aa 368-382 of Cx43) or with non-immune rabbit serum. EGF treatment, 100 ng/ml, during the last 30 min of the labeling period induced increased phosphorylation on Cx43 in the K/O cells expressing wt Cx43 (Figure 1A). In the unstimulated cells, Cx43 is present primarily in the non-phosphorylated form (designated NP) and in a phosphorylated isoform (P1) that migrates more slowly on SDS-PAGE (Fig. 1A, lane 2). Stimulating the cells with EGF increased the amount of Cx43 in the P1 isoform and in a slower migrating, more phosphorylated isoform, P2 (shown in lane 4). However, when the cells were pretreated with the MEK inhibitor (MekInh, 50 μM) for one hr prior to the stimulation with EGF, the increase in the P1 and P2 isoforms of Cx43 was not observed (lane 6). Similar results were obtained when the HeLa cells expressing wt Cx43 were treated with EGF (data not shown).

Panel B in Figure 1 shows the results of treating [$^{35}$S]-labeled cells with 50 μM LPA during the last 15 min of the labeling period. LPA stimulation increased phosphorylation on Cx43 as shown by relative increases in the P1 and P2 isoforms as compared to the unstimulated cells (Figure 1B, lanes 4 and 2). Preincubation with the MEK inhibitor prevented the increase in phosphorylation on Cx43 (Figure 1B, lane 6).

We determined the effect of the MEK inhibitor on the activation of MAP kinase directly using myelin basic protein (MBP) as a MAP kinase substrate in a kinase assay with γ-[$^{32}$P] ATP (Figure 2A). A basal level of MAP kinase activity was present in the unstimulated HeLa cells expressing wt Cx43 as shown by some phosphorylation on MBP at the zero time point. MAP kinase activity was greatly enhanced in EGF treated cells as shown at two min and five min following the addition of EGF. Preincubating the cells with the MEK inhibitor prevented the enhanced MAP kinase activity as shown in panel B for the three min time point.

Treating cells with pervanadate, a potent tyrosine phosphatase inhibitor, is reported to be sufficient to activate Raf, MEK, and MAP kinase [6]. Pervanadate treatment (100 μM for 20 min) increased the phosphorylation on Cx43 in the HeLa cells expressing wt Cx43 (Figure 3, lane 2, unstimulated cells; lane 4, pervanadate-stimulated cells).

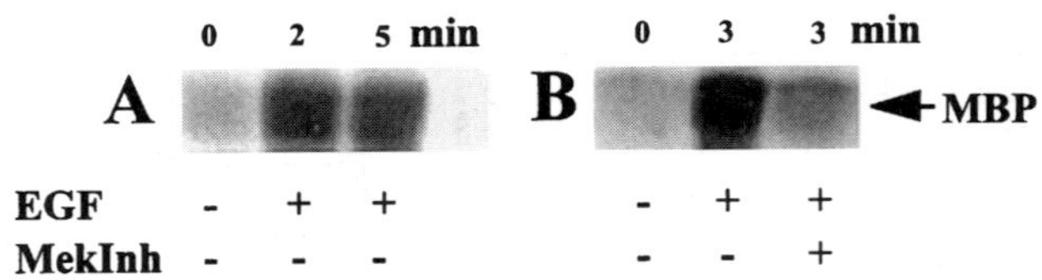


Figure 2. Autoradiograph of a SDS-PAGE gel showing that EGF treatment rapidly activates MAP kinase (panel A) as demonstrated by the MAP kinase-induced phosphorylation on myelin basic protein (MBP) at 2 min and at 5 min. Pretreating the cells with the MEK inhibitor (MekInh) blocked MAP kinase-induced phosphorylation on MBP (panel B).

Similar results were obtained with the K/O cells expressing wt Cx43 (data not shown). Pretreating the cells with the MEK inhibitor prior to pervanadate treatment appeared to only partially block the increased phosphorylation on Cx43 (see Figure 3, lane 6).

We examined the effects of EGF on GJC in the K/O and HeLa cells expressing wt Cx43 as determined by the transfer of Lucifer yellow dye. The K/O cells expressing wt Cx43 communicated on average to 4-5 neighboring cells (Figure 4). Preliminary data indicated that treatment with EGF decreased communication in these cells by ~ 40% at 30 min (n=4). The HeLa cells expressing wt Cx43 also communicated on average to 4-5 neighboring cells. GJC in these cells appeared to fall by ~25% following EGF treatment (n=5). Treating the HeLa cells expressing wt Cx43 with the MEK inhibitor (MekInh) alone for one hr appeared to have some stimulatory effect on GJC (Figure 4, average of two experiments). Treating the cells with EGF, after pretreating with the MEK inhibitor to prevent MAP kinase activation, did not induce a fall in GJC (Figure 4, average of two experiments).

## Discussion

EGF, LPA, and pervanadate all induced an increase in phosphorylation on Cx43 with less Cx43 migrating as the non-phosphorylated isoform (NP) and more Cx43 migrating in the phosphorylated P1 and P2 isoforms on SDS-PAGE (see Figures 1 and 3). Pretreating the cells with the MEK inhibitor, blocked the increase in phosphorylation on Cx43 in the EGF and in the LPA treated cells, indicating that phosphorylation was dependent on the activation of MAP kinase. EGF treatment initiated activation of the MAP kinase cascade as determined directly by the phosphorylation of the MAP kinase substrate, MBP, in kinase assays of cell lysates.

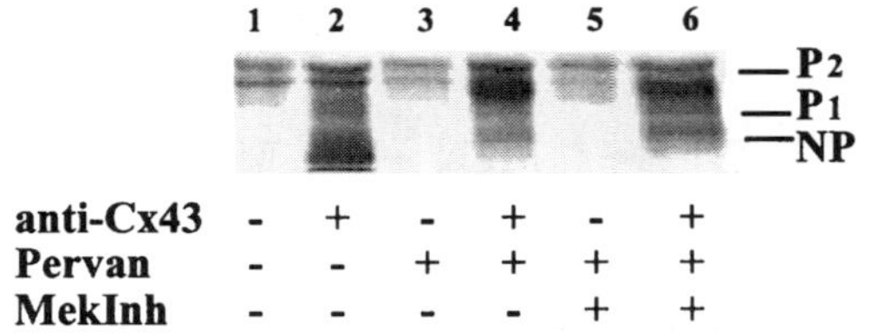


Figure 3. Autoradiograph showing a SDS-PAGE gel of pervanadate-induced increased phosphorylation on Cx43. HeLa cells expressing wt Cx43 were metabolically labeled with [$^{35}$S]-methionine and treated with pervanadate in the presence or absence of the MEK inhibitor (MekInh) as described in the text. Cx43 was immunoprecipitated with a Cx43 antiserum or non-immune rabbit serum (control).

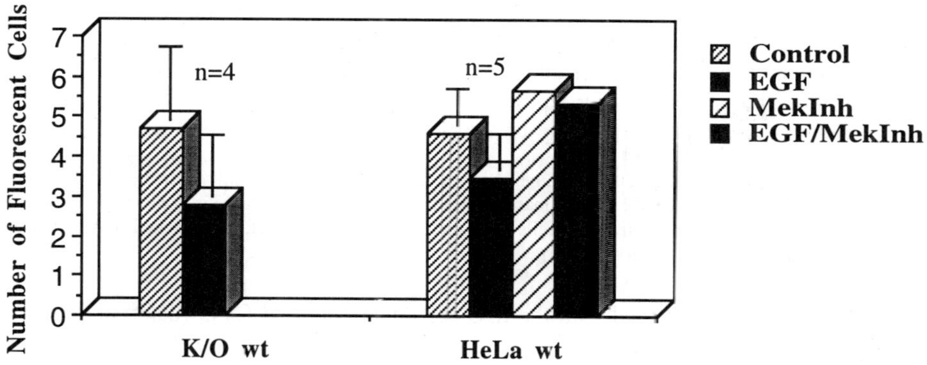


Figure 4. GJC was determined by the transfer of Lucifer yellow dye from a microinjected cell to neighboring cells. Data are shown for the K/O cells expressing wt Cx43 and treated with EGF and for the HeLa cells expressing wt Cx43 and treated with EGF alone, the MEK inhibitor alone, or with EGF following treatment with the MEK inhibitor.

The EGF-induced activation of MAP kinase was blocked in the cells pretreated with the MEK inhibitor (Figure. 2B). These data clearly indicate that EGF and LPA-induced increased phosphorylation on Cx43 are mediated through the activation of the MAP kinase signal transduction pathway and that Cx43 is a cellular substrate of activated MAP kinase. The phosphotryptic peptides obtained from Cx43 phosphorylated by MAP kinase *in vitro* comigrated with a subset of the phosphotryptic peptides obtained from Cx43 phosphorylated *in vivo* in EGF-treated cells [2,3], indicating that phosphorylation was occurring on some of the same peptides. Therefore, it is reasonable to assume that the EGF-induced activation of MAP kinase is responsible for the increased phosphorylation on Cx43 and that phosphorylation is occurring at the Ser255, Ser279, and Ser282 MAP kinase consensus phosphorylation sites identified in our earlier *in vitro* studies [3].

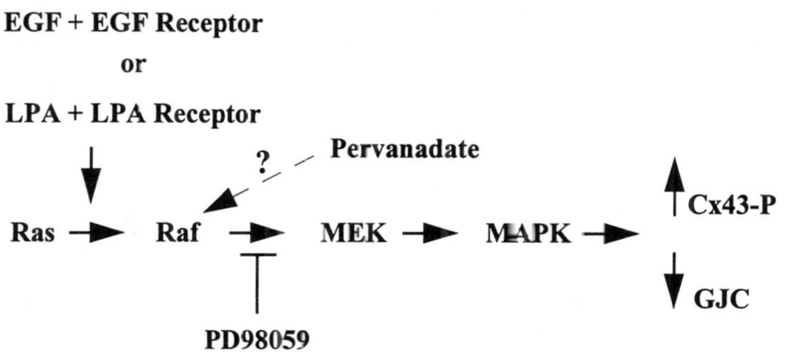


Figure 5. Schematic representation of the activation of the MAP kinase signaling pathway through ligand-induced activation of the EGF or LPA receptor. Pervanadate treatment is known to activate Raf, MEK, and MAP kinase [6]. The MEK inhibitor (PD98059) blocks the activation of MEK [7]. Activation of the Ras/Raf/MEK/MAP kinase pathway results in increased phosphorylation on Cx43 and a reduction in GJC.

Since EGF induces a fall in GJC and that fall is prevented by blocking the activation of MAP kinase, it is clear that MAP kinase-mediated phosphorylation on Cx43 at these specific serine sites is also regulating GJC. Phosphorylation on one or more of these serine residues appears to be sufficient to downregulate GJC through the Cx43 channel. Studies are currently underway utilizing Cx43 mutants with alterations of one or more of the MAP kinase consensus phosphorylation sites to isolate the effects of phosphorylation at each of these serine sites on the regulation of GJC.

## Acknowledgments

The authors thank Kendra D. Martyn, Ph.D., Cancer Research Center of Hawaii for her assistance in establishing the retroviral infection system. This research was supported by National Cancer Institute, NIH grant CA52098 (to A.F.L.) and by grant HIFW-09-95 from the American Heart Association, Hawaii Affiliate (to B.J.W.-C.).

## References

[1] A.F. Lau *et al.*, Epidermal Growth Factor Disrupts Gap-Junctional Communication and Induces Phosphorylation of Connexin43 on Serine, *Molecular Biology of the Cell* **3** (1992) 865-874.

[2] M.Y. Kanemitsu and A.F. Lau, Epidermal Growth Factor Stimulates the Disruption of Gap Junctional Communication and Connexin43 Phosphorylation Independent of 12-O-tetradecanoylphorbol 13-acetate-sensitive Protein Kinase C: The Possible Involvement of Mitogen-activated Protein Kinase, *Molecular Biology of the Cell* **4** (1993) 837-848.

[3] B.J. Warn-Cramer *et al.*, Characterization of the Mitogen-activated Protein Kinase Phosphorylation Sites on the Connexin-43 Gap Junction Protein, *Journal of Biological Chemistry* **271** (1996) 3779-3786.

[4] K.D. Martyn *et al.*, Immortalized Connexin43 Knockout Cell Lines Display a Subset of Biological Properties Associated with the Transformed Phenotype, Cell Growth and Differentiation (1997) In press.

[5] C.S.T. Hll *et al.*, Lysophosphatidic Acid Inhibits Gap Junctional Communication and Stimulates Phosphorylation of Connexin-43 in WB cells: Possible Involvement of the Mitogen-activated Protein Kinase Cascade, *Biochemistry Journal* **303** (1994) 475-479.

[6] Z. Zhao *et al.*, Activation of Mitogen-activated Protein (MAP) Kinase Pathway by Pervanadate, a Potent Inhibitor of Tyrosine Phosphatases, *Journal of Biological Chemistry* **271** (1996) 22251-22255.

[7] D.R. Alessi *et al.*, PD098059 Is a Specific Inhibitor of the Activation of Mitogen-activated Protein Kinase Kinase *In Vitro* and *In Vivo*, *Journal of Biological Chemistry* **270** (1995) 27489-27494.

*Gap Junctions*
*R. Werner (Ed.)*
*IOS Press, 1998*



# Characteristics of coupling in cardiac myocytes and CNS astrocytes cultured from wildtype and Cx43-null mice

David C. Spray[1,2], Monique J. Vink[1], Eliana Scemes[1,3], Sylvia O. Suadicani[1,4], Glenn I. Fishman[2,5], and Rolf Dermietzel[1,6]

*Departments of [1]Neuroscience, [2]Medicine and [5]Molecular Genetics, Albert Einstein College of Medicine, Bronx, NY; [3]Department of Physiology, Biosciences Institute, University of Sao Paulo, Brazil; [4]Department of Biology, Sao Judas College, Sao Paulo, Brazil; [6]Department of Anatomy, University of Bochum, Germany*

**Abstract** Connexin43 (Cx43) is the most abundantly expressed gap junction protein in ventricular myocytes of the heart and in astrocytes of the brain. In order to determine the functional importance of Cx43 in coupling between these cell types, we have compared coupling strength and junctional channel properties in cardiac myocytes and cortical astrocytes obtained from wildtype (WT) mice and littermates in which Cx43 has been deleted through homologous recombination. Because the Cx43 (-/-) mice die at birth, cultures were established perinatally from individual animals and genotypes subsequently determined from PCR of tail DNA. For cardiac myocytes from Cx43 (-/-) mice, spontaneous contractions were less frequent and more arrhythmic than in wildtype littermates; dye coupling was virtually absent in Cx43(-/-) cardiocytes, although junctional conductance ($g_j$) was 36% that of WT. Voltage sensitivity of $g_j$ and unitary junctional conductances ($\gamma_j$) were primarily Cx43-like in WT cardiocytes and Cx40-like in Cx43 (-/-) cells, although smaller channel sizes, perhaps reflecting Cx45 activity, were also prominent in event histograms prepared from recordings from cardiac myocytes of both genotypes. Conduction velocity and extent of spread of mechanically elicited $Ca^{2+}$ waves were reduced in Cx43 (-/-) cardiocytes compared to WT littermates. For astrocytes from Cx43 (-/-) mice, dye coupling was present, though less extensive than in WT glia; $g_j$ in Cx43 (-/-) astrocytes was only 5% that measured in WT littermates. Voltage sensitivity of $g_j$ was markedly stronger in Cx43 (-/-) astrocytes than in WT, in many cases exhibiting no voltage-induced substate, suggesting the presence of Cx46. Cx43 (-/-) $\gamma_j$ values were substantially lower than in WT astrocytes, although larger Cx40-like channels were also present in some cell pairs. Comparing results from individual littermates, $Ca^{2+}$ wave spread among Cx43 (-/-) and WT astrocytes was remarkably similar with regard to velocity of propagation, amplitude, and extent of spread, although significant differences in each parameter were obtained when all animals from each genotype were compared. $Ca^{2+}$ wave spread was similarly but only partially reduced in Cx43 (-/-) and WT cultures exposed to the ATP receptor blocker suramin and almost totally eliminated by heptanol treatment, indicating that under the conditions of these experiments the primary route of $Ca^{2+}$ wave spread is through junctional channels. No differences were detected in coupling strength or Cx43 mRNA expression between neonatal heterozygous and WT cardiac or glial cells, and no compensatory changes in mRNA expression of other connexins were seen in brain or cardiac tissue. We conclude from these studies that the impact of Cx43 deficiency on intercellular coupling differs markedly in cardiac myocytes and CNS astrocytes. The substantially higher $g_j$ values in cardiocytes than in astrocytes from Cx43 (-/-) mice contrasts with the lower strength of dye coupling and $Ca^{2+}$ wave spread in cardiocytes than in astrocytes from these animals and presumably reflects the different properties of channels formed by the connexins responsible for the residual coupling.

## 1. Introduction

From a functional standpoint, both brain and heart can be considered to consist of communication compartments, where connections within compartment members and from one compartment to another are established by gap junction channels [1,2]. In the heart, communication compartments are regionally organized, forming a pathway where impulses are rhythmically initiated and synchronized within sino-atrial and atrio-ventricular nodes before being conducted to the atrial and ventricular working myocardium, where the wave of propagated contraction leads to ejection of blood from the cardiac chambers. In the brain and spinal cord, communication compartments are cell-type specific, with cells of separate compartments co-mingling in different regions even though individual compartments may extend across very large distances. Glia are coupled, astrocytes more extensively that oligodendrocytes, as are leptomeningeal, ependymal and endothelial cells; at least some neurons are also coupled, most extensively during early development.

Connexin43 is the most abundant gap junction protein in the nervous and cardiovascular systems. In the mouse heart at birth, Cx43 is expressed in all regions; during the first postnatal week, Cx43 within working myocardium becomes organized into the adult pattern of localization at intercalated disks between longitudinal ends of muscle fibers [3]. In the neonatal mouse brain, Cx43 is expressed between astrocytes, and between leptomeningeal, ependymal and endothelial cells[4]; Cx43 is found between undifferentiated neuroblasts [5], and may also contribute to neuronal coupling during development and in adulthood [6]

Transgenic mice lacking Cx43 have recently been generated [7]; the homozygous null [Cx43 (-/-)] animals die at birth due to a congenital right ventricular hyperplasia that leads obstruction of the pumonary circulation. In order to explore the extent to which Cx43 contributes to functional coupling in cardiac myocytes and CNS astrocytes, we have compared the characteristics of intercellular coupling in cultured cells prepared from littermates of wildtype and Cx43 (-/-) mice. Our findings indicate that additional connexins contribute to coupling in cells from both tissues and that differences in properties of the connexins that mediate the residual coupling in cardiocytes and astrocytes from Cx43 null mice give rise to contrasting characteristics in functional coupling.

## 2. Materials and Methods

Mice heterozygous for the Cx43 knockout allele [7] were obtained from Jackson Laboratories (Bar Harbor, ME) and bred to obtain litters of Cx43-null [Cx43 (-/-)], heterozygous [Cx43 (+/-)] and wildtype [Cx43 (+/+)] offspring. Brains and hearts from each individual within more than a dozen litters were digested and highly enriched cardiac myocyte and astrocyte cultures were established using methods modified from those used to prepare similar cultures from rat tissues [8,9]. Genotype of each animal was determined subsequently using PCR primers designed to amplify 269 and 101 bp segments corresponding to wildtype and Cx43 (-) alleles, respectively.

Rates of spontaneous beating in cardiocyte cultures and $Ca^{2+}$ wave propagation in both cell types were measured using a Nikon RCM8000 real time confocal microscope equipped with UV laser in cells previously incubated for 45 min in the presence of 10 μM Indol-AM (Molecular Probes, Eugene, OR), a ratiometric $Ca^{2+}$ indicator. Analogue images were generally acquired onto an OMDR at rates of 1 or 8/sec and analysed during playback with Nikon Polygon-Star oftware. Conduction velocities were calculated from latencies to half-maximal responses determined from sigmoidal curve fitting using Origin software, with distances between cell centers measured from prints of the fields.

Dye coupling was measured in subconfluent cardiocyte and astrocyte cultures 1-2 min after iontophoretic intracellular injection of Lucifer Yellow (5% wt/vol). Junctional conductance ($g_j$) was measured in 1 day cultured cardiac myocyte cell pairs and in freshly dissociated astrocytes using the dual whole cell voltage clamp technique [10,11] with internal solution containing 135 mM CsCl [12].

Voltage sensitivity of junctional current was evaluated as previously described [13,14]. Unitary junctional conductances ($\gamma_j$) were measured under conditions where $g_j$ was spontaneously low or after reducing $g_j$ to low levels by halothane exposure [9].

## 3. Results and Discussion

Whereas gross cardiac abnormality is a hallmark feature of the Cx43 (-/-) mouse [7], we have not detected differences in overall structure, cortical lamination pattern or thickness, or expression of the astrocytic protein GFAP between brains of wildtype and Cx43 (-/-) littermates (not shown). This observation confirms a previous report [15] and suggests either that intercellular coupling within the astrocyte communication compartment is not essential for prenatal brain development and organization or that the presence of other connexins between the cells is sufficient to sustain a level of intercellular communication that is adequate for these tasks.

Cardiac myocytes contract spontaneously after 3-4 days in culture, which is presumably due to the presence within the culture of cells generating pacemaker curents. In cardiac myocytes from WT animals, such beating occurred at an average of 69 bpm and the contractions of different cells within the microscope field were highly coordinated and rhythmic. In cardiac myocytes from Cx43 (-/-) mice, beat rate was slower (38 bpm); although contractions were still coordinated throughout the culture dish, the pattern of contractions was much more arrhythmic [variance of interbeat interval 20-fold higher in Cx43 (-/-) as compared to WT cardiocytes]. This decreased rhythmicity of contractions in cardiocyte cultures lacking Cx43 is similar to that reported in cultures in which Cx43 expression was reduced following intracellular infection with *Trypanosoma cruzi*, the protozoan parasite causing Chagas' disease [16]. The mechanism for reduced rhythmicity in poorly coupled cultures may involve impaired entrainment of pacemaker cells within the population.

Table 1: Characteristics of coupling between WT and Cx43 (-/-) cardiocytes and astrocytes

| | Cardiac Myocytes | | Astrocytes | |
|---|---|---|---|---|
| Parameter | Wildtype[a] | Cx43 (-/-) | Wildtype[a] | Cx43 (-/-) |
| $g_j$ (nS) | 11.7±1.3 (42)[b] | 4.2±0.8 (62) | 16.7±2.6 (29) | 0.76±0.01 (95) |
| $V_0$ (mV) | ~60 | ~45 | ~60 | ~30 |
| $g_{min}/g_{max}$ | ~0.4 | ~0.3 | ~0.4 | <0.1 |
| $\gamma_j$ (pS) | 50,100,140,170 | 50,160 | 30,90,110 | 30,40 |
| Velocity (μm/sec) | 9.8±0.7 (40) | 7.8±0.8 (17) | 18.8±0.7(216) | 12.1±0.8 (146) |
| Attenuation (mean ratio) | 0.67±0.04 (51) | 0.9±0.07 (22) | 1.81±0.04 (216) | 1.54±0.03 (146) |
| Responding cells (%) | 88.7 | 75 | 94±3 (216) | 80±5 (146) |

[a] In cases where values for Cx43 (+/-) and Cx43 (+/+) tissues were not different, the values were combined in this table
[b] Values are given as means ± SE; numbers of experiments indicated in parentheses

Cardiac myocytes from Cx43 (-/-) mice rarely showed perceptible Lucifer Yellow coupling to any of the surrounding cells, whereas Cx43 (-/-) astrocytes generally exhibited coupling to at least one contiguous neighbor and occasionally even to second order cells. In contrast, junctional conductance of Cx43 (-/-) cardiocytes was much higher than in astrocytes (Table 1). The finding that in both astrocytes and cardiac myocytes from Cx43 (-/-) mice are coupled indicate that other connexins are either constitutively expressed or upregulated in the absence of Cx43 and suggest that gap junction proteins in addition to Cx43 participate in the coupling between both types of cells in wildtype animals.

Gap junction channels formed by different connexins display electrophysiological properties that often allow the connexin type to be inferred from physiological measurements [17]. Voltage sensitivity of macroscopic condictance and sizes of unitary junctional conductances were evaluated in both types of cell from both genotypes. Voltage sensitivity of $g_j$ was stronger in Cx43 (-/-) than in

WT cells from both tissues (see $V_0$ and $g_{min}$ values in Table 1) and some astrocyte cell pairs displayed no residual conductance, a property of Cx46 [18]. Comparison with voltage sensitivities obtained for connexins transfected into mammalian cells, Cx43 (-/-) heart cells behaved as though Cx40 was the major gap junction channel, whereas Cx43 (-/-) astrocytes appeared to functionally express Cx46 and/or Cx45. Single channel studies further support the notion that these channel types are functionally expressed. In the case of the astrocytes, most frequently encountered channel sizes were predominantly 30 and 40 pS (similar to values reported for Cx45 and Cx46: [12,18]), although larger events were also present, perhaps due to Cx40 [19,20]. In the case of cardiac myocytes, unitary conductances were primarily as expected for channels formed of Cx40 (mainstate conductance 180-190 pS, substate conductance 70 pS: 19,20), although smaller events were also present, possibly contributed by Cx45 [12]. Quantitative Northern blot analysis revealed the presence of Cx45 and Cx40 in mouse hearts and brains, without genotype-dependent expression level; in astrocytes of both genotypes, RT-PCR analysis revealed the presence of Cx46 (results not shown).

$Ca^{2+}$ or $IP_3$ injection into single cells in a monolayer leads to spread of elevated intracellular $Ca^{2+}$ in neighboring cells [21]; focal mechanical, electrical or pharmacological stimualtion also provokes such waves (for review, see 22). To evaluate the extent to which $Ca^{2+}$ wave spread in cardiac myocytes and astrocytes was affected by the absence of Cx43, intracellular $Ca^{2+}$ levels were mesured in littermate cultures as a function of time before and after one cell was pricked with a microelectrode. In the case of cardiac myocytes from WT and Cx43 (-/-) mice, such stimulation led to waves that travelled between cells with velocities that were slightly slower and with larger amplitudes in responding cardiocytes lacking Cx43 (Table 1). Fewer Cx43 (-/-) cells participated in the $Ca^{2+}$ waves than in WT cardiocytes.

$Ca^{2+}$ wave responses were also measured in cultures obtained from sibling WT and Cx43 (-/-) mice. Although paired cultures from individual litters behaved differently, significant differences were seen in each of the measured parameters of $Ca^{2+}$ wave propagation when results were pooled according to genotypes (Table 1). In order to determine the extent to which the $Ca^{2+}$ wave propagation in these experiments was through junctional channels, as distinct from extracellular signalling pathways, experiments were performed in the presence of suramin, which blocks ATP receptors, and in the presence of heptanol, which inhibits gap junctional communication. Measurements from these studies indicated that all parameters of $Ca^{2+}$ wave spread were substantially reduced in WT and Cx43 (-/-) astrocytes treated with heptanol, whereas suramin only reduced velocity and augmented attenuation in both types of littermates, without reducing the efficacy of transmission. Thus, the major route of intercellular $Ca^{2+}$ wave signalling under these conditions is through gap junction channels; similar conclusions were reached from studies of rat astrocytes by others [23,24].

## 4. Conclusions

We conclude from these studies that cardiac myocytes and CNS astrocytes from mice lacking the expression of the gene encoding Cx43 are deficient in intercellular coupling, although the degree of impairment varied for different functional assays (Table 1). Dye coupling was reduced in both cardiac myocytes and in astrocytes from Cx43 (-/-) mice, but more severely in the cardiocytes. Conversely, $g_j$, which is similar in WT astrocytes and cardiac myocytes, is about 3 to 4-fold higher in Cx43(-/-) cardiocytes than astrocytes. It seems likely that this paradox between effects on different measurements of intercellular coupling strength reflects the expression of different additional connexins in the two systems: High-conductance but less anion-permeant Cx40 channels appear to predominantly contribute the non-Cx43 conductance in cardiac myocytes, whereas low conductance but more anion-permeant Cx46 channels appear to contribute to the coupling seen in Cx43 (-/-) astrocytes. Intriguingly, both astrocytes and myocytes from Cx43 (-/-) mice appear to be connected on average by about a dozen junctional channels ($g_j/\gamma_j$).

The preservation of coordinated, although less rhythmic, contractions in Cx43 (-/-) myocytes is consistent with reports that only a few junctional channels are required for impulse synchronization in cardiac myocytes [25,26], and the maintenance of only slightly altered $Ca^{2+}$ wave propagation in Cx43 (-/-) cardiocytes and astrocytes indicate that the safety factor for this process in wildtype cells is also quite high. Although cardiac conduction in Cx43 (-/-) mice is favored by the high conductance of the residual channels, the restricted permeability of these channels might be expected to impede diffusion of anionic morphogens at developmentally relevant times. This difference in communication compartment function might conceivably contribute to the congenital malformation of the ventricular outflow tract.

## References


[1] D.C. Spray, *et al.*, Cardiovascular gap junctions: Gating properties, function and dysfunction. In: P.M. Spooner *et al.* (eds) Ion Channels in the Cardiovascular System: Function and Dysfunction. Futura Publishing, Mt. Kisco, NY, 1994, pp. 185-217.

[2] R. Dermietzel and D.C.Spray Gap junctions in the brain: What type, how many and why? Trends Neurosci 16 (1993):186-192.

[3] C. Fromaget *et al.*, Distribution pattern of connexin43, a gap junctional protein, during the differentiation of mouse heart myocytes. Differentiation 51 (1992):9-20

[4] R. Dermietzel *et al.*, Differential expression of three gap junction proteins in developing and mature brain tissues. Proc Natl Acad Sci USA 86 (1989): 10148-10152

[5] R. Rozental *et al.*, Changes in gap junction properties induced during neuronal differentiation in immortalized hippocampal progenitor cells. Submitted to J. Neuroscience.

[6] E. Simburger *et al.*, Expression of connexin43 mRNA in adult rodent brain. Histochem Cell Biol 107 (1997):127-137.

[7] A.G. Reaume *et al.* Cardiac malformation in neonatal mice lacking connexin 43. Science 267 (1995):1831-1834.

[8] R. Dermietzel *et al*, Gap junctions between cultured astrocytes: Immunocytochemical, molecular and electrophysiological analysis. J. Neurosci. 11(1991):1421-1432.

[9] J.M..Burt and D.C. Spray, Volatile anesthetics block intercellular communication between neonatal rat myocardial cells. Circ Res 65 (1989):829-837

[10] J. Neyton and A.Trautmann, Single-channel currents of an intercellular junction. Nature 317(1985):331-335.

[11] R.L. White *et al.*, Some electrical and pharmacological properties of gap junctions between adult ventricular myocytes Am J Physiol 249 (1985): C447-C455

[12] A.P. Moreno *et al.*, Properties of gap junction channels formed of connexin 45 endogenously expressed in human hepatoma (SKHep1) cells. Amer J Physiol 268 (1995):C356-C365.

[13] D.C. Spray *et al.*, Equilibrium properties of a voltage-dependent junctional conductance. J Gen Physiol 77(1981):77-93.

[14] A.L. Harris *et al.*, Kinetic properties of a voltage-dependent junctional conductance. Gen Physiol 77 (1981):95-117.

[15] J.L. Perez Velazquez *et al.*, Development of astrocytes and neurons in cultured brain slices from mice lacking connexin43. Devel Br Research 97 (1996):293-296.

[16] A.C. Carvalho *et al.*, Gap junction distribution is altered between cardiac myocytes infected with Trypanosoma cruzi. Circ Research 70 (1992):733-742.

[17] D.C. Spray, Physiological properties of gap junction channels in the nervous system. In DC Spray and R Dermietzel, (eds) Gap Junctions in the Nervous System. RG Landes Company, Austin, TX., 1996, pp.39-59.

[18] M. Chanson *et al.*, Gating characteristics of a steeply voltage-dependent gap junction channel in rat Schwann cells. J gen Physiol 102 (1993):925-946.

[19] P. Hellmann *et al.*,Properties of connexin40 gap junction channels endogenously expressed and exogenously overexpressed in human choriocarcinoma cell lines. Pflug Arch 432 (1996): 501-509.

[20] F.F. Bukauskas *et al.*, Biophysical properties of gap junction channels formed by mouse connexin40 in induced pairs of transfected human HeLa cells. Biophys J 68 (1995): 2289-2298

[21] J.C. Sáez *et al.*, Hepatocyte gap junctions are permeable to the second messenger inositol 1,4,5-triphosphate, and to calcium ions. Proc Natl Acad Sci USA 86 (1989): 2708-2712.

[22] M.J. Sanderson, Intercellular waves of communication. News Physiol Sci 11 (1996): 262-269.

[23] L. Venance *et al.*, Mechanism involved in initiation and propagation of receptor-induced intercellular calcium signaling in cultured rat astrocytes. J Neurosci 17 (1997): 1981-1992.

[24] M.O. Enkvist and K.D. McCarthy, Activation of protein kinase C blocks astroglial gap junction communication and inhibits the spread of calcium waves. Neurochem 59 (1992): 519-526.

[25] R. Weingart and P. Maurer, Action potential transfer in cell pairs isolated from adult rat and guinea pig ventricles. Circ Res 63 (1992): 72-80.

[26] D.E. Clapham *et al.*, Junctional resistance and action potential delay between embryonic heart cell aggregates. J Gen Physiol 75 (1980): 633-654.

# Section 9:

# Functional Studies of Gap Junctions in Mutants and Knockout Mice

*Gap Junctions*
*R. Werner (Ed.)*
*IOS Press, 1998*



# Identification of a Mutation in the Connexin 50 (Cx50) Gene of the *No2* Cataractous Mouse Mutant

Ernest C. Steele, Jr.[1], Mary F. Lyon[2], Peter H. Glenister[2], Pascale V. Guillot[2], and Robert L. Church[1*]

[1]*Emory Eye Center, Emory University School of Medicine, 1365B Clifton Rd., NE, Atlanta, GA 30322 [USA]*

[2]*MRC Mammalian Genetics Unit, Harwell, Didcot, Oxfordshire OX11 ORD, United Kingdom*

**Correspondence to: Robert L. Church, Ph.D.*
*Emory Eye Center, Rm. B5601*
*1365B Clifton Rd., NE*
*Atlanta, GA 30322 [USA]*
*Phone: [404] 778-4101*
*FAX: [404] 778-2232*
*Email: rlchurc@emory.edu*

**Abstract.** The purpose of this study was to screen the entire Cx50 (MP70) protein coding region of the *No2* (nuclear opacity #2) mouse mutant for a mutation that was linked to and consistent with the observed cataractous phenotype.

Genomic DNA's were prepared from both mutant and normal, parental spleens. DNA oligonucleotides were designed and used to amplify the entire Cx50 coding region, including the upstream splice acceptor site from these DNA's via PCR. PCR products were gel purified and then directly sequenced

A single A→C transversion was observed in the coding strand of the mutant DNA. This DNA substitution results in the nonconservative amino acid substitution of Ala for the normally encoded Asp in a conserved domain of the encoded polypeptide. The DNA alteration also creates a restriction site not present in the wild type DNA sequence. This was used to corroborate the sequence data.

A mutation has been identified in the Cx50 gene of the *No2* mouse mutant. This mutation is both linked to and consistent with the observed cataractous phenotype, as it is only present in mutant DNA and is predicted to alter a conserved domain of the ocular-specific membrane protein encoded by the gene. These results indicate that Cx50 plays an important role in normal lens development and maintenance and heralds its gene as a likely candidate in human congenital, hereditary cataract.

## 1. Introduction

Connexin 50(Cx50) was originally identified as an intrinsic membrane protein which localized to fiber gap junctions in the ovine ocular lens and was referred to as MP70 [1]. The Cx50 protein coding region was subsequently cloned from mouse genomic DNA and demonstrated to be a functional member of the gap junction family of proteins as well as the murine homologue of MP70 [2]. The gene structure of Cx50, like all known connexins [3] consists of a promoter and a small noncoding exon followed by a large intronic stretch and a single coding exon. This relatively simple gene structure combined with an unusually high degree of conservation at both the nucleotide and amino acid level, has facilitated the recent cloning of the Cx50 protein coding region from human [4] and sheep [5].

The chromosomal localization of the Cx50 gene has also been investigated in both the mouse and human. In the mouse, the gene, symbolized *Gja8*, was first assigned to chromosome 3 using somatic cell hybrids [6]. Subsequently, the gene was assigned to human chromosome 1 [4]. More intensive mapping experiments then allowed the subchromosomal localization to a position 11.9 $\pm$ 5.0 cM distal to the molecular marker DMit22 on mouse chromosome 3 [7]. Most recently, the gene has been narrowed down to band 21.1q on the long arm of human chromosome 1 [8]. Interestingly, this region of human chromosome 1 shows a high degree of conserved synteny with the above mentioned region of mouse chromosome 3.

At the same time as the subchromosomal localization of Cx50 in human was being carried out, linkage studies of the cataractous mouse mutant *No2* (nuclear opacity #2) were carried out. These studies indicated that the cataract mapped to the same region as the Cx50 gene on mouse chromosome 3 (M. Lyon, personal communication). The *No2* mouse, originally named ENU-326, was isolated in a study evaluating the efficacy of ethylnitrosourea (ENU) as a spermatogonial mutagen in the mouse [9]. Both heterozygous and homozygous *No2* mice display a bilateral nuclear opacity of the ocular lens, the severity being greater in the latter. The cataract is both congenital, manifesting as soon as the eyes are open, and hereditary (semi-dominant). The linkage data implicated *Gja8*, the gene encoding the mouse Cx50 protein, as an ideal candidate gene. We therefore, decided to screen the entire Cx50 protein coding region of the *No2* mutant for a mutation that was linked to and consistent with the cataractous phenotype. We report here the identification of such a mutation and propose a causative relationship between this mutation and the observed cataract in the *No2* mouse.

## 2. Methods

High molecular weight genomic DNA's were isolated from 200 mg of frozen spleens from normal parental mouse strains and heterozygous and homozygous *No2* mice according to standard methods using Proteinase K digestion of cells followed by organic extraction and precipitation of DNA. The DNA's were quantitated using a UV spectrophotometer and were analyzed on an agaraose gel to confirm high molecular weight and lack of degradation. DNA oligonucleotides were then designed and used along with a proofreading DNA polymerase to amplify the entire Cx50 protein coding region, including the upstream splice acceptor site, from these genomic DNA's. PCR products were then resolved on a preparative agarose gel, recovered from agarose, and then sequenced using Cy5-labeled DNA oligonucleotide primers and the Cy5 AuoCycle Sequencing Kit (Pharmacia Biotech, Piscataway, NJ). Sequencing reactions were electrophoresed and analyzed using an ALF express automated DNA sequencer (Pharmacia Biotech, Piscataway, NJ).

## 3. Results

PCR yielded an amplification product approximately 1.6 kilobases long from both normal parental DNAs as well as heterozygous and homozygous DNAs. This uniformity in size indicated that there were no large insertions, deletions or rearrangements in the mutant DNA. Sequencing of the amplification products identified a single A$\rightarrow$C transversion in the coding strand of the homozygous *No2* DNA (See Fig. 1). Both nucleotides were observed in the heterozygous *No2* DNA as would be expected.

```
A : 5'- GAG TTT GTG TGG GGC GAT GAG CAA TCT GAT TTT-3'
        GLU PHE VAL TRP GLY ASP GLU GLN SER ASP PHE

B : 5'- GAG TTT GTG TGG GGC GCT GAG CAA TCT GAT TTT-3'
        GLU PHE VAL TRP GLY ALA GLU GLN SER ASP PHE
```

**FIGURE 1.** Nucleotide and translated amino acid sequences of wild type (A) and *No2* mutant (B) Cx50. Altered nucleotides and amino acids are underlined.

This single nucleotide alteration results in the nonconservative substitution of alanine (Ala) for the normally encoded aspartic acid (Asp) at amino acid position 47 in the Cx50 polypeptide. In addition, the nucleotide substitution creates an additional *HhaI* restriction endonuclease recognition site that is not present in the wild type sequence. This allowed corroboration of the sequence data by digestion of the PCR products with this enzyme (See Fig. 2).

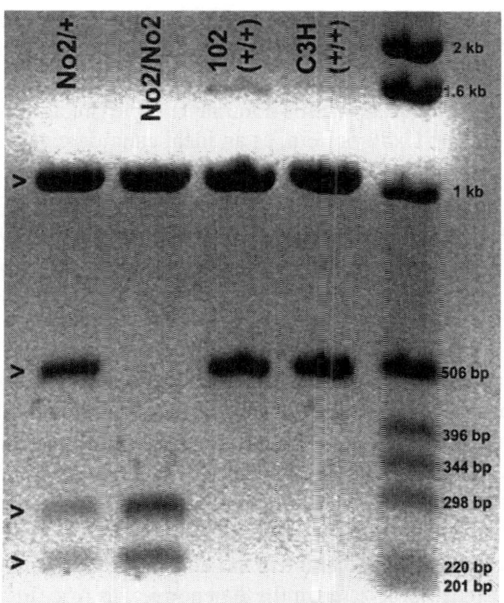


**FIGURE 2. *HhaI* - digested Cx50 PCR amplification products.** Wild type product (102 and C3H) are cut at the wild type endogenous site only, yielding fragments of ~1 kb and 500 bp. Homozygous *No2/No2* mutant product is cut twice, once at the wild type site and once at the new site created by the mutation, yielding fragments of ~1 kb, ~300, and ~200 bp. Heterozygous *No2/+* product displays all 3 bands as only half of the product contains the mutation.

The identified mutation results in an amino acid substitution within the first extracellular loop (E1) of the Cx50 polypeptide (See Fig. 3). A survey of all connexin predicted amino acid sequences in the GenBank database revealed that this domain is particularly well conserved in all connexins of all species studied to date. In particular, the Asp residue is absolutely nonvariant. A putative role in voltage gating has been inferred for this domain of the connexin from the results of structure-function studies using point mutations and chimeric connexins [10,11]. The fact that this region is so well conserved and is of functional significance leads one to believe that a mutation within this region hould have consequence.

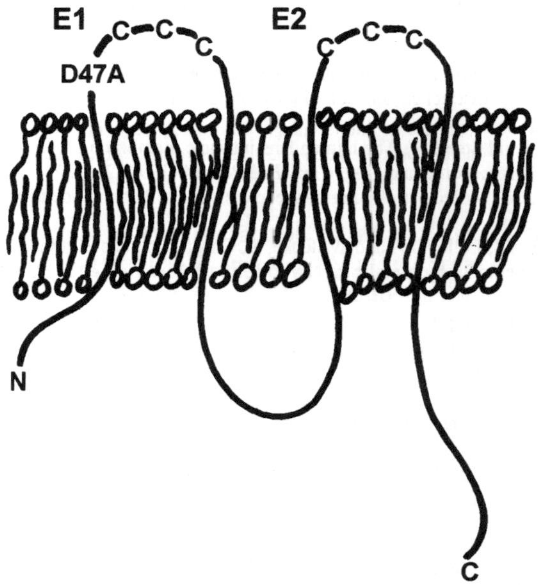


**FIGURE 3. Topological position of amino acid substitution in Cx50 polypeptide.** The Asp→Ala (D47A) mutation is located within the first extracellular loop (E1) between the membrane junction and the first cysteine residue (C) within this domain.

## 4. Discussion

Being an avascular organ, the ocular lens depends heavily upon extensive metabolic and electrochemical coupling of fiber cells for homeostasis and maintenance of lens transparency. Gap junctions and their protein components have long been suspected as players in cataractogenesis. To date, however, no evidence of a mutation within a lens connexin has been linked to cataract. Here, we have provided the first evidence that a mutation in a lens connexin may be the direct cause of a congenital, hereditary cataract. The identified mutation is only present in the cataractous mice and results in a nonconservative amino acid substitution within the highly conserved E1 domain of Cx50.

There exists a human correlate to the *No2* mouse. Zonular pulverulent (or Coppock) cataract has been linked to the same region as the human Cx50 homologue [12]. We propose that this human hereditary cataract is also the result of a mutation in the Cx50 gene. We further propose that there may be several forms of human hereditary cataract that are the result of a genetic lesion in one or more of the connexin proteins expressed within the ocular lens.

## References


[1] Kistler, J, Kirkland, B and Bullivant, S, Identification of a 70,000-D protein in lens membrane junctional domains, *J. Cell. Biol.* 101 (1985) 28-35.

[2] White, TW, Bruzzone, R, Goodenough, DA and Paul, DL, Mouse Cx50, a functional member of the connexin family of gap junction proteins, is the lens fiber protein MP70, *Mol. Biol. Cell.* 129 (1992) 711-720.

[3] Fishman, GI, Eddy, RL, Shows, TB, Rosenthal, L and Leinwand, LA, The human connexin gene family of gap junction proteins: distinct chromosomal locations but similar structures, *Genomics* 10 (1991) 250-256.

[4] Church, RL, Wang, J and Steele, E, The human lens intrinsic membrane protein MP70 (Cx50) gene : clonal analysis and chromosome mapping. *Curr. Eye Res.* 14 (1994) 215-221.

[5] Yang, DI and Louis, CF, Molecular cloning of sheep connexin 49 and its identity with MP70, *Curr. Eye Res.* 15 (1996) 307-314.

[6] Schwarz, HJ, Chang, YS, Lalley, PA and Willecke, K, Chromosomal assignments of mouse genes for connexin 50 and connexin 33 by somatic cell hybridization, *Somatic Cell and Molecular Genetics* 20 (1994) 243-247.

[7] Kerscher, S, Church, RL, Boyd, Y and Lyon, MF, Mapping of four mouse genes encoding eye lens-specific strucural, gap junction, and integral membrane proteins: *Cryba1* (crystallin βA3/A1), *Crybb2* (crystallin βB2), *Gja8* (MP70) and *Lim2* (MP19), *Genomics* 29 (1995) 445-450.

[8] Geyer, DD, Church, RL, Steele, EC, Heinzmann, C, Kojis, TL, Klisak, I, Sparkes, R, and Bateman, JB. Regional mapping of the human MP70 (Cx50) gene by fluorescence *in situ* hybridization to 1q21.1, *Mol. Vis.* (1997), in press.

[9] Favor, J, A comparison of the dominant cataract and recessive specific-locus mutation rates induced by treatment of male mice with ethylnitrosourea, *Mutat. Res.* 110 (1983) 367-382.

[10] Rubin, JB, Verselis, VK, Bennett, MVL and Bargiello, TA, A domain substitution procedure and its use to analyze voltage dependence of homotypic gap junctions formed by connexins 26 and 32, *Proc. Natl. Acad.Sci. USA* 89 (1992) 3820-3824.

[11] Rubin, JB, Verselis, VK, Bennett, MVL and Bargiello, TA, Molecular analysis of voltage dependence of heterotypic gap junctions formed by connexins 26 and 32, *Biophys. J.* 62 (1992) 183-195.

[12] Renwick, JH and Lawler, SD, Probable linkage between a congenital cataract and the Duffy blood group locus. *Ann. Hum. Genet.* 27 (1963) 67-84.

*Gap Junctions*
*R. Werner (Ed.)*
*IOS Press, 1998*

# Characterization of Immortalized Connexin43 Knockout Cell Lines Reveals Altered Biological Properties

Kendra Dean MARTYN[1,2], Wendy KURATA[1], Janis BURT[3], Bonnie WARN-CRAMER[1], Erica TENBROEK[4], and Alan F. LAU[1,2]. *Molecular Carcinogenesis[1], Cancer Research Center and the Department of Genetics and Molecular Biology[2], School of Medicine, University of Hawaii at Manoa, Honolulu, HI, 96813; the Department of Physiology[3], University of Arizona, Tucson, AZ, 85724; and the Department of Genetics and Cell Biology[4], University of Minnesota, St. Paul, MN 55108*

**ABSTRACT.** Immortalized cells from embryonic connexin43 knockout mice (Cx43$^{-/-}$) and homozygous littermates (Cx43$^{+/+}$) were cloned and characterized to determine whether the absence of connexin43 function would induce observable phenotypic changes supportive of a role for Cx43 in cell growth control and neoplastic transformation. Cells of the Cx43$^{+/+}$ clones expressed Cx43 and engaged in gap junctional communication (GJC) with 10-12 neighboring cells. The Cx43$^{-/-}$ cells were devoid of Cx43 and communicated to less than 1 cell. Electrophysiological analysis indicated that the Cx43$^{-/-}$ cells communicated through Cx45 channels from 8-80 fold less than the Cx43$^{+/+}$ clones, which appeared to communicate through Cx43 and Cx45 channels. The Cx43$^{-/-}$ clones grew at faster rates and to higher saturation densities, had more spindly morphologies, were more refractile, and adhered less well to the substratum than the Cx43$^{+/+}$ clones. Re-introducing the Cx43 gene into the Cx43$^{-/-}$ clones resulted in three subclones that communicated to 3-4 cells. Partial restoration of GJC in the 3 subclones was accompanied by reduced growth rates and saturation densities (2-fold compared to parental Cx43$^{-/-}$ clones), but no reversions in morphology or cell-substratum adhesion. The increased growth rates and saturation densities, altered morphology, and decreased cell adhesion displayed by the Cx43$^{-/-}$ clones are a subset of the properties of transformed cells [1]

## Introduction

The 43 kDa connexin protein, Cx43, is a member of a family of evolutionarily conserved integral plasma membrane proteins which form channel structures permitting the passive intercellular exchange of small molecules and ions [2], a process referred to as gap junctional communication (GJC). Cx43 is expressed in numerous mammalian tissues, and is believed to be involved in a wide spectrum of biological processes including heart development [3], synchronous contraction of the heart and uterus [4], generation of action potentials in excitable cells [5], exocrine and endocrine gland secretion [6], ovarian follicular growth [7], as well as cell differentiation, development and growth control [8-10 ].

Early evidence for the involvement of GJC in growth control came from the observation that many tumor tissues lacked GJC. Further evidence now implicates loss of GJC in the process of neoplastic transformation. For example, tumor promoters and carcinogens have been shown to disrupt GJC [8, 11,12], and the activities of various oncogenes including *v-src* [13], polyoma middle T antigen [14], *v-Fps* [15], *v-ras*, *v-mos*, SV40 large T antigen, and [16,17], *neu* [18] have been linked to loss of GJC. Moreover, the reestablishment of GJC in transformed cell lines through the incorporation of cDNA for various connexin genes, including Cx32 [19], Cx26 [20, 21], and Cx43 [21-25], have led to the reversal of transformed properties, as demonstrated by reduced growth rates and saturation densities, normalization of cell morphology, and reduced tumorigenicity. Taken together, these results suggest that connexins may represent a novel class of tumor suppressor genes.

We exploited the availability of Cx43 knockout mice to establish immortalized Cx43$^{-/-}$ and Cx43$^{+/+}$ mouse embryonic fibroblast cell lines to use as a suitable cell system for investigating whether the loss of the Cx43 gene at the onset of mouse development might initiate the process of neoplastic transformation. Here we report partial characterization of these cell lines which provides evidence supporting a role for Cx43 in growth control, and suggesting that loss of Cx43 function may be involved in the transformation of normal cells to a preneoplastic state.

## Results

We immortalized Cx43$^{-/-}$ and Cx43$^{+/+}$ fibroblast cells by continuous passage in culture medium rather than using immortalizing viruses like SV40 that have other functions which could complicate interpretation of results. Northern and immunoblot analysis was used to confirm the absence of Cx43 mRNA and protein in three Cx43$^{-/-}$ clones, and to demonstrate approximately equal levels of Cx43 mRNA and protein in three companion Cx43$^{+/+}$ clones (data not shown). Lucifer Yellow dye permeability and electrophysiological analysis using the dual whole-cell patch clamp technique demonstrated that the Cx43$^{-/-}$ clones were coupled from 8 to 30-fold less than the Cx43$^{+/+}$ clones (See Table 1). Single channel conductance studies (Table 1) and Northern blot analysis (data not shown) indicated that the Cx43$^{-/-}$ clones were likely coupled through Cx45 channels, while the Cx43$^{+/+}$ clones were coupled through Cx45 and Cx43 channels.

Microscopic analysis revealed striking morphological differences between the Cx43$^{-/-}$ and the Cx43$^{+/+}$ clones which were similar to morphological differences often noted between normal and transformed cell types. The Cx43$^{-/-}$ cells had a more spindly shape, with enhanced refractility, more disorganized growth on the culture plate, and few cell-cell contacts, whereas the Cx43$^{+/+}$ clones were very flat cells that grew in a highly organized fashion, and appeared to have many close contacts with neighboring cells.

The Cx43$^{-/-}$ clones also demonstrated a marked reduction in cell-substratum adhesion compared to that of the Cx43$^{+/+}$ clones. An assay to indirectly measure the number of cells remaining attached to the culture dish after treatment with EDTA, a reagent known to dissociate cell-cell and cell-substratum adhesions, indicated that the Cx43$^{-/-}$ clones were approximately 2.5-fold less adhesive to the culture dish than the Cx43$^{+/+}$ clones (Table 1). These results suggested that loss of Cx43 may affect focal adhesions through disruptions of normal cell-extracellular matrix interactions involving integrins and extracellular matrix proteins. These kinds of disruptions are common in transformed cells. [26].

The hallmark of cellular transformation is increased growth rates and saturation densities. Growth curve analysis clearly demonstrated that the loss of Cx43 function in the Cx43$^{-/-}$ clones was associated with higher growth rates (average of 3-fold higher; Table 1) and elevated cell saturation density (average of 3-fold higher; Table 1).

The soft agar growth assay was conducted to determine whether the Cx43$^{-/-}$ clones had lost their anchorage dependent growth. The results demonstrated no growth in soft agar of any Cx43$^{-/-}$ or Cx43$^{+/+}$ clones, whereas 35% of the v-src transformed Rat1 control cells grew under these conditions (data not shown).

Reexpression of Cx43 in the Cx43$^{-/-}$ cells by infection and transfection of mammalian Cx43 cDNA expression vectors resulted in 3 clones expressing high levels of Cx43 (data not shown), and communicating on average to 3 or 4 neighboring cells, as measured by Lucifer yellow dye transfer (Table 1). Growth curve analysis of these clones revealed reduced growth rates (2-fold; Table 1) and saturation densities (2-fold; Table 1) in each of these clones compared to the parental knockout cell line, whereas 6 out of 7 noncommunicating controls showed no reduction in these properties (data not shown). However, these Cx43 reexpressing clones failed to show reversion to that of the Cx43$^{+/+}$ condition in morphology or cell-substratum adhesions (Table I).

Table I. Biological properties of Cx43$^{-/-}$ and Cx43$^{+/+}$ cells.

| Clone | Dye Transfer[1] | Macro Conduct. (nS)[2] | Unitary Conduct. (pS)[3] | Morphology[4] | Adhesion (%)[5] | Growth rate[6] | Saturation density (x $10^6$)[7] |
|---|---|---|---|---|---|---|---|
| +/+3 | 12.5 | 4.5±1.2 | 30, 65 | F | 79.3 | 1.45 | 0.98 |
| +/+7 | 11.8 | 9.6±1.2 | 30, 65 | F | 75.9 | 0.93 | 0.52 |
| +/+8 | 8.9 | 14.5±1.7 | 30, 65 | F | 72.0 | 0.93 | 0.69 |
| -/-1 | 0.7 | 0.26±0.09 | 30 | S | 31.1 | 5.90 | 3.80 |
| -/-3 | 0.3 | 0.18±0.05 | 30 | S | 27.8 | 3.09 | 2.14 |
| -/-4 | 0.5 | 0.53±0.26 | 30 | S | 27.1 | 2.19 | 1.56 |
| 1R1 | 4.0 | ND | ND | S | 21.3 | 2.53 | 1.72 |
| 3R1 | 3.3 | ND | ND | S | 52.8 | 2.48 | 1.19 |
| 3R2 | 4.0 | ND | ND | S | 58.6 | 1.26 | 1.06 |

1R1 is a Cx43 reexpressing subclone of clone -/-1.

3R1 and 3R2 are Cx43 reexpressing subclones of clone -/-3.

[1]Dye Transfer measures the number of fluorescing cells surrounding a single cell microinjected with Lucifer yellow.

[2]Macro Conduct., the macroscopic conductance measured using the dual whole-cell patch clamp technique.

[3]Unitary Conduct., the recording of single channel conductance events.

nS , nanoSeimens; pS, picoSeimens.

[4]F, flat. S, spindly.

[5]Adhesion data represents % of 5mM EDTA-treated cells remaining attached to culture dish compared to untreated controls.

[6]Growth rate is represented by the slope of the growth curve between day 2 and 6.

[7]Saturation density is determined by counting number of cells on a 60 mm culture plate at day 8 of growth.

ND, not determined.

## Discussion

We have investigated the role that loss of Cx43-mediated GJC might play in the process of cellular transformation by characterizing immortalized cell lines from Cx43 knockout mice and from Cx43 wildtype littermates. Our analysis of these genetically altered cell lines has provided evidence that lack of Cx43 gene function may cause cells to acquire biological properties that are indicative of progression towards neoplastic transformation. These properties include altered morphological characteristics such as more spindly shape and enhanced refractility, decreased cell-substratum adhesion, and increased growth rates and saturation densities. Reexpression of Cx43 in 3 Cx43$^{-/-}$ subclones resulted in partial restoration of GJC. These three subclones demonstrated reduced growth rates and saturation densities, and clearly provided evidence supportive of a growth control function for Cx43-mediated GJC. However, the Cx43 reexpressing subclones failed to show any reversion in morphology or cell-substratum adhesion. Because these clones did not communicate at a level equal to the Cx43 wildtype cells it is not clear whether reversal of these properties required a higher level of GJC.

We did not observe loss of anchorage dependent growth, as indicated by a failure of the cells to grow in soft agar, which suggests that the Cx43$^{-/-}$ cells have not acquired the ability to form tumors in animals. Taken together, our results indicate that loss of Cx43-mediated GJC is not sufficient to induce full transformation, but suggest that loss of Cx43 function may be an important step in a multistep process of cellular transformation.

Interestingly, our analysis of GJC using dual whole-cell patch clamping techniques demonstrated that the Cx43$^{-/-}$ cells were coupled electrically at very low levels through Cx45 channels. In contrast, the coupling of the Cx43$^{+/+}$ cells was from 8 to 80 fold higher and appeared to be mediated through both Cx43 and Cx45 channels. Therefore, our Cx43$^{-/-}$ and Cx43$^{+/+}$ cells maybe similar in the selectivity of signaling through Cx45 channels, but differ by a complete lack of signaling through Cx43 channels. Cx45 channels are not permeated well by negatively charged molecules, whereas Cx43 channels are less charge selective [27] and can more easily pass negative charged molecules like the second messengers cAMP, cGMP and IP3. These results suggest that it may be the failure of cells to pass such second messengers through Cx43 channels that contributes to the acquisition of neoplastic properties.

## Acknowledgments

The authors thank Dr. J. Rossant for the generous gift of primary culture cells from the Cx43 knockout and wild type mice, Dr. M. Kanemitsu for the pBabeCx43b virus, Dr. D. Miller for retroviral packaging cells, and Dr. E. Beyer for connexin cDNAs. This research was supported by grants from National Institutes of Health to A.F.L (CA52098) and J.M.B. (HL31008) and the American Heart Association, Hawaii Affiliate to B.W.-C. (HIFW-09-95). K.D.M. was supported by NIH T32 training grant (CA61806).

## References


[1] K.D. Martyn *et al.*, Immortalized connexin43 knockout cell lines display a subset of biological properties associated with the transformed phenotype, *Cell Growth Diff.* (1997) In Press.

[2] W.R. Loewenstein, Junctional intercellular communication: the cell-to-cell membrane channel, *Physiol. Rev.* 61 (1981) 829-913.

[3] A.G. Reaume *et al.*, Cardiac malformation in neonatal mice lacking connexin43, *Science* (1995) 267: 1831- 1834.

[4] D.C. Spray and J.M. Burt, Structure-activity relations of the cardiac gap junction channel, *Am. J. Physiol.*258 (1990) C195-C205.

[5] P.R. Brink *et al.*, Gap junctions in excitable cells, *J. Bioenerg. and Biomembr.* (1996) 28: 351-358.

[6] P. Meda, The role of gap junction membrane channels in secretion and hormonal action, *J. Bioenerg. and Biomembr.* 28 (1996) 369-377.

[7] J. Wiesen and R. Midgley, Changes in expression of connexin 43 gap junction messenger ribonucleic acid and protein during ovarian follicular growth, *Endocrin.* 133 (1993) 741-746.

[8] H. Yamasaki and C.C.G. Naus, Role of connexin genes in growth control, *Carcinogenesis* 17 (1996) 1199-1213.

[9] W.R. Loewenstein and B. Rose, The cell-cell channel in the control of growth, *Cell Biol.* 3 (1992) 59-79.

[10] C.W. Lo, The role of gap junction membrane channels in development, *J. Bioenerg. and Biomembr.* 28 (1996) 379-385.

[11] J.W. Holder *et al.*, Gap junction function and cancer, *Cancer Res.* 53 (1993) 3475-3485.

[12] J.E. Trosko *et al.*, Chemical, oncogene and growth factor inhibition of gap junctional intercellular communication: an integrative hypothesis of carcinogenesis, *Pathobiol.* 58 (1990) 265-278.

[13] R. Azarnia *et al.*, The cellular src gene product regulates junctional cell-cell communication, *Science* 239 (1988) 398-401.

[14] R. Azarnia and W.R. Loewenstein, Polyomavirus middle T antigen down regulates junctional cell-to-cell communication, *Mol. Cell Biol.* 7 (1987) 946-950.

[15] W.E. Kurata and A.F. Lau, $p130^{gag\text{-}fps}$ disrupts gap junctional communication and induces phosphorylation of connexin43 in a manner similar to that of $pp60^{v\text{-}src}$, *Oncogene* 9 (1994) 329-335.

[16] M. Atkinson and J. Sheridan, Altered junctional permeability between cells transformed by v-ras, v-mos, or v-src, *Am J. Physiol.* 255 (1988) C647-C683.

[17] R. Arzania and W.R. Loewenstein, Intercellular communication and the control of growth. XII. Alteration of junctional permeability by Simian virus 40, *J. Membr. Biol.* 82 (1984) 213-222.

[18] Y.-W. Jou *et al.*, Inhibition of gap junctional intercellular communication and malignant transformation of rat liver epithelia cells by *neu* oncogene, *Carcinogenesis* 16 (1995) 311-317.

[19] B. Eghbali *et al.*, Involvement of gap junctions in tumorigenesis: Transfection of tumor cells with connexin 32 cDNA retards growth *in vivo*, *Proc. Natl. Acad. Sci. USA* 88 (1991) 10701-10705.

[20] M. Mesnil *et al.*, Negative growth control of HeLa cells by connexin genes: connexin species specificity, *Cancer Res.* 55 (1995) 629-639.

[21] K.K. Hirschi *et al.*, Gap junction genes Cx26 and Cx43 individually suppress the cancer phenotype of human mammary carcinoma cells and restore differentiation potential, *Cell Growth Diff.* 7 (1996) 861-870.

[22] P.P Mehta *et al.*, Incorporation of the gene for a cell-cell channel protein into transformed cells leads to normalization of growth, *J. Membr. Biol.* 124 (1991) 207-225.

[23] B. Rose *et al.*, Gap-junction protein gene suppresses tumorigenicity, Carcinogenesis 14 (1993) 1073-1075.

[24] D. Zhu *et al.*, Transfection of C6 glioma cells with connexin 43 cDNA: analysis of expression, intercellular coupling, and cell proliferation, *Proc. Natl. Acad. Sci. USA* 88 (1991) 1883-1887.

[25] S.-C. Chen *et al.*, Connexin43 reverses the phenotype of transformed cells and alters their expression of cyclin/cyclin-dependent kinases, *Cell Growth Diff.* 6 (1995) 681-690.

[26] S.H. Lo and L.B. Chen, Focal adhesion as a signal transduction organelle, *Cancer and Metas. Reviews* 13 (1994) 9-24.

[27] R.D. Veenstra, Size and selectivity of gap junction channels formed from different connexins, *J. Bioenerg. and Biomembr.* 28 (1996) 327-337.

*Gap Junctions*
*R. Werner (Ed.)*
*IOS Press, 1998*



# Role of Cx40 in Cardiac Conduction

Alexander M. SIMON, Daniel A. GOODENOUGH*, and David L. PAUL
*Dept. of Neurobiology and Dept. of Cell Biology*, Harvard Medical School, Boston, MA 02115*

**Abstract.** In the mouse heart, connexin40 (Cx40) is the only known connexin in gap junctions of Purkinje fibers in the His bundle and the proximal portions of the left and right bundle branches. These fibers mediate fast conduction of action potentials from the atrioventricular node to the ventricles and thus are essential for coordinated myocardial contraction. To directly test if Cx40 is required for conduction through the His/Purkinje system, we generated mice that have the Cx40 gene deleted. Surprisingly, homozygous mutant mice were viable and had normal sinus rhythm. Further studies are required to determine if Cx40 null mice have altered conduction characteristics.

## 1. Introduction

The generation of efficient pumping action by the heart requires the coordinated contraction of the myocardium. Ventricular contraction begins with the stiffening of the interventricular septum followed rapidly by the contraction of the apex and then the base of the heart. Contraction is also controlled radially; ventricles are first activated at the endocardial surface, with depolarization subsequently spreading through the myocardium to the epicardial surface. Coordinated activation of the myocardium is controlled by a specialized conduction system (Fig. 1). Rhythmic firing of the sinoatrial node activates atrial contraction and initiates impulse propagation to the atrioventricular (AV) node. Conduction through the AV node is very slow, allowing for a delay between atrial and ventricular contraction. Distal to the AV node, conduction occurs very rapidly through the bundle of His, which branches on either side of the interventricular septum, forming a network of fibers that activate the endocardial surface of the ventricles. Since the atria and ventricles are electrically insulated from each other by a fibrous ring of tissue, the AV node/His bundle pathway is the normal route for atrioventricular impulse propagation.

The cells that comprise the bundle of His are modified myocardial cells called Purkinje fibers which exhibit two features contributing to fast conduction. First, these cells have characteristic fast rising, large amplitude action potentials. Second, Purkinje fibers are well coupled by a high density of gap junctions which provide a low resistance pathway for current flow at cell-cell borders. Gap junctions are intercellular channels made up of connexin, a family of related proteins with distinct patterns of expression and functional properties. Although gap junctions are thought to be essential for rapid conduction through the His/Purkinje system, this has not been demonstrated directly.

Connexin expression in the specialized conduction system has been studied in a number of species including rodents, dogs and humans [1-7; reviewed by 8]. In all species except guinea pig, Cx40 was detected in Purkinje fibers of adult heart. Human Purkinje fibers also expressed Cx43, whereas canine fibers contained Cx43 and Cx45 in addition to Cx40. In the mouse, however, Cx40 was the only connexin detected in the His bundle and

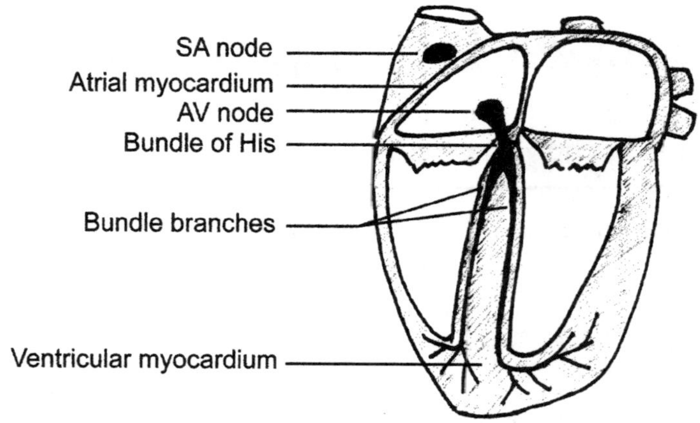


Fig. 1 Specialized cardiac conduction system

the proximal portion of its branches, providing a unique system for studying the role of gap junction communication in cardiac conduction. To directly test if Cx40 is required for conduction through the His/Purkinje system, we generated mice that have the Cx40 gene deleted. Surprisingly, Cx40 null mice were viable and had normal sinus rhythm. Our results raise the possibility of an additional connexin in the His/Purkinje system of Cx40 knockout mice, or an alternate conduction pathway, to account for the lack of complete conduction block.

## 2. Materials and Methods

### *Immunocytochemistry and histology*

Hearts were frozen unfixed in O.C.T embedding medium. Frozen sections (10 μm) were fixed in acetone at -20° C for 5 minutes and immunolabeled with anti-Cx40 antibody [9]. Antigen-antibody complexes were detected with rhodamine-conjugated goat anti-rabbit IgG (Boehringer Mannheim). To identify the conduction system tissue, adjacent sections were stained for acetylcholinesterase activity [10].

### *Gene targeting*

A rat Cx40 cDNA [11] was used to isolate genomic clones from a 129/Sv mouse genomic library (gift of Dr. En Li). A replacement targeting vector containing a *pgk-neo* cassette was designed to delete the entire coding region (Fig. 2). The vector contained 5' and 3' homology regions of 2 kb and 10 kb respectively, inserted into the pPNT vector [12] and included a *pgk-TK* casette downstream of the 3' homology region. The vector was electroporated into J1 ES cells (gift of Dr. En Li) and $neo^r$-$FIAU^r$-clones were isolated by positive/negative selection as described [13]. Homologous recombinants were identified by PCR and genomic structure was confirmed by Southern blotting using a 5' probe (probe 1) and a probe within the neo gene (probe 2).

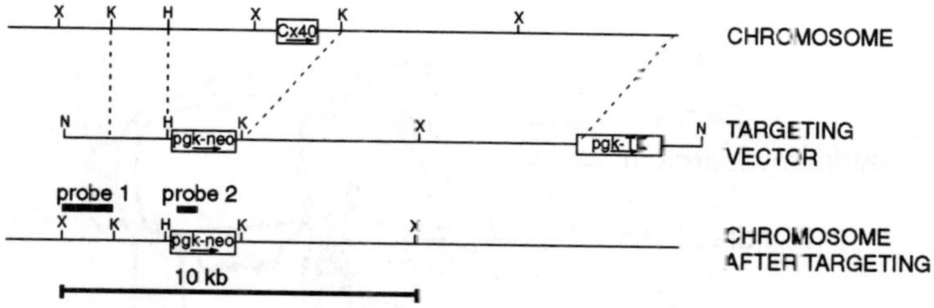


Fig. 2 Strategy used for targeted inactivation of the Cx40 gene.

### *Electrocardiography*

Animals were lightly anesthetized by intraperitoneal injection of sodium pentobarbital (~70 mg/kg) and kept warm with a heating pad to minimize muscular activity during recordings. Three-lead EKGs were measured in mice using mini test clip adaptors (Radio Shack) which were dipped in conductive jelly and attached to the limbs at the digits. Data were amplified with a Gould 4600 Biotach EKG amplifier and recorded on a Gould 2400 strip-recorder or digitally with a PC running the Axotape program (Axon Instruments). Digital data acquisition was at 2.5 kHz.

## 3. Results and Discussion

To directly test if Cx40 is required for conduction through the His/Purkinje system, we generated mice in which the Cx40 gene was deleted using the targeting vector shown in Fig. 2. Heterozygous (Cx40+/-) mice were interbred and resulting litters genotyped by Southern blotting or PCR. Surprisingly, the ratio of genotypes was close to Mendelian (1:2:1), indicating that there was no reduction in viability of homozygous (Cx40-/-) mutant mice. Immunocytochemistry was performed on frozen sections of Cx40-/- hearts to confirm that the Cx40 protein had been eliminated. Anti-Cx40 antibodies [9] labeled Purkinje fiber junctions in wildtype His bundle tissue (Fig. 3), but did not label these fibers in Cx40-/- mice (Fig. 4). Adjacent sections were stained for acetylcholinesterase activity to unambiguously identify the conduction system tissue in these sections [10]. Cx40 was also absent in atrial myocardium and vascular endothelium of Cx40-/- mice, tissues where it is normally expressed [3, 14].

Three-lead EKGs were taken on lightly anesthetized mice using surface electrodes attached to their limbs. In these recordings, the first deflection (P wave) results from atrial depolarization, the second series of positive and negative deflections (QRS complex) results from ventricular depolarization and the last deflection (T wave) is due to ventricular repolarization. Heart rates in the Cx40-/- mice and wildtype mice were similar and showed normal sinus rhythm; P waves were always followed at the same interval by a QRS complex. Thus, ventricular depolarization in Cx40-/- mice was still triggered by impulses originating in the atria and suggested that conduction through the His/Purkinje system was not completely blocked. Further analysis will be required to determine if Cx40-/- mice have altered conduction characteristics. The absence of complete conduction block in mice lacking Cx40 raises the possibility of the existence of a previously undetected connexin in the mouse conduction system or an alternative pathway for conduction.

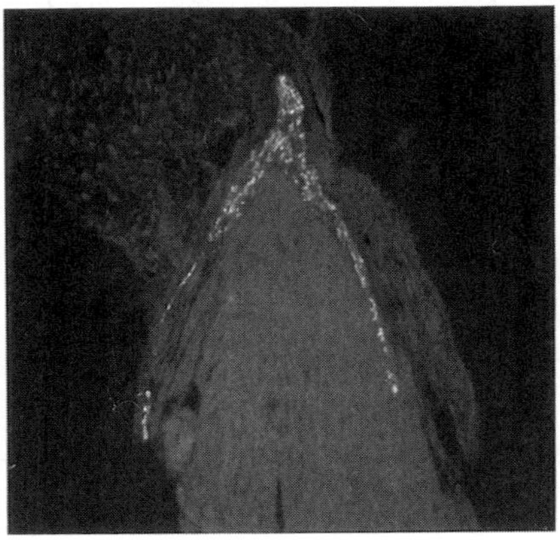

Fig. 3 Wildtype heart section immunostained with anti-Cx40 antibody. The conducting fibers are labeled.

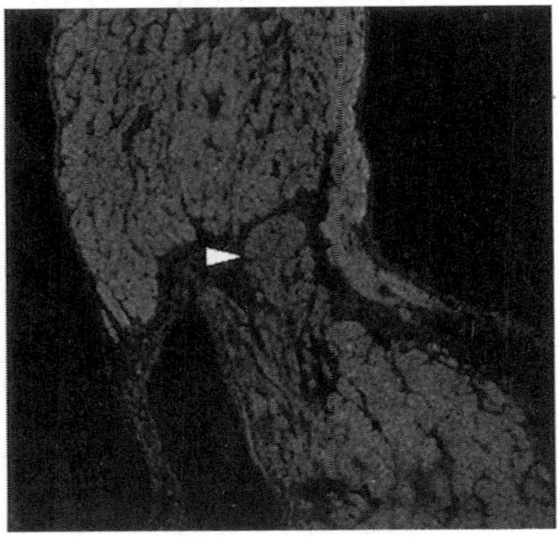

Fig. 4 Cx40-/- heart section immunostained with anti-Cx40 antibody. Cx40 is absent from conducting fibers (arrowhead).

Immunohistochemical staining of His bundle tissue from wildtype or Cx40-/- mice with antibodies against connexins 43, 45, 37, 46, 50, 32, and 26 has so far not revealed the presence of another connexin.

**Acknowledgments**

We are grateful to Joe Gabriels for preparations of affinity purified antibodies to Cx40, Cx37, and Cx43. We thank Dr. Janis Burt for assistance in preliminary electrocardiographic experiments.

## References

[1] Davis, L. M., Kanter, H. L., Beyer, E. C., and Saffitz, J. E. (1994). Distinct gap junction protein phenotypes in cardiac tissues with disparate conduction properties. Journal of the American College of Cardiology *24*, 1124-1132.

[2] Van Kempen, M. J. A. K., Vermeulen, J. L. M., Moorman, A. F. M., Gros, D., Paul, D. L., and Lamers, W. H. (1996). Developmental changes of connexin40 and connexin43 mRNA distribution patterns in the rat heart. Cardiovascular Research *32*, 886-900

[3] Delorme, B., Dahl, E., Jarryguichard, T., Marics, I., Briand, J. P., Willecke, K., Gros, D., and Theveniaurussy, M. (1995). Developmental Regulation Of Connexin 40 Gene Expression In Mouse Heart Correlates With the Differentiation Of the Conduction System. Developmental Dynamics *204*, 358-371.

[4] Bastide, B., Neyses, L., Ganten, D., Paul, M., Willecke, K., and Traub, O. (1993). Gap junction protein connexin40 is preferentially expressed in vascular endothelium and conductive bundles of rat myocardium and is increased under hypertensive conditions. Circ Res *73*, 1138-1149.

[5] Gourdie, R. G., Severs, N. J., Green, C. R., Rothery, S., Germroth, P., and Thompson, R. P. (1993). The spatial distribution and relative abundance of gap- junctional connexin40 and connexin43 c orrelate to functional properties of components of the cardiac atrioventricular conduction system. J Cell Sci *105*, 985-991.

[6] Gros, D., Jarry-Guichard, T., Ten Velde, I., de Maziere, A., van Kempen, M. J., Davoust, J., Briand, J. P., Moorman, A. F., and Jongsma, H. J. (1994). Restricted distribution of connexin40, a gap junctional protein, in mammalian heart. Circulation Research *74*, 839-851.

[7] Kanter, H. L., Laing, J. G., Beau, S. L., Beyer, E. C., and Saffitz, J. E (1993). Distinct patterns of connexin expression in canine purkinje fibers and ventricular muscle. Circ.Res. *72*, 1124-1131.

[8] Gros, D. B., and Jongsma, H. J. (1996). Connexins in mammalian heart function. Bioessays *18*, 719-730.

[9] Gabriels, J. E., and Paul, D. L. (1993) Characterization of connexin expression in rat aortic endothelial cells during in vitro wounding. Mol. Biol. Cell *4*, 329a.

[10] Carbonell, L. M. (1955). Esterses of the conductive system of the heart Histochem Cytochem *4*, 87-97.

[11] Haefliger, J. A., Bruzzone, R., Jenkins, N. A., Gilbert, D. J., Copeland, N. G., and Paul, D. L. (1992). Four novel members of the connexin family of gap junction proteins. Molecular cloning, expression, and chromosome mapping. J Biol Chem *267*, 2057-2064.

[12] Tybulewicz, V. L., Crawford, C. E., Jackson, P. K., Bronson, R. T., and Mulligan, R. C. (1991). Neonatal lethality and lymphopenia in mice with a homozygous disruption of the c-abl proto-oncogene. Cell *65*, 1153-63.

[13] Li, E., Bestor, T. H., and Jaenisch, R. (1992). Targeted mutation of the DNA methyltransferase gene results in embryonic lethality. Cell *69*, 915-26.

[14] Bruzzone, R., Haefliger, J. A., Gimlich, R. L., and Paul, D. L. (1993). Connexin40, a component of gap junctions in vascular endothelium, is restricted in its ability to interact with other connexins. Mol Biol Cell *4*, 7-20.

*Gap Junctions*
*R. Werner (Ed.)*
*IOS Press, 1998*

# Biological functions of gap junctions revealed by targeted inactivation of mouse connexin32, -26 and –40 genes

Klaus Willecke with Albrecht Buchmann[1], Christoph Bützler, Heinz-Dieter Gabriel, Andreas Hagendorff[2], Dirk Jung, Susanne Kirchhoff, Olaf Krüger, Eric Nelles, Michael Schwarz[1], Achim Temme, Otto Traub, and Elke Winterhager[3]

*Institut für Genetik, Abt. Molekulargenetik, Universität Bonn, 53117 Bonn*
*[1]Institut für Toxikologie, Universität Tübingen, 72074 Tübingen*
*[2]Medizinische Klinik (Kardiologie), Universität Bonn, 53105 Bonn*
*[3]Institut für Anatomie, Universität Essen, 54122 Essen, Germany*

**Abstract.** Using standard techniques of embryonic stem cell mediated homologous recombination, mouse null mutants were generated by targeted inactivation of the connexin32, -26 or –40 genes that code for subunit proteins of gap junction channels. Connexin32 defective mice were fertile and showed defective propagation of signals generated by sympathetic nerve stimulation in liver (cf. Nelles et al, Proc. Natl. Acad. Sci. **93** (1996) 9565-9570). In addition, one year old connexin32 deficient mice had significantly more spontaneous and chemically induced liver tumors than wild type controls. - Mutant mice, deficient for the connexin26 gene, died in utero around day 11 post coitum, presumably due to insufficient uptake of glucose and/or other nutrients through the labyrinth region of the placenta, where connexin26 was highly expressed in wild type mice. - Connexin40 deficient mice were viable, although they lacked expression of connexin40 in conductive myocardium. Homozygous connexin40 deficient mice showed prolonged P wave, PQ-, QS-, and $QT_{max}$ intervals in their electrocardiogram, compared to wild type litter mates, suggesting delayed atrial and ventricular conduction.

Our results show that different intercellular connexin channels fulfill different functions in various tissues, such as protection against hepatocarcinogenesis (connexin32), intercellular transfer of nutrients (connexin26), or efficient propagation of myocardial action potentials during heart beat (connexin40). It remains to be clarified whether these functional differences are due to intrinsic differences in the permeability of connexin channels.

## 1. Introduction

The ability to generate targeted mouse mutants in genes with known genomic sequences, has revolutionized the molecular biology of the mouse genome and has produced many mouse models for human diseases. It is now possible to delete or substitute each DNA fragment or nucleotide in the mouse genome and to study the functional consequences in the adult mouse or in the mouse embryo, if the genetic alteration turns out to be embryonic lethal. In order to generate null mutants or „knock-out" mice, one had to transfect the altered gene with flanking DNA into cultured embryonic stem cells, select for homologous recombination, inject the recombinant embryonic stem cells into mouse blastocysts and transfer these manipulated early mouse embryos into the uterus of foster mice. The chimeric progeny of foster mice were backcrossed, in order to recognize germ line transmission of the altered gene in heterozygous mice. Then these mice were inbred to generate the desired homozygous defective „knock-out" mice [cf. 1,2 for reviews].

The first connexin gene that had been reported to be inactivated by this technique was the mouse connexin43 (Cx43) gene which coded for the most abundant gap junctional protein and is strongly expressed in mammalian heart [3]. Cx43 deficient mice died shortly after birth. Since the right ventricular outflow tract of their hearts was obstructed, the blood could not be circulated and properly oxygenized in the lung. The molecular reason for the malformed heart in Cx43 defective mice is not understood. Cx43 containing gap junction channels are likely to fulfill an essential role in morphogenesis of the mouse heart during embryonic development [3]. A similar conclusion has been reached by overexpression of Cx43 in transgenic mice [4].

In our laboratory we have characterized three different connexin deficient („knock-out“) mice, generated by standard techniques of gene targeting, as briefly outlined above. Two of these connexins, connexin32 (Cx32) and connexin26 (Cx26) are coexpressed in hepatocytes, whereas the third, connexin40 (Cx40), is expressed in conductive myocardium and in endothelial cells. Below we have briefly summarized recent results and conclusions from characterization of these connexin deficient mice.

## 2. Connexin32 deficient mice develop hepatocarcinomas

Previously, we have reported that Cx32 deficient mice were viable and fertile and weighed on the average about 17% less than wild type controls. When sympathetic liver nerves in Cx32 deficient mice were electrically stimulated, 78% less glucose than in wild type control mice was released from hepatic glycogen stores into the blood stream [5]. Noradrenaline, secreted from activated sympathetic liver nerves, could bind to surface receptors on hepatocytes and trigger an intracellular increase of inositol 1.4.5 trisphosphate. Our results suggested that the intercellular propagation of this second messenger mediated signal for glucose release was impaired, due to the missing Cx32 protein in gap junctions of connexin32 deficient mice. Interestingly, the amount of the second hepatic gap junctional protein, Cx26, was decreased by about two thirds in Cx32 deficient livers [5]. The Cx32 protein may stabilize the Cx26 protein in wild type hepatocytes, possibly by interaction in heterotypic or heteromeric connexin channels.

Recently, we have noticed that one year old Cx32 deficient mice showed 20-fold (male) and 8-fold (female) more spontaneous liver tumors than control mice [6]. Incorporation of bromodeoxyuridine into Cx32 deficient liver cells was higher than in wild type liver, suggesting that there are more proliferating hepatocytes in Cx32 deficient liver. Intraperitoneal injection of the carcinogen diethylnitrosamine into young Cx32 deficient mice led, after one year, to more liver tumors and accelerated tumor growth than in control animals [6]. Apparently, Cx32 mediated gap junctional communication among hepatocytes could suppress proliferation of initiated (i.e. mutated) cells and, thus, contribute to inhibition of hepatocarcinogenesis in normal mouse liver. Since the amount of Cx26 protein was partially downregulated in Cx32 deficient liver [5], we could not exclude that the increased hepatocarcinogenesis could also be due to decreased expression of Cx26, in addition to the loss of Cx32 protein. Primary hepatocytes from Cx32 deficient liver showed about 15% electrical conduction, compared to 100% in wild type hepatocytes [V. Valiunas, H. Nießen, K. Willecke, and R. Weingart, manuscript in preparation, 1997].

Human patients harboring a defective Cx32 gene in their genome, suffer from Charcot-Marie-Tooth Disease (X Type), an inherited demyelination disorder of peripheral nerves, characterized by progressive weakness of distal muscles in the second decade of life [7]. Recently we have shown that > 8 months old Cx32 deficient mice also exhibited morphological signs of demyelination („onion bulbs“) in their peripheral nerves [8], suggesting that older Cx32 deficient mice could serve as model organisms for the human Charcot-Marie-Tooth disorder (X Type). So far, there is no report in the literature that human Charcot-Marie-Tooth patients (X Type) have an increased risk of hepatocarcinogenesis, compared to unaffected individuals. Since both hepatocarcinomas and the Charcot-Marie-Tooth (X Type) disorder occur very rarely in the Western World, it is possible that an increased coincidence has not yet been noticed, in the human population. In contrast to the rare occurrence of hepatocarcinomas in humans, wild type mice showed a higher frequency of spontaneous hepatocellular carcinomas, which was still significantly increased in Cx32 deficient mice.

## 3. Connexin26 deficient mouse embryos die in utero, probably due to placental dysfunction

When we tried to generate homozygous Cx26 deficient mice, we noticed that these embryos ceased to grow after day 10 post coitum and died at day 11 post coitum. Except of growth retardation, no other morphological alteration was detected in homozygous Cx26 deficient embryos. After day 9 post coitum, the chorioallantoic placenta started to function, where connexin26 was highly expressed between the syncytiotrophoblasts I and II in the labyrinth region of wild type mice. When we measured transplacental uptake of the radioactively labelled, non-metabolizable glucose analogue, 3-O[$^{14}$C] methylglucose, at day 10 post coitum, we found that Cx26 deficient embryos accumulated only about 40% of the glucose analogue, compared to heterozygous or wild type embryos of the same size. Apparently, this decreased uptake of glucose and presumably other essential nutrients was not sufficient to support further growth and the rapid organogenesis that took place at this stage of embryonic development [9]. The embryonic lethality of homozygous Cx26 deficient mice did not allow to study the function of Cx26 channels in the adult animal, for example in hepatocytes, where Cx32 and Cx26 proteins are coexpressed. In order to generate adult mice which lack Cx26 in the liver, one could attempt to delete the Cx26 gene specifically in hepatocytes. This may be accomplished in the future by Cre mediated, hepatocyte specific recombination of the mouse Cx26 gene, flanked by loxP sites [cf. 10]. Recently it has been reported that hereditary non-syndromic sensorineural deafness in human patients is due to mutations in the connexin26 gene [11].

## 4. Connexin40 deficient mice exhibit delayed cardiac conduction

Homozygous Cx40 defective mice that we generated by standard techniques of gene targeting, turned out to be viable and fertile, although they totally lacked expression of Cx40. Previously, it has been shown that mouse Cx40 is specifically expressed in atrium, conductive myocardium, and in endothelial cells [cf. 12]. Since propagation of action potentials along the His bundle and Purkinje fibers of the conductive myocardium is thought to be essential for the normal function of the heart, possibly another connexin participates in myocardial conduction and may compensate for the loss of Cx40 in Cx40 deficient mice. When we recorded electrocardiograms by surface electrodes, fixed to the four limbs of homozygous Cx40 defective mice, we found that the P wave as well as the PQ-, QS- and $QT_{max}$ intervals were significantly prolonged, compared to wild type mice. These results suggested that the atrial as well as ventricular conduction in the myocardium were delayed in homozygous Cx40 deficient mice [S. Kirchhoff, E. Nelles, A. Hagendorff, O. Krüger, O. Traub and K. Willecke, (1997), manuscript in preparation]. Since we occasionally observed arrhythmias in Cx40 deficient mice, but not in wild type mice, it is possible that Cx40 deficiency may increase the risk of cardiac arrhythmias in mice.The phenotype of Cx40 deficient mice is similar to that of human patients showing first degree of AV block in the presence of extended QRS complex.

## 5. Outlook

When one reviews the different phenotypes caused by defective connexin32, -26, and –40 genes in transgenic mice, it becomes obvious that the functions of connexin genes that code for the subunit proteins of gap junction channels are different in various tissues and cell types [cf. 13]. Is this only due to cell type specific expression of connexin genes, or is it due to different permeabilities of various connexin channels for second messenger molecules, other metabolites and ions? The functional relevance of the latter possibility could be verified by „knock-in“ studies [cf. 2], whereby coding regions of different connexin genes are exchanged for one another, expressed in transgenic mice and analyzed for functional complementation. If new and different phenotypes are generated under these conditions, it would strongly support the notion that the diversity of connexin channels is associated with different permeabilities of biologically relevant molecules.

## Acknowledgement

Work in the Bonn laboratory on the generation of connexin deficient mice was supported by grants of the Deutsche Forschungsgemeinschaft through SFB 284 (project C1), the Deutsche Krebshilfe, BIOMED program of the European Community, and Fonds der Chemischen Industrie to K.W.

## References

[1] W.S. Pascoe, R. Kemler and S.A. Wood, Genes and functions: trapping and targeting of embryonic stem cells, Biochem. Biophys. Acta **1114** (1992) 209-221.

[2] A. Bradley and P. Liu, Targeted practice in transgenics, Nature Genetics **14** (1996) 121-123.

[3] A.G. Reaume, P.A. de Sousa, S. Kulkarni, B.L. Langille, D. Zhu, T.C. Davies, S.C. Jujena, G.M. Kidder and J. Rossant, Cardiac malformation in neonatal mice lacking connexin43, Science **267** (1995) 1831-1834.

[4] J.L. Ewart, M.F. Cohen, R.A. Meyer, G.Y. Huang, A. Wessels, R.G. Gourdie, A.J. Chin, S.M.J. Park, B.O. Lazatin, S. Villabon and C.W. Lo, Heart and neural tube defects in transgenic mice overexpressing the Cx43 gap junction gene, Development **124** (1997) 1281-1292.

[5] E. Nelles, C. Bützler, D. Jung, A. Temme, U. Dahl, O. Traub, F. Stümpel, K. Jungermann, J. Zielasek, K.V. Toyka, R. Dermietzel and K. Willecke, Defective propagation of signals generated by sympathetic nerve stimulation in the liver of connexin32-deficient mice, Proc. Natl. Acad. Sci. **93** (1996) 9565-9570.

[6] A. Temme, A. Buchmann, H.-D. Gabriel, E. Nelles, M. Schwarz and K. Willecke, High incidence of spontaneous and chemically induced liver tumors in connexin32 deficient mice, Current Biology **7** (1997) 713-716.

[7] J. Bergoffen, S.S. Scherer, S. Wang, S.M. Oronzi, L.J. Bone, D.L. Paul, K. Chen, M.W.Lensch, P.F. Chance and K.H. Fischbeck, Connexin mutations in X-linked Charcot-Marie-Tooth disease, Science **262** (1993) 2039-2042.

[8] P. Anzini, D.H.H. Neuberg, M. Schachner, E. Nelles, K. Willecke, J. Zielasek, K.V. Toyka, U. Suter and R. Martini, Structural abnormalities and deficient maintenance of peripheral nerve myelin in mice lacking gap junction protein connexin32, J. Neuroscience **17** (1997) 4545-4551.

[9] H.D. Gabriel, D. Jung, C. Bützler, A. Temme, O. Traub, E. Winterhager and K. Willecke, Transplacental uptake of glucose is decreased in embryonic lethal connexin26 deficient mice, (1997) submitted for publication.

[10] H. Gu, J.D. Marth, P.C. Orkan, H. Nossmann and K. Rajewsky, Deletion of a DNA polymerase beta segment in T cells using cell type specific gene targeting, Science **265** (1994) 103-106.

[11] D.P. Kelsell, J. Dunlop, H.P. Stevens, N.J. Lench, J.N. Liang, G. Parry, R.F. Müller and I.M. Leigh, Connexin26 mutations in hereditary non-syndromic sensorineural deafness, Nature **387** (1997) 80-83.

[12] D.B. Gros and H.J. Jongsma, Connexins in mammalian heart function, BioEssays **18** (1996) 719-730.

[13] S.M. Nicholson and R. Bruzzone, Gap Junctions: getting the message through, Current Biol. **7** (1997) R340-R344.

*Gap Junctions*
*R. Werner (Ed.)*
*IOS Press, 1998*



# Heart Defects in Connexin43-Deficient Mice

Jing Ya, Erna BHW Erdtsieck-Ernste, Piet AJ de Boer, Marjan JA van Kempen[1], Habo Jongsma[1], Daniel Gros[2], Antoon FM Moorman & Wouter H Lamers. *Depts Anatomy & Embryology, Univ Amsterdam, Medical Physiology[1], Univ Utrecht, The Netherlands, and [2]IBDM, Fac Sciences de Luminy, Marseille, France*

**Abstract.** Cardiac malformation in CX43-disrupted mice is spatially restricted, even though CX43 is abundantly expressed elsewhere in the heart as well. To determine the underlying cause of this abnormality, we analyzed cardiac morphogenesis in immuno- and hybridohistochemically stained, and threedimensionally reconstructed serial sections of CX43-deficient embryos between ED10 and birth The major abnormalities we observed, were an abnormal looping of the arterial pole of the heart, which was seen between ED10 and ED12, and subsequently, an abnormal delamination of the anterosuperior leaflet of the tricuspid valve which became manifest by the progressive development of intertrabecular pouches at the junction of the right ventricle and the outflow tract. These pouches were first seen at ED13 and progressively expanded until birth. Apparently, the prolonged existence of the acute bend at the junction of the embryonic right ventricle and outflow tract that results from the unduly long persistence of the *A*-loop configuration of the heart loop, predisposes this myocardium to the excessive development of inter-trabecular pouches that subsequently develop in these hearts. Persistence of the A-loop configuration itself generated hearts with a typical "criss-cross" configuration, with the atrioventricular cushions rotated $90^0$ and a horizontal muscular ventricular septum. CX43 deficiency is not compensated by an upregulation of CX37, CX40 or CX45.

Congenital malformations of the heart are among the most common birth defects in man. The systematic analysis of these defects almost entirely rests on findings in human neonates with these structural anomalies. As a consequence, little direct information is available about the development of these malformations. In particular, it is usually not known to what extent a particular malformation can be traced to a single morphogenetic event or process. A case in point is the development of the malformation of the subpulmonary outflow tract of the heart in mice that are deficient for gap junctions of the CX43 type that was reported by Reaume et al. [1]. CX43 is abundantly expressed in the developing and formed heart, and gap junctions of this type are thought to be responsible for the behaviour of the myocardium as an electrical syncytium. In this connection, it was recently shown the spread of the impulse in animals heterozygous for the CX43-null mutation is markedly impaired [2]. The survival to birth of embryonic mice homozygous for the CX43-null mutation is therefore remarkable. The death of CX43-deficient mice has been associated with an obstruction of the subpulmonary outflow tract. However, the underlying cause of this abnormality and, in particular, the reason why the malformation is restricted to the subpulmonary outflow tract, whereas other parts of the heart in which CX43 is also expressed abundantly, are not affected, remains to be established. We analyzed serially sectioned CX43-gene disrupted mouse embryos between ED10 and birth. The sections were stained immuno- or hybridohistochemically to aid in delineating the myocardial and fibrous structures, and were subsequently reconstructed to visualize structural anomalies. A total of 44 homozygous mice, obtained from matings of hetero-zygotes, were available for study. These homozygotes represented 16.2% of the live animals, which is significantly below the expected 25% ($p<0.01$, Chi-square test) and confirms the earlier finding of Reaumé et al. [1]. The reduced percentage of homozy-gotes was observed at all ages investigated.

**Table 1.** Prevalence of homozygous Cx43-deficient embryos in offspring of hemizygous mice of the 129Sv strain.

| | 10 | 11 | 12 | 13 | 14 | 15 | 16 | 17 | 18 | total |
|---|---|---|---|---|---|---|---|---|---|---|
| -/- | 4 | 4 | 4 | 10 | 5 | 4 | 7 | 1 | 5 | 44 |
| total | 23 | 19 | 50 | 35 | 27 | 23 | 23 | 5 | 26 | 272 |
| % | 17 | 21 | 8 | 29 | 19 | 17 | 30 | 20 | 19 | 16.2 |

**Cardiac looping in CX43-deficient embryos**

One of the early signs of developing asymmetry in the embryo is the transformation of the midline tubular heart into an asymmetrical organ. Normally, the cranial, so called ascending loop of the tube will come to occupy the rightward position (so called dextro- or *D*-loop), but in exceptional cases, it can retain the middle position (so called anterior or *A*-loop) or even come to occupy the leftward position (so called laevo- or *L*-loop). In CX43 knock-out embryos, the *A*-loop persists up to ED11 before it changes into a D-loop like configuration as a result of the selective growth of the right ventricle, whereas in normal embryos, this process already occurs at ED9. Consequently, the right ventricle occupies an aberrant cranial rather than the normal right-lateral position in CX43-deficient ED10 and ED11 embryos. As a result, the angle between the right ventricle and the outflow tract is temporarily more acute than in control embryos. We do not know at present how this geometric distortion contributes to the subsequent developmental abnormalities at this location.

Our study of the development of the heart in CX43-disrupted mouse embryos suggests that the temporal coincidence of the formation of the cardiac loop and the appearance of gap junctions in the heart may result from a causal relation. CX43-deficiency only delays and does not abolish the establishment of the usual *D*-loop, because the growth of the right ventricle during ED12 and ED13 also favours a transition of the metastable *A*-loop into the stable *D*-loop. If this scenario is correct, it is to be anticipated that in exceptional cases the *A*-loop will persist. Indeed, in one ED12 embryo we observed the persistence of the *A*-loop configuration, while, in addition, a possible ED11 precursor was identified. We did not observe the persistence of the *A*-loop configuration at any later stage, suggesting that embryos suffering from it either are not viable or that the condition is rare (<1:34 CX43-deficient embryos studied after ED12). Persistence of the A-loop configuration generates a heart in which both atria occupy their respective left- and right-sided position, but in which the position of the more distal parts of the heart has rotated by $90^0$. Thus, the cushions in the atrioventricular canal occupy lateral positions, while the right ventricle retains its cranial position relative to the left ventricle. Accordingly, the left atrioventricular connection is dorsal and the right atrioventricular connection ventral. Furthermore, the ventricular origin of the outflow tract is cranial rather than right-sided, leading to a parallel rather than a spiral course of the endocardial ridges in the outflow tract, that will, upon further development, lead to transposition of the great arteries. Although we did not notice embryos with persisting *A*-loop beyond ED12, the observed configuration resembles that of the criss-cross heart, a rare and poorly understood cardiac malformation [3]. The hallmark of the criss-cross heart, a horizontal ventricular septum due to the ventro-dorsal topographic relation of the right and left ventricle, mostly in conjunction with atrioventricular concordance and ventriculoarterial discordance, is also present in our ED12 embryo. The criss-cross condition has been explained as a postseptational rotation of the ventricular mass around its long axis and as a displacement of the bulbus cordis with consequent alteration of the bulboventricular flange. However, our present observation shows that it more likely arises as a result of the persistence of the *A*-loop configuration. These findings argue in favour of screening human patients with criss-cross hearts for CX43 mutations.

Together with the topography of other asymmetric organs in the body, such as the lungs, the spleen and the stomach, the configuration of the cardiac loop is a marker of the sidedness of the body plan. Although more direct experimental evidence is available for the involvement of other genes in the development of left-right asymmetry, our finding of a criss-cross heart in statu nascendi in a CX43-deficient mouse embryo and the association of another disturbance in the sidedness of organs, viscero-atrial heterotaxia, with CX43 mutations in humans infants, both imply that CX43 gap junctions may be involved in the development of laterality in the body plan. The association of the inactivation of the connexin43 gene with impaired cardiac looping, that is, with a laterality problem, is particularly intriguing since the initial report that viscero-atrial heterotaxia is associated with mutations in this gene [4], has been questioned recently [5,6].

**Development of the tricuspid valve in CX43-deficient embryos**

Careful examination of serial sections and 3D reconstructions of affected and normal control embryos revealed that the increasingly overt structural malformation in CX43-deficient hearts was confined to the myocardium supporting the ventricular base of the endocardial outflow tract ridges. These shelf-shaped myocardial structures, which are involved in the formation of the anterosuperior leaflet of the tricuspid valve, had already drawn our attention when studying the development of this valve in human embryos [7]. These myocardial shelves are identifiable in both normal and malformed hearts from ED13 onwards, but can be delineated more easily in the CX43 knock-out embryos because of the development of pouches underneath these shelves. Topographically, the malformation therefore develops where, in preceding stages, the junction of the embryonic right ventricle and the outflow tract made a more acute bend in CX43-deficient embryos than in normal control embryos. In situ hybridization clearly showed that high expression of mRNA of CX43 and SERCA2 was confined to the myocytes of the right ventricle, including the shelf-like structures under which formation of pouches occurs from ED13 onwards, whereas the myocytes of the outflow tract did not express mRNA of CX43 and showed much lower SERCA2 expression. In normal hearts, these markers delineate the boundary between right ventricle and the outflow tract. Even though we have been able to ascribe the progressive cardiac malformation in CX43-deficient mice to that part of the subpulmonary myocardium that is responsible for the formation of the anterosuperior leaflet of the tricuspid valve, it is difficult to envisage how the expression pattern of CX43 itself can account for this topographical restriction of the abnormality. Our findings therefore imply that the abnormality is a late consequence of the acute bend between the right ventricle and the outflow tract that temporarily exists at this location between ED10 and ED12 or, alternatively and less likely, that CX43 gap junctions at the junction of the right ventricle and the outflow tract mediate the transfer of signal molecules that (negatively) regulate the delamination of the myocardium that accompanies valve leaflet formation.

After ED13, we observed a progressive development of the bilateral pouches at the junction of the right ventricle with outflow tract. The diameter of the pouches as well as that of the local trabeculae continues to increase toward birth. As the size of the pouches increases, the wall in the superior part of the right ventricle bulges further, surrounding the pulmonary orifice. In fact, the pronounced prenatal increase in size of the peripulmonary bulge probably already reflects the local failure of the myocardium. In

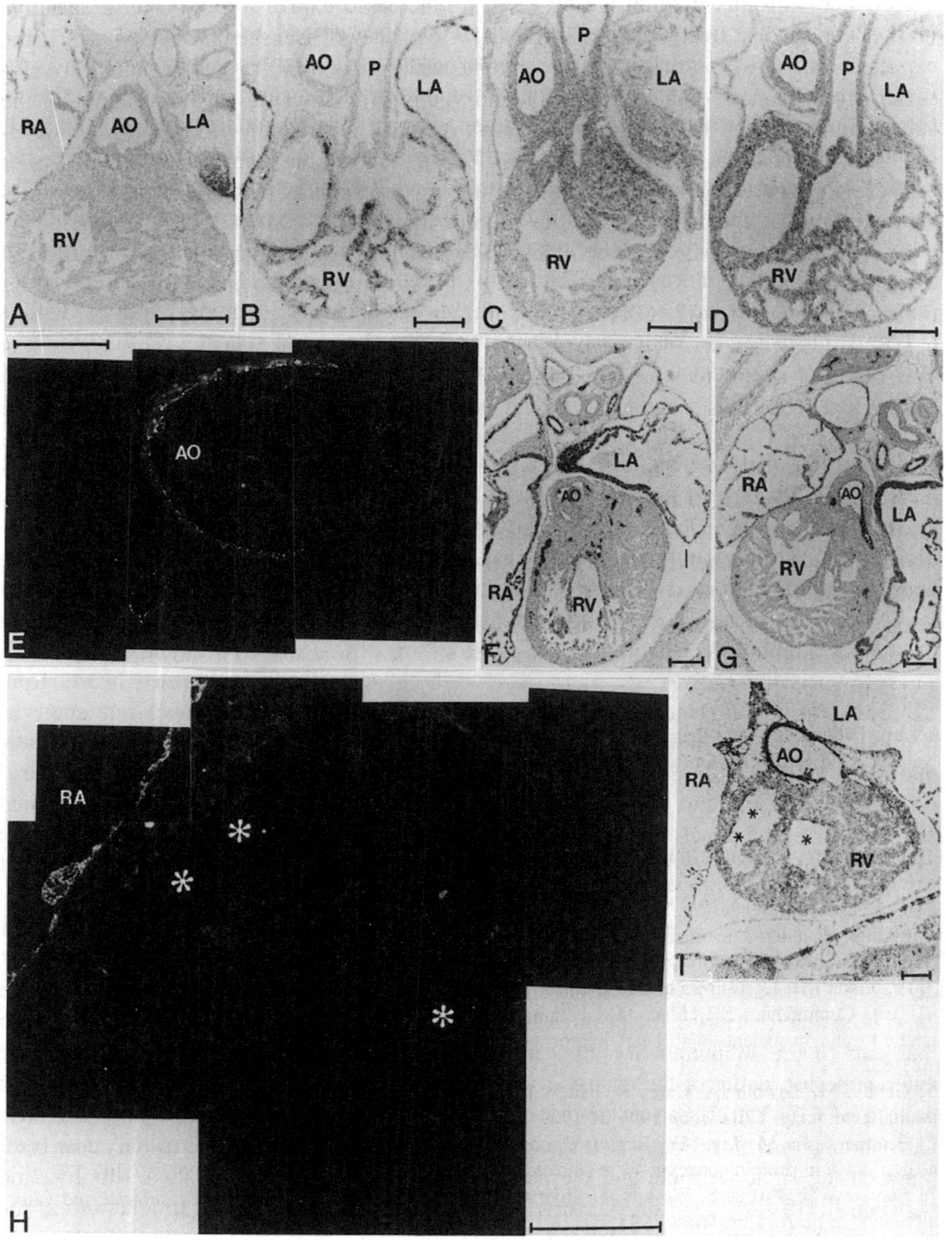


**Figure 1.** CX43-deficient hearts express ANF and show no compensatory upregulation of CX37, CX40 and CX45. The hypertrophic trabeculae in the right ventricle of CX43-deficient mice show an upregulation of ANF expression (compare panels A and B (normal and CX43-deficient ED18 hearts, respectively). CX37 (panel E) and CX40 protein (panel H) are expressed in the expected structures (vessels and atrium) in CX43-deficient ED18 hearts (panel I is a low magnification serial section identifying the aorta (panel E) and the intertrabecular pouches (asterisks; panel H)), while the distribution and staining intensity of the mRNAs for CX40 (panels F and G; ED16) and CX45 (panels C and D; ED18) in control (panels C and F) and CX43-deficient hearts (panels D and G) are similar. Ao: aorta; LA: left atrium; P: pulmonary trunk; RA: right atrium; RV: right ventricle; bar: 200$\mu$m.

accordance with this finding, we observed an increased expression of ANF, a sensitive marker for cardiac overload in adult heart. Most lines of transgenic mice in which the expression of CX43 is driven by the cytomegaly viral (CMV) promoter/enhancer, also suffer from neonatal lethality, when homozygous [8]. However, in these animals, the subpulmonary outlet is narrowed as a result of local hypertrophy. Interestingly, CX43-deficient animals that express the CMV-CX43 transgene, are temporarily rescued after birth [8], suggesting that the locally failing myocardium of CX43-deficient hearts is shored up by the hypertrophic stimulus originating from the CMV-CX43 transgene.

To find out if the absence of CX43 expression was compensated by an increased expression of CX37, CX40 or CX45, the expression of these connexins was assayed by immunofluorence (CX37, CX40) or in situ hybridization (CX40, CX45) (Fig. 1, panels C-H)). However, none of these connexins was upregulated in mutant embryos, as was also reported at this meeting by Gao and Spray at the mRNA level (RT-PCR), and by Saffitz and colleagues at the protein level (Western blots). The vessel walls were positive for CX37 and CX40, with CX40 also being abundantly expressed in the atria and the ventricular trabeculae. The marked difference in the developmental timing of the disappearance of CX40 mRNA from the trabeculae of left and right ventricles, that we previously noted in the rat, was also seen in control and mutant mice (not shown). The myocardium of the shelves at the base of the endocardial ridges of the outflow tract, that is, the structures below which the intertrabecular pouches develop in the CX43-deficient hearts, clearly expressed CX45 in both control and affected embryos (panels C and D). Elsewhere in the heart, CX45 expression was very low or absent (not shown).

**Acknowledgements**. We thank Drs DL Paul, K Willecke, K Boheler and N Carter for making cDNAs available, and Mr C Gravemeijer for his care of the photographs.

**References**

[1] Reaume AG, de Sousa PA, Kulkarni S, Langille BL, Zhu D, Davies TC, Juneja SC, Kidder GM, Rossant J: Cardiac malformation in neonatal mice lacking connexin43. *Science* 1995;267:1831-1834

[2] Guerrero PA, Schuessler RB, Davis LM, Beyer EC, Johnson CM, Yamada KA, Saffitz JE: Slow ventricular conduction in mice heterozygous for a connexin 43 null mutation. *J Clin Invest* 1997;99:-1991-1998

[3] Anderson RH: Criss-cross hearts revisited. *Pediatr Cardiol* 1982;3:305-313

[4] Britz-Cunningham SH, Shah MM, Zuppan CW, Fletcher WH: Mutations of the connexin43 gap junction gene in patients with heart malformations and defects of laterality. *N Engl J Med* 1995;332:-1323-1329

[5] Gebbia M, Towbin JA, Casey B: Failure to detect connexin43 mutations in 38 cases of sporadic and familial heterotaxy. *Circulation* 1996;94:1909-1912

[6] Penman Splitt M, Trai MY, Burn J, Goodship JA: Absence of mutations in the regulatory domain of the gap junction protein connexin 43 in patients with visceroatrial heterotaxy. *Heart* 1997;77:369-370

[7] Lamers WH, Virágh S, Wessels A, Moorman AFM, Anderson RH: Formation of the tricuspid valve in the human heart. *Circulation* 1995;91:111-121

[8] Ewart JL, Cohen MF, Meyer RA, Huang GY, Wessels A, Gourdie RG, Chin AJ, Park SMJ, Lazatin BO, Villabon S, Lo CW: Heart and neural tube defects in transgenic mice overexpressing the Cx43 gap junction gene. *Development* 1997;124:1281-1292

*Gap Junctions*
*R. Werner (Ed.)*
*IOS Press, 1998*

# Gap junction expression in lens and cornea of wildtype (WT) and Cx43 deficient (KO) mice: Cataractous changes in lens but no striking defect in cornea

**Yang Gao and David C. Spray**
*Department of Neuroscience, Albert Einstein College of Medicine, Bronx New York 10461*

**Abstract.** We have compared gap junction expression, tissue morphology and function in WT and Cx43 KO mice in order to evaluate the role of Cx43 in controlling lens development and organization and the role of junctional communication in migration and proliferation of corneal epithelial cells during wound repair. Eyes from newborn WT and Cx43 KO littermates were fixed for light and transmission electron microcopy or frozen in liquid $N_2$ for light and confocal microscopy of immunofluorescence in cryosections; corneal epithelial cells were cultured to evaluate growth rate, response to wounding, and gap junction channel properties. In WT mice, Cx43 was immunolocalized in lens epithelial cells and throughout the cornea, primarily at lateral regions of epithelial cells but also within the stroma and endothelial cells. The distribution of MIP in lens fiber cells in the bow region was similar in Cx43 KO and WT lenses. Epithelial cells and keratocytes in the cornea were connected by typical gap junctional contacts; contact areas were similar in WT and Cx43 KO mice. However, in Cx43 KO lenses, only tiny regions of contact were seen between epithelial cells, and organization of appositional membranes between lens fiber cells and between fiber and epithelial cells differed dramatically in the Cx43 KO mice. Large vacuolar spaces were apparent between fiber cells, which were largely separated from apical surfaces of epithelial cells. Corneal epithelial cells from WT and Cx43 KO mice grew at the same rate, and migration after wounding was not different. Gap junctional permeability and conductance were only slightly lower in Cx43 KO corneal epithelial cells than in WT. Junctional channels in WT corneal epithelial cells were primarily Cx43-like; channels in Cx43 KO were smaller and more voltage sensitive. Cx45 and Cx50 staining were positive in the cornea in both WT and Cx43 KO mice. These results indicate that lenses of Cx43 KO mice exhibit histological features typical of early cataract, highlighting the requirement for normal Cx43 expression in this tissue. By contrast, histology and wound repair in corneal epithelial cells were relatively normal in Cx43 KO mice, presumably due to functional expression of other connexins in this tissue.

## Introduction

Cornea and lens are key transparent elements in the visual zone of the eye, neither absorbing nor scattering significant amounts of light. Since the cornea and lens have no blood supply and most ions and nutrients are transported from the aqueous humor through epithelial layers, gap junction channels are hypothesized to provide the exchange of nutrients and metabolites between cells (1). Because Cx43 is the major gap junction protein in both epithelial cell types (2) it was of interest to evalutate lens and corneal structure and function in transgenic mice lacking Cx43.

## Methods

Mice lacking Cx43 (3) were genotyped at birth using PCR analyses of genomic tail DNA. For both Cx43 KO and WT mice from each litter, both eyes were removed after decapitation. One eyeball of each pair was frozen in liquid nitrogen for immunohistochemical studies on frozen sections, and from the other the lens was isolated with microsurgery for ultrastructural studies and the cornea was isolated for cell culture. Indirect immunofluorescence labelling of gap junction proteins in frozen sections of mouse eye tissues was carried out (4) using monospecific Cx43, Cx45, Cx46, Cx50 and MIP26 antibodies. For conventional thin section electron microscopy, lenses were fixed in 2% glutaraldehyde and 1% $OsO_4$. After staining with 0.25% uranyl acetate and dehydration in an ethanol series, the samples were embedded in Quetol 812 (5). Thin sections of the equatorial region were examined on a JEOL 1200 EX electron microscope at 80-kV accelerating voltage.

Corneal epithelial cells were cultured at low density and numbers of cells in randomly chosen fields of view were counted at 1,2,4, and 6 days of culture. Confluent cell cultures were wounded by scraping with a rubber policeman to produce a denuded area. At 10, 24, and 30 hrs thereafter, distance of the leading edge of cells migrating into the wound was measured and plotted as % coverage of the previously denuded area. Gap junction channel permeability was measured using Lucifer Yellow injection and scrape loading of confluent monolayers. Junctional conductance was measured in cell pairs using dual whole cell voltage clamp.

## Results and Discussion

Immunostaining revealed Cx43 immunoreactive sites in many parts of the eye in WT animals, including epithelia of cornea, lens and pigmental and non-pigmental epithelia (Fig.1A,B); staining was negative in Cx43 KO littermates (not illustrated). As has been described previously in mice (6), Cx43 in WT lenses showed a graded decrease in expression in the bow region, being highest in the epithelial layer and disappearing in lens fiber cells beyond the boundary. Our study of MIP26 expression showed a complementary pattern in the bow region of the outer cortex, being detected only in differentiating fiber cells (not shown). In lenses of Cx43 KO mice, MIP26 was also absent from epithelial cells; distribution of MIP26 in bow region lens fiber cells was not detectably different in lenses of WT and Cx43

KO animals.

To further investigate structural differences, we prepared Cx43 KO and WT lenses for thin section examination. Crossly similar differentiation of lens fibers in the bow region was observed at low power magnification using light microscopy of Toluidine blue-stained semithin sections; epithelium, outer and inner cortex, and the marginal region in which cell differentiation occurred were clearly discernible (Fig.1C,D). However, cells within the lenses of WT animals showed tight connections with one another, both between epithelial cells and between epithelial and fiber cells (Fig.1C), whereas lenses from Cx43 KO littermates were characterized by loosely apposed epithelial cells and by the large spaces between epithelial and fiber cells (Fig.1D). In outer cortex of lenses from both animals, fiber cell membranes were closely interlocked. However, vesicles with lipid bilayer structure and vacuoles in cytoplasm of fiber cells were much more abundant in Cx43 KO than in WT lenses (Fig. 1E,F). In the inner cortex, fiber cells of WT mice maintained membrane appositions, but those of the KO showed dilated intercellular spaces (Fig. 1G,H). Striking morphological changes were detected in the deeper inner cortex regions of Cx43 KO lenses, where large vacuolar spaces appeared between fiber cells (Fig. 1I,J). Measurements of appositional membrane showed significant reductions in all regions of Cx43 KO lenses.

The morphological changes observed in Cx43 KO lenses are similar to abnormalities observed at early stages of cataract, which is characterized by the abundance of intracellular vacuoles and widening of intercellular spaces (7).

The avascular cornea provides the eye with a smooth, transparent, refractive, and protective surface; because the cornea is exposed to the environment, it is susceptible to injury, after which healing must restore its spatial clarity. Cx43 is abundant in corneal epithelial cells (Fig.1A,B), where it may contribute to metabolic compensation among epithelial and stromal cells and may coordinate wound healing. Similarities between corneal and lens epithelia in terms of their common ectodermal origin and expression of Cx43 led us to examine whether Cx43 KO mice exhibit pathological corneal structure and whether corneal epithelial cells display altered wound healing.

Electron microscopic examination of Cx43 KO and WT corneas showed that superficial cells contacted one another along their entire lengths, whereas basal cells were in looser contact (not shown). Bundles of collagen fiber lamellae were arranged tangentially and perpendicularly to each other, surrounding keratocytes that were in contact with each other at their ends in the corneal stroma. Growth rates of corneal epithelial cells cultured from Cx43 KO, heterozygous and WT littermates, were the same independent of genotype (Fig. 1K). The rate of migration into the denuded area following scrape injury was also similar for all animals (Fig. 1L). Scrape loading in the presence of Lucifer Yellow showed that at three days after wounding, corneal epithelial cells from all genotypes were coupled equally strongly in wounded and unwounded regions (Fig 1M,N). Lucifer Yellow dye coupling was present in both Cx43 KO and WT cells, although it was stronger in corneal epithelium cells from WT mice.

These data indicating moderately strong coupling in Cx43 KO corneal epithelial cells reveal the presence of one or more additional connexin(s). Electrophysiological studies on macroscopic junctional conductance revealed stronger voltage-sensitivity in Cx43 KO than WT cells; single channel conductances were about 30-40 pS for Cx43 KO and 100 pS for WT

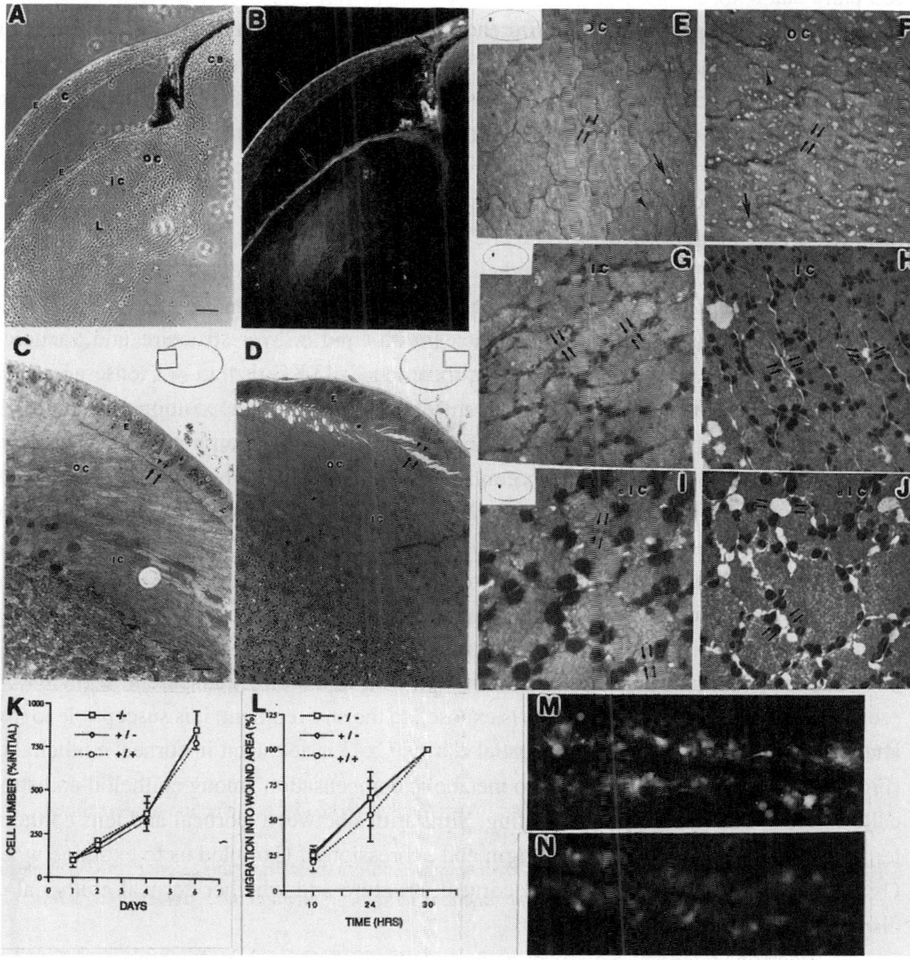

Fig.1. Comparison of lens structure and corneal cell function in WT and Cx43 KO mice. **A,B.** Phase constrast and fluorescence views of eyeball sections of WT mouse stained with Cx43 antibody Cx43 (arrows) is present in corneal (C), and lens (L) and pigment epithelium as well as in the iris (I) and cilliary body (CB). **C,D.** In lens of WT mouse, epithelilal cells are tightly apposed to one another and to fiber cells (arrowheads and arrows), in KO lens apposition between epithelial cells is looser and large spaces are present between epithelial and fiber cells (arrowheads and arrows). **E-J.** Ultrastructural images of lens cortical regions. Sections to the left show outer cortex (E), inner cortex (G) and deeper inner cortex (I) of WT and sections to the right show corresponding areas (F,H,J) of Cx43 KO lenses. **K.** Growth rates of corneal epithelial cells cultured from Cx43 KO (-/-), HT (+/-) and WT (+/+) mice were similar. **L.** Rates of migration of cultured corneal epithelial cells were similar for all genotypes. **M,N.** Scrape loading in the presence of Lucifer Yellow on day 3 after wounding indicated moderately strong coupling in both WT and KO corneal cells.

cell pairs (data not shown). These data are consistent with Cx43 being the major gap junction protein in WT, whereas voltage-sensitivity and small channel sizes are consistent with properties of Cx45 (8). Immunostaining of WT and KO corneas with antibodies against Cx45 and Cx50 showed scattered punctate staining for Cx45 in epithelial cells and keratocytes, with Cx50 being expressed only in epithelial cells (not shown).

## Summary and Conclusions

Cx43 is a major component of epithelial gap junctions in lens and cornea. In Cx43 KO mice, the differentiation of lens fiber cells in the bow region appears normal, indicating that the expression of Cx43 is not required for this process. However, loss of Cx43 expression in epithelium of KO mice leads to wider membrane separations and abundant vacuoles; both differences are expected to increase light scattering and optical opacity of the lens, consistent with changes seen in other cataractous conditions. By contrast, comparison of corneas of Cx43 KO and WT mice revealed no functional or morphological differences, including contact areas and junctions between superficial and basal epithelial cells or keratocytes, cell growth rate, and cell migration in response to wounding. Because strength of dye coupling was also not greatly affected by the lack of Cx43, and properties of gap junction channels in Cx43 KO mice showed lower conductance and smaller channel sizes, expression of additional connexins by corneal epithelium appears to compensate to prevent functional disturbance in the cornea.

## References


[1] DA Goodenough, The crystalline lens. A system networked by gap junctional intercellular communication. *Sem. In Cell. Biol.* (1992) 3:49-58.

[2] EC Beyer, J Kistler, DL Paul, DA Goodenough, Antisera directed against connexin43 peptides react with a 43-kD protein localized to gap junctions in myocardium and other tissues. *J. Cell Biol.* (1989) 108:595-605.

[3] AG Reaume, PAD Sousa, S Kulkarni, BL Langille, D Zhu, TC Davies, JC Juneja, GM Kidder, J Rossant, Cardiac malformation in neonatal mice lacking connexin43. *Science* (1995) 267:1831-1834.

[4] R Dermietzel, JA Kessler, EL Hertzberg, DC Spray, Gap junction between cultured astrocytes: immunocytochemical, molecular and electrophysiological analysis. *J. Neurosci.* (1991) 11:1421-1432.

[5] T Yamamoto, A Ochalski, EL Hertzberg, JI Nagy, LM and EM immunolocalization of the gap junctional protein connexin43 in rat brain. *Brain Res* (1990)508:313.

[6] Y Dong, M Roos, T Gruijiters, P Donaldson, S Bullivant, E Beyer, J Kistler, Differential expression of two gap junction proteins in corneal epithelium. *Eur. J. Cell Biology* (1994) 64:95-100.

[7] JS Friedenwald JS, ID Ryetl, Contribution to the histopathology of cataract. *Arch. Opthalmol.* (1955) 53:825-831.

[8] AP Moreno, JG Laing, EC Beyer, DC Spray, Properties of gap junction channels formed of connexin45 endogenously expressed in human hepatoma (SKHep1) cells. *Am.J. Physiol.* (1995) 268:C356-C365.

# Section 10:
# Connexin Gene Regulation

*Gap Junctions*
*R. Werner (Ed.)*
*IOS Press, 1998*



# Cloning of A Potential Regulatory Protein Involved In Connexin43 Gene Expression

**Elisa J. Oltra and Rudolf Werner**
***Dept. of Biochemistry and Molecular Biology. University of Miami School of Medicine***
***Miami, FL 33101***

**Abstract.** A rat cDNA clone coding for a potential transcription factor was isolated from a uterine λ Uni-ZAP expression library by its ability to bind to a double-stranded DNA probe containing the sequence of the rat connexin43 promoter corresponding to nucleotides –34 to –71 with respect to the transcription start site. *In vivo* DNA footprinting analysis of this region had shown different patterns of protection in heart as compared to uterus of rats treated with estrogen, indicating tissue- and hormone-dependent differences in protein binding throughout this area of the connexin43 promoter. The isolated clone codes for a 12 kDa protein containing three potential zinc-finger DNA-binding motifs. This protein shows no significant homology to any known sequence deposited in Genbank. It is highly homologous, however, to mouse and human sequences in the EST database, suggesting that this clone codes for a novel transcription factor involved in regulating connexin43 gene expression.

## Introduction

In the heart, the connexin43 gene is expressed constitutively, whereas in the uterus the same gene is under the control of steroid hormones. In the myometrium, gap junctions appear just before the onset of labor and quickly disappear after parturition. Previous studies on myometrial smooth muscle in animal and human models have suggested an important role for gap junctions in uterine contractility [1,2]. Transcription of the connexin43 gene in the myometrium has been shown to be induced by estrogen [3,4] and inhibited by progesterone [5,6]. In contrast, in the heart the expression of connexin43 does not change during labor or following steroid treatment [7]. In an effort to study the mechanism of this differential regulation, the rat connexin43 promoter was linked to the reporter gene for luciferase and tested for its responsiveness to estrogen and progesterone treatment in transiently transfected Hela cells. Luciferase activity was induced by estrogen, but also by progesterone [8]. In view of this unexpected result, we decided to study the connexin43 promoter by *in vivo* DNA footprinting analysis. Using this technique, we were able to identify regions of the rat connexin43 promoter that are involved in the regulation of gene expression. This analysis was followed by DNA mobility shift experiments which led to the isolation of a potential transcription factor that appears to be binding to this region.

## Results and Discussion

*In vivo* DNA footprinting analysis of approximately 500 bp of the connexin43 promoter immediately upstream of the transcription start site was used to investigate possible binding sites for transcription factors. *In vivo* DNA footprinting analysis identifies individual nucleotides engaged in protein interactions in intact cells [9,10]. These residues are visualized as bands of weaker intensity in the Ligation-Mediated PCR (LMPCR) autoradiogram. Sometimes, however, a bound protein actually enhances methylation, leading to a more intense signal in DNA from chromatin than in naked DNA. In both cases the observed banding patterns show clear differences between DNA that is involved in protein interactions and DNA that is not.

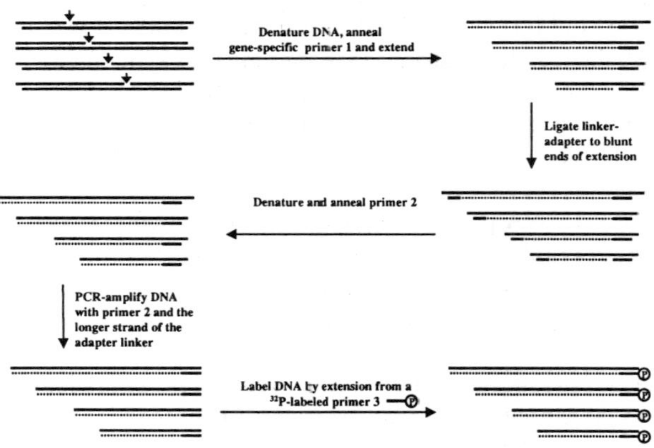


**Figure1. Schematic of *in vivo* DNA footprinting.** Tissue is pulverized in liquid nitrogen and then treated with dimethylsulfate (DMS). DMS methylates purines (preferentially guanines) that are accessible in the DNA, but bases involved in DNA-protein interactions are protected from methylation. This selectively methylated DNA is treated with piperidine, which cleaves DNA at the methylated bases. The cleaved DNA is then used as template in a single PCR extension reaction from a sequence-specific primer (primer 1). Synthetic double-stranded DNA linkers are subsequently ligated to the blunt-end of the extended fragments. A second set of primers, one from the linker sequence and the other specific to the region being analyzed and nested in relation to primer 1 (primer 2), is then used to amplify the fragments by PCR. This series of PCRs is called LMPCR (Ligation-Mediated PCR). Before the fragments can be visualized they must be PCR-amplified with a third sequence-specific primer. This third primer, nested to primer 2, is $^{32}$P-labeled at its 5' end, which allows the visualization of the final PCR products in a DNA sequencing gel.

Distinct patterns of methylation were observed in DNA isolated from heart, uterus and estrogen-induced uterus in the area of the connexin43 promoter between nucleotides –34 and –71 with respect to the transcription start site. This region includes a sequence (nucleotides –57 to –67) that has been proposed to act as a cis-activator on the mouse cx43 promoter, as shown by deletion analysis in transiently transfected cells [11]. A summary of the protection patterns is shown in figure 2.

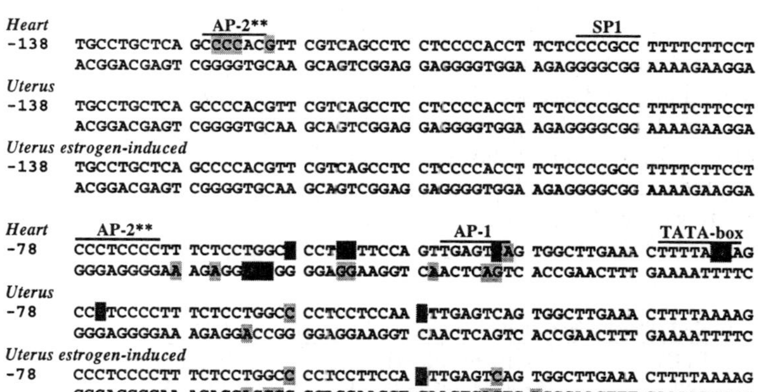


**Figure 2. Summary of connexin43 promoter *in vivo* footprinting analysis:** The rat connexin43 sequence from nucleotide –138 to –19 is shown. Known consensus binding sites are indicated. (** indicates AP-2 sites that have been reported functional in the rat tyrosine aminotransferase gene.) Tissue source of DNA is indicated on the left. Protected nucleotides are lightly shaded. Hypermethylation sites are darkly shaded.

This heavily protected region, located between nucleotides -34 and -71, was then selected to generate a double-stranded DNA probe to be used in DNA mobility shift analysis. When this double-stranded probe was incubated with nuclear extracts from estrogen-induced uterus, the formation of three different complexes was indicated by the retarded migration of the labeled probe. These protein-DNA interactions appear to be sequence-specific, since only unlabeled probe DNA could compete with the labeled probe, while other non-related sequences could not (data not shown). Our probe contains an AP-1 consensus binding site. Thus, it was of interest to check if *fos* and *jun* proteins (which are known to bind to AP-1 sites) could be part of these complexes. Competition studies with unlabeled double-stranded DNA probes corresponding to the left (-48 to -71) or the right part (-34 to -54) of the full probe showed that this was not the case, since only the left part of the probe was capable of effectively competing with complex formation but the right part was not (data not shown). This suggests that proteins other than *fos* and *jun* bind to this area of the connexin43 promoter.

In order to isolate proteins that bind to this area of the connexin43 promoter, an expression library was constructed in λ Uni-ZAP vector (Stratagene) from uterine-mRNA isolated from pregnant rats at day 18 undergoing premature labor, which had been induced by injection of the antiprogesterone drug onapristone (3 mg per animal) followed by intravaginal application of prostaglandine $E_2$ [12]. The library was screened with a synthetic $^{32}$P-labeled double-stranded DNA probe containing the sequence between nucleotides –34 and –71. ( The binding buffer used in the screening was the same as that used in the mobility shift analysis: 12mM Hepes pH 7.9, 12%glycerol, 2mM $MgCl_2$, 100mM NaCl and 0.5mM DTT ). Since the protein expressed in the λ Uni-ZAP vector was fused in frame to the β-galactosidase gene (37 amino acids downstream from the ATG), its coding region must be in frame with that of β-galactosidase. This open reading frame of our clone contained four cysteine residues arranged so that they can form a zinc finger DNA-binding motif. The fact that there was no methionine present in this open reading frame suggested that the clone was incomplete. The clone also contained the complete 3'UTR consisting of 523 nucleotides. In order to find the missing amino-terminal part of this putative transcription factor, the λ Uni-ZAP library was re-screened by PCR with a specific primer located within the coding region of the isolated clone and a second primer containing flanking vector arm sequence. A single amplified band was obtained and sequenced. Its sequence contained a region overlapping our isolated clone by 87 bp and additional sequence coding for 55 more amino acids toward the amino-terminus of the protein. A starting methionine, however, still was not found. The PCR sequence contained 2 additional zinc finger motifs: one constituted by 4 cysteines and the other by 3 cysteines and 1 histidine. A blast search in the Genbank database using the 109-amino acid sequence of our clone produced no hits. When the search was performed in the EST (Expressed Sequence Tags) database, however, a long list of highly related sequences was obtained. Among them a mouse embryo sequence and a human pregnant uterus sequence matched amazingly well with our rat clone. As summarized in Table1, the mouse clone was 94% identical and the human clone was 91% identical to our rat clone at the nucleotide level.

**Table 1**. Comparison of clone amino acid sequence with EST database

| Source of DNA | Percent identity At nucleotide level | Percent identity at amino acid level (similarity) |
|---|---|---|
| Mouse embryo | 94% | 100% |
| Human uterus | 91% | 100% |
| Rice plant (O. sativa) (partial sequence) | 78% | 92% (98%) |

When translated into protein, however, all these mismatches encoded the same amino acid, making these three sequences 100% identical at the protein level. Surprising as well was the finding of a plant sequence (*Oryza sativa*) 92% identical to our sequence at the protein level (only a partial sequence was available for the comparison). This high conservation across species suggests an important function for this new potential transcription factor. This sequence alignment provided us with another piece of information: the position of the first methionine in frame with our sequence which was only 3 nucleotides upstream of our sequence. This methionine most likely is the missing start codon since the surrounding nucleotides match the Kozak consensus sequence for a translation start site, and because in the alignments no clone contained more than 40 nucleotides upstream of our sequence. Furthermore, comparison of the sequences upstream of this methionine in different clones from the ETS database showed that the homology decreased dramatically, supporting the hypothesis that the sequence upstream of this methionine may be part of the 5'UTR region of the gene. The estimated size of the mRNA for our clone is 1 to 1.2 kb as determined by Northern blot analysis (data not shown), suggesting a short 5' UTR. The complete coding sequence of our clone is shown in figure 3.

MAKHHPDLIFCRKQAGVAIGRLCEKCDGKCVICDSYVRPCTLVRICD
ECNYGSYQGRCVICGGPGVSDAYYCKECTIQEKDRDGCPKIVNLGSS
KTDLFYERKKYGFKKRStop

**Figure 3. Sequence of potential transcription factor**. Shades indicate residues that could form zinc-finger motifs. Boxes indicate residues that could form a leucine-zipper domain. Proposed NLS (Nuclear Localization Signal) sequence is underlined.

The two most amino-terminal zinc-finger domains in our clone resemble a RING-finger motif. The consensus sequence for the RING-finger motif is : Cys-$X_2$-Cys-$X_{(9-39)}$-Cys-$X_{(1-3)}$-His-$X_{(2-3)}$-Cys-$X_2$-Cys-$X_{(4-48)}$-Cys-$X_2$-Cys where X can be any amino acid [13,14 and 15]. Our protein differs by having the histidine in position 1 instead of 4, and by having 5 amino acids between the histidine and the neighboring cysteine. The RING motif is found in a family of proteins ubiquitously expressed in organisms ranging from plants to viruses, all of which have putative functions involving some aspect of transcriptional regulation [14]. Its three-dimensional structure is similar to the LIM domain and to the nuclear hormone receptor DNA-binding domains [16,17]. It appears that the RING motif is involved in mediating protein-protein interactions and in some cases multi-protein complexes, which might depend on the presence of other proteins and/or domains [18]. In addition to the RING-finger like domain, our protein contains a region that resembles a leucine zipper (amino acids are framed in figure 3) that may function to form homodimers or to allow it to interact with other proteins. The leucine zipper occurs in a number of transcription factors such as *fos* and *jun*. It is often located near the carboxy-terminal region of the proteins and is normally preceded by a region containing DNA-binding properties, as could be the case of the carboxy-terminal zinc-finger domain in our protein.
Transcription factors need to be transported to the nucleus to be able to exert their function. Often proteins targeted to the nucleus contain an NLS (Nuclear Localization Signal) sequence in their primary structure. There is no single, strict consensus NLS, but there are some general rules. Nuclear localization signals are: (1) typically short, usually not more than 8-10 amino acids; (2) contain a high proportion of positively charged amino acids (lysine and arginine); (3) are not located at specific sites within the protein; (4) are not removed following localization; and (5) can occur more than once in a given protein [19]. It is possible that the sequence underlined in figure 3 acts as an NLS for our protein.

Even though molecules smaller than 40 kDa theoretically do not require an NLS for nuclear entry [20], some small proteins such as histone 2B contain functional NLSs. Possession of NLSs by small molecules implies that they may be important to target larger molecular weight complexes to the nucleus.

In summary, our protein appears to have all the properties required of a transcription factor. The fact that its amino acid sequence is highly conserved across species suggests that this protein plays an important role in the cell. It remains to be determined, however, "if" and "how" this novel transcription factor functions in the regulation of the connexin43 gene expression.

## References

[1] Garfield R.E. (1984) Control of myometrial function in pre-term versus term labor Clin. Obstet. Gynecol. 27 (3): 572-591

[2] Garfield R.E., Sims S.M. and Daniel E.E. (1977) Gap junctions: their presence and necessity in myometrium during parturition . Science 198: 958-960

[3] Dahl G., Arzania R. and Werner R. (1981) Induction of cell-cell channel formation by mRNA Nature 289: 683-685

[4] Petrocelli T. and Lye S.J. (1993) Regulation of transcripts encoding the myometrial gap junction protein connexin43 by estrogen and progesterone Endocrynology 133: 284-290

[5] MacKenzie L.W. and Garfield R.E. (1985) Hormonal control of gap junctions in the myometrium Am. J. Physiol. 248 (3 Pt 1): C296-C308

[6] Lye S.J., Nicholson B.J., Mascarenhas M., MacKenzie L. and Petrocell T (1993) Increased expression of connexin43 in the rat myometrium during labor is associated with an increase in the plasma estrogen:progesterone ratio. Endocrinology 132: 2380-2386

[7] Lang L.M., Beyer E.C., Schartz A. L. and Gitlin J.D. (1991) Molecular cloning of a rat uterine gap junction protein and analysis of gene expression during gestation. Am. J. Physiol. 260: E787-E793

[8] Yu W., Dahl G. and Werner R. (1994) The connexin43 gene is responsive to oestrogen Proc. R. Soc. Lond. 255: 125-132

[9] Mueller P.R. and Wold B. (1989) *In vivo* footprinting of a muscle-specific enhancer by ligation mediated PCR. Science 246: 780-786

[10] Mueller P.R. and Wold B.(1991) Ligation-mediated PCR: applications to genomic footprinting Methods 2: 20-31

[11] Chen Z.Q., Lefebvre D., Bai X.H., Reaume A., Rossant J. and Lye S.J. (1995) Identification of two regulatory elements within the promoter region of the mouse connexin43 gene J. Biol. Chem. 270: 3863-3868

[12] Rechberger T., Abramson S.R. and Woessner F. (1996) Onapristone and prostaglandin $E_2$ induction of delivery in the rat in late pregnancy: a model for the analysis of cervical softening Am. J. Obstet. Gynecol. 175 (3 Pt 1): 719-723

[13] Freemont P.S., Hanson I. M. and Trowsdale J. (1991) A Novel Cysteine-Rich Sequence Motif. Cell 64:483-484

[14] Saurin A.J., Borden K.L., Boddy M.N. and Freemont P.S. (1996) Does this have a familiar RING? TIBS 21(6): 208-214

[15] Border K.L. and Freemont P.S. (1996) The RING finger domain: a recent example of a sequence-structure family. Curr. Op. Struct. Biol. 6(3): 395-401

[16] Yammerstrom A., Berndt K.D., Sillard R., Adermann K. and Otting G. (1996) Solution structure of a naturally-occurring zinc-peptide complex demonstrates that the N-terminal zinc-binding module of the Lasp-1 LIM domain is an independent folding unit. Biochemistry 35 (39): 2723-32

[17] Schwabe J. W., Neuhaus D. and Rhodes D.. (1990) Solution structure of the DNA-binding domain of the oestrogen receptor. Nature 348: 458-61

[18] Borden K. L. B., Boddy M. N., Lally J., O'Reilly N. J., Martin S., Howe K., Solomon E. and Freemont P. S. (1995) The solution structure of the RING finger domain from the acute promyelocytic leukaemia proto-oncoprotein PML EMBO J. 14: 1532-41

[19] David A. J. (1994) Nuclear signaling pathways for polypeptide ligands and their membrane receptors? 8:841-47

[20] Miller M., Park M. Y. and Hanover J. A (1991) Nuclear pore complex: structure-function and regulation. Physiol. Rev. 71: 909-949

*Gap Junctions*
*R. Werner (Ed.)*
*IOS Press, 1998*

# Regulation of cx26 and cx43 Transcripts by Progesterone and 17-β-Estradiol

R. Grümmer[1], B. Ströbl, O. Traub[2], and E. Winterhager[1]
[1]*Institut für Anatomie, Universitätsklinikum, 45122 Essen,*
[2]*Institut für Genetik, Universität Bonn, 53117 Bonn, Germany*

**Abstract.** During preimplantation in the rat the ovarian hormones progesterone (P) and estradiol ($E_2$) lead to changes in differentiation pattern of uterine cells to prepare the endometrium for embryo implantation. This differentiation process is accompanied by regulation of connexin(cx)26 in the uterine epithelium and of cx43 in the stromal compartment. During preimplantation both connexins are suppressed by an increase in maternal progesterone. This could be evidenced since application of the antigestagen onapristone on day 3 of pregnancy abolished the suppression of the connexin genes. In contrast, injection of $E_2$ induced expression of both connexins in the rat endometrium. This induction was clearly dependent on $E_2$ as evidenced by simultaneous application of the antiestrogene ICI 182780. This regulation of connexin mRNA as well as protein expression is very sensitive to the relation of $E_2$:P concentration. P had to be given at least in a physiological concentration comparable to pregnancy to suppress connexin transcripts. Time course of connexin expression after injection of $E_2$ demonstrated an early gene response of cx26 within 3h, whereas start of cx43-transcription was detected within 14h. To follow up the molecular pathway of this hormonal regulation we investigated if transactivation of the connexin genes was directly mediated by estrogen receptors. Application of cycloheximide prevented induction of connexin expression and supported the idea of a newly synthesized transcription factor. Though c-jun and c-fos were found to be upregulated after $E_2$-injection in the rat endometrium prior to connexin induction, no AP-1 site could yet be detected in the promoter of rat cx26 within the sequenced 500 bp upstream. This study provides evidence that transcription of the both connexin genes shows different sensitivity to $E_2$ and that at least expression of cx26 is induced by a newly synthesized transcription factor.

## 1. Introduction

Implantation of the mammalian blastocyst involves a complex series of precisely synchronized physiological and cell biological events which lead to a differentiated, receptive endometrium. This differentiation process is strongly under the control of the ovarian steroid hormones 17-ß-estradiol ($E_2$) and progesterone (P) [1-3]. Intercellular communication via gap junctions, which is known to be involved in cell differentiation, tissue organization and development [4-7] seems to play an important role in preparing the endometrium for embryo implantation.

We could show that in the endometrium connexin genes are regulated by $E_2$ and P in a well coordinated and regulated spatial and temporal pattern during periimplantation and appear to play an important role in regard to uterine receptivity [8-10]. From all connexins investigated (cx26, cx32, cx37, cx40, cx43, cx45) only cx26 and cx43 seem to play a role in the rat endometrium during early pregnancy. In the nonpregnant uterus transcripts of both connexins were found in all cyclic phases. During cycling only a moderate hormonal regulation with an increased expression during the $E_2$-dominated proestrus / estrus phase and an reduced expression during the P-dominated metestrus / diestrus phase has been observed; however, the corresponding proteins were rarely found [11,12]. During transformation of the endometrium into the receptive phase in early pregnancy the increase in maternal

progesterone completely suppresses both cx26 in the epithelium of the implantation chamber as well as cx43 in the stroma. This dependency of connexin suppression on progesterone has been evidenced by application the antigestagen onapristone (ZK 98.299, Schering AG, Berlin) which inhibited the decline in cx26 and cx43 expression observed during preimplantation in pregnant rats [10].
Such cell specific regulatory mechanisms of connexin gene expression are not very well understood. Only few promoter studies have been performed until now, e. g. for the cx43 and cx32 gene [13], but tissue specific regulatory elements have not been proved yet.

## 2. Results

By supplying ovarian steroid hormones to ovariectomized rats we could confirm that connexin gene expression is hormonally regulated in the rat endometrium. Cx26 as well as cx43 were induced by application of $E_2$, whereas P led to a suppression of both connexins. This hormonal regulation of connexin expression is very sensitive to the $E_2$:P-ratio. Only application of these hormones in a physiological ratio which is usually used to maintain pregnancy in rats (0.1 µg $E_2$ + 4 mg P / day / rat) led to a suppression of both connexin genes comparable to the situation during preimplantation (Fig. 1). A tenfold increase of $E_2$ (1 µg $E_2$ + 4 mg P / day / rat) abolished the suppressive effect of P on cx26 expression. In contrast to cx26, cx43 seemed to be less sensitive to $E_2$ as it was still suppressed by P despite of the increase in $E_2$-concentration. These experiments indicated that only a hormonal profile adequate to conditions during pregnancy with a high amount of P in combination with low $E_2$ levels was able to suppress cx26 transcription. Time course experiments demonstrated that the cx26 gene is more sensitive to regulation by $E_2$ which was able to induce cx26 transcripts in an completely uncoupled endometrium within 3 hours whereas cx43 induction was found within 14 hours (Fig. 2). Induction of both connexins could be abolished by simultaneous application of ICI 182780. The time course experiments revealed that the expression of cx26 represents an early gene response to $E_2$ treatment and thus may be able to react very fast to changing physiological requirements.
Application of $E_2$ in combination with the protein synthase inhibitor cycloheximide to ovariectomized rats for 4 hours prevented $E_2$-induced cx26 expression (Fig. 3). This led to

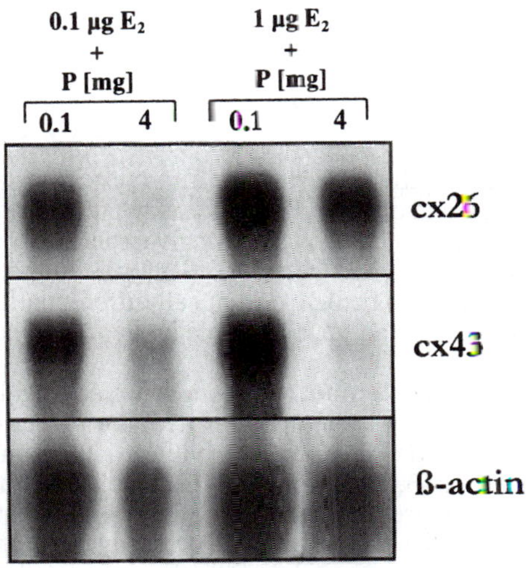


Fig. 1. Dependency of connexin-mRNA expression on $E_2$:P-ratio. Application of a physiological $E_2$:P-ratio as found in pregnancy (0.1 µg $E_2$ + 4 mg P / day / rat) led to suppression of cx26 and cx43. A tenfold increase in $E_2$-concentration (1 µg $E_2$) abolished the suppressive effect of P on cx26, whereas cx43 was still sensitive to P-mediated suppression.

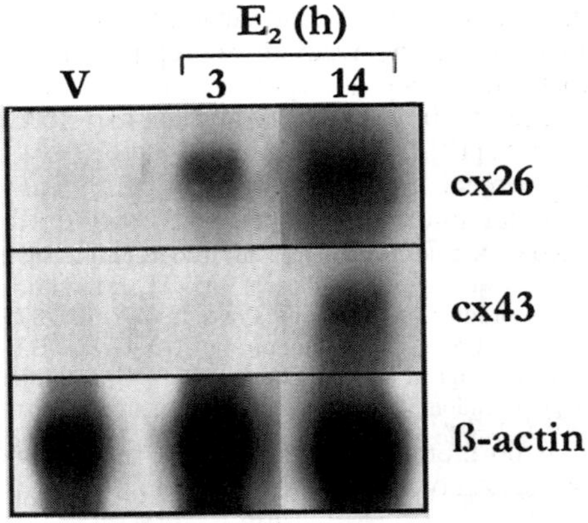


Fig. 2. Expression of cx26-mRNA is induced in ovariectomized rats within 3 hours, cx43 within 14 hours after $E_2$ injection.

the conclusion that a newly synthesized transcription factor is necessary for induction of the cx26 gene, which seems not to be directly regulated by the estrogen receptor. In addition to connexin induction $E_2$ appplication led to an induction of c-fos and an increase in c-jun transcript expression within 1 hour after application. Such an $E_2$-mediated induction of c-jun and c-fos in addition to cx43 could already be shown in the myometrium [14,15]. Since the cx43 promoter contains an AP-1 site [16] possibly the cx43 gene can be regulated via the transcription factors c-jun and c-fos. A functional proof, however, is still missing. In the putative promoter region of the rat cx26 gene we could neither identify an AP-1 site nor an estrogen responsive element within 500 bp upstream of the exon1, but a binding site for the transcription factor NFkB. The different time course of $E_2$ mediated induction of cx26 and cx43 as well as the missing AP1 site in the putative cx26 promoter points to different regulative mechanisms for cx26 and cx43, respectively, which may be related to different functions of these connexins during periimplantation in the rat.

## 3. Conclusions

The dynamic cell type specific modulation and the steroid hormone dependent regulation of connexin genes seems to play an important role for embryo implantation. In the rat endometrium cx43 and especially cx26 are able to react with a very high plasticity to hormonal changes. The signaling pathway involved in this transcriptional regulation of connexin genes is still under investigation. Consequently functional promoter studies are

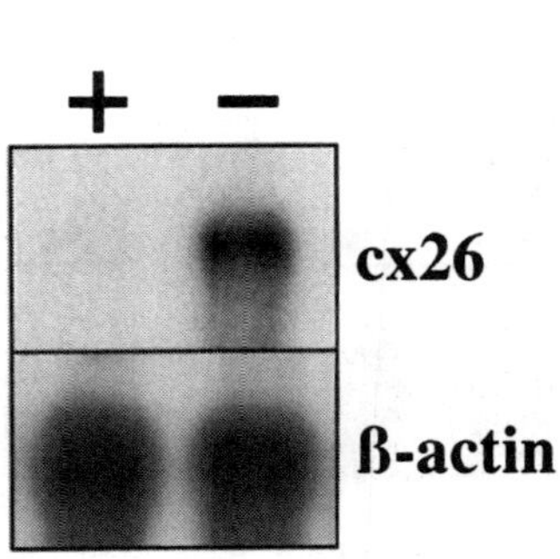


Fig. 3. Application of $E_2$ in combination with the protein kinase inhibitor cycloheximide did not induce cx26 expression within 4 hours (+) compared to application of $E_2$ alone (-).

required to determine single steps in the cascade leading to this coordinated and cell type specific regulation properties of cx26 in the uterine epithelium and cx43 in the stromal compartment of the endometrium.

## References


[1] S.R. Glasser and J.H. Clark, A determinant role for progesterone in the development of uterine sensitivity to decidualization and ovoimplantation. In: C.L. Merkert. J. Papaeonstantinou (eds), Developmental Biology of Reproduction. Academic Press, New York, 1975, pp. 311-345.

[2] A. Psychoyos, Endocrine control of egg implantation. In: R.O. Greep, E.B. Astwood (eds), Handbook of Physiology. American Physiological Society, Washington DC, 1992, pp. 187-215.

[3] H.W. Denker, Cell biology of endometrial receptivity and of trophoblast-endometrial interaction. In: S.R. Glasser *et al.* (eds), Endocrinology of Embryo-Endometrial Interactions. Plenum Press, New York, 1994, pp. 17-32.

[4] E. Hertzberg and R. Johnson, Gap Junctions. Modern Cell Biology Vol. 7, Alan R. Liss Inc., New York, 1988.

[5] F. Miragall *et al.*, Expression of connexins in the developing olfactory system of the mouse, *J. Comp. Neurol.* 325 (1992) 359-378.

[6] C.P. Ruangvoravat and C.W. Lo, Restrictions in gap junctional communication in the Drosophila larval epidermis, *Dev. Dynamics* 194 (1992) 261-281.

[7] E. Dahl *et al.*, Expression of gap junction genes, connexin40 and connexin43, during fetal mouse development, *Anat. Embryol.* 191 (1994) 267-278.

[8] E. Winterhager *et al.*, Gap junction formation in rabbit uterine epithelium in response to embryo recognition, *Dev. Biol.* 126 (1988) 203-211.

[9] E. Winterhager *et al.*, Spatial and temporal expression of connexin 26 and connexin 43 in rat endometrium during trophoblast invasion, *Dev. Biol.* 157 (1993) 399-409.

[10] R. Grümmer *et al.*, Regulation of connexin26 and connexin43 in rat endometrium by ovarian steroid hormones, *Biol. Reprod.* 51 (1994) 1109-1116.

[11] B. Risek *et al.*, Modulation of gap junction transcripts and protein expression during pregnancy in the rat, *J. Cell Biol.* 110 (1990) 269-282.

[12] E. Winterhager *et al.*, Expression of different connexin genes in rat uterus during decidualization and at term, *Eur. J. Cell Biol.* 55 (1991) 133-142

[13] R. Bruzzone *et al.*, Connections with connexins: the molecular basis of direct intercellular signaling, *Eur. J. Biochem.* 238 (1996) 1-27.

[14] G.M. Stancel *et al.*, Interactions of estrogens, proto-oncogenes and growth factors. In: Cellular Signals Controlling Uterine Function, L.A. Lavia (ed), Plenum Press, NY, (1991), pp. 49-61.

[15] C. Chiappetta *et al.*, Estrogen regulates expression of the jun family of proto-oncogenes in the uterus, *J. Steroid Biochem. Mol. Biol.* 41 (1992) 113-123.

[16] Z.Q. Chen *et al.*, Identification of two regulatory elements within the promoter region of the mouse connexin43 gene, *J. Biol. Chem.* 270 (1995) 3863-3868.

*Gap Junctions*
*R. Werner (Ed.)*
*IOS Press, 1998*

# Upregulation of Connexin 32 in the PNS of the Shiverer Mouse

Candra Smith-Slatas and Elisa Barbarese
*Dept. Neurology, University of Connecticut Health Center, Farmington, CT 06030*

**Abstract.** Connexin 32 (CX32) is the main gap junction protein expressed in adult Schwann cells (Sc), the myelinating cells of the peripheral nervous system (PNS). Cx32 hexamers form hemi-connexons which pair with others to connect adjacent layers of the myelin sheath. These gap junctions are located in the Schmidt-Lantermann incisures (SLI) which are cytoplasmic channels running through the compact myelin sheath. The *shiverer* (*shi*) mutant mouse is a null allele for myelin basic protein (MBP), a component of compact myelin. In *shi* PNS myelin, the number of SLI is three times that of wild type (wt). There is a corresponding increase in the level of CX32 protein which is appropriately localized to SLI. Expression of CX32 in the liver, a tissue that does not express MBP, is similar in *shi* and *wt*. In vitro translation of Cx32 mRNA is inhibited by MBP polypeptide suggesting that the regulation of Cx32 expression in nervous tissues may take place, in part, at the translational level.

## 1. Introduction

Cx32 is a gap junction protein found in many tissues, including liver and nervous tissue [1]. In the PNS Cx32 is present in the Sc and its myelin [2]. During the course of normal development in the mouse, expression of Cx32 follows that of myelin specific components [3], so that there is an increase in Cx32 protein as myelin accumulates. The bulk of Cx32 is localized in a specialized compartment of myelin, the SLI where it forms gap junctions. SLI are cytoplasmic channels running through the compact myelin sheath [4]. Their postulated function is the maintenance of myelin by transport of nutrients and removal of toxic metabolites. Gap junctions presumably enhance the efficacy of this process.

The shi mutation of the mouse is a null allele for MBP, a specific myelin component [5]. Homozygous shi mice elaborate PNS myelin which, in most respects, is identical to that of wt [6]. However, shi PNS myelin contains more than 3 times the wt number of SLI [7]. The relationship of MBP to Cx32 is not understood, but in vitro binding experiments indicated that MBP could bind to a specific hexanucleotide RNA sequence [8]. A data bank search revealed that several repeats of this sequence were present in Cx32 mRNA. This study was therefore designed to determine if the level of Cx32 polypeptide was elevated in shi mice (as suggested by the supernumerary SLI), to determine if only MBP containing tissues were affected, and to determine if MBP could affect the translation of Cx32 mRNA.

## 2. Materials and Methods

*Animals and tissue preparation*: Sciatic nerves and liver from adult (greater than 90 day old) wt and shi mice were dissected out and kept frozen at -75 C until use, or fixed in 4% paraformaldehyde. Frozen tissues were processed for Cx32 analysis according to the protocol of Hertzberg [9].

*Western Blot*: Extracts were electrophoresed on polyacrylamide gels, blotted unto nitrocellulose, and probed with antibodies according to standard protocols. An enhanced chemiluminescence protocol was used to detect immune complexes.

*Immunofluorescence*: Fixed tissues were processed according to standard protocols, and examined by confocal fluorescence microscopy.

*In vitro translation*: A chimeric Cx32/green fluorescent protein (GFP) mRNA was used for translation in a rabbit reticulocyte lysate system according to the manufacturer's protocol. Bovine MBP and cytochrome C (cyto C) were obtained commercially.

## 3. Results

MBP is expressed exclusively in nervous tissues in adult animals where it is found in compact myelin [10]. In contrast, Cx32 which is present in the cytoplasmic, uncompacted domains of the myelin sheath, is expressed in several tissues, including liver [1]. Therefore, the expression of Cx32 was analyzed in sciatic nerves and liver obtained from age matched wt and shi adult mice by Western blot (qualitative) and modified Western blot (quantitative) (not shown). Cx32 protein appeared as a polymeric series in all the tissues examined. The relative amount of Cx32 was the same in wt and shi liver, but was 4 times greater in shi nerves than in wt (Fig. 1). The amount of P0, the major structural component of compacted PNS myelin [11], was identical in both wt and shi (Fig.1). MBP was present in wt sciatic nerves but absent from shi sciatic nerves (Fig. 1) as shown previously [6].

Immunofluorescence of whole sciatic nerves showed Cx32 to be similarly localized in SLI, identified by the presence of filamentous actin, Sc bodies, and at the nodes of Ranvier (not shown) in shi and wt tissues (Fig. 2). The distribution of Cx32 appeared

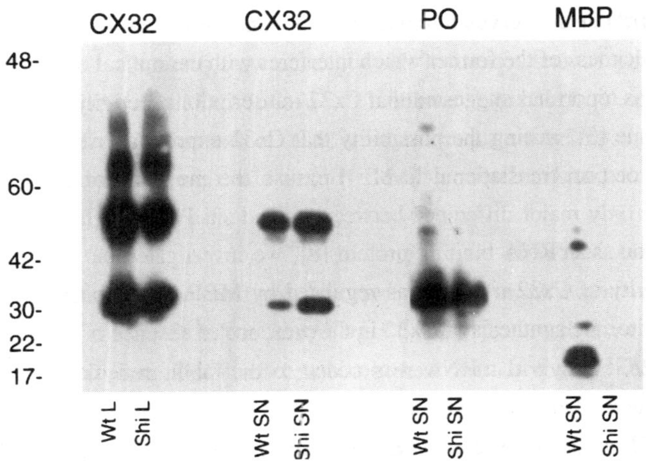


Figure 1: Western blot of Cx32 in sciatic nerves and liver from adult wt and shi mice. Wt (8μg) and shi (8μg) sciatic nerves (SN), wt (2μg) and shi (2μg) livers (L), probed with mouse anti-Cx32 (1:1000), rabbit anti-P0 (1:1500), and rabbit anti-MBP (1:500) antibodies. Molecular weight markers (kD) are indicated on the left.

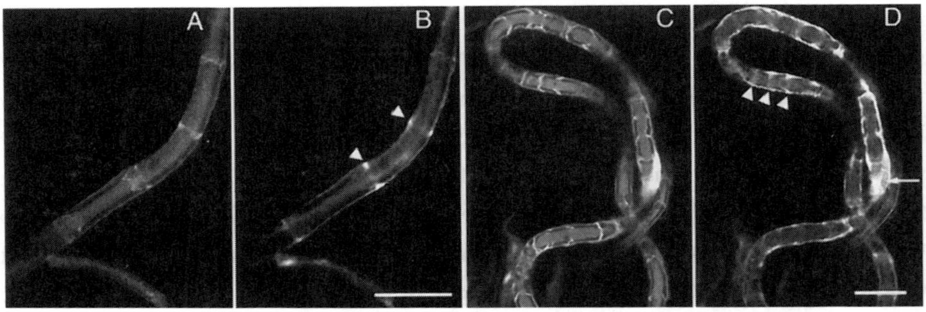


Figure 2: Confocal microscopy of Cx32 in mouse tissues. Wt (A, B) and shi (C, D) sciatic nerves were incubated with fluorescein-conjugated phalloidin (A, C) and Cx32 antibodies (1:75) with an appropriate Texas red-conjugated secondary antibody (B, D). Schwann cell body (arrow); SLI (arrowheads). Scale bar = 25 μm.

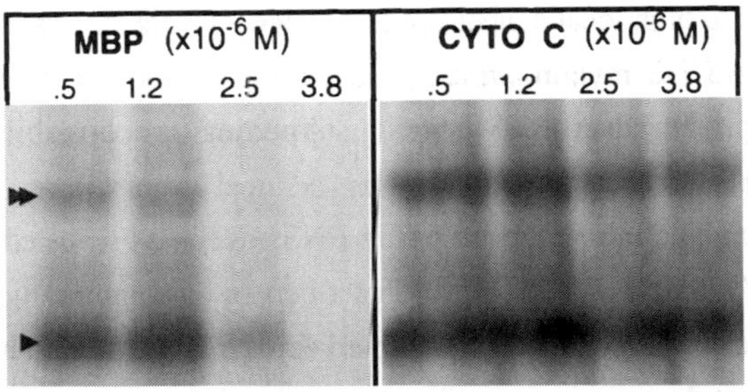


Figure 3: In vitro translation of Cx32 mRNA. Cx32 mRNA (50 ng) was added to each vial containing rabbit reticulocyte lysate and various amounts of MBP or cyto C at the final concentration indicated above each lane. Single and double arrowheads indicate Cx32 monomer and dimer, respectively.

smooth and uniform in the nerves in contrast to its punctate pattern in liver [12], possibly because of the thickness of the former which interferes with the optical resolution.

A previous report has suggested that Cx32 transcriptional activity was similar in wt and shi PNS tissue [3], raising the possibility that Cx32 expression might be regulated at the translational or post-translational level. Because the presence of MBP polypeptide appears to be the only major difference between wt and shi PNS myelin [6], and because MBP may function as an RNA binding protein [8], we investigated the possibility that the translational activity of Cx32 mRNA was regulated by MBP. A rabbit reticulocyte lysate was used to monitor the synthesis of Cx32 in the presence or absence of MBP. A constant amount of Cx32/GFP hybrid mRNA was added to the rabbit reticulocyte lysate in the absence or presence of increasing amount of MBP or cyto C as indicated in Fig. 3. The newly synthesized radioactive protein was identified by autoradiography after SDS PAGE. MBP at a final concentration of 3.8 x $10^{-6}$M was sufficient to abolish the translation of MBP mRNA (Fig. 3). Cyto C did not inhibit translation at the same or higher concentration.

## 4. Discussion

The increased amount of Cx32 protein found in PNS myelin of adult shi mice may be due to enhanced translational activity and/or protein stability. In vitro translation of Cx32 mRNA in the presence of MBP showed that the latter has an inhibitory effect on the translation of Cx32 mRNA suggesting that overexpression of Cx32 in shi may be due to the absence of MBP in this mutant. Liver, a tissue that expresses Cx32 but does not contain MBP, shows no differences in its level of Cx32 protein when wt and shi were compared. Three observations suggest that translational repression may be operative in vivo. First, MBP and Cx32 are simultaneously expressed in sciatic nerve indicating that interactions can take place; second, MBP mRNA is clustered at sites of myelin assembly which may give rise to high local concentrations of MBP protein necessary for translational inhibition; and third, CNS myelin, which contains higher amounts of MBP, has a lower content of Cx32 protein than PNS myelin supporting the possibility of an inverse correlation between these two components.

Cx32 is assembled into gap junctions in PNS myelin as revealed by freeze fracture microscopy [13]. Visualization of Cx32 in intact nerves by immunofluorescence does not show a prominent clustering pattern as observed for the liver [12]. This could be due to the fact that the clusters are too numerous to resolve by light microscopy, or that the thickness of the tissue and its geometry give rise to diffraction of the fluorescence signal. However, fluorescence microscopy clearly indicates that Cx32 is spatially restricted to the cytoplasm-containing domains of the myelin sheath, and that this distribution is unchanged in shi mice where the excess number of SLI appear to accommodate the higher level of Cx32 protein.

## Acknowledgements

This work was supported by NIH grant #NS19943 (E.B.), and by a graduate fellowship from the U. of Ct. Health Ctr. (C.S.-S). We thank Dr. J. H. Carson, Dr. Y. Gao, S. Kwon, and H. Tedford for sharing their unpublished findings and for providing reagents, Frank Morgan for his help with fluorescence image analysis, and Dr. M. Filbin for antibodies to P0.

## References

[1] D.L. Paul, New functions for gap junctions. Curr. Op. Cell Biol. 7:665-672 (1995).

[2] J. Bergoffen *et al.*, Connexin mutations in X-linked Charcot-Marie-Tooth disease. Science 262:2039-2042 (1993).

[3] S.S. Scherer *et al.*, Connexin32 is a myelin-related protein in the PNS and CNS. J. Neurosci. 15:8281-8294, 1995.

[4] S.M. Hall and P.L. Williams, Studies on the 'incisures' of Schmidt and Lanterman. J. Cell Sci. 6:767-791, 1970.

[5] A. Roach *et al.*, Characterization of cloned cDNA representing rat myelin basic protein: absence of expression in brain of shiverer mutant mice. Cell 34:799-806, 1983.

[6] D.A. Kirschner and A.L. Ganser, Compact myelin exists in the absence of basic protein in the shiverer mutant mouse. Nature 283:207-210, 1980.

[7] R.M. Gould *et al.*, The number of Schimdt-Lanterman incisures is more than doubled in shiverer PNS myelin sheaths. J. Neurocytol. 24:85-98, 1995.

[8] Y. Gao and J.H. Carson, Myelin basic protein is a sequence specific RNA binding protein. Mol. Biol. Cell 7:653a (1996).

[9] E.L. Hertzberg, A detergent-independent procedure fro the isolation of gap junctions from rat liver. J. Biol. Chem. 259:9936-9943 (1984).

[10] S.R. Cohen and M. Guarnieri, Immunochemical measurement of myelin basic protein in developing rat brain: An index of myelin synthesis. Dev. Biol. 49:294-299 (1976).

[11] S.W. Brostoff *et al.*, Isolation and partial characterization of the major proteins of rabbit sciatic nerve myelin. Brain Res. 86:449-458 (1975)

[12] J.-T. Zhang and B.J. Nicholson, The topological structure of connexin 26 and its distribution compared to connexin 32 in hepatic gap junctions. J. Membrane Biol. 139:15-29 (1994).

[13] E. Reale *et al.*, Freeze-fracture faces of the perineurial sheath of the rabbit sciatic nerve. J. Neurocytol. 4:261-270 (1975).

*Gap Junctions*
*R. Werner (Ed.)*
*IOS Press, 1998*



# Structural and Functional Analysis of the Rat Connexin31 Promoter

B. Ströbl, P. Hellmann, R. Büttner* and E. Winterhager
*Department of Anatomy, University of Essen, Germany*
**Department of Pathology, University of Regensburg, Germany*

**Abstract.** Connexin31 (cx31) expression seems to play an important role during trophoblast development in rodents. During preimplantation cx31 is coexpressed with cx43 in the trophoblast cells as well as in the inner cell mass of the blastocyst. Following implantation, compartmentalization of both connexins occurs: cx31 is expressed in the proliferative and invasive trophoblast and cx43 in the embryo proper. Trophoblast differentiation into a functional placenta is indicated by a switch from cx31 to cx26 and cx43 [1]. Such a switch from cx31 to cx43 accompanied by a differentiation process can be induced in rat trophoblast cells by retinoic acid [2]. To further elucidate the trophoblast specific regulation properties of cx31, we have isolated the cx31 gene and performed structural and functional promoter analysis. The isolated rat cx31 gene contains two exons separated by an intron of 2.6 kb. Exon 1 comprises 105 and 139 bp, respectively due to two transcription start sites. Within 935 bp of the 5′ flanking region of exon 1, four putative binding sites for GATA trancription factors, furthermore a NF-κB element, a CAAT-box and an E-box were identified. For functional promoter analysis two cell lines, rat choriocarcinoma cells and mouse keratinocyte cells, which both express cx31 were chosen. HeLa cells were used as a negative control. A construct including 880 bp of the promoter region, exon 1 and the intron linked to a luciferase reporter gene showed activity in transient transfection experiments in both cell lines. Deleting the 880 bp construct to 420 bp led to a decrease in promotor activity. Even without promotor the intron itself conferred the same basic activity as found for the 100 bp construct. Deletion of 2.3 kb from the 3' end of the intron resulted in a loss of promoter activity. Cell specific differences could be observed within the 5' region of the promoter. All constructs tested did not show any activity in HeLa cells. In conclusion an enhancer seems to be located between 880 bp and 420 bp in the promoter region and the intron plays a role in conferring activity.

## 1. Introduction

Trophoblast cells are responsible for embryo implantation and development in utero by establishing a feto-maternal exchange organ, the placenta. Acquisition of these functions is accompanied by a highly regulated cell differentiation process along a multilineage pathway into different trophoblast cell types (labyrinth, spongiotrophoblast, giant cells). In previous studies we could demonstrate that this differentiation process is correlated to a spatially and temporally regulated connexin expression pattern [3]. The highly invasive and proliferative trophoblast of the implantation phase is characterized by expression of cx31. The differentiation phase is indicated by a switch from cx31 to cx26 gene expression in the labyrinth and from cx31 to cx43 in the spongiotrophoblast and giant cells [1]. The availability of rat trophoblast cell line Rcho-1 [4] has provided a useful model to study cell type specific connexin gene expression since this cell line represents the invasive trophoblast and expresses cx31. In addition, Rcho-1 cells can be manipulated to differentiate into giant cells, a process which is accompanied by a switch from cx31 to cx43 gene expression [2]. Therefore this cell line is appropriate to study regulation of connexin genes during trophoblast differentiation.

In this study we wanted to identify elements in the promoter of the cx31 gene responsible for its trophoblast specific expression and regulation.

## 2. Results and Discussion

### *2.1 Isolation and characterization of the rat cx31 gene*

To isolate a genomic clone for rat cx31 a rat lambda phage library was screened with a probe representing the complete coding region of rat cx31 [5]. Using stringent conditions six positive clones were isolated and characterized by restriction mapping and Southern blotting. When the restriction pattern was compared to that of the rat cx31 genomic clone published by Hoh et al., one clone was identified which contained the coding region and additional 9 kb of upstream sequence. This clone was chosen for further analysis. The phage insert was cut into EcoRI fragments and subcloned into pBluescript vector. A 4.4 kb EcoRI fragment hybridised to the above probe on a Southern blot and was sequenced from the 3' end. Sequence comparison confirmed that this clone represents the cx31 gene.

To determine the size of the intron of rat cx31 and to localize exon 1 we performed a PCR reaction using phage DNA and rat genomic DNA as template. Using the upper primer (position 42-63) of mouse exon 1 [6] and the lower primer located 340 bp 3' of the start of exon 2 of rat, an amplification product of 3 kb was obtained. From this result it could be concluded that rat cx31 contains an intron, 2.6 kb long.
Sequence comparison of exon 1 of rat and mouse revealed that they differ in size. Rat exon 1 (105 bp) is 18 bp larger than mouse exon 1 (86 bp) [6]. Not taking into account the additional 18 bp in rat exon 1 the DNA sequences show 93% homology.

The exon-intron boundaries were verified by RT-PCR using RNA isolated from Rcho-1 cells as a template. By comparison of the obtained cDNA with the genomic DNA a divergence in the sequences indicates the beginning and the end of the intron. The assumed splice sites which match exactly the acceptor and donor consensus splice sites described in the literature could be confirmed [7].

To map the transcription start site a primer extension analysis was performed. A 25 bp antisense primer overlapping the 3' end of exon 1 and the 5' end of exon 2 was annealed to RNA from Rcho-1 cells and extended with AMV reverse transcriptase. This reaction yielded two products corresponding to position +1 and -34 of the genomic sequence (Fig. 2). The transcription start site at position +1 corresponds to the 5' end of mouse cx31 cDNA shifted one base pair upstream [6].

Summarizing the above results we could compose a gene structure for rat cx31 as shown in Fig. 1. Similar to all the other connexin genes cx31 is build up of two exons separated by an intron. Exon 1 is 105 respectively 139 bp in size and exon 2 contains the complete coding region (810 bp). The intron of the cx31 gene is 2.6 kb long and represents the smallest intron of all rat connexins identified.

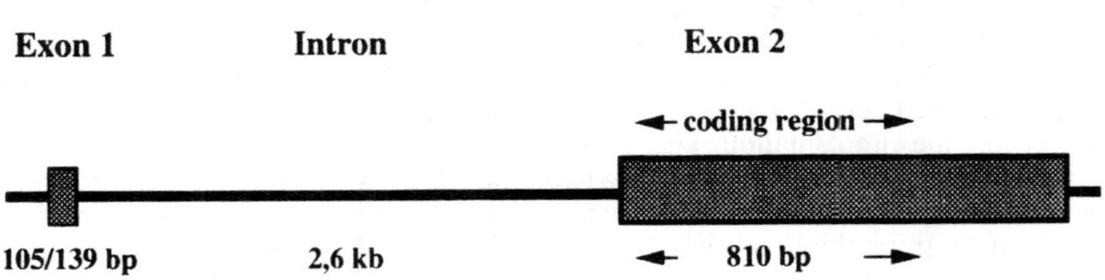


Fig. 1. Gene structure of rat cx31. Exons are indicated as boxes.

```
-935           AG CTATACCTTT GGGCCAGCCA CCATGCTGGT AGAAANAGAT CGACGGTTGG

-883 GTCCTGCAGC TGGATAAAAA CTGGATCAAA AAAATAATAA TAATTGTTCT GGTAGCTGAA

-823 GTTGGGGGGC AACAGGGATC CAGAACCGGC TCCACCAGTC CCCAGCTCGG GCTAGGCGGG

-763 AAATTCTTTT CCGGTGAGTC TATCTCAGGG TGTGTCTTAG AAGGGACTAG GCAGCCTGCA

-703 CCAAAATGGG GAACCTCAGA CGGCTATCCT TAGAGGGTGC CAGGTGCAAG GAACCCCTCT

-643 CCTCCCCTTC ACTCTCCTAG GAAGAGCACT CCTAGGAAGA GCAGCCTGGG GAGGGAACAG
                                       GATA
-583 TTGAAGATTC TTGAAATAAT CCAAGATAAG AGATAAGGAA CAGGCTTAGC TACATTGCAG

-523 AATTCTCTTG GATGTCTCCC AGCTAGTAAT CCTGGCAGTT CCTGGGGGTC AGAATTCAGG

-463 TGCTAATCTT TGGCCTTCCC TGGGACTTTT AGAATATGCA ACGGAACCCA TGCGTGTCGT
              NF-κB
-403 CACCCACTGG GGCTTCCCAG GCAGAACTGA GACTCTCAAA AGACCACCCT GTGAAAGGCA

-343 GCCATACAGC ACTCCGGGNC TATACTGGCC CTCTGCCTTT CTGCCCAGCC CTCTCCGAGA
                                                   GATA
-283 ACTAGGCTCT CTTGTGCCTG AATTGAGTGC CTATTGTATG CTGATACCAA ACTTGAAGCA
                            CCAAT-Box
-223 GCCAGCAAGT CAAGTAGGCA TGTTAATTGG ACAGACAGGT GTTAAATGAG AACTCAAACA

-163 AACAAACAAA CAAATGACCC AGGCTTAAAT TGCAAATTGA GATTGAGGCT ACTAAAACAA
                                             E-Box
-103 GGACAAAGCA GGAACAGACT GAGCACAGCA GGTTCAGATG CAAAGTAGTCT TGAGAGCAGA
          Exon I GATA                       GATA          +1 Exon I
-43 CCCTAAGGG↓GATA)(TTGAGAG CAGACCCTAA GGGGATA)TCG GAC↓ CAAGATG CCTCCTTAAT

18 GAGTAGGGTC TGGCCACTTA AGGCTCTAGA GAAAGAGCAT TCCAGGCGGG AGGAATCAAG
                            splice donor site
78 GCGAGGACTG AGCAGAGAGA GAGGCAAG↓ GTAAGACAGACA GACAGAGAGA GAGAGAGAGA
```

Fig. 2. DNA sequence of the 5' flanking region of the rat cx31 gene. Included are the first exon and a portion of the intervening sequence. Numbering is relative to the lower transcription initiation site which is designated as +1. Nucleotide sequences related to binding sites for known DNA binding proteins are underlined. Transcription start sites and the splice donor site are indicated by an arrow.

935 bp upstream of the transcription start site were sequenced and screened for putative transcription factor binding sites using a transcription factor data base (Fig. 2). No canonical TATA-box was found within the first 200 bp upstream of exon 1. Remarkably, a direct repeat of 14 bp is present upstream of the first start site. Four different putative transcription factor binding sites are present: an E-box, a CCAAT-box on the antisense strand and binding sites for the transcription factors GATA-2/-3 and NF-κB.

Four consensus sites for GATA-2 and GATA-3 transcription factors could be identified. These transcription factors are of special interest as they are known to be coexpressed with cx31. Similar to cx31, GATA-2 and GATA-3 are present in the invasive trophoblast cells of the mouse as well as in Rcho-1 cells. These GATA factors are known to regulate a trophoblast specific gene, prolactin-I [8]. The E-box motif (CANNTG) could be identified at position -68. A couple of transcription factors belonging to the basic helix-loop-helix transcription factor family can bind to the E-box. Some of them as Hxt, Id-1 and Mash-2 could play a role in trophoblast specific expression of cx31 [9, 10, 11, 12]. Hxt and Mash-2 have been shown to be expressed in the invasive trophoblast and the giant cells of mouse placenta [9, 10]. In addition, Hxt stimulates the expression of an early trophoblast specific gene, interferon-τ and the expression of prolactin-I in undifferentiated Rcho-1 cells [9].

## 2.2 Functional promoter studies

To identify cis-acting regulatory elements in the cx31 promoter, fragments of the promoter were linked to a luciferase reporter gene. Transient transfection experiments were performed using three different cell lines: Rcho-1, Hel-37 a mouse keratinocyte cell line also expressing cx31 [13] and HeLa cells as a negative control. The relative activity conferred by the contructs was measured in relation to a control vector containing a SV40 promoter and enhancer which was transfected in parallel.

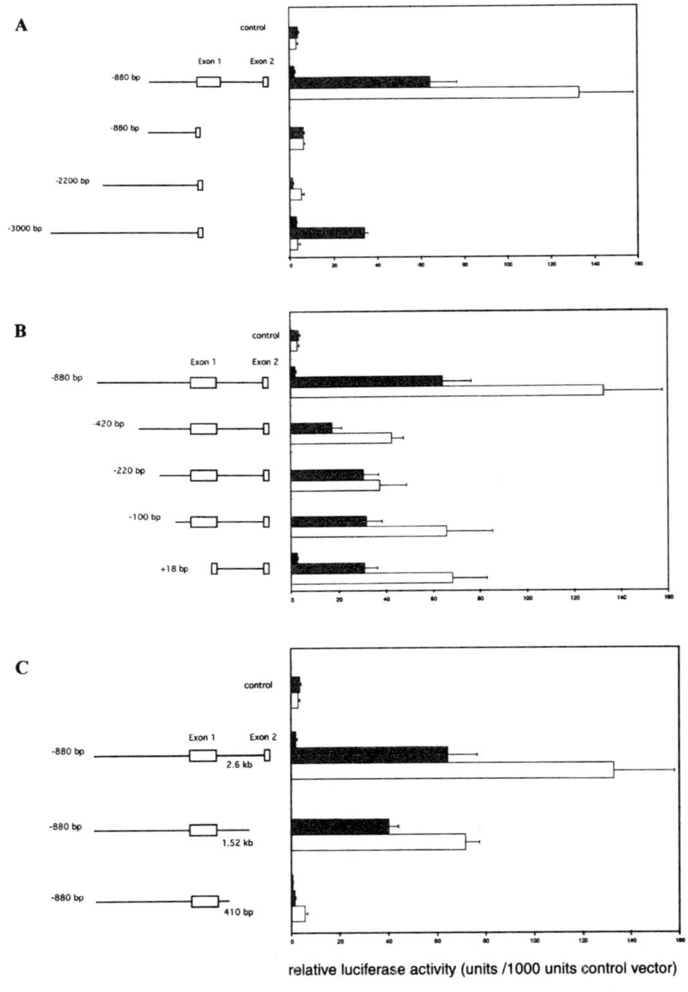


Fig. 3. Transient transfection experiments of promoter constructs. (A) 5' deletion analysis of the promoter region without the intron. (B) 5' deletion analysis of the promoter region including the intron. (C) 3' deletion analysis of the intron region. The recombinant Luciferase constructs including DNA fragments from the cx31 promoter region are shown on the left. The respective promoter activities are shown in the histograms on the right. The activities are expressed relative to the activity of the control vector containing the SV40 promoter and enhancer. White, grey and black bars present results obtained from Rcho-1, Hel-37, and HeLa cells, respectively.

As shown in Fig. 3A the construct containing sequence from -880 to +17 did not confer activity. When the 5' region was enlarged to 3 kb, the resulting construct -3000 to +17 was active only in Hel-37 cells. This activity is situated within a 800 bp fragment because the shorter construct -2200 to +17 was inactive. Interestingly, when the intron was included in the -880 to +17 construct, luciferase activity could be induced. In the following, a series of 5' deletion constructs -420, -220, -100 and +18 including the intron were tested (Fig. 3B). A reduction in activity points to a positive regulatory element between -880 and -420. Even the +18 construct, which contains exclusively the intron produced luciferase activity comparable to the -100 construct. Intron activity could be mapped to the region between +1.52 and +410 by constructing deletions -880 to +1.52 and -880 to +410 (Fig 3C). All constructs tested were inactive in HeLa cells. In future experiments we intend to understand the cell specific difference in the promoter activity of the 5' flanking region and the role the intron plays in conferring activity.

## 3. Summary and Conclusion

The rat cx31 gene shares the common structure of all other connexin genes investigated. It contains two exons separated by a short intron of 2.6 kb and has two transcription start sites. The putative promoter region contains no TATA-box but several interesting transcription factor binding sites which could be involved in trophoblast specific expression or regulation of this gene. Functional promoter analysis in trophoblast and keratinocyte cell lines reveals that part of the intron is necessary to confer basic activity. This activity is enhanced by enlarging the 5' region from 420 to 880 bp. The region between -2200 and -3000 mediates cell type specific expression of cx31 in keratinocytes compared to trophoblast cells.

## References


[1] B. Reuss *et al.*, Connexins and E-Cadherin are differentially expressed during trophoblast invasion and placenta differentiation in the rat, *Developmental Dynamics* 205 (1995) 172-182

[2] R. Grümmer *et al.*, Regulation of connexin31 gene expression upon retinoic acid treatment of rat choriocarcinoma cells, *Exp. Cell Res.* 227 (1996) 23-32

[3] R. Grümmer *et al.*, Expression pattern of different gap junction connexins is related to embryo implantation, *Int. J. Dev. Biol.* 40 (1996) 361-367

[4] T. N. Faria and M. J. Soares, Trophoblast cell differentiation: Establishment, characterization, and modulation of a rat trophoblast cell line expressing members of the placental prolactin family, *Endocrinology* 129 (1991) 2895-2906

[5] J. H. Hoh *et al.*, Molecular cloning and characterization of a new member of the gap junction gene family, connexin-31, J. *Biol. Chem.* 266 (1991) 6524-6531

[6] H. Hennemann *et al.*, Characterization of gap junction genes expressed in F9 embryonic carcinoma cells: molecular cloning of mouse connexin31 and -45 cDNAs, *Eur. J. Cell. Biol.* 57 (1992) 51-58

[7] M. B. Shapiro and P. Senapathy, RNA splice junctions of different classes of eukaryotes: sequence statistics and functional implications in gene expression. *Nucleic Acids Res.* 15 (1987) 7155-7157

[8] Y. K. Ng *et al.*, GATA factor activity is required for the trophoblast-specific transcriptional regulation of the mouse placental lactogen I gene, *Development* 20 (1994) 3257-3266

[9] J. C. Cross *et al.*, Hxt encodes a basic helix-loop-helix transcription factor that regulates trophoblast cell development, *Development* 121 (1995) 2513-2523

[10] F. Guillemot *et al.*, Essential role of Mash-2 in extraembryonic development, *Nature* 371 (1994) 333-336

[11] P.-Y. Desprez *et al.*, Suppression of mammary epithelial cell differentiation by the helix-loop-helix protein Id-1, *Mol. Cell. Biol.* 15 (1995) 3398-34004

[12] R. Benezra *et al.*, The protein Id: a negative regulator of helix-loop-helix DNA binding proteins, *Cell* 61 (1990) 49-59

[13] N. E. Fusenig *et al.*, Growth and differentiation characteristics of transformed keratinocytes from mouse and human skin in vitro and in vivo, *J. Invest. Dermatol.* 91 (1983) 168s-175s

Gap Junctions
R. Werner (Ed.)
IOS Press, 1998

# Expression of Connexin 32-Green Fluorescent Protein in Oligodendrocytes

H. W. Tedford, Hongyi Cui and John H. Carson
*University of Connecticut Health Center, Farmington, CT 06030*

**Abstract.** mRNA encoding a translational fusion between connexin 32 (Cx32) and green fluorescent protein (GFP) was microinjected into mouse oligodendrocytes in culture. The distribution of Cx32-GFP fusion protein expressed in the injected cells was visualized by confocal microscopy. Cx32-GFP appeared as numerous small clusters distributed throughout the perikaryon, processes and veins but excluded from the myelin compartment. Three dimensional reconstruction revealed that some of the clusters were embedded in the plasma membrane while others were intracellular. Time lapse imaging revealed that some of the clusters exhibited directed movement within the cell and others exhibited periodic fluctuations of fluorescence intensity.

## 1. Introduction

Connexin 32 (Cx32) is a gap junction protein expressed in myelinating glial cells where it is believed to form reflexive gap junctions at cytoplasmic incisures between adjacent layers in the myelin sheath (1). The intracellular trafficking pathway that targets Cx32 to these regions in myelin is not understood. Previous studies on intracellular trafficking of gap junction proteins have utilized heterologous systems such as oocytes or fibroblasts where endogenous Cx32 is not expressed. If particular aspects of Cx32 trafficking are peculiar to myelinating glial cells it may be misleading to analyze expression of this protein in such heterologous systems. In this study we have examined the intracellular distribution of a Cx32-GFP fusion protein expressed in oligodendrocytes which are the myelinating glial cells in the CNS.

Oligodendrocytes in primary culture have several features that make them particularly suitable for studies of intracellular trafficking of Cx32. First, the metabolic and physiological properties of oligodendrocytes in culture are presumably more similar to myelinating glial cells in vivo than the heterologous cell systems that have been used previously. This means that intracellular trafficking in oligodendrocytes in culture may more accurately reflect the trafficking pathway in vivo. Second, the morphology of the oligodendrocyte in culture lends itself to high resolution microscopic analysis of intracellular trafficking. The cell is extremely thin and flat. The various subcellular compartments (nucleus, perikaryon, processes, veins and myelin compartment) are spatially resolved in the X/Y axes and intracellular components are displayed within a thin layer of cytoplasm, enhancing resolution in the Z axis by minimizing the contribution of out-of focus material from above and below the plane of focus. These optical properties are ideal for confocal microscopy and also facilitate three dimensional analysis of Cx32 distribution from serial confocal optical sections. Third, exogenous Cx32-GFP can be expressed in oligodendrocytes in culture by microinjecting the encoding mRNA directly into the perikaryon. After translation of the injected mRNA the expressed protein can be visualized by fluorescence microscopy permitting time lapse imaging for temporal resolution of Cx32 trafficking in live cells. In this study we have taken advantage of these features to investigate intracellular trafficking of Cx32-GFP in oligodendrocytes.

## 2. Materials and Methods

The coding region of rat Cx32 cDNA was amplified by PCR from plasmid pfCRL obtained from Dr. Rudolf Werner (University of Miami, Miami, FL). The sense primer (5'TCGTCTAGAGCCGCCACCATGAACTGGACAG3') contained a consensus Kozak sequence flanked by an XbaI restriction site. The antisense primer (5'GCTTCTAGACAGGCTGAGCATCGG3') also contained an XbaI site. The amplified DNA was digested with XbaI and cloned into plasmid pHJ2 containing GFP cDNA followed by a polyA tail. The resulting plasmid (pATDP) was linearized with BsaW1 and transcribed with SP6 polymerase to generate capped Cx32-GFP mRNA for microinjection. Cx32-GFP mRNA (1 ug/ul) was coinjected with Texas Red conjugated dextran into murine oligodendrocytes in culture as described previously (2). After 6-24 hours the cells were examined by dual channel confocal microscopy. GFP fluorescence revealed the intracellular distribution of Cx32-GFP in the injected cells. Texas Red fluorescence indicated the amount of RNA injected and also provided a cytoplasmic volume marker. Three dimensional reconstruction from serial confocal optical sections was performed as described previously (3). The surface of the oligodendrocyte was rendered based on the isosurface of the Texas Red dextran-filled compartment. Isosurfaces of Cx32-GFP containing structures were rendered into the same coordinates to visualize their three dimensional intracellular distribution within the oligodendrocyte. Time lapse imaging was performed by collecting confocal images at 1 minute intervals.

## 3. Results

A confocal microscopic image of two oligodendrocytes microinjected with Cx32-GFP mRNA is shown in Figure 1. The coinjected Texas Red dextran is uniformly distributed throughout the cells. Cx32-GFP fusion protein expressed in the injected cells

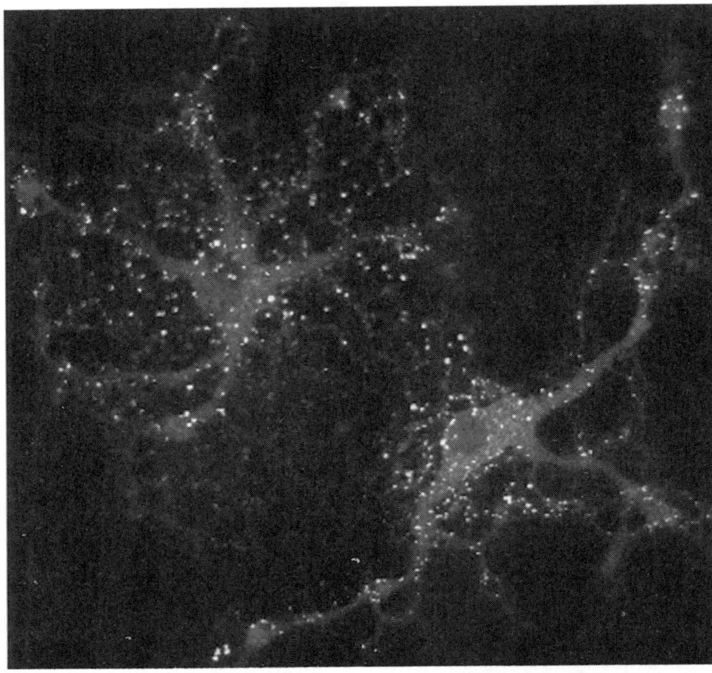

Figure 1 - Confocal image of Cx32-GFP expression in microinjected oligodendrocytes. Cells were coinjected with Cx32-GFP mRNA and Texas Red dextran. The dextran (dark grey) fills the cytoplasm. The Cx32-GFP (light grey) appears as numerous small clusters throughout the perikaryon and processes.

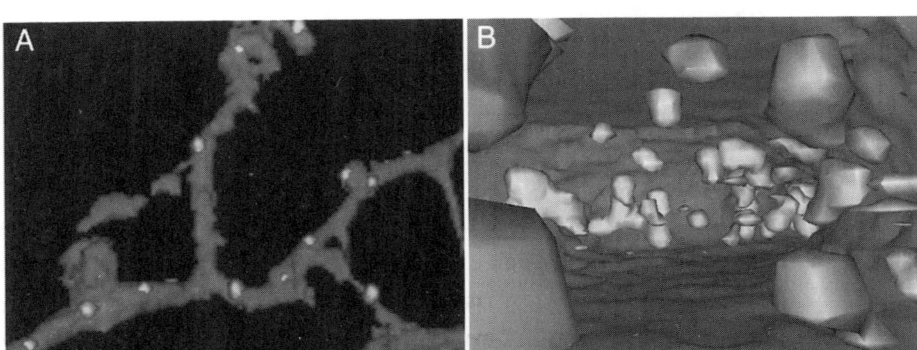


Figure 2 - 3-dimensional reconstruction of Cx32-GFP distribution in microinjected oligodendrocyte. Panel A - Exterior view - Cx32-GFP clusters (light grey) are embedded in the cell surface (dark grey). Panel B - Interior view - Cx32-GFP clusters (light grey) are distributed throughout the perikaryon.

appears as numerous small clusters throughout the perikaryon, processes and veins of the injected cells. The clusters are excluded from the interstices between the processes and veins which is the region of the cell corresponding to the lamellar myelin compartment in vivo. When two injected cells contacted each other the clusters were more concentrated in regions of cell-cell contact (data not shown).

Three dimensional reconstruction of Cx32-GFP clusters in an injected oligodendrocyte is shown in Figure 2. When the cell is viewed from the exterior (Figure 2A) some clusters appear to be embedded in the cell surface. When the cell is viewed from the interior (Figure 2B) many clusters appear to be entirely intracellular, not intersecting the cell surface at any point. Some of the intracellular clusters are congregated around the nucleus.

Time lapse images of Cx32-GFP clusters in an injected oligodendrocyte are shown in Figure 3. Most of the clusters are immobile with stable fluorescent intensity. However some clusters exhibit directed movement within the cell (data not shown) and other clusters (indicated by arrows) exhibit periodic fluctuations in intensity.

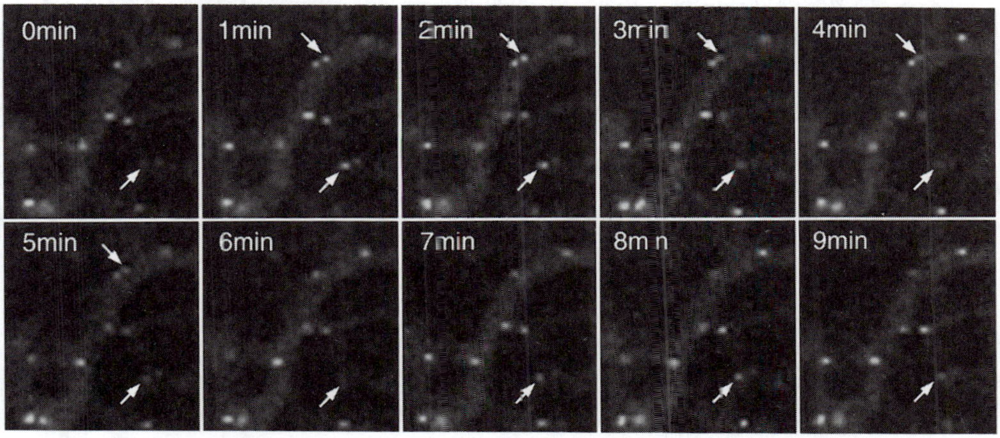


Figure 3 - Time lapse confocal imaging of Cx32-GFP clusters in a microinjected oligodendrocyte. Cx32-GFP clusters (light grey) indicated by arrows exhibit periodic fluctuations in intensity.

## 4. Discussion

The main finding in this study is that Cx32-GFP is expressed in microinjected oligodendrocytes and assembles into clusters that can be visualized by confocal microscopy. The clusters are dispersed throughout the perikaryon, processes and veins of the oligodendrocye but are excluded from the compartment corresponding to lamellar myelin. This distribution is consistent with the localization of gap juntions to cytoplasmic incisures in the myelin sheath in vivo, indicating that intracellular trafficking and targetting of Cx32-GFP in the injected cells provides a useful model for understanding gap junction targetting and assembly in myelinating glial cells in vivo.

The relation of the Cx32-GFP clusters in microinjected oligodendrocytes to the various intermediates in gap junction assembly and turnover that have been described previously remains to be determined. The fluorescent intensity and the diameter of the clusters indicate that each individual cluster must contain hundreds of Cx32-GFP molecules. The clusters embedded in the cell surface may correspond to hemi-channels in the oligodendrocyte plasma membrane. The observation that clusters tend to be concentrated in regions of cell-cell contact raises the possibility that hemi-channels on one cell may be capable of docking with hemi-channels on another cell to form true gap junctions. This could be tested by measuring the permeability properties of the injected cells. The intracellular clusters may represent trafficking intermediates in the pathway of translocation of Cx32-GFP to the cell surface, degradation intermediates in the pathway of Cx32 turnover or ectopic gap junctions between intracellular membranes. It may be possible to distinguish among these possibilities by electron microscopy.

The periodic fluctuations in fluorescence intensity exhibited by some clusters is indicative of dynamic changes in either the structure or environment of the Cx32-GFP molecules within the cluster. Some possible explanations include: periodic assembly and disassembly of Cx32-GFP molecules into clusters, periodic conformational or structural changes of the Cx32-GFP molecules within the cluster leading to fluorescence quenching and/or dequenching effects, spectral diffusion caused by periodic changes in the local environment in the vicinity of the cluster. More detailed analysis of the kinetic and spectral features of the periodic intensity fluctuations may distinguish among these possible explanations. The intensity fluctuations may provide a sensitive probe for dynamic structural or environmental changes affecting gap junction proteins in living cells.

## Acknowledgements

This work was supported by NIH grant #NS15190 to J.H.C., and UCHC pre- and post- doctoral fellowships to H.W.T. and H.C., respectively. We thank Dr. Elisa Barbarese for providing oligodendrocyte cultures andFrank Morgan for help with three dimensional reconstruction and time lapse imaging.

## References

[1] D.L. Paul, New functions for gap junctions. Curr. Op. Cell Biol. 7:665-672 (1995).

[2] Ainger, K., Avossa, D., Morgan, F., Hill, S.J., Barry, C., Barbarese, E. and Carson, J.H. (1993) Transport and Localization of Exogenous Myelin Basic Protein mRNA Microinjected into Oligodendrocytes. J. Cell Biol. 123: 431-441.

[3] Morgan, F., Barbarese, E. and Carson, J.H. (1992) Visualizing Cells in Three Dimensions Using Confocal Microscopy, Image Reconstruction, and Isosurface Rendering: Application to Glial Cells in Mouse Central Nervous System. Scanning Microscopy 6: 345-358.

# Section 11:

# Gap Junctions in Development and Cancer

*Gap Junctions*
*R. Werner (Ed.)*
*IOS Press, 1998*



# Modulation of Gap Junctional Intercellular Communication during Human Breast Stem Cell Differentiation and Immortalization

Kyung-Sun Kang[1,2], Chia-Cheng Chang[2] and James E. Trosko[2]

[1]*Division of Cellular and Molecular Toxicology, National Institute of Health Sciences, 1-18-1 Kamiyoga, Setagaya-ku, Tokyo 158, Japan*
[2]*Department of Pediatrics and Human Development, College of Human Medicine, Michigan State University, Michigan 48824, USA*

**Abstract.** The gap junctional exchange of ions or small molecules, including second messenger such as cAMP and $Ca^{2+}$ between adjacent cells, plays an essential role in the maintenance of homeostasis, coordination of secretion, morphogenesis, cell differentiation, and growth control in multicellular organs. Aberrant gap junctional intercellular communication (GJIC) has been implicated in tumor promotion, neuropathy and teratogenesis. We have isolated and characterized two types of normal human breast epithelial cells with stem cell characteristics (Type I) or basal epithelial cell characteristics (Type II). Furthermore, we have demonstrated that Type I normal human breast epithelial cells (HBEC) with stem cell characteristics were gap junctional intercellular communication (GJIC)-deficient, while Type II HBEC with basal epithelial characteristics were GJIC-efficient.

Using cholera toxin, we have also shown that human breast stem cells could differentiate into basal epithelial cells. This study showed that Type I HBEC could differentiate into Type II HBEC using cholera toxin or forskolin treatment. These differentiated cells have GJIC activity. In addition, these differentiated cells from Type I HBEC expressed connexin 43 (Cx43). We have developed SV40 immortalized cell lines from a normal human breast epithelial cell (HBEC) type with basal epithelial cell characteristics, and we examined the modulation of GJIC during immortalization. GJIC activity was reduced in the extended-life span and immortalization state of Type II HBEC. The phosphorylation pattern of Cx43 was altered during immortalization. Taken together, these results indicated that the modulation of GJIC might be crucial in human breast differentiation and carcinogenesis.

## Introduction

Homeostatic control of the many functions of cells within tissues and organs of virtually all metazoans is controlled, in part, by intercellular channels between contiguous cells, called gap junctions (1).

These gap junction channels allow the transfer of ions and small molecules to synchronize their functions in electrotonic cells and to metabolically cooperate in non-excited cells (2). Gap junctions consist of two membrane-bound hemi-channels of two neighboring cells, called connexons (3). Each connexon is composed of a hexameric composite of a highly evolutionarily-conserved family of proteins called connexins (4). The direct transfer of ions, second messages, nutrients and macromolecular substrates between cells via gap junctions has been correlated with the entrainment of synchronous activity of excitable cells, control of pattern formation, cell proliferation, differentiation, apoptosis, secretion, and the "bystander effect" (5-7).

Dysfunction of gap junctional intercellular communication (GJIC) has been linked to teratogenesis, carcinogenesis, specifically the tumor promotion and progression phases, reproductive dysfunction, neurological anomalies and many other disease states (8). Tumor promoting chemicals, various oncogenes, as well as anti-sense connexin genes, appear to be associated with the down regulation of GJIC, whereas the anti-tumor promoting chemicals, tumor suppressor genes or connexin-transfected tumor cells restore GJIC in cells having deficient GJIC (9,10). Various transgenic and connexin knock-out mice, together with various connexin mutants in human beings have further demonstrated the critical roles gap junctions play in maintaining health (11-14)

Recently, we have developed a method to culture normal human breast epithelial cells which do not express GJIC, but expresses the estrogen receptor(15), has luminal epithelial cell markers and stem cell characteristics (i.e., they have the ability to differentiate into another cell type and to form a budding/ductal organoid). This study was designed to characterize the potential roles of gap junctions during the differentiation and immortalization of a normal human breast epithelial cell strain.

## Results and Discussion

### Differentiation of the Breast Stem Cell to its Differentiated Daughter by cAMP - Inducing Agents

Based on the hypothesis that totipotent and/or pluripotent stem cells do not express their connexin genes and on previous observations that our human presumptive breast epithelial stem cells (Type I cells) are deficient in GJIC (16), we wished to examine the hypothesis that induction of differentiation of Type II cells from the Type I or stem cell would be accompanied by the induction of GJIC. Therefore, after 8 days treatment of Type I cells with cholera toxin or forskolin, the Type II cells which appeared in culture were examined for their ability to perform GJIC. The scrape loading dye transfer assay was used to detect GJIC in these cells. The type II or differentiated cells exhibited GJIC while the Type I cells had no GJIC (Figure 1,H). These results suggest that cAMP -inducing agents can induce cell differentiation in these human breast epithelial cells, possibly though a protein kinase A pathway. The question remains whether the differentiation preceded the induction of connexin proteins and GJIC or vice-versa.

## Characterization of GJIC during the Blockage of Differentiation ("immortalization") of Human Breast Epithelial Stem Cells

One concept of the process of carcinogenesis is that a normal, "mortal" cell must first be "immortalized" before it can be neoplastically transformed. Our view is that the normal human totipotent and pluripotent stem cells are, by definition and reality, "immortal". It is when they are induced to terminally differentiate that they become "immortal". In reality, then, the first step of carcinogenesis is to block "mortalization" of a normal "immortal" stem cell. Exposing stem cells to chemicals, radiation or "immortalizing" viruses could mutate or inactivate genes or gene products involved in the "mortalization" of stem cells. Transfecting our human breast epithelial stem (Type1) cells with SV40 viral DNA, clones of cells exhibiting first extended life span, then "immortalization" or lack of differentiation of Type II characteristics, were examined for their ability to perform GJIC. Results in Figure 2 show that these cells do not exhibit any functional GJIC as measured by dye transfer. This result is consistent with the hypothesis that "immortalization" of Type I or stem cell is, in reality, the blockage of "mortalization" of an immortal, undifferentiated stem cell. Moreover, it is consistent with the hypothesis that differentiation is dependent on the ability of stem cells to perform GJIC and not vice-versa.

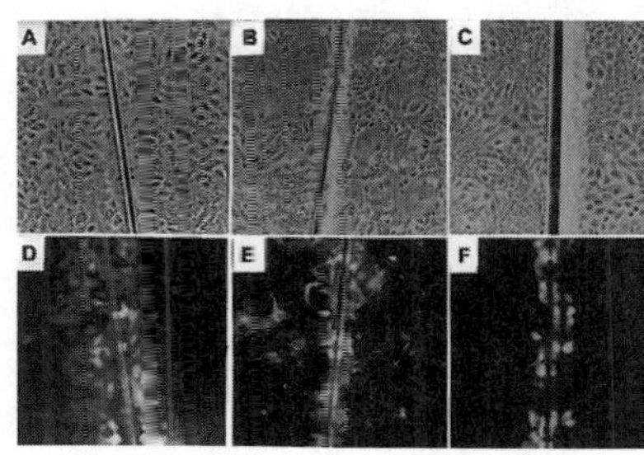

Figure 2. Scrape loading dye transfer images. Panels A and D, normal HBEC Type II; B and E, Extended life span; C and F, Immortalized HBEC Type II.

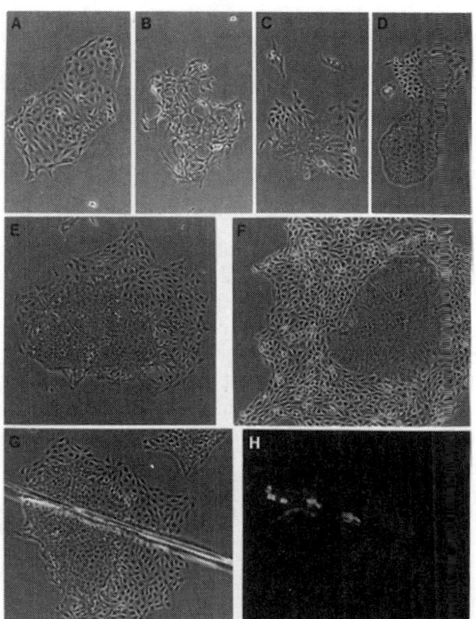

Figure1. Differentiation from Type1 HBEC to Type II HBEC by forskolin (10nM). Normal HBEC; Panels: A, Control, non-treated; B, 4 hrs. treatment; C, 3 days; D, 4 days; E, 7 days; F' 8 days; G and H, Epifluorescent image of scrape loading -dye transfer after 7 days treatment

## Characterization of Localization and Phosphorylation Patterns of Cx43 during "Immortalization" of Human Breast Epithelial Stem Cells

In the differentiated human breast epithelial, Type II, differentiated cells, membrane -localized gap junction plaques were observed (Figure 3). Interestingly, antibodies directed to Cx43 in the SV-40 extended-life span and "immortal" HBEC cells were observed to be localized in the nucleus. These results indicate that the blockage of differentiation by the SV-40 large T antigen might affect the trafficking of the connexin protein to the membrane, rather than the transcriptional or translational function of the connexin gene or mRNA.

To examine if the phosphorylation of connexin 43 is changed, thereby, altering its cellular localization, we performed western blot analyses of the Cx43. In figure 4, two major bands ($P_0$ and $P_1$) and a weak band ($P_2$) were detected in the normal Type II, while only a $P_0$ band was observed in both extended -life span and "immortalized" HBEC. It, then, appears that during the blockage of differentiation or "mortalization" of the human breast epithelial stem cell by SV-40 large T antigen, hypophosphorylation of the Cx43 protein occurred as compared to the normal human breast epithelial, Type II cell. Taken together, it appears that functional GJIC is important in the differentiation of human breast stem cells and that blockage of differentiation or prevention of "mortalization" of these cells depends on the prevention of functional GJIC. Since these cells are not tumorigenic, it also suggests that while the blockage of GJIC is a characteristic of most, if not all, tumor cells, it is a necessary, but insufficient, step in carcinogenesis.

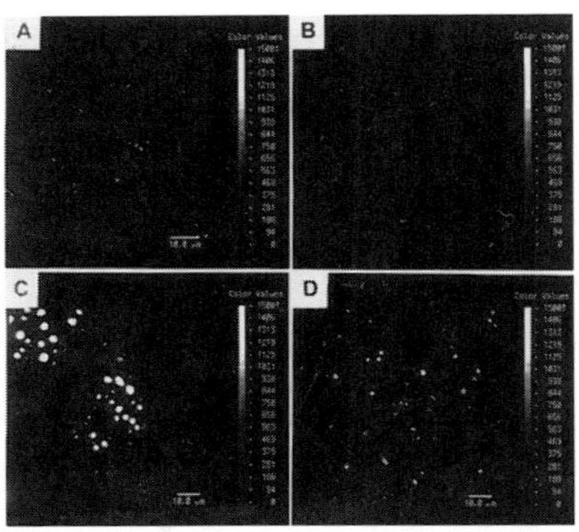


Figure 3. Immunofluorescent staining for connexin 43. A, Normal HBEC Type II; B, No primary Antibody; C, Extended-life span; D, Immortal HBEC Type II. Antibody for Cx 43 is anti-rabbit polyclonal antibody from Zymed Co. Confocal scaning was performed by Ultima laser confocal microscope from Meridian Co.

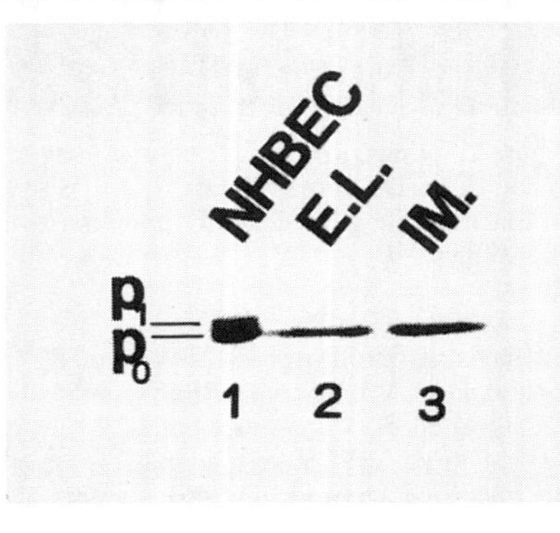


Figure 4. Western blot analysis. Lane 1, NHBEC Type II; Lane 2, Extended-life span; Lane 3, Immortal HBEC Type II. NHBEC, normal human breast epithelial cell ; E.L., extended life-span; IM., Immortal. Antibody for Cx 43 is anti-rabbit polyclonal antibody from Zymed Co. Detection was performed by ECL kit from Amersham Co.

**Acknowledgement**

This work was supported in part by a grant from the National Cancer Institute (CA21104) to J.E.T. and by a grant from the National Institute of Environmental Health (ES07256) to C.C.C. And this first author(K.S. K) thank Dr. Tohru Inoue, chairperson at K-S K's visiting address and supported him to join international gap junction meeting.

## References


[1] Bruzzone,R., White, T.W., and Paul, D.L.,Connections with connexins: the molecular basis of direct intercellular signalling. Eur. J. Biochem. 238:1-27, 1996.

[2] Loewenstein, W.R. and Rose, B. The cell -cell channel in the control of growth. Semin.Cell Biol. 3:59-79,1992.

[3] Kumar, N.W. and Gilula, N.B. The gap junction communication channel. Cell 84: 381-388, 1996.

[4] Kanter, H.L., Saffitz,J.E., and Beyer,E.C. Cardiac myocytes express multiple gap junction proteins. Circ. Res. 70: 438-444, 1992.

[5] Bruzzone, R., and Meda, P. The gap junction: a channel for multiple functions? Eur. J. Clin. Invest. 18:444-453, 1988.

[6] Trosko, J.E., and Goodman, J.A., Intercellular communication may facilitate apoptosis: Implications for tumor promotion." Molecular Carcinogen. 11: 8-12,1994.

[7] Mesnil, M., Piccoli, C., Tiraby, G., Willecke, K., and Yamasaki, H., Bystander killing of cancer cells by herpes simplex virus thymidine kinase gene is mediated by connexins. Proc. Natl. Acad. Sci. USA 93: 832-835, 1966

[8] Trosko, J.E., Madhukar, B.V. and Chang, C.C., Endogenous and exogenous modulation of gap junctional intercellular communication : toxicological and pharmacological implications. Life Sci. 53: 1-19, 1993

[9] Trosko, J.E., Chang, C.C., Madhukar, B.V. and Dupont E., Oncogenes, tumor suppressor genes and intercellular communication in the "Oncogeny as partially blocked ontogeny" hypothesis. In: New Frontiers in Cancer Causation. Iversen, O.H., ed., Taylor and Francis,Publishers, Washington, D.C., pp. 181- 197, 1993.

[10]Ruch, R.J., Guan, X., and Sigler, C., Inhibition of gap junctional intercellular communication and enhance ment of growth in BALB/c3T3 cells treated with connexin 43 antisense oligonucleotides. Molec. Carcinogen. 14:269-274, 1995.

[11] Reaume, A.G., de Sousa, P.A., Kulkarni, S., Langille, B.L., Zhu, D., Davies, T.C., Juneja, S.C., Kidder, G.M., and Rossant, J. Cardiac malformation in neonatal mice lacking connexin 43. Science 267: 1831-1834, 1995.

[12]Simon, A.M., Goodenough, D.A., Li, En, and Paul, D.,L., Female infertility lacking connexin 37. Nature 385: 525-529,1997.

[13]Britz- Cunningham,S.H., Shah, M.M., Zuppan, C.W., and Fletcher, W.,H., Mutations of the connexin43 gap junction gene in patients with heart malformations and defects of laterality. New Engl. J. Med. 332: 1323-1329,1995.

[14]Bergoffen, J., Scherer, S.S., Wang, S., Oronzi-Scott, M., Bone, L.J., Paul, D.L., Chen, K., Lensch, M.W., Chance, P.F., and Fischbeck, K.H., Connexin mutations in X-linked Charcot-Marie -Tooth disease. Science 262:2039-2042, 1993.

[15] Kang, K.-S., Morita, I., Cruz, A., Jeon, Y. J., Trosko, J. E., and Chang, C. C. Expression of estrogen receptors in a normal human breast epithelial cell typorita, I., Cruz, A., Jeon, Y. J., Trosko, J. E., and Chang, C. C. Expression of estrogen receptors in a normal human breast epithelial cell type with luminal and stem cell characteristics and its neoplastically transformed cell lines. Carcinogenesis, 18:251-257, 1997.

[16] Kao, C-Y., Nomata, K., Oakley, C.S., Welsch, C. W., and Chang, C.C., Two types of normal human breast epithelial cells derived from reduction mammoplasty: phenotypic characterization and response to SV40 transfection. Carcinogenesis 16: 531-538, 1995.

*Gap Junctions*
*R. Werner (Ed.)*
*IOS Press, 1998*

# Connexin Expression During *In Vitro* Cardiac Differentiation of Mouse Embryonic Stem Cells

Masahito OYAMADA[1], Yumiko OYAMADA[2], Kiyoshi KOMATSU[2], Tetsuro TAKAMATSU[1] and Michio MORI[2]

[1]*Department of Pathology and Cell Regulation, Kyoto Prefectural University of Medicine, Kawaramachi Hirokoji, Kamigyo-ku, Kyoto 602, Japan*

[2]*Department of Pathology, Sapporo Medical University School of Medicine, South 1, West 17, Chuo-ku, Sapporo 060, Japan*

**Abstract.** Embryonic stem (ES) cells can differentiate into various cell lineages *in vitro*, which provides a good model system to study early embryogenesis. Using an *in vitro* system for differentiation of ES cells into cardiac myocytes, we found that differential modulation of Cx40, Cx43 and Cx45 occurs during early stages of cardiogenesis.

## 1. Introduction

Gap junctional intercellular communication (GJIC) has long been considered to play a major role in regulation of cell differentiation and early developmental events in animal bodies [1]. A plausible hypothesis is that cells differentiating into a similar phenotype can communicate easily with each other, whereas communication becomes restricted between cell populations with different phenotypes. There exists evidence supporting possible roles of GJIC in cell differentiation. First, injection of antibodies to gap junction protein or gap junction antisense RNA into amphibian embryos and mouse preimplantation embryos results in developmental abnormalities [2]. Second, gap junctional intercellular communication-deficient murine ES cells have limited differentiation capacity compared with the parental cell line [3]. Third, Cx43 knockout mice die at birth, as a result of a failure in pulmonary gas exchange caused by a swelling and blockage of the right ventricular outflow tract from the heart [4].

Embryonic stem (ES) cells are totipotent lines derived from the inner cell mass of the preimplantation blastocyst [5, 6]. The establishment of ES cell lines has enabled us to make many new experimental approaches in the field of mammalian developmental biology, including the generation of knockout mice. Besides various applications *in vivo*, differentiation of ES cells *in vitro* provides a good model system to study early events in morphogenesis [7, 8]. This is because ES cells can differentiate into various cell types, including the hematopoietic, endothelial, muscle and neuronal lineages, under appropriate conditions in culture and because early precursors of mature lineages are much more accessible *in vitro* than *in vivo*. The *in vitro* differentiation of ES cells follows the normal developmental program of the embryo. For example, ES cell-derived cardiac myocytes developmentally express cardiac-specific genes such as cardiac isoforms of myosin heavy chain, a ventricular isoform of myosin light chain, atrial natriuretic factor, a cardiac-specific homeobox (Csx) and all major heart-specific ion channels [9-22]. We analyzed the expression of connexin (Cx) genes by RT-PCR during cardiomyocytic differentiation of ES cells *in vitro* to learn [illegible] changes in the expression of multiple Cx genes occur during the early stage of heart development.

## 2. Connexin expression during cardiomyocytic differentiation of ES cells *in vitro* [23]

Mouse embryonic stem cell (ES cell) line J1 was used. For differentiation of ES cells into cardiac myocytes, the hanging drop culture method was used, as shown in Figure 1. Cultures were monitored every day under a microscope to detect the presence of spontaneously contracting cardiomyocytes. The expression of cardiac Cx genes, Cx40, Cx43 and Cx45, and cardiac-specific genes such as α and ß cardiac isoforms of myosin heavy chain (MHC α and ß), a ventricular isoform of cardiac regulatory myosin light chain (MLC-2v), was examined by RT-PCR. Detailed information about RT-PCR can be found in our paper [23].

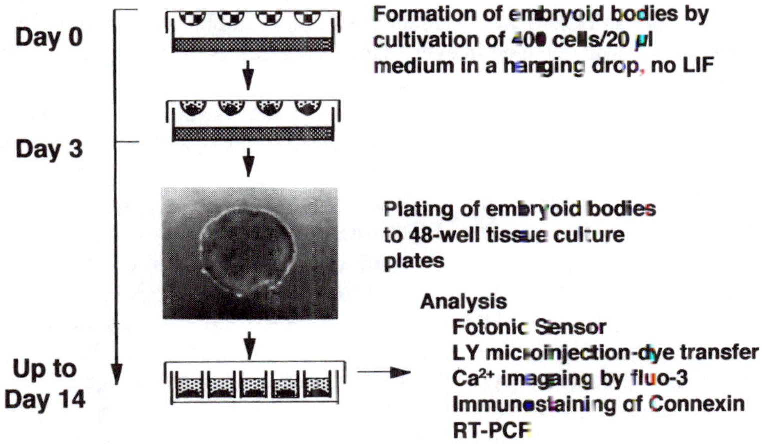


Figure 1. Experimental procedure for cardiomyocytic differentiation of ES cells *in vitro*

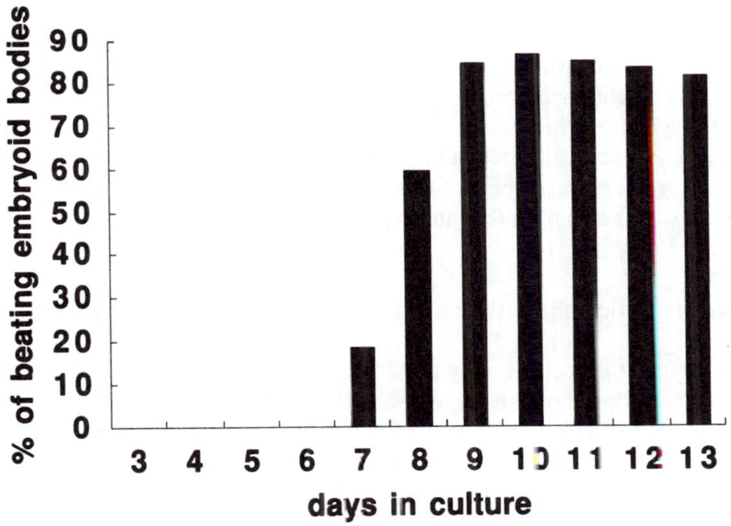


Figure 2. Appearance of spontaneously beating embryoid bodies during *in vitro* differentiation of ES cells

Figure 2 shows percentages of embryoid bodies containing spontaneously beating cardiomyocytes during *in vitro* differentiation of ES cells. At day 7 of culture, beating cells were first observed in about 20% of the plated embryoid bodies. At day 9, more than 80% of the embryoid bodies showed beating and maintained this level during the observation period (up to 14 days).

Figure 3 demonstrates expression of heart-specific genes and connexins during ES cell differentiation into cardiomyocytes *in vitro* measured by RT-PCR. Transcripts for heart-specific genes such as α and β cardiac isoforms of MHC, and myosin light chain 2v were detected on and after the appearance of beating embryoid bodies (day 7 and later), and showed single bands of the expected sizes of 302, 205 and 499 base pairs, respectively. Cx40 transcripts were not observed in undifferentiated ES cells, and were barely detectable in 3- and 5-day-old embryoid bodies. However, Cx40 transcripts significantly increased with the appearance of beating cells at day 7. On the other hand, transcripts of the major cardiac gap junction genes Cx43 and Cx45 were present even in undifferentiated embryonic stem cells (day 0 of the culture) and detected in every sample regardless of the appearance of beating cardiomyocytes.

Like heart-specific genes such as MHC α and β and MLC-2v genes, Cx40 gene expression corresponded to the appearance of beating cardiomyocytes, whereas Cx43 and Cx45 were expressed in undifferentiated ES cells and their expression continued throughout the developmental period examined. Immunofluorescence for Cx43 also showed punctate staining on cell-cell borders not only among ES cell-derived cardiomyocytes but also surrounding nonbeating cells in embryoid bodies. Concerning Cx43 expression in undifferentiated ES cells, our results agreed with those reported previously by Nishi *et al.* [24]. Concerning Cx40, Delorme *et al.* [25] reported the expression of Cx40 transcripts and protein in the heart at 11 days post-coitum and suggested that mouse Cx40 is not only the marker of the conduction system in adults but is also a potential marker for "embryonic" unorganized conduction tissue. The appearance of Cx40 transcripts coincided with that of beating cells and expression of cardiac-specific genes, suggesting that the expression of Cx40 is closely related to cardiomyocytic differentiation of ES cells.

Concerning GJIC function during cardiomyocytic differentiation of ES cells *in vitro*, Lucifer Yellow dye transfer assay showed that GJIC was present among undifferentiated ES cells, prebeating cells of embryoid body outgrowth and ES cell-derived beating cardiomyocytes, and that when dye was injected into a beating cell at the border, dye spread was restricted to neighboring beating cells.

These data indicate that the expression of multiple connexins is differentially regulated during the cardiomyocytic differentiation of ES cells *in vitro*. A paper on ultrastructure and cell-cell coupling of cardiac myocytes differentiating in embryonic stem cell cultures written by Westfall *et al.* was also recently published [26].

## 3. Cardiomyocytic differentiation of Cx43-deficient ES cells *in vitro*

Cx43-deficient ES cells (kindly provided by Dr. J. Rossant) showed little GJIC measured by Lucifer Yellow dye transfer assay, but there was cardiomyocytic differentiation *in vitro*, which was not significantly different from that of wild-type ES cells. RT-PCR revealed that the amounts of Cx40 and heart-specific gene transcripts significantly increased with the appearance of beating cells as with wild-type ES cells, whereas Cx43 transcripts were not detected in the Cx43-deficient cells. These data indicate that Cx43 is not essential for cardiomyocytic differentiation of ES cells *in vitro* and that expression of other connexins such as Cx40 and Cx45 is not influenced by the absence of Cx43 expression.

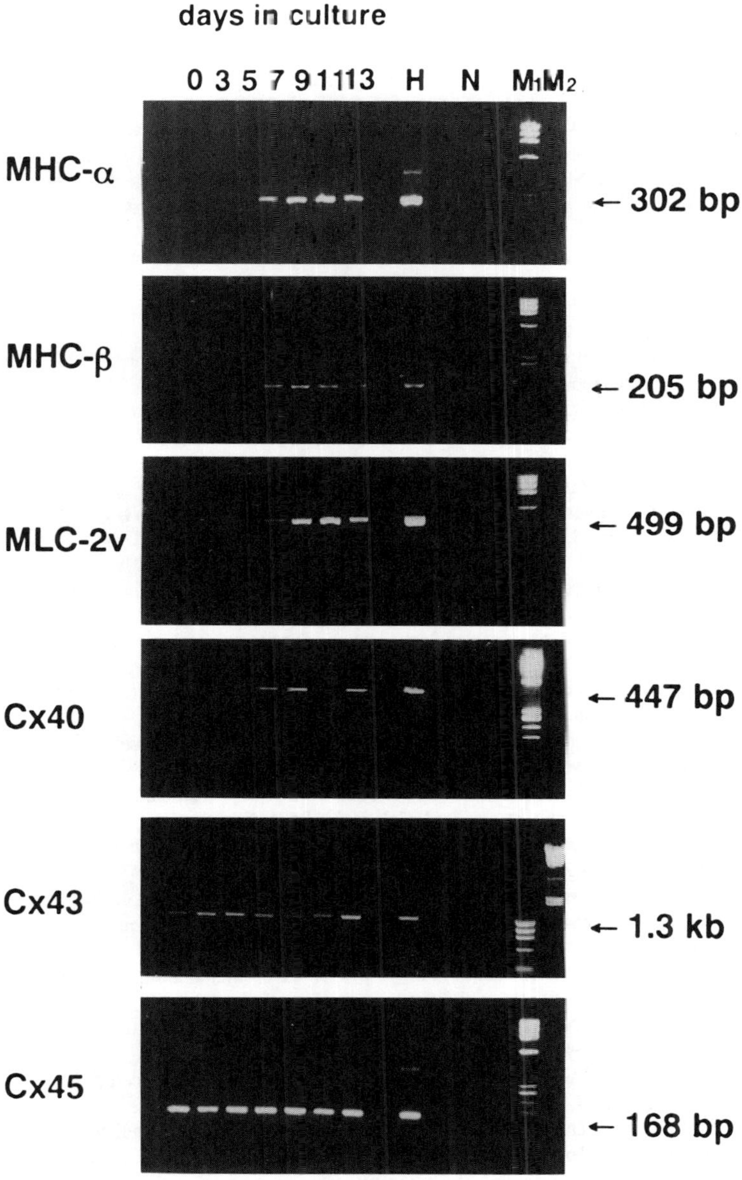


Figure 3. RT-PCR analysis for expression of heart-specific genes and connexins during ES cell differentiation into cardiomyocytes *in vitro*. 0.5 μg of the total RNA from each sample was reverse transcribed and PCR amplified with sequence-specific oligonucleotide primers. The gene products were analyzed by agarose gel electrophoresis. H, sample from mouse heart as a positive control; N, negative control. Sample with no RNA was reverse transcribed and PCR amplified; M1, molecular weight marker. Product of φX 174 plasmid DNA cut by Hae III; M2, molecular weight marker. Product of λphage DNA cut by Hind III .

## References


[1] S.C. Guthrie and N.B. Gilula, Gap junctional communication and development. *Trends Neurosci.*, **12** (1989) 12-16.

[2] A. Bevilacqua, *et al.*, Abnormal development and dye coupling produced by antisense RNA to gap junction protein in mouse preimplantation embryos. *Proc. Natl. Acad. Sci. USA*, **86** (1989) 5444-5448.

[3] S.A. Sheardown and M.L. Hooper, A relationship between gap junction-mediated intercellular communication and the *in vitro* developmental capacity of murine embryonic stem cells. *Exp. Cell Res.*, **198** (1992) 276-282.

[4] A.G. Reaume, *et al.*, Cardiac malformation in neonatal mice lacking connexin43. *Science*, **267** (1995) 1831-1834.

[5] M. Evans and M. Kaufman, Establishment in culture of pluripotential cells from mouse embryos. *Nature*, **292** (1981) 154-156.

[6] G. Martin, Isolation of a pluripotent cell line from early mouse embryos cultured in medium conditioned by teratocarcinoma stem cells. *Proc. Natl. Acad. Sci. USA*, **78** (1981) 7634-7638.

[7] T.C. Doetschman, *et al.*, The *in vitro* development of blastocyst-derived embryonic stem cell lines: formation of visceral yolk sac, blood islands and myocardium. *J. Embryol. Exp. Morphol.*, **87** (1985) 27-45.

[8] G.M. Keller, *In vitro* differentiation of embryonic stem cells. *Curr. Opin. Cell Biol.*, **7** (1995) 862-869.

[9] J. Robbins, *et al.*, Mouse embryonic stem cells express the cardiac myosin heavy chain genes during development *in vitro*. *J. Biol. Chem.*, **265** (1990) 11905-11909.

[10] A.M. Wobus, *et al.*, Pluripotent mouse embryonic stem cells are able to differentiate into cardiomyocytes expressing chronotropic responses to adrenergic and cholinergic agents and $Ca^{2+}$ channel blockers. *Differentiation*, **48** (1991) 173-182.

[11] A. Sanchez, *et al.*, Myosin heavy chain gene expression in mouse embryoid bodies. An *in vitro* developmental study. *J. Biol. Chem.*, **266** (1991) 22419-22426.

[12] I. Komuro and S. Izumo, Csx: a murine homeobox-containing gene specifically expressed in the developing heart. *Proc. Natl. Acad. Sci. USA*, **90** (1993) 8145-8149.

[13] V.A. Maltsev, *et al.*, Embryonic stem cells differentiate *in vitro* into cardiomyocytes representing sinusnodal, atrial and ventricular cell types. *Mech . Dev.*, **44** (1993) 41-50.

[14] J.M. Metzger, *et al.*, Transition in cardiac contractile sensitivity to calcium during the *in vitro* differentiation of mouse embryonic stem cells. *J. Cell Biol.*, **126** (1994) 701-711.

[15] V.A. Maltsev, *et al.*, Cardiomyocytes differentiated *in vitro* from embryonic stem cells developmentally express cardiac-specific genes and ionic currents. *Circ. Res.*, **75** (1994) 233-244.

[16] A.M. Wobus, *et al.*, Development of cardiomyocytes expressing cardiac-specific genes, action potentials, and ionic channels during embryonic stem cell-derived cardiogenesis. *Ann. N.Y. Acad. Sci.*, **752** (1995) 460-469.

[17] G. Weitzer, *et al.*, Cytoskeletal control of myogenesis: a desmin null mutation blocks the myogenic pathway during embryonic stem cell differentiation. *Dev. Biol.*, **172** (1995) 422-439.

[18] J.M. Metzger, *et al.*, Myosin heavy chain expression in contracting myocytes isolated during embryonic stem cell cardiogenesis. *Circ. Res.*, **76** (1995) 710-719.

[19] M.G. Klug, *et al.*, DNA synthesis and multinucleation in embryonic stem cell-derived cardiomyocytes. *Am. J. Physiol.*, **269** (1995) H1913-1921.

[20] M.V. Westfall, *et al.*, Troponin I isoform expression is developmentally regulated in differentiating embryonic stem cell-derived cardiac myocytes. *Dev. Dyn.*, **206** (1996) 24-38.

[21] J.M. Metzger, *et al.*, Vital staining of cardiac myocytes during embryonic stem cell cardiogenesis *in vitro*. *Circ. Res.*, **78** (1996) 547-552.

[22] M.O. Sowell, *et al.*, Targeted inactivation of alphai2 or alphai3 disrupts activation of the cardiac muscarinic $K^+$ channel, $IK^+Ach$, in intact cells. *Proc. Natl. Acad. Sci. USA*, **94** (1997) 7921-7926.

[23] Y. Oyamada, *et al.*, Differential regulation of gap junction protein (connexin) genes during cardiomyocytic differentiation of mouse embryonic stem cells *in vitro*. *Exp. Cell Res.*, **229** (1996) 318-326.

[24] M. Nishi, *et al.*, Developmental regulation of gap junction gene expression during mouse embryonic development. *Dev. Biol.*, **146** (1991) 117-130.

[25] B. Delorme, *et al.*, Developmental regulation of connexin 40 gene expression in mouse heart correlates with differentiation of the conducting system. *Dev. Dyn.*, **204** (1995) 358-371.

[26] M.V. Westfall, *et al.*, Ultrastructure and cell-cell coupling of cardiac myocytes differentiating in embryonic stem cell cultures. *Cell Motil. Cytoskeleton*, **36** (1997) 43-54.

*Gap Junctions*
*R. Werner (Ed.)*
*IOS Press, 1998*



# Connexins and Carcinogenesis: Upregulated Expression of Connexin 43 by Cancer Preventive Agents or by Gene Transfer Decreases Proliferation and Expression of Neoplasia

Timothy J. King, Laurie H. Fukushima, and John S. Bertram

*Cancer Research Center of Hawaii, University of Hawaii, 1236 Lauhala Street, Honolulu, Hawaii 96813, USA*

**Abstract.** Cancer preventive retinoids and carotenoids are antiproliferative and upregulate connexin43 expression and gap junctional communication (GJC). To determine if effects on GJC are central to actions of these compounds we have created HeLa cell clones containing a tet-inducible Cx43 gene. Clones respond rapidly to induction by doxycycline with Cx43 synthesis and incorporation into junctional plaques. In dense cultures, induction results in decreased proliferation. Clonal heterogeneity among HeLa cells with regard to Cx43 expression and neoplastic phenotype was also discovered. These results support the hypothesis of growth control via GJC.

**1. Introduction**. The role of gap junctional communication in growth control and neoplasia, was first proposed by Loewenstein and Kanno in the 1960s [1]. This evidence stemmed from observations that neoplastically transformed cells in culture were deficient in junctional communication. The central tenet of this hypothesis was that normal cells form communicating compartments, the size of which is monitored via junctionally transferred signals. Cells would proliferate until the magnitude of the signal increased beyond a certain threshold, determined by the cell number and degree of communication, when proliferation would cease. In this model the loss of communication by neoplastic cells would cause unregulated growth [2]. This hypothesis was strengthened by the later discovery that tumor promoting agents of diverse structure strongly inhibited junctional communication [3], and that restoration of communication between tumor cells and normal cells restored growth control [4].

Since this hypothesis was first proposed, a central role for proliferation in the process of carcinogenesis has become evident i.e. in the genesis of the neoplastic cell, not merely in the expression of neoplasia itself. Proliferation is required because cancer in humans is a multi-step process requiring mutations of many genes in a single cell [5]. Mutations arise as a consequence of replication of damaged genes, thus proliferation is a prerequisite for mutation. Moreover, in order that there is a finite probability of multiple mutations in a single cell, this cell must be capable of forming a population containing at least $10^6$ cells. For a detailed discussion see [6].

As an extension of these arguments, it would follow that agents or procedures that inhibited aberrant proliferation would inhibit the carcinogenic process, i.e. be chemopreventive. In the 10T/1/2 cell system of mouse fibroblasts, this prediction has been fulfilled for two important classes of cancer preventive agents: the retinoids and the carotenoids. The retinoids can reduce the incidence of cancers in high-risk subjects: in head and neck cancer 13-cis retinoic acid caused dramatic decreases in second primary cancer incidence [7]. This expression was reversible however on drug removal [8]. In 10T/1/2 cells, application of retinoids or carotenoids after exposure to a carcinogenic stimulus caused a partial to complete inhibition of formation of neoplastically transformed cells [9]; a scenario similar to the clinical studies which were conducted in a population of smokers. This effect was again reversible upon drug removal. We demonstrated that this action of retinoids was strongly correlated with their ability to upregulate junctional communication [10]; an action that was due to increased expression of connexin43 (Cx43) at the message and protein level [11]. Retinoids active as inhibitors of neoplastic transformation also strongly decreased the proliferation of cultured cells. This action correlated strongly with their ability to upregulate communication and their ability to suppress transformation. However this was only seen in confluent cells, not in cells plated sparsely and deprived of cell-cell contacts and thus the ability to form functional junctions [12]. In clinical studies retinoic acid has been shown to upregulate expression of Cx43 in the suprabasal cells of human epidermis [13].

Thus retinoid-induced junctional communication could contribute to chemopreventive activity by suppressing cell growth. However it should be noted that these are only associations and do not prove cause and effect. Because of these concerns we have initiated studies to examine the properties of cells in which Cx43 is upregulated utilizing molecular techniques.

## 2. Results and Discussion

*Studies utilizing Cx43 transfected cells: Production of cell lines containing Cx43 under the control of the tetracycline-regulated expression system.* The *tet* expression system, comprised of *tet* operator sequences proximal to a CMV minimal promoter and a *tet* receptor/viral transactivator fusion protein derived from a prokaryotic antibiotic resistance mechanism, was first developed and described by Bujard and colleagues in Heidelberg, Germany [14]. Cells are first stably transfected with the tetracycline receptor followed by antibiotic selection. A second stable transfection is then carried out with a plasmid containing the gene of interest under the control of the *tet* operator/CMV promoter sequences followed by a second round of antibiotic resistance selection. Cells are subsequently screened for the ability of tetracycline, or more commonly the tetracycline analog doxycycline, to regulate expression of the gene of interest. A major advantage of this system is that tetracycline or doxycycline are without side-effects at concentrations sufficient to activate the *tet* expression system in mammalian cells.

*Discovery of clonal heterogeneity amongst HeLa cells.* At the time we initiated these studies, all published reports of the use of the *tet* system had utilized HeLa cells. Since HeLa cells have been previously studied in connexin gene transfer experiments and are known to form connexons [15], we decided to employ this cell line. HeLa cells were obtained from the ATCC, and subjected to the first round of transfection utilizing the tet receptor and a geneticin (G418) resistance marker. Although previous reports indicated that HeLa cells did not express Cx43 mRNA and protein [15], we decided to characterize G418-resistant clones. We were thus surprised to discover clones that exhibited high, medium or undetectable levels of Cx43 protein expression. The extent of this clonal heterogeneity is shown in Figure 1a. We next determined if the subclones differed in their phenotypic properties. Selected clones exhibiting a range of Cx43 expression were assessed for 1/ their growth in monolayer culture, 2/ their growth in soft agar and 3/ their ability to grow on a monolayer of growth inhibited 10T/1/2 cells. As can be seen in Table 1, there was a strong correlation between Cx43 expression and possession of a less malignant phenotype as indicated by these surrogate markers. Of particular interest was the behavior of these clones in soft agar; as shown in Figure 1b, whereas all clones formed colonies with approximately equal efficiency, only those clones which were null for Cx43 were capable of producing large colonies. This is the expected behavior in the growth control model of junctional communication. The *in vivo* tumorigenicity of these clones is not known.

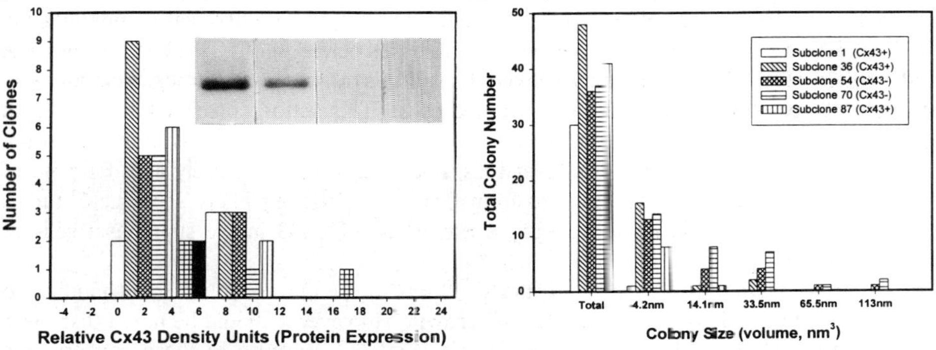


**Figure 1a-b. Left panel: 1a. HeLa subclones differ in their expression of Cx43.** Cx43 expression was determined by Western blotting of equal amounts of total protein derived from cell extracts of HeLa clones. We used previously described rabbit polyclonal antibody to the C-terminal region of Cx43 [11]. Results are expressed in arbitrary units after digital image analysis. Inset: Western blots of representative clones. Lanes 1-4, left to right: Clone#1; clone#36; clone#54; clone#70. **Right panel: 1b. Anchorage independent growth of HeLa subclones.** Clones were seeded at 200 cells/dish in 0.4% agarose/media on top of a soldified 0.8% agarose/media layer. Colonies were incubated for 4 weeks and stained with nitroblue tetrazolium salt.

| Clone | Cx43 Protein | Cx43 mRNA | Immuno-fluorescence (a) | GJIC by Scrape-load | Saturation Density (b) | Growth in Co-culture (c) | Soft Agar Colony Size (d) |
|---|---|---|---|---|---|---|---|
| 1 | +++ | ++ | ++ | ++ | ND | + (42) | + |
| 36 | ++ | ++ | + | +/- | – (100) | + (69) | ++ |
| 87 | ++ | ++ | + | +/- | ND | ND | + |
| 54 | --- | --- | - | -- | –+ (113) | +++ (133) | +++ |
| 70 | --- | --- | - | -- | –+– (136) | –+ (119) | +++ |

**Table 1. Characteristics of clones derived from HeLa culture.** Clones were isolated and characterized for Cx43 expression, localization, function and growth control. (a), presence of immunoreactive plaques in regions of cell-cell contact; (b), final cell density expressed as a (%) of clone 36; (c), HeLa cells were plated on confluent monolayers of 10T1/2 cells or alone and colonies counted after 7 days. Numbers represent a ratio of these values; (d), for details see Fig 2.

It must be stressed that these Cx43 positive clones express endogenous Cx43 mRNA (3 kb transcript, not shown) as well as protein and were not transfected with a Cx43-containing plasmid (2.2 kb transcript). All clones possessed HeLa marker chromosomes and had the appropriate G6PD alloenzyme (data not shown). Had we transfected Cx43 under the control of a constitutive promoter and subsequently analyzed cells for connexin expression and their phenotypic properties, we may have reached some misleading conclusions concerning the role of this transfected gene. These studies emphasize the requirement for inducible promoters in order to analyze the effects of transfected genes.

*Development of HeLa Cx43 tetracycline-inducible cell line.* We next took a HeLa cell line, negative for connexin expression and positive for the tet-on receptor, and performed the second-stage transfection with a plasmid containing the Cx43 gene [15] downstream from the tet operator/CMV promoter sequences and a hygromycin selectable marker. A total of 44 clones were assessed for connexin inducibility after treatment with doxycycline 1.0μg/ml/48 hours, of these approximately 16 percent (7/44) were strongly inducible and exhibited little expression in the absence of doxycycline.

In Figure 2a is shown Western blots of four such clones. Digital analysis indicated up to a 30 fold induction of Cx43 protein. This protein was functional as indicated by scrape loading and microinjection and was assembled into junctional plaques as indicated by immunofluorescence using a rabbit polyclonal antibody to Cx43 (Figure 2b). Time-course and dose-response studies with doxycycline have revealed that optimum induction occurs with 1 μg/ml after 24 hours and that increased expression can be detected by Western blot within 8 hours of treatment.

**Figure 2a,b. Transfected HeLa cell clones are inducible for Cx43 by doxycycline. Fig. 2a: Western blotting:** Cells were cultured in the presence (+) or absence (-) of 1.0 μg/ml doxycycline for 48 hours, harvested for protein and immunoblotted with a Cx43 rabbit antibody. Clockwise from top left: clones 114,122,89,100. **Fig. 2b: Immunofluorescent micrographs:** Clone #100 cultured in the presence (+) or absence (-) of 1.0 μg/ml doxycycline for 48 hours as in Fig. 2a. Note the many reactive junctional plaques in the induced cells.

*Characterization of tetracycline-inducible cell lines.* We have conducted preliminary studies using flow cytometry, to determine if connexin induction modifies cell cycle transit times. HeLa cell cultures were treated with doxycycline under two conditions: first, while sparsely plated and in logarithmic phase; second, under conditions of high cell density. When in logarithmic phase, when minimum cell contacts occurred, addition of doxycycline did not influence cell cycle parameters as measured by propidium iodide staining of DNA (data not shown). However when cultures were treated under conditions of high cell density, we detected decreased numbers of cycling cells.

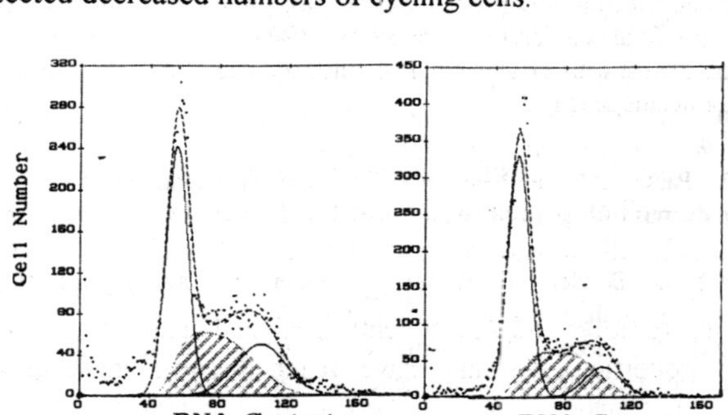


**Figure 3. Cell cycle distribution of Cx43 inducible HeLa clone.** HeLa clone #122 was seeded at high density in the presence (+) or absence (-) of 1.0 μg/ml doxycycline for 48 hours. Cultures were then stained for DNA with propidium iodide and analyzed by FACS using a Coulter EPICS system.

As shown in Figure 3, the population of G1 cells increased from 44% to 56% while the population of G2 cells decreased from 21% to 12% of total cells. These studies indicate that induced cells were partially G1 arrested and were delayed in their progression through S-phase. These studies need repeating with other clones and under more varied conditions of growth and treatment.

In conclusion, we have attempted to more definitively link effects on communication with effects on proliferation and transformation and have utilized molecular techniques to upregulate Cx43 expression without exposure to chemopreventive agents. The tet-on expression system has been successfully employed in HeLa cells resulting in clones which strongly express functional Cx43 after doxycycline treatment. Preliminary data indicates that connexin induction in cell cultures at high cell density decreases proliferation. The behavior of these clones in immunosuppressed mice has yet to be evaluated. These studies also revealed clonal heterogeneity in HeLa cultures and indicate that great care must be taken in interpreting results obtained utilizing constitutive promoters.

## References


1. Loewenstein, W.R. and Kanno, Y. Intercellular communication and the control of tissue growth: lack of communication between cancer cells. *Nature, 209* (1966) 1248-1249.

2. Loewenstein, W.R. Junctional intercellular communication and the control of growth. *Biochim Biochimica et Biophysica Acta, 560* (1979)1-65.

3. Yamasaki, H. Gap junctional intercellular communication and carcinogenesis. *Carcinogenesis, 11* (1990) 1051-1058.

4. Mehta, P.P., Bertram, J.S., and Loewenstein, W.R. Growth inhibition of transformed cells correlates with their junctional communication with normal cells. *Cell, 44* (1986) 187-196.

5. Kinzler, K.W. and Vogelstein, B. Lessons from hereditary colon cancer. *Sci, 87* (1996) 159-170.

6. Ames, B.N., Shigenaga, M.K., and Gold, L.S. DNA lesions, inducible DNA repair, and cell division: Three key factors in mutagenesis and carcinogenesis. *Environ. Health Perspect. 101 Suppl. 5* (1993) 35-44

7. Hong, W.K., Lippman, S.M., Itri, L.M., Karp, D.D., Lee, J.S., Byers, R.M., Schantz, S.P., Kramer, A., Lotan, R., Peters, L.J., Dimery, I.W., Brown, B.W., and Goepfert, H. Prevention of second primary tumors with isotretinoin in squamous-cell carcinoma of the head and neck. *N. Eng. J Med. 323* (1990) 795-800.

8. Papadimitrakopoulou, V.A., Hong, W.K., Lee, J.S., Martin, J.W., Lee, J.J., Batsakis, J.G., and Lippman, S.M. Low-dose isotretinoin versus b-carotene to prevent oral carcinogenesis Long-term follow-up. *J. Nat. Cancer Instit. 89* (1997) 257-258.

9. Merriman, R. and Bertram, J.S. Reversible inhibition by retinoids of 3-methylcholanthrene-induced neoplastic transformation in C3H10T1/2 cells. *Cancer Res. 39* (1979) 1661-1666.

10. Mehta, P.P., Bertram, J.S., and Loewenstein, W.R. The actions of retinoids on cellular growth correlate with their actions on gap junctional communication. *J. Cell Biol. 108* (1989) 1053-1065,.

11. Rogers, M., Berestecky, J.M., Hossain, M.Z., Guo, H.M., Kadle, R., Nicholson, B.J., and Bertram, J.S. Retinoid-enhanced gap junctional communication is achieved by increased levels of connexin 43 mRNA and protein. *Mol. Carcinog. 3* (1990) 335-343.

12. Hossain, M.Z. and Bertram, J.S. Retinoids suppress proliferation, induce cell spreading, and up-regulate connexin43 expression only in postconfluent 10T1/2 cells: Implications for the role of gap junctional communication. *Cell Growth & Differentiation, 5* (1994) 1253-1261.

13. Guo, H., Acevedo, P., Parsa, D.F., and Bertram, J.S. The gap-junctional protein connexin 43 is expressed in dermis and epidermis of human skin: differential modulation by retinoids. *J. Invest. Dermatol. 99* (1992).460-467.

14. Gossen, M., Freundlieb, S., Bender, G., Müller, G., Hillen, W., and Bujard, H. Transcriptional activation by tetracyclines in mammalian cells. *Sci, 268* (1995).1766-1769.

15. Mesnil, M., Krutovskikh, V., Piccoli, C., Elfgang, C., Traub, O., Willecke, K., and Yamasaki, H. Negative growth control of HeLa cells by connexin genes: Connexin species specificity. *Cancer Res. 55* (1995) 629-639.

16. Beyer, E.C., Paul, D.L., and Goodenough D.A. Connexin 43: A Protein from Rat Heart Homologous to a Gap Junction Protein from Liver. *J. Cell Biol. 105* (1987)2621-2629.

*Gap Junctions*
*R. Werner (Ed.)*
*IOS Press, 1998*

# Role of Altered Gap Junctional Intercellular Communication in Lung Neoplasia

Randall RUCH[1], Kimberley CESEN-CUMMINGS[1], Marie PIECHOCKI[1], and Alvin MALKINSON[2]
[1]*Medical College of Ohio, Toledo, OH 43699 and* [2]*University of Colorado Health Sciences Center, Denver, CO 80262, USA*

**Abstract.** The role of gap junctional intercellular communication (GJIC) in lung neoplasia is poorly understood. Here we review our studies on tumor promoter modulation of lung epithelial cell GJIC, the effects of ß-carotene, dibutyryl cAMP, and dexamethasone on lung epithelial cell GJIC and growth, the frequency of altered GJIC in human and rodent lung carcinoma cells, and the ability of connexin cDNA transfection to restore lung carcinoma GJIC and reduce growth and tumorigenicity.

## 1. Introduction

Lung cancer is the leading cause of cancer-related death in the United States and is also a major cause of death in other countries; approximately 150,000 deaths were attributable to lung cancer in the United States in 1990 [1]. The vast majority of lung cancers are due to tobacco smoking, so this high mortality rate could be reduced by decreasing smoking. Despite extensive educational and smoking cessation programs, however, the incidence of smoking in the United States has decreased only slightly in men and is rising in women. In other countries, the incidence of tobacco smoking is increasing.

Gap junctional intercellular communication (GJIC) is involved in growth control and is frequently defective in neoplastic cells [reviewed in 2]. Gap junctions are often decreased in size and number in neoplastic cells compared to normal cells. This is due to reduced expression of gap junction proteins (connexins), improper localization of connexins in the cells, inability to assemble connexins into junctional channels, and to defects in cell-cell adhesion which prevent docking of gap junction hemichannels (connexons) between adjacent cells [reviewed in 3]. Some neoplastic cells have numerous, functional gap junctions between themselves (homologous GJIC), but not with normal cells (heterologous GJIC) [4,5]. Many growth factors, tumor promoters, and oncogene products reduce GJIC, whereas growth inhibitors, differentiating agents, and anticarcinogens enhance GJIC [reviewed in 6]. Antisense inhibition of connexin expression in normal cells reduces GJIC and causes abnormal growth whereas restoration of connexin expression and GJIC in neoplastic cells reduces growth and tumor-forming ability [6]. Genetic "knockout" of the connexin32 (Cx32) gene, the major connexin expressed by hepatocytes [7,8], increased spontaneous and chemically-induced liver cancer formation in mice [9]. Thus, most neoplastic cells have defective GJIC and this is involved in their abnormal phenotypes.

Despite the high frequency and mortality of lung cancer, few studies have investigated the possible role of GJIC in lung neoplasia. Our group has studied the effects of tumor promoters on GJIC in lung epithelial cells [10,11], the ability of ß-carotene and $N^6$,2'-*O*-dibutyryladenosine 3':5'-cyclic monophosphate (DB-cAMP) to enhance GJIC and suppress growth in lung epithelial cells [12,13], the frequency of altered GJIC in lung carcinoma cells [14], and the effect of connexin transfection on lung carcinoma cell growth and tumorigenicity [15]. Other groups have shown that human lung carcinoma cells have few intercellular junctions compared to normal lung epithelial cells [16,17], that TGF-ß and EGF can modulate GJIC in human bronchial epithelial cells [18,19], and that the phorbol ester, 12-*O*-tetradecanoylphorbol-13-acetate (TPA), and airborne particulate matter can reduce GJIC in primary cultures of rat alveolar Type II cells [20]. In this manuscript, we review our studies on the role of GJIC in lung neoplasia.

## 2. Alterations of Lung Epithelial Cell GJIC by Tumor Promoters

Tumor promoters increase the formation of neoplasms, but are not themselves carcinogenic. Many tumor promoters do this by stimulating the growth and/or inhibiting the apoptotic death of "initiated" cells, i.e., mutated cells that have not yet developed into neoplasms. The growth-enhancing and apoptosis-inhibiting effects of tumor promoters may be related to their abilities to block GJIC [21].

Yotti et al. [22] and Murray and Fitzgerald [23] first showed that a tumor promoter, TPA, blocked GJIC in cultured cells. Since then, many tumor promoters have been reported to do this in cultured cells and in the intact animal [reviewed in 24]. Few studies have examined the effects of tumor promoters on GJIC in cells of the lung or conducting airways. As noted above, TPA and airborne particulate matter reduced GJIC in primary cultures of rat Type II pneumocytes [20].

We reported that the tumor promoters, butylated hydroxytoluene (BHT), TPA, and *p,p'*-dichlorodiphenyltrichloroethane (DDT), but not phenobarbital (PB), blocked GJIC in C10 cells, a cell line derived from mouse Type II pneumocytes [10,11]. The effects of these agents on GJIC were assayed by Lucifer Yellow dye microinjection and were dose-dependent, time-dependent, and reversible following washout of the compound. Furthermore, a tumor promoting metabolite of BHT, 6-t-butyl-2-(hydroxy-t-butyl)-4-methylphenol (BHTOH), blocked GJIC whereas a non-promoting metabolite, 2,6-di-t-butyl-4-hydroxymethylphenol (BHTBzOH) did not affect GJIC in C10 cells [11].

These data suggest a direct correlation between the inhibition of GJIC by a chemical agent and its ability to promote lung tumor formation. The lung tumor promoters, BHT and BHTOH, blocked lung epithelial cell GJIC. PB, which is a liver tumor promoter but not a lung tumor promoter in mice [25], did not block lung epithelial cell GJIC, but does inhibit mouse hepatocyte GJIC [26]. The lung tumor-promoting abilities of TPA and DDT have not been reported.

We have also examined gap junction number and connexin expression in these cells. Cultured lung epithelial cells express connexin43 (Cx43) as their predominant gap junction protein [14]. Treatment of C10 cells with BHT and TPA caused the rapid internalization (within 15 min) of Cx43-immunopositive gap junction plaques [11]. These plaques reappeared within 24 h in TPA-treated cells, but disappeared completely in BHT-treated cells. Western blot analyses revealed that Cx43 was expressed as 42 kDa nonphosphorylated (Cx43-NP) and 44 and 46 kDa phosphorylated (Cx43-P1 and Cx43-P2) forms [11]. TPA induced a 47 kDa phosphorylated form (Cx43-P3) coincident with gap junction internalization whereas BHT caused the loss of Cx43-P2 along with the loss of Cx43 immunostaining. Although preliminary, the data suggest that TPA activates protein kinase C (PKC) which alters Cx43 phosphorylation leading to gap junction internalization; this is reversed when further TPA exposure downregulates PKC. In contrast, BHT induces gap junction internalization without PKC activation; the internalized plaques are probably degraded in lysosomes as we have reported for the GJIC inhibitors, DDT and lindane [27].

## 3. Effects of ß-Carotene, DB-cAMP, and Dexamethasone on Lung Epithelial Cell GJIC and Growth

Carotenoids and retinoids, cyclic AMP agonists and analogues, and steroid hormones can modify GJIC and growth [6]. We have examined the relationship between altered growth and effects on GJIC of ß-carotene, DB-cAMP, and dexamethasone in normal and neoplastic mouse and human lung epithelial cells. ß-Carotene did not affect GJIC or growth in normal C10 cells, three murine lung carcinoma cell lines (E9, 82-132, and PCC4), and several human lung carcinoma cell lines (NCI-H125, NCI-H446, and NCI-H460) [12 and unpublished data]. DB-cAMP enhanced GJIC and Cx43 expression and suppressed cell growth in a dose-dependent (0.05-1 mM) and time-dependent (1-5 d) manner in C10, 82-132, and E9 cells [13]. Dexamethasone did not affect GJIC or Cx43 expression in E10 and E9 cells [unpublished data], although the steroid inhibits growth in E10 cells [28]. Thus, DB-cAMP and dexamethasone inhibit the growth of lung epithelial cells while ß-carotene does not. For the cAMP analogue, this inhibition was related to enhanced GJIC and Cx43 expression.

## 4. Altered GJIC and Connexin43 Expression in Neoplastic Lung Epithelial Cells

Because other cancer cells often have reduced GJIC and connexin expression [2,3], we examined this in lung carcinoma cells. GJIC and connexin expression were analyzed by Northern blotting, Western blotting, and immunohistochemistry in seventeen human and five murine lung carcinoma cell lines [14]. To investigate nontumorigenic lung epithelial cells, we utilized normal human bronchial epithelial cells (NHBEC) and normal human small airway epithelial cells (NHSAEC), both obtained from Clonetics Corp. (San Diego, CA), and C10 and E10 murine Type II lung epithelial cell lines.

GJIC was assessed by Lucifer Yellow dye microinjection. Normal human and murine lung epithelial cells exhibited high dye-coupling percentages (>80%) whereas most of the human lung carcinoma cell lines (11 of 17) and all of the murine lung carcinoma cell lines (5 of 5) exhibited dye-coupling percentages <40%.

Northern and Western analyses indicated that Cx43 was expressed by normal lung epithelial cells and that Cx26, Cx32, Cx37, and Cx40 were not expressed by these cells. Cx43 mRNA and protein content were generally higher in normal cells than in carcinoma cells and the levels of these correlated well with dye-coupling.

Immunostaining for Cx43 also revealed differences between normal and carcinoma cells. In normal cells, Cx43 was localized predominantly at regions of cell-cell contact and in discrete loci characteristic of gap junction plaques. In the carcinoma cells, Cx43 staining was decreased or was localized abnormally as evidenced by pronounced cytoplasmic and nuclear staining.

## 5. Connexin Transfection Induces Phenotypic Changes in Lung Carcinoma Cells

The above data indicate that reduced GJIC and Cx43 expression occur frequently in lung carcinoma cells. To test whether this is involved in the neoplastic phenotype of these cells or is an epiphenomenon associated with neoplastic transformation, we transfected a poorly communicating, low Cx43-expressing murine lung carcinoma cell line (E9) with Cx43 and Cx32 cDNA expression vectors [15]. We obtained several stably transfected clones of Cx43-transfected cells (E9-2, E9-3, E9-5, and E9-25) and one stable Cx32-transfected clone (E9/32-9). E9 cells were also transfected with empty expression vectors to produce several stable clones (E9-41, E9-42, and E9/32-12) that were used as controls for DNA transfection. We then compared GJIC, Cx43 expression, growth *in vitro*, and tumorigenicity of these clones versus parental E9 cells and normal E10 cells.

E9 cells exhibited a low dye-coupling percentage (22%) compared to E10 cells (94%) and Cx43 expression followed a similar pattern. Connexin transfection increased dye-coupling in E9-2 (82%), E9-3 (72%), E9-5 (52%), E9-25 (79%), and E9/32-9 (76%). Cx43 and Cx32 expression were also increased accordingly. E9-41, E9-42, and E9/32-12 vector control clones had dye-coupling and Cx expression levels similar to E9 cells.

E9 cells grew faster and to a higher saturation density than E10 cells and the Cx43 and Cx32 transfected clones exhibited growth rates and saturation densities intermediate between the two cell lines. The vector control cells grew like E9 cells. These data indicate that restoration of GJIC only partially normalizes growth in E9 cells. This is not surprising considering that E9 cells have several defects in growth regulatory mechanisms [29].

We also examined the tumorigenicity of these cells by injecting $5x10^6$ cells subcutaneously into male Balb/c mice. E9 cells were tumorigenic in all test mice (6/6), with palpable tumors appearing within 14 d; E10 and E9-2 cells were not tumorigenic (0/10 mice each). E9-3, E9-5, E9-25, and E9/32-9 cells were tumorigenic in 4/10, 6/8, 7/8, and 7/8 mice, respectively, and these tumors appeared later (17-125 d) than E9 cell tumors. These results indicate that Cx43 transfection reduced the tumor incidence and increased the tumor latency of E9 cells. When tumors arising from E9-3, E9-5, and E9-25 cells were immunostained for Cx43, little protein was detected. This suggests that the exogenous gene was lost or its expression was suppressed in these cells *in vivo*. Additionally, immunostaining of tumors arising from E9/32-9 cells revealed high levels of Cx32 expression, indicating that the transfected gene was expressed *in vivo*. This suggests that the tumorigenicity of E9 cells cannot be suppressed by a nonhistiotypic connexin.

## 6. Conclusions

These studies imply that altered GJIC is involved in the expression of the neoplastic phenotype in human and rodent lung carcinoma. Enhancement of GJIC might inhibit the formation and progression of lung cancer.

## References


[1] C. Boring et al., Cancer statistics. *CA Cancer Journal* 44 (1994) 7-13.

[2] W. Loewenstein, Junctional intercellular communication and the control of growth. *Biochimica et Biophysica Acta* 560 (1979) 1-65.

[3] H. Yamasaki and C. Naus, Role of connexin genes in growth control. *Carcinogenesis* 17 (1996) 1199-1213.

[4] V. Krutovskikh et al., Altered homologous and heterologous gap-junctional intercellular communication in primary human liver tumors associated with aberrant protein localization but not gene mutation of connexin 32. *International Journal of Cancer* 56 (1994) 87-94.

[5] S. Garber et al., Altered gap junctional intercellular communication in neoplastic rat esophageal epithelial cells. *Carcinogenesis* 18 (1997) 1149-1153.

[6] R. Ruch, The role of gap junctional intercellular communication in neoplasia. *Annals of Clinical Laboratory Science* 24 (1994) 216-231.

[7] D. Paul, Molecular cloning of cDNA for rat liver gap junction protein. *Journal of Cell Biology* 103 (1986) 123-134.

[8] N. Kumar and N. Gilula, Cloning and characterization of human and rat liver cDNAS coding for a gap junction protein. *Journal of Cell Biology* 103 (1986) 767-776.

[9] K. Willecke et al., Biological functions of gap junctions revealed by targeted inactivation of mouse connexin32 and -26 genes. *1997 International Gap Junction Conference* (1997) p. 105.

[10] R. Chaudhuri et al., Gap junctional intercellular communication in mouse lung epithelial cell lines: Effects of cell transformation and tumor promoters. *Cancer Letters* 71 (1993) 11-18.

[11] X. Guan et al., Down-regulation by butylated hydroxytoluene of the number and function of gap junctions in epithelial cell lines derived from mouse lung and rat liver. *Carcinogenesis* (1995) 2575-2582.

[12] R. Banoub et al., Lack of growth inhibition or enhancement of gap junctional intercellular communication and connexin43 expression by ß-carotene in murine lung epithelial cells in vitro. *Cancer Letters* 108 (1996) 35-40.

[13] R. Banoub et al., Enhancement of gap junctional intercellular communication by dibutyryl cyclic AMP in lung epithelial cells. *Anticancer Research* 16 (1996) 3715-3719.

[14] K. Cesen-Cummings et al., Frequent reduction of gap junctional intercellular communication and connexin43 expression in human and mouse lung carcinoma cells. *Carcinogenesis* (in press).

[15] K. Cesen-Cummings et al., Altered growth properties and decreased tumorigenicity of mouse lung carcinoma cells stably transfected with connexin43. *Carcinogenesis* (in press).

[16] D. McDonagh et al., Intercelluar junctions and tumor behavior in lung cancer. *Modern Pathology* 4 (1991) 436-440.

[17] R. Vollmer, Intercellular junctions and tumor behavior in lung cancer. *Modern Pathology* 5 (1992) 96-7.

[18] C. Albright et al., Intercellular communication in bronchial epithelial cells: review of evidence for a possible role in lung carcinogenesis. *Toxicological Pathology* 18 (1990) 324-343.

[19] C. Albright et al., Cell-to-cell communication: a differential response to TGF-β in normal and transformed (BEAS-2B) human bronchial epithelial cells. *Carcinogenesis* 12 (1991) 1993-1999.

[20] G. Heussen and G. Alink, Inhibition of gap-junctional intercellular communication by TPA and airborne particulate matter in primary cultures of rat alveolar type II cells. *Carcinogenesis* 13 (1992) 719-722.

[21] J. Trosko and J. Goodman, Intercellular communication may facilitate apoptosis: Implications for tumor promotion. *Molecular Carcinogenesis* 11 (1994) 8-12.

[22] L. Yotti et al., Elimination of metabolic cooperation in Chinese hamster cells by a tumor promoter. *Science* 206 (1979) 1089-1091.

[23] A. Murray and D. Fitzgerald, Tumor promoters inhibit metabolic cooperation in cocultures of epidermal and 3T3 cells. *Biochemical and Biophysical Research Communications* 91 (1979) 395-401.

[24] I. Budunova and G. Williams, Cell culture assays for chemicals with tumor promoting or inhibiting activity based on the modulation of intercellular communication. *Cell Biology and Toxicology* 10 (1994) 71-116.

[25] M. Pereira et al., Effect of subsequent treatment of chloroform or phenobarbital on the incidence of liver and lung tumors initiated by ethylnitrosurea in 15 day old mice. *Carcinogenesis* 6 (1985) 203-207.

[26] R. Ruch and J. Klaunig, Kinetics of phenobarbital inhibition of intercellular communication in mouse hepatocytes. *Cancer Research* 48 (1988) 2519-2523.

[27] X. Guan and R. Ruch, Gap junction endocytosis and lysosomal degradation of connexin43-P2 in WB-F344 rat liver epithelial cells treated with DDT and lindane. *Carcinogenesis* 17 (1996) 1791-1798.

[28] K. Droms et al., Altered dexamethasone responsiveness and loss of growth control in tumorigenic mouse lung cell lines. *International Journal of Cancer* 53 (1993) 1017-1022.

[29] A. Malkinson et al. Mouse lung epithelial cell lines - tools for the study of differentiation and the neoplastic phenotype. *Toxicology* (in press).

# The Balance Between Birth and Death: Apoptosis, Mitosis, and Gap Junctions

Melinda R. Wilson and James E. Trosko
*Michigan State University, College of Human Medicine, Department of Pediatrics and Human Development, East Lansing, Michigan 48824.*

**ABSTRACT**

The balance between cell birth and cell death (apoptosis is critical to homeostasis of all organisms. Defective regulation of apoptosis may contribute to the etiology of various diseases, including cancer. It has been hypothesized that gap junctions may facilitate the signal for programmed cell death. Support of this hypothesis include observations that chemicals and genes, which induce apoptosis, also enhance gap junctional intercellular communication (GJIC). Conversely, oncogenes and tumor promoting chemicals, which block apoptosis, reduce GJIC. Previous results indicated that there was an increase in GJIC and augmentation of connexin 43 phosphorylation during induction of apoptosis in rat liver epithelial cell populations. Since apoptosis is a single cell event, i.e. only a percentage of cells within a population will initially undergo apoptosis, we wanted to determine which morphologically distinguishable cells within the population were altered in their ability to perform GJIC. Using the FRAP (fluorescent recovery after photobleaching) assay, GJIC was assessed in the predominant sub-populations of cells. Morphological criteria was used to categorize sub-populations. Stages of cell division (prophase, metaphase, anaphase, telophase) were easily distinguished in a phase contrast microscope and subjected to the FRAP assay. In addition, apoptotic bodies (late stage apoptosis), and cells with crescent shaped chromosome condensation (early stage apoptosis) were subjected to FRAP analysis. As mitosis progressed there was a dramatic decrease in dye transfer. Metaphase and anaphase cells performed GJIC at reduced levels. There was a large variation in the levels of GJIC within individual telophase cells. In general, late telophase cells had higher proficiency in GJIC than early telophase cells. In addition, we assessed the frequency of mitosis during induction of apoptosis. The level of mitosis *increased,* coincident with alteration of GJIC. The highest level of GJIC was observed in cells displaying crescent chromosomal condensation (early apoptotic) and normal, non-dividing cells. Early apoptotic bodies were able to transfer dye from neighboring normal cells. The augmentation of GJIC and thus subsequent enhancement of apoptosis in cell populations deficient in their ability to undergo normal programmed cell death (i.e. cancer cells) may aid in development of treatment strategies for various diseases.

## *1. Introduction*

Apoptosis is an important mechanism for regulating cell numbers within a population to maintain the balance between birth and death during development and cell population homeostasis. Defective regulation of apoptosis contributes to the emergence of various diseases, including cancer [1]. The biological sequence of events resulting in apoptosis must involve mechanisms within single cells combined with interactions among cells within a population. Which cells divide and which cells die must be regulated, to some degree, by the organized microenvironment that cells live within. Thus, to understand that which regulates apoptosis, interactions that occur within a cell population during apoptosis must also be understood. Disruption of physical connections between the extracellular matrix and normal epithelial cells will induce apoptosis [2]. In addition, epithelial cells interact with one another via connections such as adherens junctions, desmosomes, and tight junctions [3]. The way in which these junctions physically connect cells to one another, serve as important regulators of cells within a microenvironment [2].

When gap junction channels are disrupted, single cells can escape the normal regulatory signals passing within a group of cells [4, 5, 6, 7].

## 2. *Hypothesis*

Since most normal adherent cells are coupled via gap junctions, it seems reasonable to postulate that gap-junctional intercellular communication (GJIC) will be altered during the course of apoptosis. Previous observations indicate that gap junctions appear to decrease during advanced stages of apoptosis [8]. Based on observations that tumor promoters inhibit apoptosis [9] and GJIC [10], it has been hypothesized that gap junctions may facilitate the signal [11] for programmed cell death pathways. To explore this hypothesis, we characterized the relationship between GJIC and apoptosis within a homogeneous epithelial cell population and on a single cell basis.

**Table 1.** Quantitation of Apoptosis During Serum Deprivation of WB-F344 Cells Using FACS Analysis.

| Hours after serum withdrawal | % Apoptotic Cells |
|---|---|
| 0 | not detected |
| 3 | 3 |
| 7 | 8 |
| 12 | 18 |

## 3. *Results and Discussion*

Apoptosis was induced by growth factor withdrawal by depriving cells of serum. Internucleosomal DNA fragmentation [12], morphological changes, and Fluorescence Activated Cell Sorting (FACS) analysis [13] were used to monitor changes in the level of

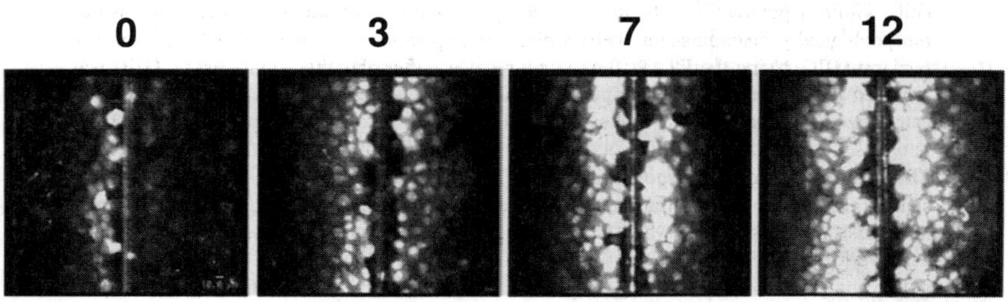


**Figure 1.** Changes in population GJIC using the SL/DT assay. Images are representative laser cytometer digitized scans of Lucifer Yellow dye transfer in WB-F3344 monolayers during serum deprivation. Hours after serum withdrawl are indicated above each image.

apoptosis in serum-deprived rat liver epithelial cell (WB-F344) cultures. The Scrape Load/dye Transfer assay (SL/DT) [14] was used to monitor the extent of dye transfer in multiple cells simultaneously within a population. DNA ladder fragmentation appeared 7 hours after serum withdrawal (data not shown). FACS analysis indicated that the number of cells undergoing apoptosis increased (Table 1) as functional GJIC increased (Figure 1). Since phosphorylation of CX43 is directly correlated with gap junction function [15] we performed western blot analysis using monoclonal antibodies specific for CX43. An increase of total CX43 protein amount as well as an increased level of the phosphorylated species was observed during serum deprivation of WB cells (Figure 2). Augmentation of CX43 protein phosphorylation and synthesis may reflect a response of non-apoptotic cells to maintain gap junction function during serum deprivation.

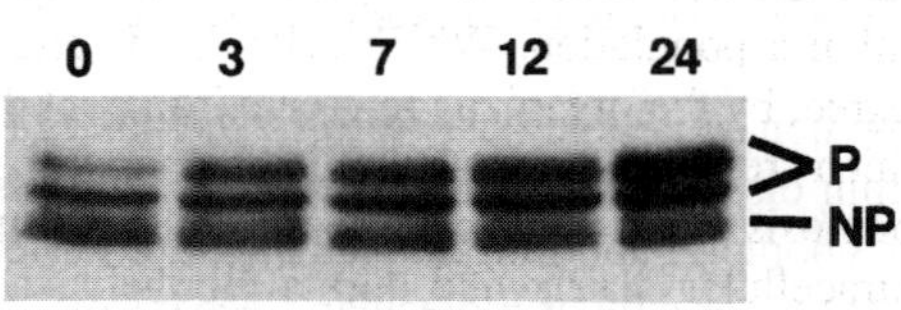


**Figure 2.** Amount and Phosphorylation of CX43. Hours after serum withdrawl is indicated above each lane. P: phosphorylated CX43 species, NP: non-phosphorylated CX43.

Results obtained from population GJIC assays and CX43 protein analysis represent an average of the majority of cells in a population. Since the proportion of apoptotic cells in the population reached less than 20% within the first 12 hours of serum deprivation, it is possible that the data collected may represent a "reaction" of mostly non-apoptotic cells. Thus, these data may reflect the response of a *population* to induction or regulation of apoptosis, not necessarily what is happening in individual apoptotic cells. Consequently, it was not apparent if apoptotic cells within the population followed the same pattern of altered GJIC. Therefore, measurement of GJIC within single cells undergoing apoptosis and mitosis was investigated using the Fluorescent Redistribution After Photobleaching (FRAP) assay [16]. As mitosis progressed there was a dramatic decrease in dye transfer between mitotic and resting cells (Table 2). Metaphase and anaphase cells communicated at an approximate 4-fold reduced level than prophase cells. Late telophase cells had higher proficiency in GJIC than early telophase cells. The highest level of GJIC was observed in cells displaying crescent chromosomal condensation (early apoptotic), prophase (early mitotic) and normal, non-dividing cells. Since early mitotic cells displayed relatively high levels of GJIC, we quantitated the level of mitosis during serum deprivation. The level of mitosis increased, coincident with augmentation of GJIC and apoptosis (Figure 3) and was statistically significant ($p<0.05$, ANOVA, Tukey test). Hence, the increased GJIC at the population level may be due to higher levels of GJIC from a combination of higher levels in prophase cells, cells containing crescent chromosome condensation, and growth arrested "resting" cells. Although late stage apoptotic bodies did not perform GJIC, early apoptotic bodies were able to transfer dye from neighboring normal cells. These results indicated that apoptotic bodies were still coupled to healthy cells via gap junctions (Figure 4) and that *uncoupling* of gap junctions was not an essential requirement prior to programmed cell death.

**Table 2.** Assessment of GJIC in sub-populations of WB-F344 cells using the FRAP assay.

| Morphology Of Chromosome Condensation | % Fluorescence Recovery / minute Average | +/- SD |
|---|---|---|
| **Control** (resting cell) | 16.0 | 2.0 |
| **Mitosis** | | |
| Prophase | 12.4 | 2.1 |
| Metaphase | 6.3 | 2.3 |
| Anaphase | 4.1 | 1.9 |
| Telophase | 4.0 | 2.3 |
| **Apoptosis** | | |
| Crescent | 12.2 | 6.1 |
| Circular | 4.1 | 1.8 |
| Apoptotic body | 3.7 | 5.3 |

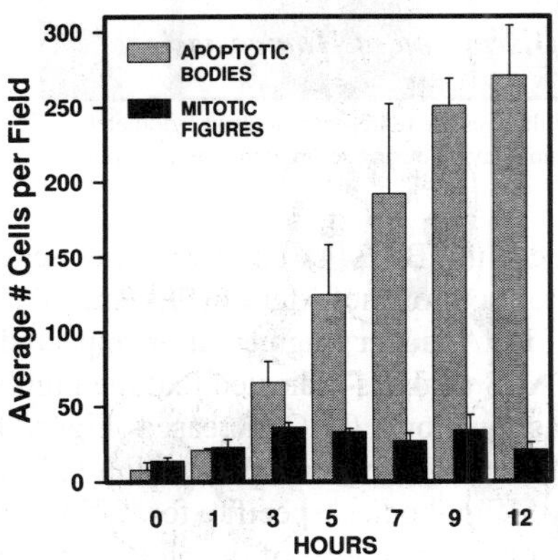


**Figure 3.** Quantitation of Apoptosis and Mitosis. Multiple fields of multiple WB monolayer cultures were scored for the number of cells undergoing mitosis and apoptosis during serum

Previous observations indicate that there are common features in the regulation of the cell cycle and apoptosis. Regulatory proteins such as fos, myc, jun, cdc2, and phosphorylation of Rb, which are activated during early cell cycle traverse, are also increased during induction of apoptosis [17]. In addition, apoptosis and mitosis share some common morphological features such as cell shrinking detachment from the substratum, lamin disassembly, and chromosome condensation [18]. Perhaps the common features of the apoptotic and mitotic pathways are not only that both pathways may be regulated by the same or similar molecular mechanisms, but that a sub-population of cells *responds to an apoptosis inducer by initiating mitosis*. The increased phosphorylation of CX43 would indicate activation of signal transduction pathways involving kinases that act on CX43. These pathways may also lead to initiation of mitosis in some cells and apoptosis in the majority of cells.

Although there are a variety of molecular mechanisms that may play a role in the interactions described in this report, it is unclear as to how the molecular decision is made to die or divide. This decision may be initiated by neighboring cells and facilitated through gap junctions. For example, in developing neurons, calcium wave fronts cross at gap junctions from a dying cell to an adjacent healthy one. The healthy neuron will initiate growth only in the axon that contacts the dying cell [19].

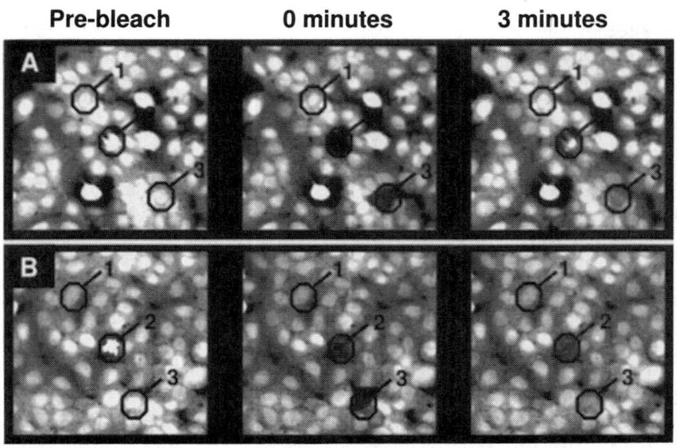


**Figure 4.** FRAP Assay of Early (A) and Late (B) Apoptotic Bodies. Cell#1: unbleached control, #2: apoptotic body, #3: control. Pre-bleach: baseline scan, 0 minutes: immediately after photobleaching, 3 minutes: delay time to allow fluorescence recovery.

The decision to die or divide may rely on the Neighborhood Coherence Principal (NCP) (Figure 5). According to the NCP, a population of cells can produce and maintain a spatial organization by virtue of cell-cell communication through gap junctions [20]. The NCP is a concept of a control mechanism whereby a cell would be told by its neighbors whether to enter a division cycle [21]. This decision depends in part on intrinsic variability of a cell's "state" and extrinsic interactions of a consensus of cells. SimCan, a mathematical model designed around the NCP [21], would predict that if apoptosis is transiently induced, there would be an accompanying increase in mitosis after which both events would return to a steady state homeostasis. If the apoptotic inducer was continuous, there may be two potential outcomes; the entire population eventually dies or the mitotic sub-population increases in cell number causing eventual resistance to the initial apoptotic inducer, and possible progression towards an abnormal phenotype. In a population of cancer cells, the reverse may be true: possible progression towards

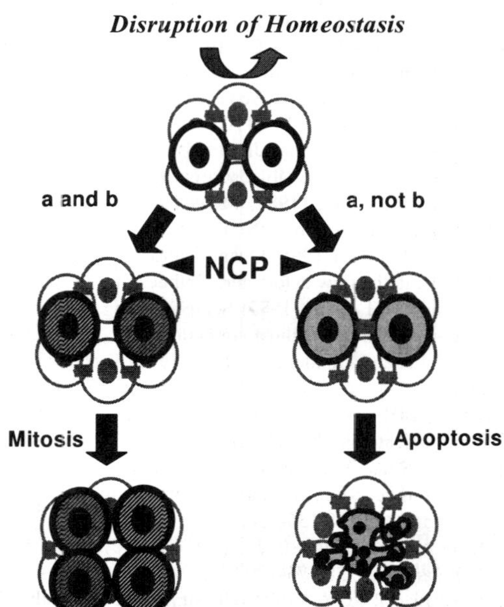


**Figure 5.** The Balance Between Birth and Death Through Gap Junctions. a: activation of regulatory pathways, b: gap junction uncoupling, NCP: neighborhood coherence principle, ■: gap junction channel.

a normal phenotype. The final outcome would most likely depend upon strength and duration of an extrinsic apoptotic inducer, and intrinsic state of cell variability.

Results presented here suggest that a common element in the early regulatory stage of apoptosis and mitosis may be a relatively high level of GJIC. These results may be useful in developing strategies to control diseases, such as cancer, involving dysregulated gap junctions, mitosis, and apoptosis.

***Acknowledgments***

We gratefully acknowledge L. King for his expertise with the FACS technique; T. Close, N. DeoCampo and B. Upham for helpful discussions. This work was supported in part by NCI grant #CA21104, Air Force grant #95NL271, awarded to J.E. Trosko, and a NIEHS service award to M.R. Wilson.

***References***

[1] Thompson CB, (1995) Apoptosis in the pathogenesis and treatment of disease. Science 267:1456-1462.

[2] Frisch SM and Hunter F, (1994) Disruption of epithelial cell-matrix interactions induces apoptosis. J. Cell. Biol. 124(4):619-626.

[3] Farquhar M and Palade G, (1963) Junctional complexes in various epithelia. J. Cell Biol. 17:375-412.

[4] Loewenstein WR and Kanno Y, (1966) Intercellular communication and the control of tissue growth: Lack of communication between cancer cells. Nature 209: 1248-1249.

[5] Yotti LP *et al.*, (1979) Elimination of metabolic cooperation in Chinese hamster cells by a tumor promoter. Science 206:1089-1091.

[6] Trosko JE *et al.*, (1983) A mechanism of tumor promotion: Potential role of intercellular communication. Cancer Invest. 1:511-526.

[7] Yamasaki H and Naus CG, (1996) Role of connexin genes in growth control. Carcinogenesis 17(6):1199-1213.

[8] Watanabe H and Tonosaki A, (1995) Gap junctions in the apoptosis: TEM observations of membrane-granulosa cells of mouse ovarian follicle. *In* Progress in Cell Research, vol. 4, Kanno Y, Kataoka K, Shiba Y, Shibata Y, and Shimazu T (eds.), Elsevier Science BV, Amsterdam, pp.37-40.

[9] Bursch F *et al.*, (1992) Cell death by apoptosis and its protective role against disease. Trends Pharmacol. Sci. 13:245-251.

[10] Trosko JE and Chang CC, (1988) Nongenotoxic mechanisms in carcinogenesis: Role of inhibited intercellular communication. *In* Banbury Report 31: Carcinogen Risk Assessment: New Directions in Qualitative and Quantitative Aspects, Hart RW and Hoerger FD (eds), Cold Spring Harbor Laboratory Press, New York, pp. 139-170.

[11] Trosko JE and Goodman JI, (1994) Intercellular communication may facilitate apoptosis: Implications for tumor promotion. Molec. Carcinogenesis. 11:8-12.

[12] Sambrook J *et al.*, (1989) Molecular Cloning, A Laboratory Manual, Second Edition,Cold Spring Harbor Laboratory Press, New York).

[13] Fraker PJ *et al.*, ( 1995) Quantification of apoptotic events in pure and heterogeneous populations of cells using the flow cytometer. Methods in Cell Biol. 46:57-76

[14] El-Fouly MH *et al.*, (1987) Scrape-loading and dye transfer: A rapid and simple technique to study gap junctional intercellular communication. Exp. Cell Res. 168:428-430.

[15] Musil LS and Goodenough DA, (1991) Biochemical analysis of connexin43 intracellular transport, phosphorylation, and assembly into gap junctional plaques. J. Cell Biol. 115:1357-1374.

[16] Wade MH, *et al.*, (1986) A fluorescence photobleaching assay of gap junction-mediated communication of human cells. Science 236:175-180.

[17] Pandey S and Wang E, (1995) Cells en route to apoptosis are characterized by the upregulation of c-foe, c-myc, c-jun, cdc2, and RB phosphorylation, resembling events of early cell-cycle traverse. J. of Cellular Bioch. 58:135-150.

[18] King KL and Cidlowski JA, (1995) Cell cycle and apoptosis: Common pathways of life and death. J. Cellular Bioch. 58:175-180.

[19] Wolszon LR *et al.*, (1994) Calcium wave fronts that cross gap junctions may signal neuronal death during development. J. Neurosci. 14(6):3437-3438.

[20] Phipps M *et al.*, (1990) Carcinogenic implications of the neighborhood coherence principal. Medical Hypotheses 31:289-301.

[21] Phipps M *et al.*, (1997) How the neighborhood coherence principal (NCP) can give rise to tissue homeostasis: a cellular automaton approach. J. Theor. Biol. 185:475-487.

*Gap Junctions*
*R. Werner (Ed.)*
*IOS Press, 1998*

# Concanavalin A-activated Lymphocytes Form Gap Junctions that Increase their Rate of DNA Replication

[1]Juan C. SAEZ, [1]Manuel A. SEPULVEDA, [1]Roberto ARAYA , [2]Claudia G. SAEZ and [1]Francis PALISSON
*[1]Departamento de Ciencias Fisiológicas, Facultad de Ciencias Biológicas and [2]Departamento de Oncología y Hematología, Facultad de Medicina, Pontificia Universidad Católica de Chile, Santiago, Chile*

**Abstract.** Lymphocytes from lymph nodes (mouse) or peripheral blood (human) were used. Cx37 and Cx43 were detected (immunofluorescence) in resting and Con-A-activated lymphocytes. After Con-A stimulation extensive transfer of Lucifer yellow occurred and was reversibly blocked by octanol. Moreover, gap junctional communication was prevented with synthetic peptides homologous to the extracellular loop 1 of Cx32 or Cx37, but unrelated peptides to this Cx domain were ineffective. Active peptides also induced a 70-80% reduction in BrdU incorporation. Thus, activated lymphocytes form gap junctions that may allow for cell-to-cell transfer of metabolites and/or cell signals that control cell proliferation.

## 1. Introduction

It has been shown that lymphocytes treated with a mitogen agent express a low resistance intercellular pathway that allows transfer of electrical stimuli [1]. Electron microscope views of PHA-stimulated human or rabbit peripheral blood lymphocytes have revealed membrane areas of adjacent cells compatible with gap junctions [2,3]. The gap junction subunit(s), termed connexins (Cxs), and their functions in circulating and lymph node lymphocytes are unknown.

The membrane topology of Cxs consist of 4 transmembrane segments, 3 cytoplasmatic domains, and 2 extracellular loops. The extracellular loop 1 is one of the most conserved domain among Cxs. A synthetic peptide with sequence corresponding to that of a segment of the extracellular loop 1 of Cx32 inhibits cell-cell channel formation [4], presumably by blocking binding sites to Cxs present in the plasma membrane of adjacent cells. The cell-cell diffusion of Lucifer yellow mediated by gap junctions could also be blocked, although less specifically, by various lipophilic agents, including octanol [5].

This work was undertaken to identify Cxs expressed by lymphocytes, demonstrate that concanavalin A- (Con A)-activated lymphocytes establish gap junctional communication and to elucidate the role of gap junctions in the proliferative response of lymphocytes.

## 2. Methods

Adult balb/c mice were used. Lymph nodes were dissected, disrupted on a metallic sieve and the resulting cell suspension was washed with PBS and resuspended in RPMI 1640 medium supplemented with 10% FCS. The dissociated cells were plated on tissue culture dishes for 1 h at 37°C and the non-adherent cells were recovered and washed once with RPMI 1640. Lymphocytes were also isolated from blood drawn from human volunteers. Mononuclear cells were isolated using Histopaque 1077 and centrifugation. Adherent cells (monocytes) were eliminated as described above and lymphocytes were retrived.

Previously characterized rabbit anti-Cx26, mouse anti-Cx32, rabbit anti-Cx33 and rabbit anti-Cx43 antibodies were used. Rabbit anti-Cx37, 40, 45 and 50 antibodies were generated using synthetic peptides homologous to non-conserved regions of the C-terminal of each Cx.

Lymphocyte smears were fixed in 70% ethanol for 20 min at -20°C and then Cxs were detected by indirect immunofluorescence.

Functional coupling between Con A-activated lymphocytes was evaluated by microinjecting one cell of a cluster with Lucifer yellow (5% LY in 150 mM LiCl). The incidence of dye coupling was scored as the percentage of injections that resulted in dye transfer to more than one neighboring cell.

Replication of lymphocyte DNA was determined by measuring the incorporation of bromo deoxybromo uridine (BrdU) using a kit commercially available from Boehringer Mannnheim.

## 3. Results

Freshly isolated lymphocytes showed a diffuse intracellular immunoreactivity to Cx37 and Cx43 (Fig. 1a and c). After 60 min treatment with Con A different size cell aggregates were frequent and the immunoreactivity to either Cx37 or Cx43 was predominantly found at cell interfaces (Fig. 1b, d). No reactivity to anti-Cx26, 32, 33, 40, 45 and 50 antibodies was detected either in freshly isolated or in Con A-treated lymphocytes. Similar results were obtained in human lymphocytes.

After about 1 h treatment with Con A (10 µg/ml) followed by a brief and gentle centrifugation, mouse or human lymphocytes remained adhered to each other forming multicellular aggregates. LY microinjected into one cell of a cluster spread to several

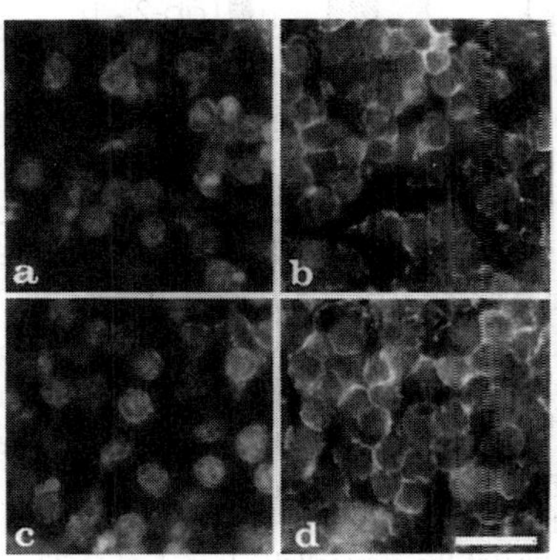


Figure 1. Immunoflurescence detection of Cx37 (c, d) and Cx43 (a, b) in resting (a, c) and 1 h after Con A addition (b, d). Bar: 25 µm.

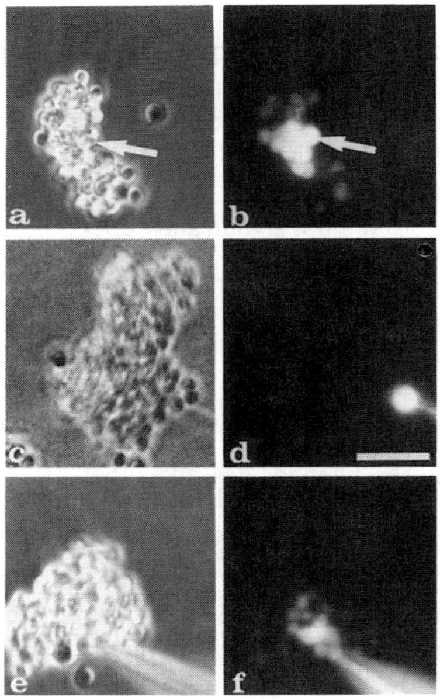


Figure 2. (b) Dye coupling in murine lymphoctes activated with Con A for 1 h. (d). Lymphocytes activated with Con A for 1 h in the presence of Lab-2 peptide were uncoupled. (f) Lymphocytes activated with Con A for 1 h in the presence of Lab-4 peptide were well coupled. a, c and e are phase views of b, d and f, respectively. Bar: 25 μm.

neighboring cells (Fig. 2b). In Con A-treated cells, the spread of LY to adyacent cells was completed blocked by 5 min treatment with 500 $\mu$M octanol. Cell coupling was completely recovered after octanol was washed out (not shown).

In mouse lymphocytes activated with Con A in the presence of 100 μM synthetic peptides homologous to the extracellular loop 1 of Cx32 (52-63, Lab-1), Cx37 (54-66, Lab-2) or Cx43 (53-66, Lab-3) LY did not spread to neighboring cells (Fig. 2d). On the contrary, dye coupling between cells activated with Con A in the presence of peptides homologous to an intracellular domain (C-terminus) of Cx37 (261-276, Lab-4) or Cx40 (339-356, Lab-5) was comparable to that observed in Con A-activated cells (Fig. 2f). Similar results were

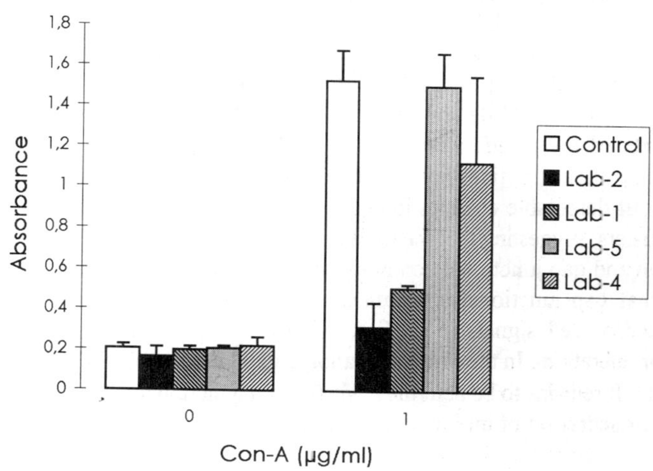


Figure 3. Effect of synthetic peptides homologous to specific regions of Cxs on the BrdU-incorporation into Con A-treated lymphocytes. Mesurement of the absorbance at 370 nm was directly related to the amount of BrdU incorporated. Lab-1 and Lab-2 are homologous to a region of the extracellular loop 1 of Cx32 and Cx37, respectively. Lab-4 and lab-5 are homologous to a C-terminal region of Cx37 and Cx40, respectively.

obtained in purified murine T- or B-cells (not shown). None of the synthetic peptides effected the cell viability, as well as, the number and size of cell clusters induced by Con A.

Under control condition, Con A (1 µg/ml) induced a 7-8 fold increase in BrdU-incorporation into mouse lymphocytes. Nonetheless, the BrdU-incorporation into cells treated with Con A in the presence of Lab-1 or Lab-2 peptide was drastically reduced as compared to control condition (Fig. 3). Lab-4 or Lab-5 did not significantly effect the Con A-induced BrdU-incorporation (Fig. 3). None of the peptides affected the cell viability as tested by trypan blue exclusion.

## 4. Discussion

Formation of gap junctions between lymphocytes and other cell types has been suggested by various independent groups. During antigen presentation *in vitro* [7] or *in vivo* [8], physical interactions between T-lymphocytes and Langerhans cells reveal gap junction-like structures. In addition, transfer of a fluorescent dye between endothelium and lymphocytes, suggest that functional gap junctions are formed between these two cell types [9]. Recently, Cx43 gap junctions were detected in follicular dendritic and lymphoendothelial cell from lymph nodes [10]. Moreover, dye transfer between thymic epithelial cells and thymocytes has been described [11]. Therefore, it is likely that Cx37 and Cx43, found in this work, to be expressed by lymphocytes (T and B) allow formation of gap junctions with homologous, as well as, heterelogous cell types. Both Cxs are known to form functional gap junction channels and are compatible [12].

The mechanism by which lymphocytes cooperate to proliferate is cell contact-dependent [13]; [$^{14}$C]uridine incorporation in mitogen-treated lymphocytes occurs only in cell clusters and is drastically depressed in cells maintained dispersed. Interestinly, we found that lymphocytes forming clusters were well coupled. Moreover, we found that formation of functional gap junction channels between murine lymphocytes was prevented by synthetic peptides identical to a domain of the first extracellular loop 1 of three different Cxs, including one not expressed by lymphoctes (Cx32). The inhibition induced by the Cx32 (Lab-1) peptide could be explained by its high homology with the corresponding region of Cx37 and Cx43.

By other way, there are evidences that cell contact mediated by cell adhesion molecules (CAM) is required for normal immune response [13]. Using mixed lymphocyte reaction, Lukacs et al. [14] demonstrated that the addition of anti-VCAM-1 monoclonal antibodies inhibited the proliferative response by over 70%. The gap junction inhibition induced by the synthetic peptides homologous to a region of the extracellular loop 1 of Cxs occured without detectable changes in cell adhesion, as judged by the number and size of lymphocyte clusters, suggesting that the used experimental strategy allows to dissect the role of cell adhesion and gap junctional communication during the immune response. Therefore, we suggest that gap junctional communication is essential for intercellular exchange of metabolites and/or cell signals (e.g., Ca2+) known to permeate gap junctions [15]that promote cell proliferation . In the thymus octanol, a gap junction blocker, depresses thymulin production [11]. It remains to be determine whether gap junctional communication affects the synthesis and/or secretion of interleukins (e.g., IL-2).

**Acknowledgements** This work was partially supported by a grant of Centro de Investigaciones Medicas PG-13/96 (to F.P.) and a FONDECYT grant 1960559 (to J.C.S.).

## References


[1] D.F. Hüsler and J.H. Peters. Contact cooperation in stimulated lymphocytes. Electrophysiological investigations on intercellular communication. Exp. Cell Res. 74 (1972) 319-326.

[2] G.M. Oliveira-Castro et al., Intercellular communication in stimulated human lymphocyte. J. Immunology. 111 (1973) 1616-19.

[3] G.M. Oliveira-Castro and G. Dos Reis. Cell communication in the immune response. In: W.C. De Mello (ed.), Intercellular communication. Plenum Publishing Coorporation, New York. 1977, pp 201-230.

[4] M.L. Kapsenberg and W. Leene. Formation of B type gap junctions between PHA-stimulated rabbit lymphocites. Exp. Cell Res. 120 (1979) 211-222.

[5] G. Dahl et al., Attempts to define functional domains of gap junction proteins with synthetic peptides. Biophys. J. 67 (1994) 1816-1822.

[6] M.V. Johnston et al., Interaction of anesthetics with electrical synapses. Nature 286 (1979)-498-500.

[7] M. Concha et al., Caorsi I. Physical interaction between Langerhans cells and T-lymphocytes during antigen presentation in vitro. J. Invest. Dermatol. 100 (1993) 429-34.

[8] C.U. Brand et al., Activated immunocompetent cells in human skin lymph derived from irritant contact dermatitis: an immunomorphological study. Br. J. Dermatol. 132 ( 1995) 39-45.

[9] S. Guinan et al., Cytoplasmic transfer between endothelium and lymphocytes: Quantification by flow cytometry. Am. J. Pathol. 132 (1988) 406-409.

[10] T. Krenacs and M. Rosendaal. Immunohistological detection of gap junctions in human lymphoid tissue: connexin43 in follicular dendritic and lymphoendothelial cells. J. Histochem. Cytochem. 43 (1995) 1125-37.

[11] L.A. Alves et al., Functional gap junctions in thymic epithelial cells are formed by connexin 43. Eur. J. Immunol. 25 (1995) 431-7.

[12] R. Bruzzone et al., Connections with connexins: the molecular basis of direct intercellular signaling. Eur. J. Biochem. 238 (1996) 1-27.

[13 ] Springer, T. Adhesion receptors of the immune system. *Nature* 1990. 346: 425-434.

[14] N.W. Lukacs et al., VCAM-1 influences lymphocyte proliferation and cytokine production during mixed lymphocyte responses. Cell Immunol. 154 ( 1994) 88-98.

[15] J.C. Sáez et al., Hepatocyte gap junctions are permeable to the second messengers inositol 1,4,5-trisphosphate and calcium ions. Proc. Natl. Acad. Sci. USA. 86 ( 1989) 2708-2712.

*Gap Junctions*
*R. Werner (Ed.)*
*IOS Press, 1998*



# Dominant-negative Inhibition of Connexin-mediated Tumor Suppression by Mutant Connexin Genes

Yasufumi OMORI, Agnès DUFLOT-DANCER, Marc MESNIL and Hiroshi YAMASAKI
*Unit of Multistage Carcinogenesis, International Agency for Research on Cancer, 150 cours Albert-Thomas 69372 Lyon Cedex08 France*

***Abstract.*** Evidence is accumulating that connexin genes form a family of tumor-suppressor genes. Thus, overexpression of Cx43 in rat C6 glioma cells, and of Cx26 in HeLa cells, suppresses growth both *in vitro* and *in vivo*. Previously, we showed that the mutant Cx32 proteins, found in X-linked Charcot-Marie-Tooth disease patients, abolish the GJIC restored in HeLa cells by the wild-type Cx32 gene in a dominant-negative fashion. However, it was not clear whether such a dominant-negative effect of mutant connexins could lead to biological phenomena such as aberrant growth control. We have, therefore, examined the dominant-negative effects of mutant Cx43 constructs in C6 cells, and of mutant Cx26 constructs in HeLa cells, on cell growth control exerted by the wild types. When two mutant Cx43 constructs - L160M (Leu 160 to Met) and A253V (Ala 253 to Val) - were transfected into Cx43-transfected C6 cells (clone C6-13), they restored anchorage-independent growth capacity and reinforced the tumorigenicity of C6-13, meaning that these two mutants can inhibit growth-suppressive function of wild-type Cx43 in a dominant-negative manner. Neither of the mutants seemed to affect subcellular localization of Cx43 proteins. In HeLa cells transfected with wild-type Cx26, the mutant constructs P87L (Pro 87 to Leu) and R143W (Arg 143 to Trp) restored tumorigenicity in nude mice after their second transfection, without showing any effect on subcellular localization of Cx26 proteins. However, there was no clear correlation between the level of GJIC and the tumorigenicity of these double transfectants. These results show that some mutant forms of Cx43 and Cx26 can inhibit the tumor-suppressive function of their wild-type counterparts, and suggest that the level of GJIC, as revealed by Lucifer Yellow transfer, is not necessarily the determinant of growth control.

## 1. *Introduction*

It has long been believed that gap junction contributes to maintenance of normal cell growth and its disruption results in aberrant cell growth and finally tumors [1], and currently, there is accumulating evidence suggesting the role of gap junction in tumor suppression [2]. Abnormality of gap junction or connexin protein is one of the most common phenomena seen in tumor tissues. Such anomalies include loss or reduction of expression of connexin and aberrant localization of connexin or connexon [2]. Another major line of evidence for the role of gap junctional intercellular communication (GJIC) in negative cell growth arises from transfection of connexin cDNAs into non-communicating tumor cells. A tumor-suppressive effect of Cx43 has been demonstrated in rat C6 glioma cells [3, 4], in mouse transformed fibroblasts MCA-10 [5] and in human mammary carcinoma cells [6]. Similarly, growth delay of a human hepatoma cell line was observed *in vivo* after Cx32 transfection [7]. We have shown that expression of Cx26 eliminates the tumorigenicity of HeLa cells [8]. The tumor-suppressive activity of the Cx26 gene in human breast cancer was also predicted and has recently been confirmed by transfection of the Cx26 gene into human mammary carcinoma cells [6].

Considering that mutations in connexin genes cause abnormalities of GJIC, we have searched for such mutations in tumor tissues. However, only a few mutations were found in chemically-induced rat tumors [9, 10], and so far no connexin mutations have been found in human tumor tissues [11-13].

In our previous study, when several Cx32 mutants, associated with X-linked Charcot-Marie-Tooth disease, were transfected into HeLa cells which had already been transfected with wild-type Cx32 gene, the mutants abolished GJIC restored by the wild type in a dominant-negative manner [14]. It seemed likely that this dominant-negative effects was due to the formation of non-functional heteromeric connexons, rather than heterotypic gap junction, composed of wild-type and mutant connexin molecules [14].

However, it was not clear whether such a dominant-negative effect of mutant connexins could lead to biological function such as aberrant growth control; the Cx32 gene is not tumor-suppressive in HeLa cells [unpublished data]. We, therefore, have examined the dominant-negative effect of mutant Cx43 constructs in C6 cells, and of mutant Cx26 constructs in HeLa cells, on both *in vitro* and *in vivo* cell growth.

## 2. Dominant-negative effects of mutant Cx43 constructs on growth of C6 cells

As illustrated in Figure 1, two mutant forms of Cx43 - L160M (Leu at codon 160 to Met) and A253V (Ala at codon 253 to Val) - were generated from wild-type Cx43 cDNA by site-directed mutagenesis and were inserted into pRc/RSV vector. These two mutant Cx43 and wild-type constructs, along with vector only as a negative control, were transfected into Cx43-transfected rat C6 glioma cells (clone C6-13) [3]. Among G418-resistant clones, only the clones that expressed enough amount of transfected genes were used for the following experiments.

First, the anchorage-independent growth capacity of each clone was examined. Trypsinized cells were seeded into soft agar media and ten days later the number of formed colonies was counted. As shown in Table 1, the clones C6-13 and wtCx43/vector indicated approximately thirty percent of the number of colonies formed by non-transfectant C6 cells. After the second transfection of wild type (wtCx43/wt), the colonies decreased further in number. On the other hand, the mutants L160M and A253V increased the number of colonies given by the clone C6-13, suggesting that these two mutants competed with wild-type Cx43 and down-regulated negative growth control driven by the wild type, resulting in the partial restoration of anchorage-independent growth capacity.

Table 1. Growth in soft agar of C6 cells transfected with both wild-type and mutant Cx43 constructs

| | Colonies growing in soft agar (%) |
|---|---|
| non-transfectant C6 cells | 100 |
| parental clone C6-13 | 30 |
| wtCx43/vector | 28 |
| wtCx43/wt | 11 |
| wtCx43/L160M | 64 |
| wtCx43/A253V | 65 |

Second, the mutants L160M and A253V were examined in their dominant-negative

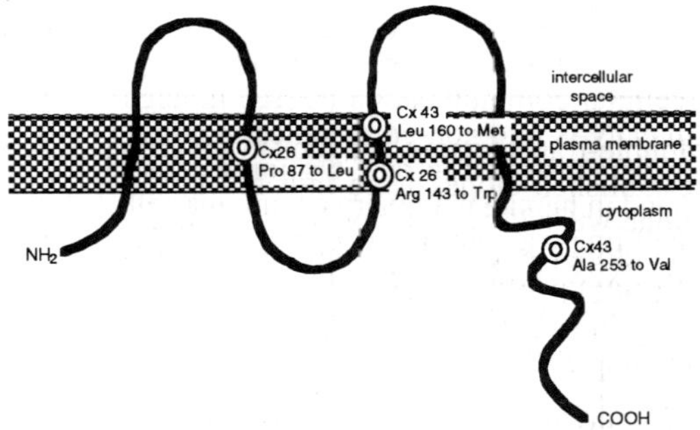


Figure 1. Schematic presentation of the topology, relative to the plasma membrane, of the examined mutations of connexin genes.

effects in tumorigenicity in nude mice. Transfected clones (wtCx43/L160M, wtCx43/A253V and wtCx43/vector) and non-transfectant C6 cells were subcutaneously injected into the backs of nude mice and the size of developed tumors was measured. In the mice injected with the clone wtCx43/vector, tumors appeared much later and grew more slowly than those in the mice injected with non-transfectant C6 cells (Table 2). To the contrary, the clone wtCx43/L160M expressing both wild-type and mutant L160M developed tumors earlier and faster than those in the clone wtCx43/vector. As for the mutant A253V, the double transfectant wtCx43/A253V restored tumorigenicity very efficiently and this dominant-negative effect was more obvious than that of L160M.

Although there were different efficiencies in their dominant-negative effects among clones, they were correlated to the levels of expression of transfected mutant genes in each clone. Therefore, it is concluded that mutant Cx43 constructs - L160M and A253V - inhibit Cx43-mediated growth control of rat C6 glioma cells.

Table 2. Tumorigenicity assay of transfected clones in nude mice

| | Tumor volume ($cm^3$) |
|---|---|
| non-transfectant C6 cells | 4.23[a] |
| wtCx43/vector | 0.56[a] |
| wtCx43/L160M | 1.85[a] |
| wtCx43/A253V | 2.16[a] |
| non-transfectant HeLa cells | 4.54[b] |
| wtCx26/vector | 0.05[b] |
| wtCx26/P87L | 2.52[b] |
| wtCx26/R143W | 1.45[b] |

[a]Tumor size was measured after 19 days growth.
[b]Tumor size was measured after 29 days growth.

### *3. Dominant-negative effects of mutant Cx26 constructs on growth of HeLa cells*

HeLa cell growth in soft agar is suppressed when the wild-type Cx26 gene is transfected [8]. To examine whether Cx26 mutant constructs can reverse this growth-suppressive effect of wild-type Cx26, Cx26 transfected-HeLa cells were seeded into soft agar media after re-transfection of Cx26 mutant constructs (Fig. 1), P87L (Pro at codon 87 to Leu) and R143W (Arg at codon 143 to Trp) prepared by site-directed mutagenesis, along with vector only as a control. As shown in Table 3, P87L-transfected population significantly recovered an ability to grow in soft agar (62%) compared with vector-transfected population (29%), suggesting a dominant-negative effect of mutant Cx26 P87L on the loss of anchorage-independent growth capacity of Cx26-transfected HeLa cells. R143W-transfected population also had a higher ability to grow in soft agar (44%) than the vector-transfected one, showing dominant-negative effect of this mutant on *in vitro* cell growth control.

Table 3. Growth in soft agar of HeLa cells after transfection of both wild-type and mutant Cx26 constructs

| Transfected constructs | Colonies growing in soft agar (%) |
|---|---|
| none | 100 |
| wtCx26 + vector | 29 |
| wtCx26 + P87L | 62 |
| wtCx26 + R143W | 44 |

For cloning of double-transfectants, we picked up colonies formed in sort agar as described above. The established clones were injected into nude mice to examine their tumorigenicity. While clone wtCx26/vector developed no tumor in any of the four injected nude mice (Table 2), the clones expressing both wild-type and mutant Cx26 constructs (wtCx26/P87L and wt Cx26/R143W) produced tumors.

These results clearly show that mutant Cx26 constructs P87L and R143W are capable of eliminating the growth-suppressive effects of wild-type Cx26 [15].

In tumor samples, aberrant subcellular localization of connexin proteins is one of the most common abnormalities [2]. Thus, we examined localization of Cx26 in the transfectants described above by indirect immunofluorescent staining [15]. In both wtCx26/vector (Fig.

2*left*) and wtCx26/P87L (Fig. 2*right*) cells, Cx26 spots were present only at cell-cell contact areas and no aberrant intracytoplasmic or perinuclear localization was observed.

## 4. Possible mechanism of the observed dominant-negative effects

How do mutant connexin proteins inhibit the function of the wild types? We consider that the function of the wild-type connexin is down-regulated by formation of heteromeric (chimeric) hexamers (connexons) with mutated connexin proteins produced by the subsequently transfected mutant gene [14]. Connexin proteins form hexamers called connexons, which are transferred and inserted into the cell membrane to form gap junctions [16]. When mutant proteins keep the capacity for oligomerization but have lost the biological functions including GJIC and tumor-suppression, they can inhibit or alter the function of the wild-type connexin proteins. This type of dominant-negative effect could, in principle, be achieved very efficiently. When equal numbers of wild-type and mutant connexin molecules are present, the probability of the formation of a connexon composed of exclusively wild type is as low as one in $2^6$. Thus, it is possible to reduce the number of intact homomeric connexons by 98.5%, if one allele of the connexin gene is mutated in a give cell. Thus, a point mutation on one allele of the connexin gene may have a much more dramatic effect on GJIC than other changes, such as complete deletion of one allele.

## 5. Discrepancy between the tumor-suppressive function and the GJIC capacity of connexin

As summarized in Table 4, the double-transfectants established above did not always show a good correlation between the tumor-suppressive function and the GJIC capacity. For example, the clone wtCx43/L160M had no GJIC capacity and showed a high tumorigenicity. In contrast, while the clone wtCx43/A253V had a high GJIC capacity, it manifested a high tumorigenicity. This does not seem consistent with the long-standing hypothesis that down-regulation of GJIC leads to development of tumors. Therefore, connexin may not only be a mediator of GJIC but may also control cell growth or tumor development in a GJIC-independent manner. Another possible explanation for the discrepancy between GJIC and growth control capacity is that the tracer molecule Lucifer yellow may not represent the molecules which pass through a gap junction thus are involved in tumor-suppression. In this scenario, Lucifer yellow may be able to pass through the gap junction formed by the clone wtCx43/A253V while the molecules responsible for growth control may not. Further studies must be addressed to this phenomenon.

Table 4. Comparison of double-transfectants in their characters

| | GJIC capacity | Tumorigenicity in nude mice |
|---|---|---|
| wtCx43/vector | ++ | + |
| wtCx43/L160M | - | ++ |
| wtCx43/A253V | ++ | +++ |
| wtCx26/vector | ++ | - |
| wtCx26/P87L | ++ | +++ |
| wtCx26/R143W | +++ | - |

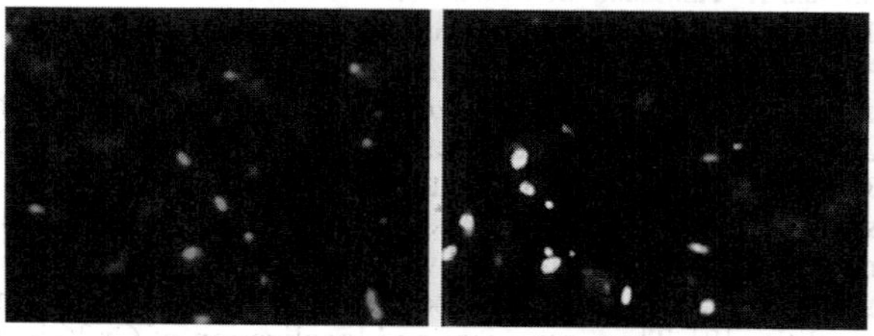

Figure 2. Indirect immunofluorescence staining of representative transfected clones. *left*, the HeLa cells expressing only wild-type Cx26; *right*, the HeLa cells expressing both wild-type and mutant Cx26 P87L.

## 6. Biological significance of the observed dominant-negative effects

As indicated above, some mutated forms of Cx43 and Cx26 proteins can abrogate cell growth control capacity of the wild types when both wild types and mutants co-exist in a cell. This fact is suggestive when one thinks of the involvement of altered connexins, whether by mutation or post-translational modification, in carcinogenesis. It is well known that connexin proteins are very sensitive to structural alteration caused not only by mutation but also by chemical, biochemical or physical factors, including phosphorylation, calcium concentration and pH [17, 18]. When such a microenvironmental change affects only a partial population of connexin molecules (i.e., mutated ones) in a cell, probably altered molecules should inhibit normal cell growth control driven by intact ones in a dominant-negative manner.

It has been reported that heteromeric connexons between Cx26 and Cx32 [19] and between Cx46 and Cx50 [20] are formed. It may be possible that mutant connexin molecules affect the function of other species of connexins. Therefore, dominant-negative inhibition of connexin-mediated cell growth control by altered connexin molecules might constitute an important phase of carcinogenesis.

## Acknowledgments

We are grateful to Mrs. Colette Piccoli for her technical assistance and to Dr. John Cheney for editing the manuscript. This work was partially supported by grants from the US National Institutes of Health (R01-CA-40534), the European Union (EVSV-CT940421), the Association for International Cancer Research (Grant no. 95-27), and l'Association pour la Recherche sur le Cancer.

### References

[1] W.R. Loewenstein, *Physiol. Rev.* **61** (1981) 829-913.
[2] H. Yamasaki and C.C.G. Naus, *Carcinogenesis (Lond.)* **17** (1996) 1199-1213.
[3] D. Zhu, S. Caveney, G.M. Kidder and C.C.G. Naus, *Proc. Natl. Acad. Sci. USA* **88** (1991) 1883-1887.
[4] C.C.G. Naus, K. Elisevich, D. Zhu, D.J. Belliveau and F. Del Maestro, *Cancer Res.* **52** (1992) 4208-4213.
[5] B. Rose, P.P. Mehta and W.R. Loewenstein, *Carcinogenesis (Lond.)* **14** (1993) 1073-1075.
[6] K.K. Hirschi, C. Xu, T. Tsukamoto and R. Sager, *Cell Growth Differ.* **7** (1996) 861-870.
[7] B. Eghbali, J.A. Kessler, L.M. Reid, C. Roy and D.C. Spray, *Proc. Natl. Acad. Sci. USA* **88** (1991) 10701-10705.
[8] M. Mesnil, V. Krutovskikh, C. Piccoli, C. Elfgang, O. Traub, K. Willecke and H. Yamasaki, *Cancer Res.* **55** (1995) 629-639.
[9] Y. Omori, V. Krutovskikh, N. Mironov, H. Tsuda and H. Yamasaki, *Carcinogenesis (Lond.)* **17** (1996) 2077-2080.
[10] T. Saito, A. Barbin, Y. Omori and H. Yamasaki, *Cancer Res.* **57** (1997) 375-377.
[11] V. Krutovskikh, G. Mazzoleni, N. Mironov, Y. Omori, A.-M. Aguelon, M. Mesnil, F. Berger, C. Partensky and H. Yamasaki, *Int. J. Cancer* **56** (1994) 87-94.
[12] N.M. Mironov, A.-M. Aguelon, G.I. Potapova, Y. Omori, O.V. Gorbunov, A.A. Klimenkov and H. Yamasaki, *Cancer Res.* **54** (1994) 41-44.
[13] V. Krutovskikh, N. Mironov and H. Yamasaki, *Carcinogenesis (Lond.)* **17** (1996) 1761-1763.
[14] Y. Omori, M. Mesnil and H. Yamasaki, *Mol. Biol. Cell* **7** (1996) 907-916.
[15] A. Duflot-Dancer, M. Mesnil and H. Yamasaki, *Oncogene* (1997) in press.
[16] E.C. Beyer, D.L. Paul and D.A. Goodenough, *J. Cell Biol.* **105** (1987) 2621-2629.
[17] R. Bruzzone, T.W. White and D.L. Paul, *Eur. J. Biochem.* **238** (1996) 1-27.
[18] V.A. Krutovskikh, M. Mesnil, G. Mazzoleni and H. Yamasaki, *Lab Invest.* **72** (1995) 571-577.
[19] K.A. Stauffer, *J. Biol. Chem.* **270** (1995) 6768-6772.
[20] J.X. Jiang and D.A. Goodenough, *Proc. Natl. Acad. Sci. USA* **93** (1996) 1287-1291.

# Author Index